Nanomedicine and Nanotoxicology

Series editor

V. Zucolotto, São Carlos, Brazil

More information about this series at http://www.springer.com/series/10620

Gil Gonçalves · Gerard Tobias
Editors

Nanooncology

Engineering Nanomaterials for Cancer Therapy and Diagnosis

Springer

Editors
Gil Gonçalves
Institut de Ciència de Materials de Barcelona (ICMAB-CSIC)
Bellaterra
Spain

Gerard Tobias
Institut de Ciència de Materials de Barcelona (ICMAB-CSIC)
Bellaterra
Spain

ISSN 2194-0452 ISSN 2194-0460 (electronic)
Nanomedicine and Nanotoxicology
ISBN 978-3-030-07885-0 ISBN 978-3-319-89878-0 (eBook)
https://doi.org/10.1007/978-3-319-89878-0

Printed on acid-free paper

This Springer imprint is published by the registered company Springer International Publishing AG part of Springer Nature
The registered company address is: Gewerbestrasse 11, 6330 Cham, Switzerland

Contents

Gold Nanoparticles for Imaging and Cancer Therapy

Marc-André Fortin, Teresa Simão and Myriam Laprise-Pelletier

Abstract Gold nanoparticles have properties useful in biomedical imaging and cancer therapy. This biocompatible metal has been used for centuries in medicine. In the last 20 years, the rapid developments in nanotechnology have revealed several applications of nanosized gold, which are now being evaluated for clinical procedures. For instance, gold nanoparticles can be used to develop sensors due to their optical properties; they also make possible the development of new hyperthermia and drug delivery treatments. However, gold nanoparticles could find more immediate and direct applications in medical physics procedures, such as X-ray imaging and radiotherapy. First, this chapter provides an overview of the different synthesis routes for the production of biomedical gold nanoparticles. Then, an overview of the physical principles of photon–matter interactions, that are fundamental to the concept of X-ray attenuation in biological tissues, is presented. The properties of gold nanoparticles as contrast agents for X-ray and computed tomography (CT) imaging are reviewed, along with the principles of the radiosensitization effect useful in medical physics and oncology. The main mechanisms leading to dose enhancement, to cell damage and to cell death, are described in the light of the specific interactions taking place between ionizing photons and high-*Z* materials such as gold (Au) when these are distributed in biological tissues such as tumours. Finally, the performance of gold nanoparticles as CT contrast agents and radiosensitizers in oncology is discussed, in the perspective of their consideration for clinical applications.

Keywords Gold nanoparticles · Radiotherapy · Brachytherapy · Computed tomography · Radiosensization · Cancer treatment · Theranostics

---

M.-A. Fortin (✉) · T. Simão · M. Laprise-Pelletier
Centre de Recherche du Centre Hospitalier, Universitaire de Québec-Université Laval (CR CHU de Québec), axe Médecine Régénératrice, Québec G1L 3L5, QC, Canada
e-mail: Marc-Andre.Fortin@gmn.ulaval.ca

M.-A. Fortin · T. Simão · M. Laprise-Pelletier
Department of Mining, Metallurgy and Materials Engineering, Université Laval, Québec, QC G1V 0A6, Canada

M.-A. Fortin · T. Simão · M. Laprise-Pelletier
Centre de Recherche sur les Matériaux Avancés (CERMA), Université Laval, Québec, QC G1V 0A6, Canada

G. Gonçalves and G. Tobias (eds.), *Nanooncology*, Nanomedicine and Nanotoxicology, https://doi.org/10.1007/978-3-319-89878-0_1

1 Introduction

Over the last decades, gold nanoparticles (Au NPs) have emerged as a new promising material for a large variety of medical applications [13, 41, 51, 52, 55, 78, 95, 97, 100, 120, 131, 142, 176, 178, 190, 197, 199, 200]. Au NPs bear several physico-chemical properties useful for the development of diagnosis tools as well as for imaging contrast agents. They can also be used as radiosensitizers: these are products that, when they are injected in biological tissues, have the capacity to increase the impact of radiotherapy [25, 39, 45, 76, 111, 204, 208]. Finally, nanosized gold is resistant to oxidation in biological media and overall, it is accepted by the cells at relatively high concentrations [64, 95, 113, 173].

A very large number of studies and reviews have been written about the properties of Au NPs used as 'plasmonic' materials [57, 186, 200]. Plasmons are oscillations of free electrons taking place at the surface of conductor nanomaterials (such as gold). The surface electrons of Au NPs can couple with electromagnetic radiation of certain wavelengths that are far larger than the particle. Plasmonic nanoparticles exhibit interesting scattering, absorbance and coupling properties, and several proof-of-concept studies useful to the biomedical field have been made by exploiting the resonant properties of Au NPs irradiated by visible and near-infrared light. For instance, new sensor technologies have been developed relying on the principle that molecules reacting with a specific molecular function present at the surface of Au NPs, induce slight changes to specific resonance peaks. These can be sensitively detected with appropriate spectroscopic tools. The resonant properties of Au NPs under visible and infrared light can also be exploited to increase the temperature of cells, a phenomenon referred to as 'hyperthermia' [53, 182].

Although the plasmonic characteristics of Au NPs appear very useful for in vitro applications, their potential for in vivo procedures has always been somewhat limited to skin diseases and to superficial cancer (e.g. skin cancer). In fact, visible and near-infrared photons diffuse strongly in the biological tissues. Diffusion limits the depth of penetration of such low-energy photons in biological tissues to a few millimetres only. Therefore, to be useful for a large variety of applications in vivo, Au NPs must preferably be irradiated with mid-to-high energy electromagnetic radiation—mainly photons—that have the capacity to penetrate deep enough in the body.

In fact, biomedical imaging and cancer treatment by radiotherapy are two areas of medicine where the injection of Au NPs in vivo could be recommended for clinical diagnoses and therapies [45, 70, 71, 76]. High-energy photons (from 1 keV and above) are already exploited in medicine. This is precisely the range of energies useful to medical physics. Irradiated with photons in the ionizing energy range, Au NPs can be used as effective contrast agents for X-ray computed tomography (CT), as well as for 'boosting' the radiation dose delivered in radiotherapy treatments [25, 84, 131].

Therefore, in vivo applications of Au NPs based on their interactions with higher energy photons, have emerged in modern medicine and in particular for the treatment of cancer. Au NPs can be used to enhance the differences of density between biological tissues (e.g. using X-ray and CT imaging). They can also be used in radiosensitizing procedures to induce more damages to cells irradiated with radiotherapy beams or with internal radioactive sources. In fact, Au NPs are a very dense material; they can attenuate high-energy photons in the biological tissues even if they are distributed at relatively moderate concentrations. Also, photons interacting with high-*Z* materials such as gold generate many by-products such as low-energy photons and electrons. These 'secondary products' have the capacity to increase the therapeutic impact of the primary photons beams, as well as the radioactive sources used in radiotherapy. In medical physics, Au NPs are often referred to as 'radiosensitizers', i.e. products that have the capacity to enhance radiotherapy treatments.

This chapter begins with an introduction to the different synthesis routes leading to the production of Au NPs, as well as to their surface functionalization. Then, an overview about the physical mechanisms by which photons (~10–500 keV) interact with the atoms present in biological tissues is given. Photons of energy higher than 10 keV can penetrate deep enough in the biological tissues (e.g. sub-visible light wavelength range), and therefore can be exploited either for imaging or therapeutic applications. Finally, the applications of Au NPs as contrast agents for CT (X-ray imaging), and as radiosensitizers in radiotherapy (medical physics in oncology), are presented in the context of their ongoing evaluation towards clinical applications.

2 Synthesis of Gold Nanoparticles for Biomedical Applications

Colour change is the most evident signature of colloidal gold. Depending on the size of the nanoparticles, gold presents either an intense red colour for small to medium-sized particles, purple for larger particles and blue for aggregates [52, 78]. Because the properties and the applications of Au NPs so closely depend on their size and shape, the colour of aqueous suspensions of Au NPs provides very efficient indications to the chemist at every step of the nanoparticle preparation.

The first experiments for the synthesis of gold sols under controlled conditions were reported by Michael Faraday in the 1850s [58, 152]. The scientist accidentally generated a ruby red solution while mounting pieces of gold leaves onto microscope slides. He referred to this solution as 'activated gold'. Then, he used phosphorus to reduce a solution of gold chloride. Already interested in the properties of light and matter, Faraday further investigated the optical properties of these colloidal gold solutions. For a long time, the composition of 'ruby' or 'activated' gold was unclear, but the colour-particle size relationship was already acknowledged. The studies on Au NPs did not make significant advances until the end of the twentieth century, with the exponential development of advanced analytical technologies, such as atomic force microscopy and electron microscopy. Due to their relatively easy synthesis routes, high stability and high density per particle, colloidal gold has also been used in the development of physical separation techniques such as ultracentrifugation, as well as electron microscopy dyes.

Gold nanoparticles produced for biomedical applications are usually generated from precursor solutions containing chloroauric acid ($H[AuCl_4]$). After $H[AuCl_4]$ is dissolved in water, the solution is rapidly stirred while a reducing agent is added. To prevent the particles from aggregating, a stabilizing agent is usually added for coating the nanoparticle surface. A comprehensive review of the different classes of Au NPs synthesis was written by Daniel et al. [48]. A selection of the most significant ones is reviewed here.

2.1 Direct Reduction: The Turkevitch Method

The chemical reduction of $AuCl_4$ by citric acid in a hot aqueous solution was investigated in 1951 [185]. In this synthesis, citrate acts as the reducing agent, then as the capping agent that stabilizes the Au NPs through electrostatic interactions. The Turkevitch method allows the growth of NPs with many different shapes and diameters usually ranging from 10 to 50 nm (Fig. 1). A significant increase in the particle diameters is observed at lower citrate concentration, leading to the generally accepted conclusion that the total particle surface area is determined by the number of citrate ions available to cover it [65].

More recently, an alternative approach showed that a high concentration of Cl^- ions also leads to larger sizes for Au NPs reduced by citric acid, since it decreases their surface charge and thereby promotes coarser particle size and even aggregation [205]. A similar result was obtained by increasing the pH of the NP solution just after initiating the reduction step. An increase of the pH limits the nucleation process and decreases the final number of grown Au NPs [167]. Although simple, the direct chemical reduction synthesis route usually leads to the production of NPs of relatively large sizes, showing anisotropy and a certain degree of size polydispersity.

2.2 Seed-Mediated Growth for Smaller and Narrower Particle Size Distributions

The constantly increasing demand for Au NPs in photonics, biology and medicine, has led to the development of new synthesis routes for achieving better control over size distributions. One of the most promising approaches in this goal is the seed-mediated growth technique. In this concept, small nanoparticle seeds of a typical size smaller than 15 nm, act as catalytic nucleation centres for the growth of larger NPs [86]. The nucleation of very small NP seeds is usually made through the reduction of gold ions by sodium borohydride ($NaBH_4$), a relatively toxic reducing agent that is tolerated as an initiator of Au NPs seeds. However, it is not indicated in the final synthesis steps of biomedical Au NPs. Shape uniformity in the seed suspension is of paramount importance to achieve both shape and size uniformity of the final NPs [141].

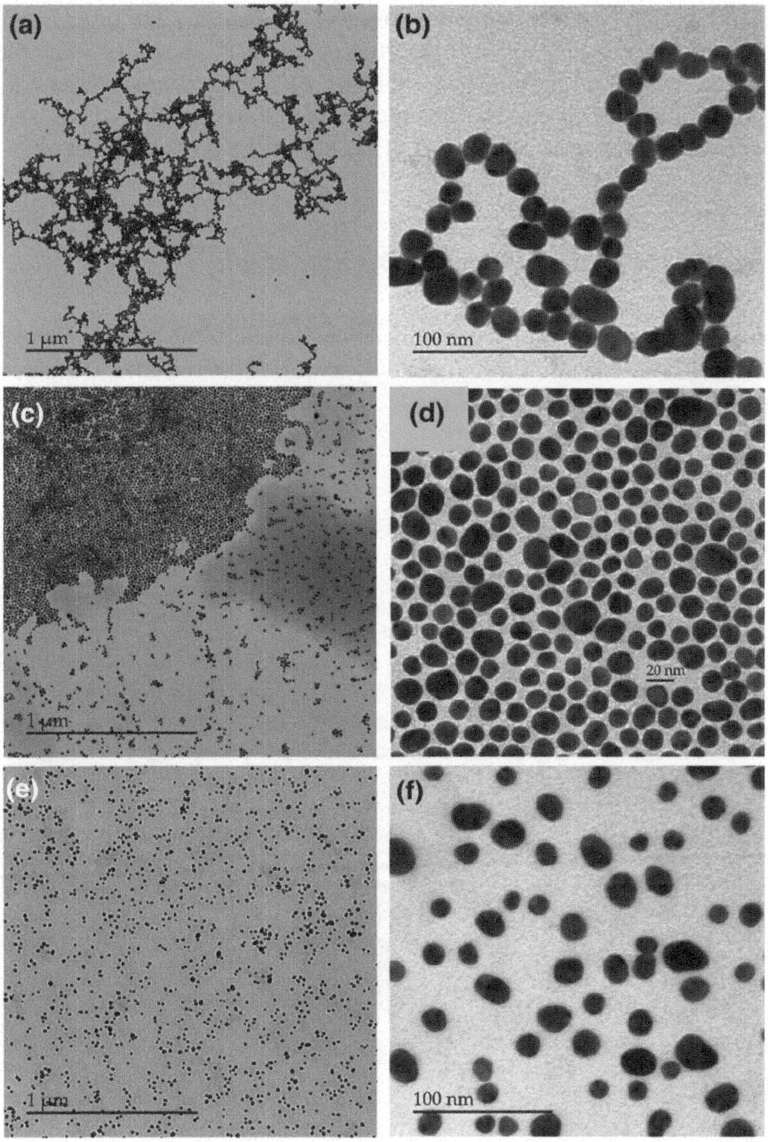

Fig. 1 Gold nanoparticles synthesized by the Turkevitch method in citrate: **a**, **b** as-synthesized; **c**, **d** grafted with polyethylene glycol PEG 1000 mw; and **e**, **f** grafted with polyethylene glycol PEG 5000 mw. The longer the chains, the larger the spacing between the particles. *Source* The authors, following procedures adapted from [185] including the use of PEG-thiol as a surfactant

Several articles have been published on the topic of seed-mediated growth of Au NPs. One of the most comprehensive ones is a three-step process that was used to produce uniform Au NPs with a diameter up to 300 nm (Fig. 2) [209]. The main limitation of this method is that an important population of smaller NPs often remains in solution additionally to the grown seeds, requiring further purification to achieve very narrow particle size distributions. Perrault et al. overcame this problem by using hydroquinone as a reducing agent [147]. Hydroquinone reduces selectively the gold ions that are located in the immediate vicinity of Au^0 seed nanoclusters, leaving isolated gold ions unreduced. This chemical route allows a selective growth of the seeds while avoiding further nucleation of Au clusters in the reaction fluid.

2.3 Syntheses Taking Place in Organic Media

Nanoparticle synthesis routes performed in aqueous conditions are easy to implement, and as a major advantage, they avoid the introduction of potentially toxic solvents into the chain of NP production. Unfortunately, the size distribution

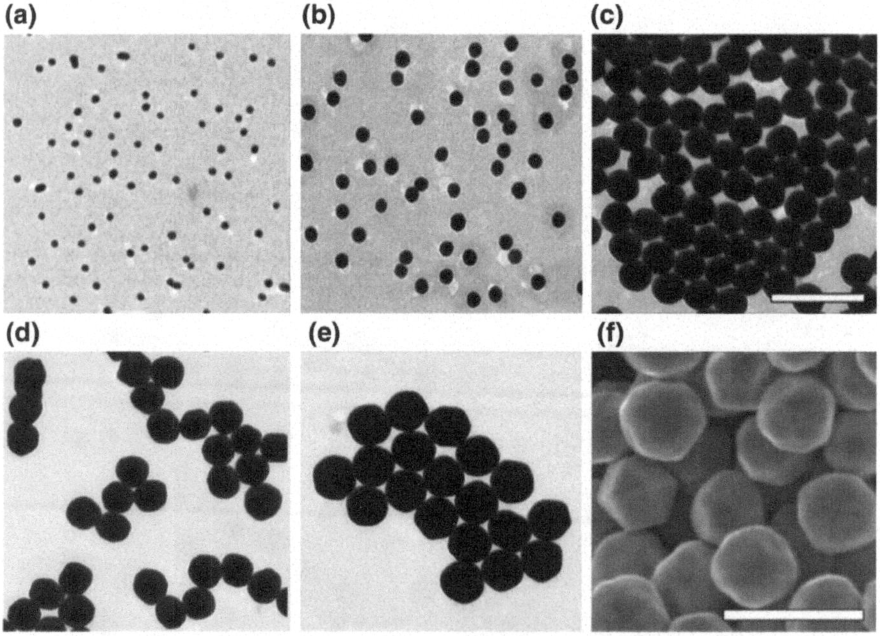

Fig. 2 TEM images of **a** 15 ± 2 nm seeds, **b** 31 ± 3 nm, **c** 69 ± 3 nm, **d** 121 ± 10 nm, and **e** 151 ± 8 nm and SEM image of **f** 294 ± 17 nm gold nanoparticles. Scale bars are 200 nm for parts **a–c** and 500 nm for parts **d–f**. Reprinted with permission from [209]. Copyright 2011 American Chemical Society

of Au NPs synthesized in aqueous media is relatively difficult to keep to narrow and small values. It is generally well established and acknowledged in the field of nanotechnology, that syntheses performed in organic media often result in much smaller and narrower particles size distributions. The Brust–Schiffrin method is one of the first reported to synthesize Au NPs in organic media, and it is until now one of the most widely performed to synthesize small particles of narrow distributions (Fig. 3) [22]. Typically, it consists of using tetraoctylammonium bromide (TOAB) to transfer the gold salts into an organic phase (e.g. toluene, ethanol and ethylene glycol). Then, NPs are nucleated through a reduction step through thiol ligands and $NaBH_4$. This approach produces very small and stable NPs ranging from 1.5 to 5 nm that can be easily functionalized with other ligands. Recent developments related to the Brust–Schiffrin method include mostly fundamental studies of mechanisms such as charge transfer, nucleation and growth during the synthesis process [54, 146].

Another class of synthesis procedures taking place in an organic solvent is the so-called polyol process. It allows the synthesis of monodisperse Au NPs, or alternatively to particles of various well-controlled geometries [60]. In this procedure, gold salts are dispersed in high boiling point alcohols (e.g. ethylene glycol), which also act as a reducing and as a capping agent. By using a polymeric stabilizer, highly symmetrical polyhedron-shaped Au NPs with a narrow size distribution can be obtained in a wide variety of sizes. Xia et al. have used a polyol process with polyvinylpyrrolidone (PVP) as the stabilizer to synthesize very uniform silver nanocubes with 50–115 nm of edge length, and used them as sacrificial templates in a solution of $HAuCl_4$ for the production of gold nanoboxes with perfectly smooth surfaces [179]. Song et al. dissolved $HAuCl_4$ and PVP in 1,5-pentanediol, instead of ethylene glycol for its higher boiling point, and added small concentrations of $AgNO_3$ [165]. Depending on the concentration of silver ions, octahedral (ratio Ag/Au = 1/600) or cubic (Ag/Au = 1/200) shape nanostructures were obtained, among others. This chemical

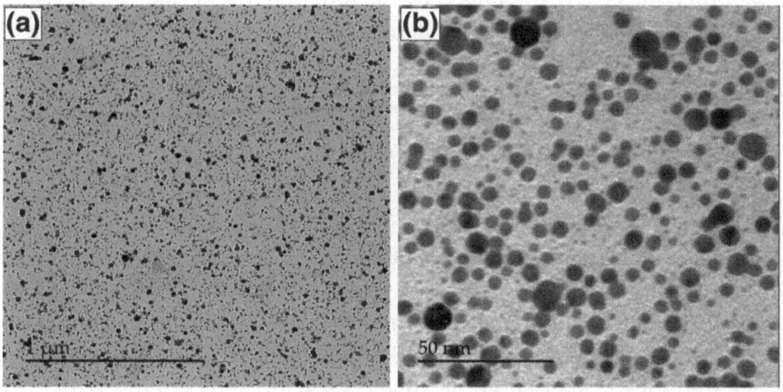

Fig. 3 Gold nanoparticles synthesized by the Brust–Schiffrin method. Among the different colloidal synthesis techniques, this one usually leads to very small particle size distributions. *Source* The authors, following procedures described in [22]

behaviour was attributed to the selective deposition of silver species on the seed surface during the reaction. Overall, the polyol process is a convenient and low-cost technology for the large-scale production of highly symmetrical Au NPs of homogeneous sizes. However, the complete elimination of the solvent residues, as well as the presence of potentially toxic chemicals, necessitates the introduction of tedious filtration, dialysis or chromatography procedures that are not necessarily easy to upscale.

2.4 Au NPs Purification Prior to Coatings and Functionalization

The purification and functionalization steps to remove potentially toxic reagents and to cover the NPs surface with biocompatible ligands are very important in the development of stable as well as functional Au NP formulations for biomedical applications. The purification is usually performed by dialysis, chromatography or centrifugation. Those methods are generally sufficient to remove the majority of the contaminants. As mentioned above, the majority of Au NPs used for biomedical applications are synthesized either by the Turkevitch [56] method, by the seed-mediated growth approach [150] or by the Brust–Schiffrin [42] technique. In the first case, citrate molecules cover the surface of the Au NP. In the second one, hexadecyltrimethylammonium bromide (CTAB) is usual, whereas alkanethiols are employed in the Brust–Schiffrin method. For biomedical applications, the Au NPs must be dispersed in aqueous solutions, and they must be free of even low traces of contaminants (e.g. organic solvents, reduction agents and excess of surfactants). They must be stable in physiological media, which are rich in diverse ions, proteins and many other molecules. Thus, in order to preserve the colloidal stability and assure biocompatibility, the Au NPs obtained by the previously described methods must be functionalized with biocompatible molecules.

2.5 Ligand-Free Au NP Suspensions

The conventional Au NP synthesis approaches do have several limitations in terms of toxicity risks induced by surfactants and chemical residues. For instance, sodium borohydride ($NaBH_4$) is a harsh reagent that must be entirely removed from the solutions [146]. Then, the gold ions used in most Au NP synthesis techniques usually come from chloroauric acid, a potent acid which must be entirely cleared from the biomedical solutions [5]. In addition, ligand exchange procedures involve several manipulation steps that lead to material loss, as well as to agglomeration. Several alternatives to the more conventional chemical synthesis approaches have emerged recently to synthesize purer Au NPs, generally free from ligands or reducing agents, and therefore ready to be efficiently functionalized. These synthesis techniques could represent a major advantage in the quest for compounds that limit the toxicity risks related to the presence of residues in Au NPs compounds.

2.5.1 Pulsed Laser Ablation in Liquid (PLAL)

The synthesis technique that possibly enables the production of the smallest and the purest Au NPs solutions, is pulsed laser ablation in liquid (PLAL). This synthesis method was introduced by Cotton and Henglein in 1993 [63, 138]. It is performed by immersing a gold metal target in a fluid, followed by irradiation of the target surface for a certain time with a pulsed laser [153]. The target absorbs the laser pulse energy, resulting in heating and photoionization of the irradiated area. Solid, liquid and vaporized materials are emitted in the form of a plasma plume, which expands outwards in a confining liquid [8, 9, 81]. Then, the plasma plume starts to cool down and a cavitation bubble is formed [8]. At this moment, the supersaturation point is achieved due to the concentration of metal ions, to the high pressure and to the temperature attained because of liquid confinement. This causes NPs to nucleate: metal atoms condense and coalesce in the form of NP nuclei. The newly formed NPs start to diffuse into the expanding cavitation bubble, where further growth, coalescence or aggregation happen. At last, when the cavitation bubble collapses, all the NPs are released into the surrounding liquid [8]. A schematic representation of PLAL for the synthesis of Au NPs is shown in Fig. 4. The ultra-small Au NPs colloids synthesized by this technique can be stabilized by PEG grafting to enhance their stability in biomedical media [169].

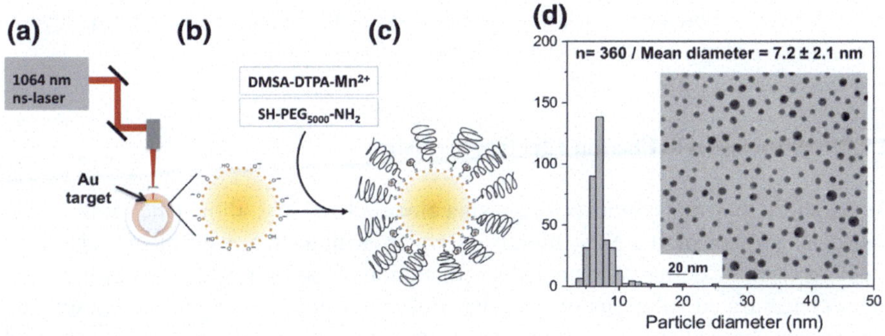

Fig. 4 Pulsed laser ablation in liquid for the nucleation of Au NPs in ultrapure water (**a**), produces Au colloids of chemically pure surfaces that are stabilized by electrostatic forces (**b**). To increase their colloidal stability in biological fluids, the Au NPs are stabilized by polyethylene glycol. Additionally, the Au NP@PEG was functionalized with DMSA-DTPA-Mn^{2+} to be applied as a contrast agent for MRI (**c**). PLAL synthesis followed by PEG and DMSA-DTPA-Mn^{2+} grafting enables the production of very small particle size distributions (**d**). Reproduced from [169] with permission from the Royal Society of Chemistry

The laser wavelength, the laser fluence, the repetition rate, the ablation time and the composition of the liquid solution can be tuned to achieve a specific concentration of NPs and size distribution [8, 153]. The most suitable wavelengths to synthesize Au NPs are in the NIR region (e.g. 1064 nm) [184]. Subsequently, post-irradiation with wavelengths coinciding with plasmon absorption or interband transition of Au NPs can be used to tune the size and to reduce polydispersity [124].

PLAL allows the production of Au NPs in water or organic solvents [10, 180]. However, organic solvents are prone to pyrolysis during laser ablation and the degraded molecules can adsorb on the surface of the NPs, which may raise biocompatibility issues. In pure water, PLAL-synthesized Au NPs show a size range of 10–40 nm; they are electrostatically stabilized by negative charges that result from the formation of $Au–O^-$ species [8, 153, 180]. To improve the stability and to control the Au NPs size in water, it is possible to use low salt concentration. The anions of the salts adsorb on the surface of the Au NPs and increase the negative charge density, thus enhancing the repulsive forces between the NPs. This effect prevents both the growth of the NPs by avoiding nuclei coalescence and improves the colloidal stability prior to further grafting of the nanoparticles with biocompatible molecules (e.g. to create steric hindrance) [130, 153].

In a biomedical context, the main advantage of PLAL over conventional Au NP synthesis colloidal chemistry routes is the production of NPs directly in water without the use of reducing or stabilizing chemicals (e.g. $NaBH_4$) [153]. As a result, the surface of the NPs is free of ligands and it is thereby readily available for the conjugation with other molecules [153]. Moreover, many reducing chemicals used in colloidal Au NP synthesis routes can be toxic, and this hazard is totally eliminated in laser ablation. Thus, the ligand-free surface allows to avoid extensive purification processes and aggregation associated with ligand exchange procedures [194]. The main limitations of PLAL for Au NP synthesis are the necessity to use very expensive lasers as well as low concentrations of solutions synthesized, which require several concentration steps in order to reach biomedical applicability.

2.5.2 Atmospheric Plasma Electrochemistry

Atmospheric plasma synthesis, or plasma electrochemistry, allows the rapid nucleation and growth of Au NPs directly in water, without the need of any chemical reducing agent. Compared to pulsed laser ablation, it has the advantage of producing more concentrated solutions of Au NPs while requiring less expensive equipment. Plasma is a charged gas containing free electrons, positive and negative ions, neutral species in the ground or excited state, and photons. This mixture of charged species gives plasmas unique physico-chemical properties. The most common and easily attained method for forming plasma is to apply an electrical voltage between two electrodes, solid or liquid, separated by a gas. With aqueous solution, plasma electrochemistry must be operated near atmospheric pressure and at low temperature to avoid solvent evaporation.

A great variety of chemical reactions take place at the plasma–liquid interface. In the presence of nitrogen as the main plasma component or as a contaminant from the presence of air, the formation and dissolution of nitric acid can cause a significant decrease in pH [108]. The electrolysis of water $\left(2H_2O + 2e^- \rightarrow 2OH^- + H_2\right)$ at the plasma cathode and the formation of chlorine gas ($2Cl^- \rightarrow Cl_2 + 2e^-$) at the anode were also reported [158]. The formation of reactive ions H^+, H^-, O^-, OH^-, radicals $H^\cdot$, $O^\cdot$ and $OH^\cdot$ and H_2O_2 is also a very important factor in the cascade of chemistry events taking place in the nucleation and growth of nanoparticles [66, 125].

The exact mechanism for the nucleation of Au NPs by plasma–liquid electrochemistry is still not entirely understood. The electrons generated in the plasma are accelerated to the surface of the liquid by the electric field between the electrodes and pass into solution in the form of solvated electrons [1, 125]. Several reactive oxygen and nitrogen species generated by the plasma treatment act as metal ion reducers. These 'fugitive' or metastable reducing agents make possible the nucleation of Au NPs while avoiding the use of external toxic chemicals (e.g. $NaBH_4$) for the synthesis of water-dispersed Au NPs. Gold ions in solution can reach the plasma–liquid interface by the combined effects of the electric field, the convection forces and the concentration gradient [2]. When an ion reaches the interface, it is reduced into atomic gold directly by the electrons: $[AuCl_4]^- + 3e^- \rightarrow Au^0 + 4Cl^-$. Once reduced, the gold atoms collide and form clusters, which diffuse away from the plasma–liquid interface. These clusters act as nucleation sites of NPs [2]. Size control is usually performed by adjusting only the concentrations of metal salt and surfactant in solution. Increasing the current can also lead to the production of smaller NPs since it allows a higher nucleation rate [79].

In the last 10 years, new 'cold plasma' reactors operating at atmospheric pressure brought the possibility to synthesize NPs from metal salts directly in water. The first type of atmospheric plasma reactor that was used to nucleate Au NPs was the microplasma [104, 160], and the most commonly used geometry of microplasma is the microhollow cathode [20, 104, 155]. It consists of a gas stream flowing in a cylindrical hollow cathode placed a few millimetres above a liquid solution containing metal ions. A DC voltage, typically several thousand V/cm, is applied between the cathode and the solution. The plasma discharge generated at the output of the microhollow cathode thus extends to the surface of the liquid, where the reduction of the metal ions and the nucleation of NPs take place.

Richmonds and Sankaran demonstrated the possibility to synthesize gold and silver NPs from the dissolution of a metal sheet in a slightly acidic aqueous solution [155]. Using fructose as a stabilizing agent, the reduction by argon plasma has formed 10 nm diameter NPs for both metals. This device shows the conventional concept of an electrochemical cell, where the metal film acts as a sacrificial anode, and the microplasma as a cathode [160]. This concept was adapted several times for the synthesis of gold and silver NPs of different sizes and morphologies, with and without stabilizer [127, 188, 189]. Unfortunately, the microplasma technologies have a rather limited surface coverage, and other types of plasma must be considered to allow the development of technologies enabling the synthesis of large volumes of solutions.

A promising alternative to the microplasma is the dielectric barrier discharge (DBD). Through the DBD technology, plasma–liquid interface area of several cm^2 can be generated [16]. This technology was recently employed to synthesize radioactive Au NPs for applications in oncology [19]. The main limitation of Au NPs synthesized by plasma electrochemistry remains the relatively high polydispersity of particles, requiring further separation steps. Further work aimed at understanding the exact electrochemical mechanisms behind the reduction of Au ions should help reaching a better control over size.

2.6 Surface Treatment of Au NPs

The size, shape and surface properties of the Au NPs are determinant factors for their successful biomedical applications [4, 35, 40, 55, 59, 95, 134]. To be used in vivo, the particles should be dispersible in water; they should form a stable colloid in biological media, which are aqueous solutions with significant ionic strengths and rich in proteins; they should not induce cytotoxicity; they should not adsorb too many proteins on their surfaces to avoid uptake by the mononuclear phagocytic system (MPS) cells; and they should be easily conjugated with specific ligands for targeting studies [35, 59, 172]. Therefore, the Au NPs produced by the previous techniques should undergo ligand exchange processes for their surface to be coated with different types of biocompatible, antifouling and/or biologically active molecules [119, 159].

The surface of Au NPs can be functionalized using molecules containing thiols, amines, carboxylic acids and phosphines [140]. The most resistant coatings are formed by making use of ligands that have the highest affinity to Au. Therefore, thiol modification is usually preferred, as the sulphur atoms form a coordinate covalent bond with the metal surface [119]. The surface of Au NPs stabilized with citrate or CTAB can be easily functionalized by exchanging these molecules with thiol containing ligands [159]. The initial stabilizing agent is quickly replaced during the adsorption of sulphur atoms, followed by a slower reorganization and packing of the incoming molecules [52, 119]. The surface of Au NPs stabilized by alkanethiols can be modified by thiol–thiol exchange [159]. This process requires higher molar excess of the incoming ligands [119]. In addition, this method can also be used to produce mixed organic monolayers containing different types of ligands at specific ratios [159]. Finally, polyethylene glycol (PEG) is a commonly used antibiofouling ligand for Au NPs [52]. When this highly hydrophobic polymer is attached at the surface of Au NPs, the long chains prevent the aggregation of the nanoparticles by steric repulsion. PEG also prevents the adhesion of proteins at the surface of the NPs.

2.7 *Cell Toxicity of Au NPs: In Vitro Studies*

Several studies and reviews have reported on the viability of a wide selection of cell types incubated with Au NPs [64, 90, 95, 113, 173]. Several sizes of Au NPs were reported, as well as different surface charges and molecular coatings. The concentration thresholds for which cytotoxicity effects are detected with the main types of Au NPs used in biomedicine can be found in this literature.

3 Principles of Physical Interactions Between Photons and High-*Z* Elements

High-energy photons (X or γ rays) in the range typically used for computed tomography and medical physics, can penetrate long distances in the biological tissues. By interacting with the different chemical elements present in vivo, they undergo specific interactions which makes them useful for whole body medical imaging and therapy. Photons used in medical physics and imaging usually have an energy included between 10 keV and 1 MeV. In that range, photons can be attenuated by four major types of interactions upon colliding with materials: (1) the photoelectric effect, which dominates at the lower energies; (2) the Compton effect, which increases at higher energies and for higher atomic numbers; (3) Rayleigh scattering; and finally (4) pair production, which occurs only at very high energies. A schematic representation of each one of these phenomena is represented in Fig. 5.

Attenuation coefficients calculated for each element, at different energies, and for each one of the interaction mechanisms, allow a comprehensive quantitation of the photon–matter interaction process. The cross section is a measure of the probability for an interaction to occur, and it depends on the atomic number (*Z*). For high-*Z* elements, the probability of interaction is generally higher due to their larger cross section. The total mass attenuation coefficient (μ/ρ) corresponds to the sum of the individual attenuation coefficients (Eq. 1). It represents the probability of interaction per mass unit of a given material (in cm^2/g):

$$\frac{\mu}{\rho} = \frac{\tau}{\rho} + \frac{\sigma}{\rho} + \frac{\kappa}{\rho} + \frac{\sigma_R}{\rho} \tag{1}$$

where ρ is the density of the material, and τ, σ, κ and σ_R are the cross sections for the photoelectric effect, the Compton effect, pair production as well as Rayleigh scattering, respectively.

The photoelectric effect occurs when a photon collides with an electron from one of the inner orbitals of the impacted atom (Fig. 5). The photon is absorbed and the electron is ejected (i.e. the ‘photoelectron’) with a kinetic energy (E_k) corresponding to the difference between the energy of the incident photon ($h\upsilon$) and the binding energy of the electron (E_B). When the photoelectron is ejected from the inner-shell, an electron of an upper shell fills the vacant place, which leads to the emission of either characteristic X-rays or Auger electrons. Characteristics X-rays are emitted from electronic transition rearrangement and their energy is specific to each element. The Auger effect occurs when the energy involved in the transition serves to eject an electron (i.e. the 'Auger electron') present on a higher orbital.

Figure 6a, b illustrates the mass attenuation coefficients (μ/ρ) for each one of the processes occurring in gold and as well as in water. For gold, the photoelectric effect is very strong at energies of 200 keV and below, whereas similar photoelectric interaction rates are reached in water at photon energies of 30 keV and below. The probability of the photoelectric effect to occur is very high for materials made of high-Z elements: in fact, it is proportional to ~Z^4 [14, 89]. Figure 6c illustrates the relative importance of each mechanism for different atomic numbers and photon energies. For conventional external beam therapy with 100–250 keV photons, the Compton effect dominates in the biological elements (C, O, N), whereas the photoelectric effect is preponderant in the interactions with gold. At lower energies, however, the photoelectric effect dominates.

The Compton effect occurs when a photon collides with an electron of the outer shell and ejects it from the atom (Fig. 5). The photon loses part of its energy and is scattered at an angle ϕ. Either it continues its course, or it undergoes further interactions such as Compton, photoelectric or Rayleigh scattering. The ejected electron is referred to as 'Compton electron', and travels in the surrounding material where

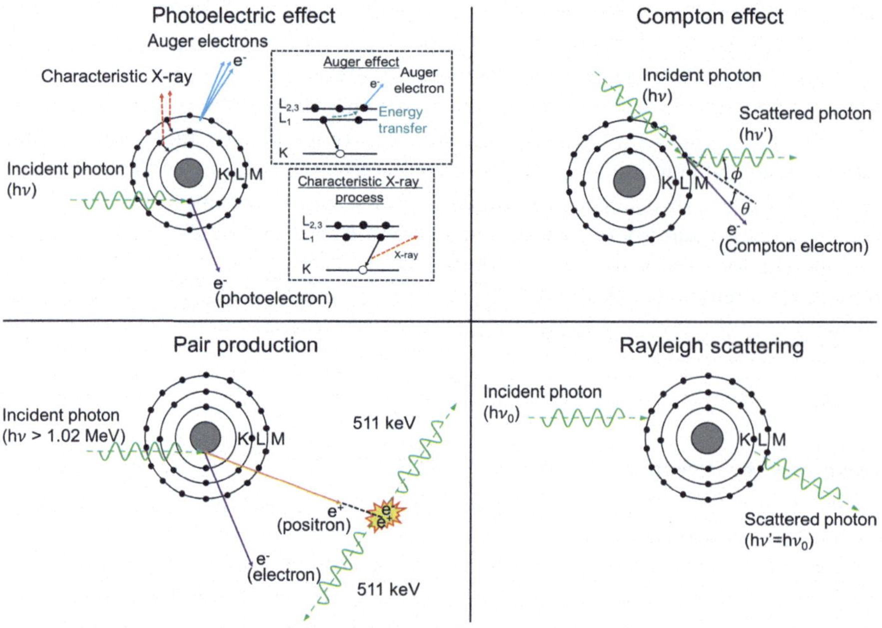

Fig. 5 Illustration of interactions of photons with the atoms that constitute materials and tissues. *Source* The authors (Review, in press at Advanced Healthcare Materials, with permission from Wiley-Verlag)

it may cause subsequent excitation and ionization of atoms. The incidence of the Compton effect is significant at photon energies ranging from 100 keV to 10 MeV (Fig. 6a, b). It is almost independent of Z, and it predominates at intermediate therapeutic energies [89].

In Rayleigh scattering or coherent scattering, the incoming photon interacts with the whole atom. The electrons oscillate in phase and release an extra quantity of energy in the form of a secondary photon having the same energy as the incident photon, however emitted in a slightly different direction (Fig. 5) [14, 89]. Overall, this interaction does not contribute to the dose enhancement in radiotherapy, since no energy is transferred to orbital electrons, and therefore, ionization does not occur. Usually negligible at high energies, Rayleigh scattering is more probable with high-Z

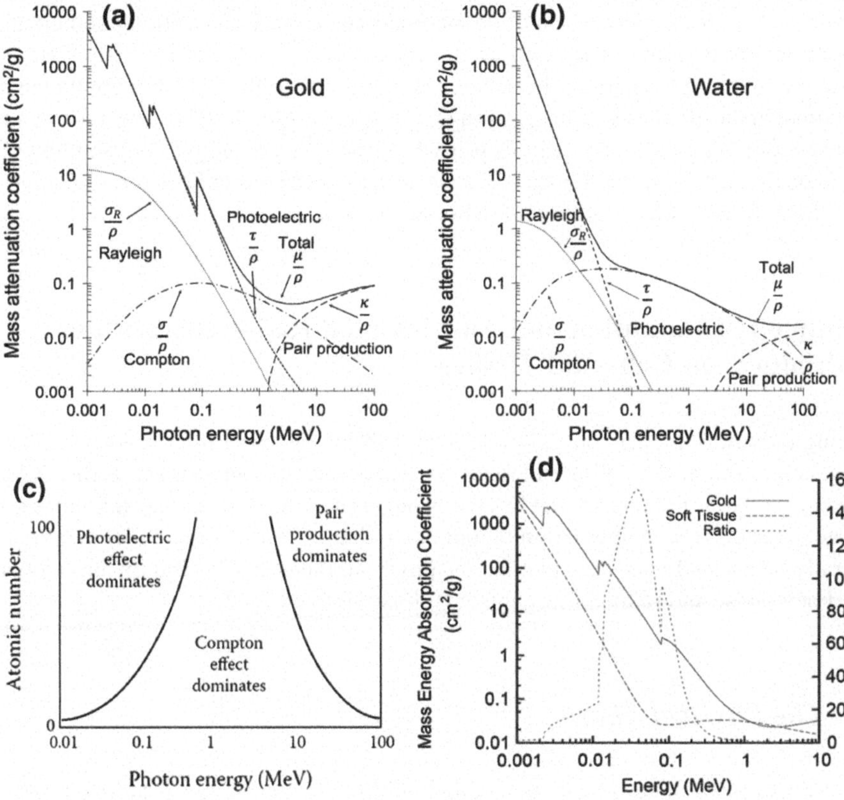

Fig. 6 **a**, **b** Mass attenuation coefficients for gold and water according to the photon energy (data from the NIST) [136]. **c** Diagram of the different photon–matter interactions taking place depending on the atomic number (Z) and the energy of photons. Reproduced from [14] with permission from Wiley-VCH Verlag. **d** Comparison of the photon mass energy absorption coefficients for gold and soft tissue. The ratio of the mass energy absorption coefficients is shown as a function of energy. Reproduced from [25], based on data originally from [80], with permission from the Royal Chemical Society

materials such as gold, as well as for low and very low-energy photons (Fig. 6a, b) [89].

Finally, pair production arises from the trajectory of a high-energy photon (>1.02 MeV) passing near the nucleus of the atom. In this specific situation, the energy of the photon is converted into mass and an electron–positron pair is created (Fig. 5) [14, 89]. The energy in excess (E = 1.02 MeV) is equally distributed in the form of kinetic energy (e.g. 511 keV) between the positron and the electron, and this energy is lost via ionization as the particles travel through matter. When the positron comes to rest, it interacts with an electron of the material, resulting in the emission of two annihilation photons of 511 keV, travelling in opposite directions. Pair production interactions are dominant at high energies (>10 MeV, e.g. external beam radiotherapy) and are more likely to happen for high-Z elements, since the cross section for this kind of interaction varies according to Z^2 (Fig. 6a, b) [14, 89].

Figure 6d shows a comparison of the photon mass energy absorption coefficients for gold and soft tissue. The ratio of the mass energy absorption coefficients is shown as a function of energy. The difference of mass absorption energy coefficients is maximal in the 10–200 keV energy range, the same as used for CT imaging and for the vast majority of radiotherapeutics procedures used in oncology. The coupling of these modalities with Au NPs is discussed later in this chapter: in Sect. 5 for imaging, and in Sect. 6 radiosensitization and radiooncology.

4 Impact of Radiation and Au-Mediated Radiosensitization Products on Cells and Tissues

Ionizing radiation impacts on the biological systems by generating several defects in organelles and at the cell membranes. In particular, the impairment to the DNA molecules is by far the impact that has the strongest and the most lasting influence on the fate of the cells. The interaction of ionizing radiation with biological systems can generally be divided into three phases: the physical, the chemical and the biological [76]. Each phase and related mechanism is illustrated in Fig. 7.

4.1 The Physical Phase

The physical phase takes place upon collision of the photon with the atoms present in the cells and more globally in the tissues (e.g. water, proteins, lipids and DNA). Through these interactions, energy from the photons is deposited in the tissues, causing thereby several direct damages to the molecules. Low-energy electrons and free radicals can also be generated through the interaction of the primary photons with the tissues [74, 76]. The interaction of photons with high-Z elements such as gold atoms results in the emission of further secondary species: photoelectrons, Auger

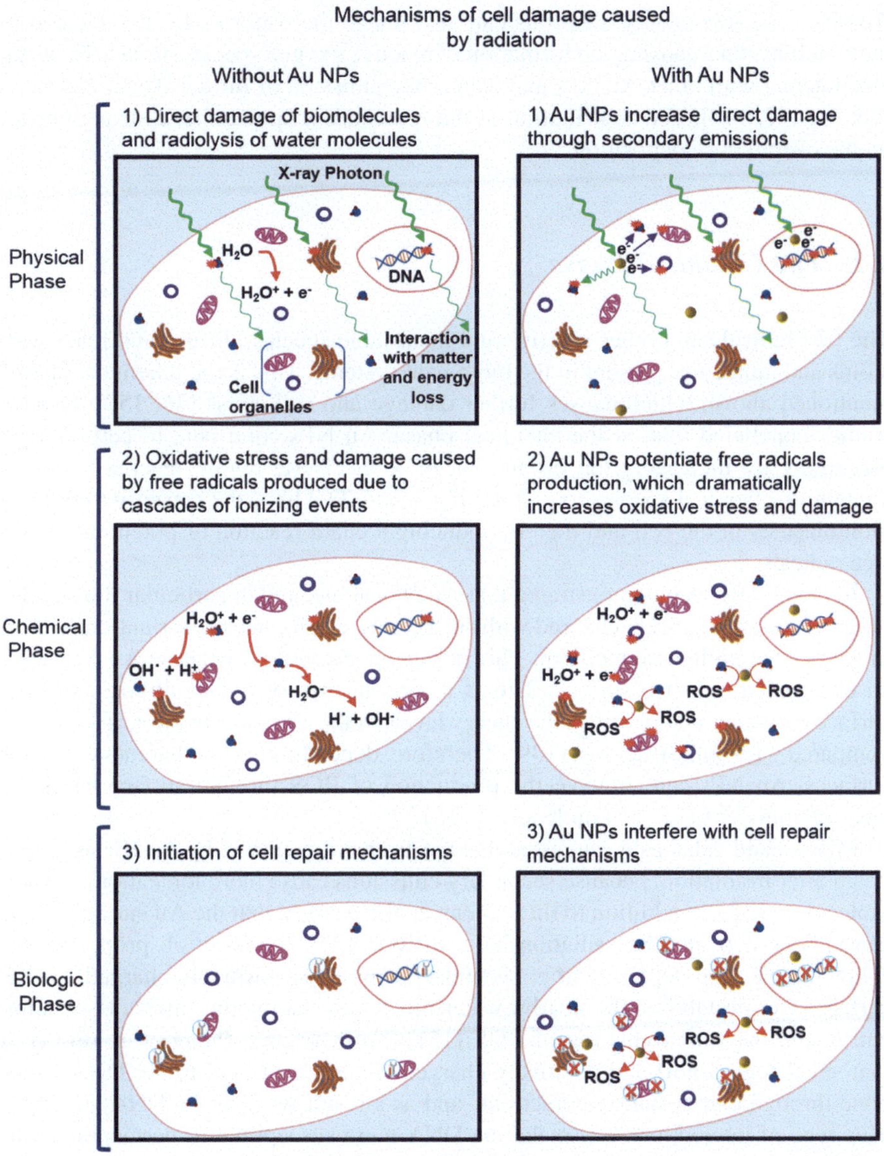

Fig. 7 Representation of the different steps of the impact of ionizing radiation on the biological tissues and cells: the physical, the chemical and the biological phases are depicted. Illustrations are inspired by descriptions provided in [74, 76, 157]

electrons and characteristic X-rays [76]. Thus, in addition to the damage created by the energy transferred from the incident photon, the DNA, proteins and lipids of the irradiated cells, can be further injured by the secondary emissions coming from the

Au NPs. The low-energy electrons can also ionize the water molecules located in their vicinity, thus causing the formation of reactive oxygen species (ROS). They can also interact with other Au NPs and lead to the emission of further Auger electrons [26, 74, 102, 143]. The contribution of these secondary species has a strong impact on the overall dose deposition.

4.2 The Chemical Phase

The free radicals and other reactive and metastable species, diffuse and react with atoms and molecules present in the biological systems. This is the 'chemical phase' mentioned above, which causes further damage and cell stress [76, 157]. Several studies concluded that, at the chemical phase, Au NPs contribute to cell damage essentially by increasing the production of ROS. These highly reactive oxygen-containing chemical species (e.g. $^{\bullet}OH$, H_2O_2 and $^{\bullet}O_2$) have the capacity to oxidize biomolecules in the cell and thereby inducing a chain reaction of potentially toxic free radicals.

In fact, it has been demonstrated that Au NPs alone, and in particular the smaller ones (<5 nm) not entirely covered with surfactants, can induce the formation of ROS in water or in biological medium. This occurs spontaneously because Au NPs have an electronically active surface. This is a consequence of the small size and high surface curvature that modifies the energy levels of the electrons in the surface atoms compared with bulk structures [139]. Therefore, depending on the 'bareness' of their surfaces, Au NPs can catalyze the production of ROS through surface-mediated electron transfer to oxygen molecules [76, 139].

Misawa and Takahashi suggested that the production of ROS by Au NPs is potentiated after irradiation, because secondary emissions cause more ionization of water molecules [132]. In addition to this, Cheng et al. observed that the Au surface atoms can also be activated by radiation-induced ROS [37]. Porcel et al. proposed that following electron depletion after secondary emissions, positively charged Au NP surfaces can destabilize the nearby water molecules, facilitating their dissociation and increasing the production of ROS [151]. Furthermore, Zheng et al. observed that small (e.g. 5 nm) and positively charged Au NPs can bind to the DNA backbone through electrostatic interactions, and as a result weaken the DNA structure. This type of interactions can make the DNA more susceptible to permanent damage, in particular when it is impacted by low-energy electrons produced through the interaction of photons with Au NPs [76, 208].

4.3 *Measurement of Physical and Chemical Impact of Au NPs on DNA*

It is difficult to clearly dissociate the impact of physical damages, from the impact of purely chemical damages caused by impacting photons to the biological tissues. The mechanisms leading to increased DNA damage and to cell death, after irradiation in the presence of Au NPs, are not completely understood. However, Au NPs have a critical effect in each phase of radiation-induced cell damage. DNA impairment is the most important parameter to take into account when evaluating the efficacy of radiotherapy [74, 76, 157]. Overviews of the mechanisms of radiobiology can be found in comprehensive books on the topic [73]. Au NPs used as radiosensitizers increase dose deposition and localize their impact more precisely into targeted tissues such as in tumours [74]. Several experimental studies indicated that dose enhancement in the vicinity of Au NPs is much higher than initially predicted by theoretical simulations [88, 111, 129]. Such observations led to a number of investigations aiming to understand the mechanisms behind the radiosensitizing effect induced by Au NPs and its influence on the radiobiological mechanisms.

Traditionally, the most direct means for quantifying the impact of ionizing radiation on cells was by measuring the frequency of chromosomal aberrations in function of the absorbed dose [73]. Nowadays, the damage caused by ionizing radiation to cells in presence of Au NPs is generally quantified by clonogenic assays [27, 39, 47, 88, 123, 129, 166, 181, 207]. This analysis provides information about cell survival after irradiation, and it allows a quantification of the sensitizing effect generated by Au NPs [27, 123, 181].

However, more analytical tests are necessary to fully quantify the impact of the radiosensitizing effect on the cells. Double-strand breaks to the DNA can be quantified by using the γ-H2AX or the 53BP1 foci formation assays [27, 39, 47, 129, 166, 181, 207]. The presence of these proteins is an indicator that irreparable damage was caused to the DNA. The quantification of ROS production is usually performed using the dichloro-dihydro-fluorescein diacetate (DCFH-DA) assay [27, 123]. By combining the complementary information provided by different assays, Butterworth et al. demonstrated that irradiation of PC3 cells containing Au@DTDTPA NPs induced a 1.7-fold increase in residual DNA double-strand breaks (e.g. 24 h after irradiation) compared with cells treated with radiation only. In addition, the authors observed that incubation of PC3 cells with Au@DTDTPA NPs did not cause a statistically significant increase in the production of ROS, compared with non-treated cells [27]. Overall, these observations indicate that the natural DNA repair mechanisms can be impeded when increasingly complex damages are caused to the helixes. However, this effect seems to be cell-dependent [27]. On the other hand, after X-ray irradiation, Ma et al. observed a 1.6-, 1.2- and 1.1-fold increase of ROS concentration in KB cells incubated with Au NPs, Au nanospikes and nanorods, compared with the control group. These studies suggested that increased ROS production is one of the possible mechanisms behind Au NPs radiosensitization in the chemical phase [123].

4.4 The Biological Phase

The biological phase begins when repair mechanisms are triggered in response to the physical and to the chemical damages caused to vital structures (DNA, mitochondria and other organelles) [76, 157]. In particular, the increased ROS production and the indirect DNA damage initiated during the chemical phase can (1) induce oxidative stress, (2) disrupt the cell cycle and (3) delay or inhibit DNA repair [76, 123, 157]. Oxidative stress can induce damage to almost all organelles, and therefore lead to cell death in a variety of ways. Although the exact impact of Au NPs on oxidative stress induction is not entirely well elucidated at the moment, Au NPs can lead to an increased production of ROS through catalytic reactions. High intracellular concentrations of ROS impair mitochondrial function, and the presence of Au NPs in cells can inhibit the mechanisms of proteins implied in cellular oxidation homeostasis [76, 116, 157, 187].

The cell cycle also plays an important role in cell sensitivity and survival to radiation. It has been demonstrated that cells are more sensitive to radiation-induced damage when they are at the end of cycle step G2 and during mitosis. On the other hand, cells in late S phase are more resistant to radiation [76, 157]. The analysis of cell cycle is often performed by flow cytometry, after DNA staining with propidium iodide and 5-bromo-2′-deoxyuridine [47, 123, 207]. This method quantifies the DNA content per cell, and with this information, one can identify the phase of the cell cycle where the impact of radioactivity is the highest. Roa et al. observed that Au NPs (e.g. glucose-coated Au NPs) can synchronize and stop cells in the G2/mitosis phase. This mechanism, in turn, increases cell sensitivity to radiation damage [193]. After exposing cells to different types of Au NPs, Ma et al. observed a slight increase on the percentage of cells in the G2/M phase for Au NP-treated cells unexposed to radiation; after X-ray treatment, the percentage of cells in the G2/M phase appeared to be significantly increased [123]. However, the Au NP-mediated ‘cell cycle synchronization effect’ reported in the previously cited papers is being debated, since other research groups have not detected any cell cycle change after treatment with Au NPs [47, 207]. One can argue that the variation in the results between the studies can be attributed to different experimental conditions, such as the physico-chemical properties of Au NPs and the cell lines [76, 157]. A systematic and more comprehensive methodology should clearly help to identify and to quantify this effect in the near future.

When DNA is affected by physically or chemically induced defects, repair mechanisms are activated that, in principle, help the cell to recover its vital functions. However, as seen in the previous sections, the presence of Au NPs in the cells can multiply the impacts on the DNA, potentially leading to a delay or to the inhibition of the DNA repair mechanisms. Usually, counting of double-stranded DNA damage takes place immediately after (e.g. about 1 h) and at 24 h post-irradiation. If the damage persists or even increases at 24 h, the residual damage reflects an inhibition or a delay of the repair process. Chithrani et al. observed higher concentrations of γ-H2AX and 53BP1 foci at 4 and 24 h after irradiation of HeLa cells incubated with

Au NPs [39], whereas Cui et al. and Butterworth et al. did not observe increase in the double-strand DNA damage immediately after irradiation (e.g. 30 min to 1 h). Instead, they detected an increase in the residual DNA damage at 24 h post-irradiation [27, 47]. Chen et al. detected a 2.02 and 1.95-fold increase in the amount of γ-H2AX foci at 2 and 4 h after irradiation of cells incubated with BSA-coated Au NPs [36]. The concentration of double-stranded DNA damage was similar at 24 h after irradiation, suggesting that Au NPs did not interfere with the repair mechanism [36]. Jain et al. also concluded that Au NPs neither increase radiation-induced double-stranded DNA damage formation nor inhibit DNA repair after irradiation of cells at clinically relevant MV X-ray energies [83]. Thus, further studies are necessary to clarify several aspects of the influence of Au NPs on the DNA repair mechanisms [76, 157].

5 Performance of Au NPs as X-Ray Computed Tomography Contrast Agents

X-ray computed tomography (CT) is a common tool in modern medicine. It is essential for the diagnosis of bone pathologies and trauma, to identify the calcification of tissues, as well as for visualizing several lung diseases [191]. In addition, the relatively short acquisition times typical of CT imaging enables a variety of cardiac and vascular diagnostic procedures [92, 101]. In the preclinical field, microCT acquires images of higher spatial resolutions than clinical CT (<50 μm), making the imaging modality an essential tool in the development of small-animal models of various pathologies [96, 191].

CT imaging is based on the attenuation of high-energy photons (typically in the range 25–140 keV) by biological tissues. In this energy range, the main photon–matter interactions taking place are the photoelectric and the Compton effects. Photons of a given energy irradiating a material, are attenuated a certain linear rate (per depth of penetration in this substance). This rate is the linear attenuation coefficient (μ), and it is proper to each compound [24]. This characteristic is intrinsic to each tissue, depending on its elemental composition. During CT acquisition, X-rays are projected at different angles across the body, and a detector collects the transmitted photons. Then, the numerical analysis allows the 3D reconstruction of images representing the attenuation of X-rays by the biological tissues, which are largely guided by their respective attenuation coefficients (μ). The differential X-ray attenuation maps are presented in Hounsfield units; these are obtained by normalizing the total μ measured in each voxel to the μ of a known reference (e.g. water) using Eq. (2) [69, 92].

$$HU = 1000\left(\frac{\mu - \mu_{\text{water}}}{\mu_{\text{water}}}\right) \tag{2}$$

The resulting map is a 3D representation of the various volumes presenting different densities in the biological system (e.g. bone, muscle, fat, etc.).

5.1 Attenuation of X-Rays by High-Z Elements

The attenuation of X-rays by the different biological tissues in the body, or by exogenous materials such as gold, is dictated by the probability of interaction of each element in function of the energy of X-rays. Attenuation of X-rays is quantified by the mass attenuation coefficient (ρ) which is often normalized to the density of the material (μ/ρ) and expressed in units of cm^2/g. Figure 8 illustrates the mass attenuation coefficients of several elements, in function of the energy of X-ray photons. Compared with carbon, calcium has a higher mass attenuation coefficient at photon energies common for CT imaging (e.g. 40–100 keV). This is the main reason why bones and calcified tissues, appear strongly contrasted in CT imaging.

In order to provide strong contrast enhancement in the blood, or for targeting certain types of diseases that are not characterized by strong differences of tissue density, it is necessary to rely on contrast agents. These are made with elements that have the capacity to strongly attenuate X-ray radiation. Any element with a high atomic number (Z) is a good X-ray attenuator and can be used as a contrast agent for CT, given that it is not too toxic for biological applications [121]. In the energy range of CT imaging, attenuation by high-Z elements mainly results from the photoelectric absorption effect. Moreover, the probability of an X-ray photon to interact with a given element varies as $(Z/E)^3$, where Z is the atomic number and E is the photon energy [24]. Just above the binding energy of the core electrons, the absorption of photons by the atoms is significantly higher [24, 69]. These 'jumps' in the mass attenuation coefficient graphs (Fig. 8) are referred to as the 'k-edges'. The presence of such discontinuities must be taken into account in the selection of a photon energy that maximizes the contrast effect. For instance, the k-edge of gold being 80.7 keV; performing a CT-imaging scan at an energy of 81 keV multiplies by at least a factor of 4, the mass attenuation coefficient compared to a scan performed at an energy of

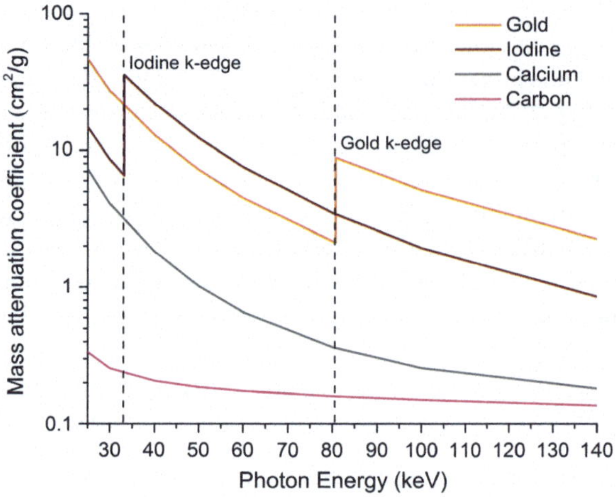

Fig. 8 X-ray mass attenuation coefficients for gold, iodine, calcium and carbon within the X-ray energy range used for CT imaging (data from [136])

80 keV. At 81 keV, the mass attenuation coefficient of Au is much higher than that of iodine, which is the most widely used element for the fabrication of clinical CT contrast agents nowadays.

5.2 Conventional Iodinated Contrast Agents

CT contrast agents are usually made of iodinated hydrosoluble molecules [114]. Iodine is an element that is endogenous to the body, and it strongly attenuates the photons (Fig. 8). Among most common formulations approved by the health authorities, figure monomeric or dimeric water-soluble tri-iodinated derivates of benzoic acid [106, 175]. After intravenous injection, these small molecules diffuse to all vascularized tissues through the fenestrations of blood vessels. Then, they are removed by the blood and they are eliminated mostly by the kidneys [106]. As these contrast agents have a short circulation time (e.g. 15–40 s), the images must be acquired quickly after the injection. This is a significant limitation in preclinical studies, because the small animals eliminate these products faster and the acquisition times in microCT usually take longer than with clinical CT scanners [96]. The main strategy to overcome this limitation is through the administration of multiple or long, continuous injections of the contrast agents in the small animals [96, 174]. This unfortunately raises toxicity concerns. To maintain the image quality as the contrast effect fades away, the X-ray exposure could also be increased. However, the dose of a single anatomical image is around 0.1 Gy. Higher imaging doses of 1.5 Gy have been reported for cardiac gated imaging in mice but should never exceed 6 Gy even in multiple sequential scans procedures. This threshold is considered lethal for small rodents [156].

5.3 Blood-Pool Contrast Agents

In order to prolong the contrast-enhancement effect, blood-pool contrast agents were developed [114]. Blood-pool contrast agents cannot go across the gaps between the endothelial cells of normal vessels, and therefore they remain for longer times in blood circulation, thus providing an improved time window for visualization of the vascular system. By increasing the size of contrast agents up to the 50–100 nm, renal elimination through glomerular filtration is reduced. The extended blood half-life of these molecules is also a result of a delayed recognition by the cells of the mononuclear phagocyte system (MPS), due to the use of well-selected surfactants [148]. Different methods and materials have been used to produce blood-pool contrast agents, and many of them are commercially available for preclinical research [96, 106, 175]. They are usually made of macromolecules, or of oil emulsions containing iodine-modified lipids. They can also be made of iodine-encapsulating micelles and liposomes [183]. Blood-pool CT contrast agents can also be based on inorganic NPs: bismuth-sulphide, barium, alkaline-earth metals or gold [96, 121]. Among these,

Au-based nanoparticles have provided the most promising proofs-of-concepts for the development of a new generation of blood-pool contrast agents for CT imaging.

5.4 Au-Based Blood-Pool Agents

Among all elements considered for CT contrast media applications, gold shows good biocompatibility and has higher X-ray attenuation coefficients than iodine at the energy range of both CT and microCT [67, 82, 85]. As mentioned in the previous section, the photoelectric effect is inversely related to the energy of the photons ($1/E^3$), except for discontinuities (i.e. absorption edges) close to the binding energies of the inner-shell electrons, at which the probability of interaction increases significantly [82]. For example, Au has a k-edge absorption at 80.7 keV [67, 82], thus when peak voltages (kVp) from 100 to 140 are used, the mass attenuation of Au (μ/ρ) increases up to 8.9 $cm^2\ g^{-1}$. In comparison, the mass attenuations of C, Ca and I in the same energy range are 0.161, 0.366 and 3.51 $cm^2\ g^{-1}$, respectively [80]. In that range, gold attenuates X-rays approximately 55 times more than C (present in soft tissues), 24 more than calcium (present in bones) and at least 2.5 times more than iodine (Fig. 8). As a consequence, in this range of energies, Au NPs can be administered at lower elemental concentrations than the iodine-based contrast agents, for a similar level of contrast enhancement [67, 137]. Au NPs also attenuate X-ray photons more efficiently by Compton scattering than soft tissues (e.g. carbon), bone (e.g. calcium and phosphorus) and iodine-based contrast agents, because of higher electron density.

In 2006, Hainfeld and colleagues were the first to report the use of Au NPs as contrast agents for vascular imaging [71]. In this study, 1.9 nm Au NPs were injected in a mouse model, and provided evidence of longer retention times and superior contrast to iodine with resolution of vessels as small as 100 μm [71] Despite high initial blood concentrations (10 mg/ml blood), no haematological nor biochemical abnormalities were detected at 11 and 30 days post-injection. Au NPs were found to be excreted by the kidneys.

Since then, several Au NPs-based formulations have been developed and used in preclinical studies [93, 98]. A variety of synthesis techniques have been employed: Turkevich [56], Brust–Schiffrin [42] and seed-mediated growth [150]. Galper and colleagues demonstrated that the maximum contrast enhancement for Au NPs dispersed in water (core diameter: 3.6 ± 1.1 nm), was achieved at 120 kVp (Fig. 9) and it was 1.9 times higher than with a commercial iodine contrast agent for an equivalent elemental concentration [67]. Jackson et al. showed that Au NPs provide better contrast over iodine at peak voltages lower than 50 kVp. However, no significant improvement was found between 70 and 90 kVp [82].

The development of fine Au NPs of well-controlled size and targeting ligands opened the possibility to improve image contrast by increasing the concentration of Au NPs in disease tissues [56]. NPs made of high-density lipoprotein (HDL) [42, 43] and low-density lipoprotein (LDL) [6] containing Au NPs also showed promising results. More specifically, Au-HDL NPs were used to specifically target and image

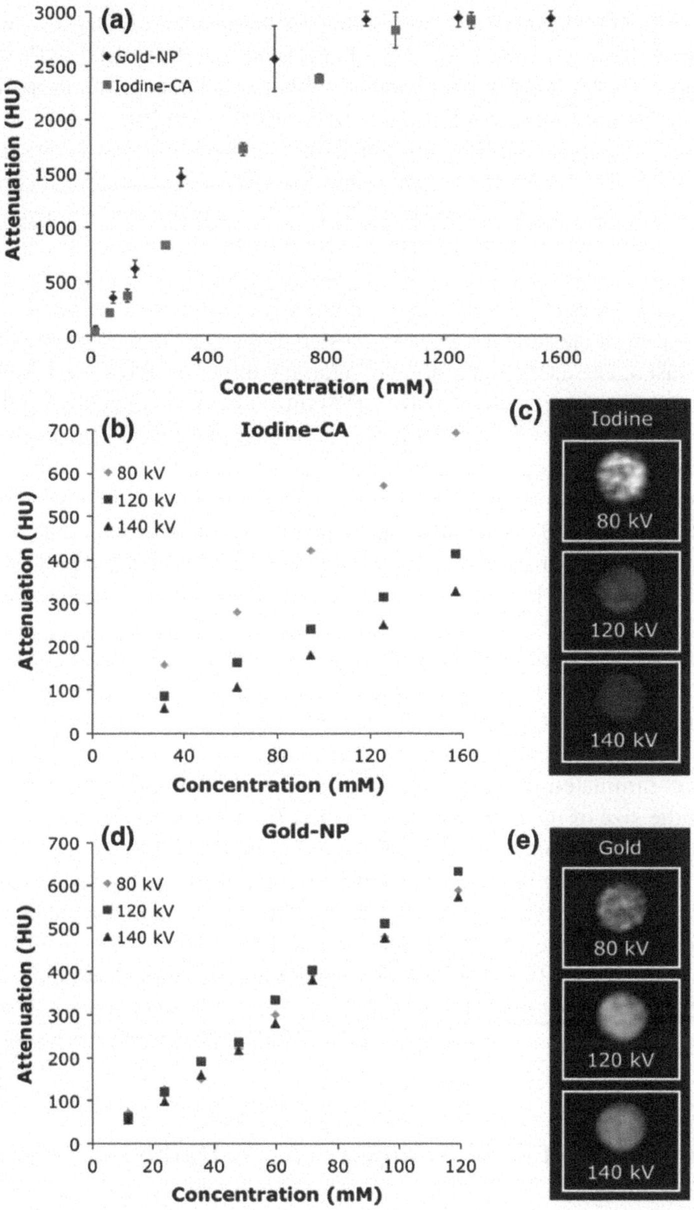

Fig. 9 **a** Graph (in Hounsfield units: HU) of attenuation versus concentration of Au NP and iodine-CA scanned in air at 100 mA and 140 kV. Error bars are standard deviations; where not seen, the error bars are hidden by the data point. Attenuation of **b** iodine-CA and **c** Au NP scanned in water with a calcium phosphate matrix at various tube voltages. CT images of **d** 157 mM of iodine-CA and **e** 119 mM of Au NP (windowing, 800Y3000 HU). Data from phantom 2. Error bars omitted for clarity. From [67], reproduced with permission from Lippincott, Williams and Wilkins

macrophage content in atherosclerotic plaque. In this system, the targeting was mediated by the apolipoprotein A–I [42, 43]. On the other hand, Au-LDL NPs were used to target tumour associated macrophages. In this case, the targeting properties were attributed to the apolipoprotein B100 and to the EPR effect [6].

Dendrimer-entrapped Au NPs (2–4 nm diameter cores) have also been synthesized and tested in vivo [103, 145, 196]. Ultra-small Au NPs (~2.4 nm diameter core) coated with dithiolated polyaminocarboxylate (DTDTPA), and injected intravascularly in the mouse model, were found massively excreted by the urinary ways (>60% of injected initial dose after 24 h) [7]. A significant fraction of the product was also found in the faeces after 72 h (2–3% cumulative, on initial dose injected).

One of the strategies that has been suggested to increase the sensitivity of Au NP-based contrast agents is the integration of a large number of ultra-small Au NP cores (e.g. 1.9 nm in diameter such as Aurovist™) into polymeric micelles ('gold-loaded polymeric micelles', or GPMs) [3]. It was demonstrated that GPMs in the form of micelles containing a large number of ultra-small Au NPs (1.9 nm diam.), remain in the blood for several hours (Fig. 10a, b) and accumulate at high concentrations in tumours (Fig. 10c, d). They also show much higher liver and spleen retention compared with their ultra-small 1.9 nm diameter individual counterparts (Fig. 10d). This could be a strong limitation to their approval for clinical applications by the health authorities.

Overall, Au NPs used as CT contrast agents, and in particular, very small ones (e.g. ~2 nm) provide longer long blood half-lives than the small iodinated molecules conventionally used in CT imaging. Few evidences of toxic reactions in the mouse model have been found in preclinical studies until now [28, 71, 137]. The smallest Au NPs are eliminated by glomerular filtration in the kidneys within minutes. By increasing the size of the contrast agents, either by using a polymer backbone, dendrimers or a core of inorganic material, the blood half-life of the contrast agent can be longer. However, the largest NPs cannot be filtered by the kidneys and therefore represent a potential toxicity risk. Indeed, their hydrodynamic diameter is usually larger than the pore size of the glomerular wall [35, 134]. This is one of the main reasons why the majority of blood-pool contrast agents, and in particular, the ones based on particles larger than 10 nm in diameter, have not been approved for clinical applications until now.

6 Performance of Gold Nanoparticles as Radiosensitizers in Oncology

As described in the previous sections, gold has a great potential for X-ray imaging, at least in preclinical applications and thanks to the fact that is chemically inert in biological media [182]. In addition, the size and shape of Au NPs are relatively easy to control, and ligands can be readily attached to their surface [52]. In addition to imaging, Au NPs can also be used as radiosensitizing agents for the radiotherapy of different tumours [25, 44, 70, 102]. The strong attenuation of photons by high-Z materials generates secondary products (e.g. Auger electrons, characteristic X-rays and photoelectrons) that contribute to enhancing the dose deposited by the primary photons.

In radiooncology, the treatments are performed by irradiating tumours with a beam of photons at a specific energy. The photons interact with the atoms of the tissues through the particle–matter interaction mechanisms described in Sect. 3. In particular, secondary emissions of various natures (electrons and photons) are ejected from the atoms of the irradiated materials, and these account for the most important fraction of the deposited dose in the tissues [14, 89]. The secondary emissions subsequently interact with atoms present in the surrounding environment, producing cascades of ionizing events [14, 89]. Generally speaking, radiotherapy treatments that could accommodate the use of Au NPs are mainly divided in two variants: (1) external beam

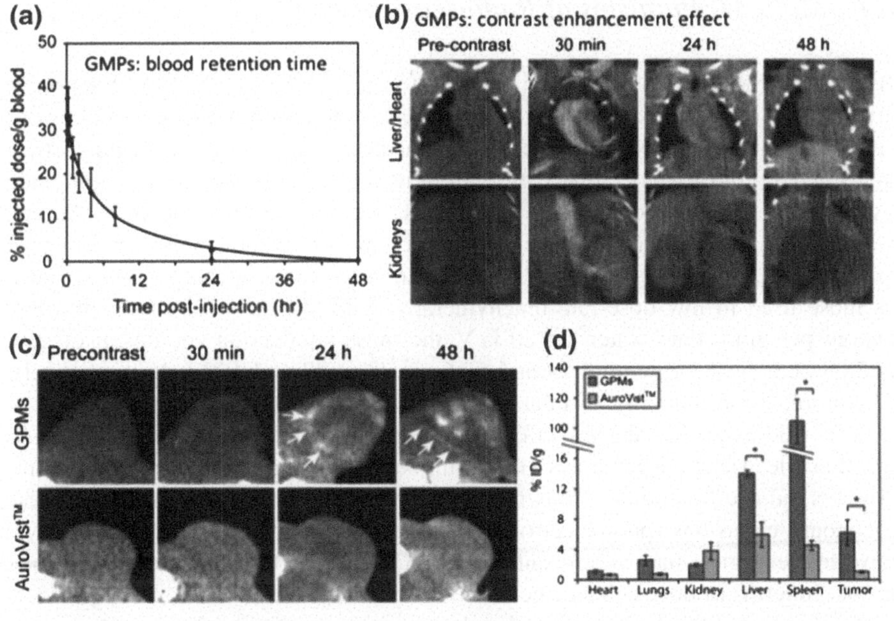

Fig. 10 Blood clearance profile of gold-loaded polymeric micelles (GMPs), measured by elemental analysis ($n = 3$). **b** Serial CT coronal views of a mouse following retro-orbital injection of 200 μL of GPM solution (650 mg/kg). Coronal views of heart and liver (top) and inferior vena cava and kidneys (bottom) are shown. A very strong contrast is seen at $t = 30$ min. **c** In vivo CT images of nu/nu nude mice with HT1080 flank tumours, injected with GMPs, as well as the ultra-small Au NPs Aurovist (1.9 nm diameter NPs, also enclosed in the GMPs compound). Representative CT images in the axial plane prior to injection (precontrast) and 30 min, 24 and 48 h post-injection of GPMs ($n = 3$) or AuroVist ($n = 3$). Tumour boundaries are indicated by white arrows. **d** Elemental analysis in excised and dissolved organs of gold distribution at 48 h following the administration of GPMs or AuroVist. The asterisk indicates statistical significance ($p < 0.05$). Reprinted from [3] with permission. Copyright 2014, the American Chemical Society

radiotherapy, where the radiation source is external to the patient, and which exploits a photon energy range between ~0.1 and 8 MeV; and (2) low-dose brachytherapy, in which radioactive seeds are inserted directly in the cancer tissues, for a photon energy range between 20 and 28 keV [161].

Over the recent years, several strategies have been investigated to maximize the dose enhancement generated by Au NPs distributed in tissues; simulation studies and experimental approaches have been tested (e.g. external beam radiation therapy combined to Au NPs, brachytherapy combined to Au NPs and radioactive NPs). In vivo experiments have been performed to demonstrate the therapeutic effect caused by the combination of Au NPs with various radiation sources on xenograft tumours. Here, we review the comprehensive aspect and the main outcomes related to the use of Au NPs as radiosensitizers in radiation therapy.

6.1 Main Mechanisms of Radiosensitization

By using a radiosensitizer, the sought effect is a maximization of the fraction of attenuated photons and consequently, an increase in the overall quantity of electrons cascading in the tissues. The mechanisms leading to the radiosensitization effect generated by gold atoms irradiated by photons in a biological tissue can be highlighted by a careful examination of Fig. 6a, b (mass attenuation coefficients of Au and water). First, at an energy of 20 keV, the total mass attenuation coefficients of gold and water differ by a factor of 97. This means that, for low-energy photons such as those used in low dose-rate brachytherapy, gold attenuates substantially more energy per gram than water. At 20 keV, the mass attenuation coefficient of gold is 78.83 cm^2/g; it drops to 2.19 and 5.16 cm^2/g at 80 and 100 keV, respectively. Therefore, the probability of interaction of photons with gold is much higher at 20 keV, where the photoelectric effect is the predominant interaction mechanism. As the photoelectric effect generates a large number of secondary electrons and photons, Au NPs radiosensitization is maximized with low-energy photons. Although these secondary emissions (photoelectrons, Auger electrons and characteristics X-rays) have low energies, they significantly contribute to dose (energy) deposited in the tissue. In terms of dose enhancement, the lower energies of the secondary products are compensated by their relative abundance in comparison to the flux of primary photons.

Energy deposition occurs when secondary electrons gradually lose their kinetic energy through many 'collisions' with orbital electrons present in the biological matter (Fig. 11). The collisional behaviour causes ionizations and excitations along the path of the electron, and these events can be defined as points in space where energy is transferred to the medium (in the biological tissues: mainly water). The amount of energy transferred per unit volume is defined as the energy lost by the electrons following each interaction ($\Delta E = E_1 - E_2$). The electron continues to travel in the medium where it interacts with other atoms. At each collision, it transfers a small amount of energy until it comes entirely to rest ($E_{\text{kin}} = 0$).

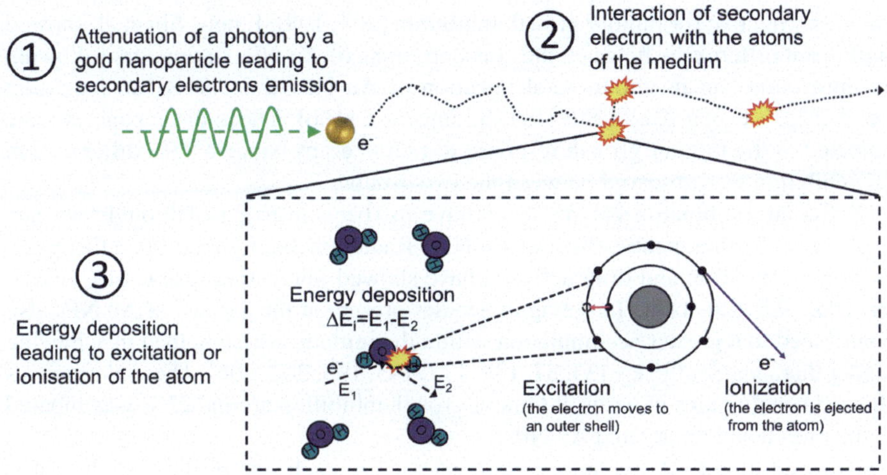

Fig. 11 Schematic representation of the mechanisms leading to energy deposition in biological tissues. First, an incident photon (1) interacts with the Au NPs and cause the emission of a secondary electron (1). This electron travels until it collides with an orbital electron of an atom in the medium (2). At this point, a fraction of the kinetic energy (E_{kin}) of the electron is transferred to an orbital electron of this atom (energy deposition). This results in electron ejection (ionization) or migration to a higher shell (excitation) (3). *Source* The authors (Review, in press at Advanced Healthcare Materials, with permission from Wiley-Verlag)

6.2 Au NPs as Radiosensitizers in External Beam Therapy

The use of Au NPs as radiosensitizers during radiotherapy treatments has emerged as one of the most promising applications of gold colloids. By directly injecting Au NPs in cancer tissues, followed by adequate irradiation by mid-to-high energy photons, it is possible to generate a large range of low-energy products (electrons and photons) in the immediate vicinity of the Au NPs [25]. In turn, these low-energy electrons and photons lead to dose enhancement in the surrounding tissues, to the production of reactive oxygen species inside or in the close vicinity of the cells. Overall, these products cause significant further damages to the tissues [25, 37, 76, 128, 132, 139, 151, 208]. It is generally assumed that the internalization of Au NPs brings them closer to the cell nucleuses, with potentially stronger impact on the DNA.

The first experimental evidences of the radiosensitization effect of Au NPs on tumoural tissues have been exhibited by Hainfeld et al. [72]. The experiment was conducted with external beam radiation (EBRT), where mice bearing xenograft tumour (breast) were injected intravenously with Au NPs. A delay in the tumour growth was observed for a group of animals treated with a combination of Au NPs and external radiation. Since then, many studies using external beam source have been reported in the literature.

Among the main parameters and conditions investigated, the Au NPs size and administration routes showed interesting results. Au NPs of 15–30 nm seems to be the most effective for tumour growth inhibition [204, 198]. Lately, Shi et al. showed significant differences between the concentrations of Au NPs in tumours following an intravenous or an intratumoral injection of Au NPs (i.t.: 496 ± 106 μg Au/g; i.v.: 5.27 ± 1.17 μg Au/g), 8 h after the injection [166]. Those discrepancies were reflected in the tumour growth response to radiotherapy, since a 25% difference in the inhibition was observed between the two groups.

Different combinations of treatments have also been investigated through the years to enhance the therapeutic effect of Au NPs. Radiation and hyperthermia therapies, as well as radiation and drug delivery, have showed interesting results [46, 75, 87, 99, 105, 122, 126, 144]. Targeting molecules present at the surface of Au NPs also contributed to a greater accumulation within the tumour, which helped in achieving higher therapeutic efficacy [33, 87, 149, 177, 195, 198, 202, 206]. These alternatives showed benefit gains in terms of tumour growth inhibition around 25% as compared to the ones using untargeted Au NPs.

However, due to the various conditions used in each one of these studies, it is still difficult to have definitive conclusions about the optimal conditions allowing the highest dose enhancement. Among the drawbacks related to the combination of EBRT and Au NPs, there is the energy range of the EBRT that is not suitable to benefit the radiosensitization effect, as well as the potential toxicity issues related to the Au NPs concentration that need to be injected to observe therapeutic effect.

6.3 Dose Simulations Have Confirmed a Good Match Between Au NPs and Brachytherapy Sources

One of the most important aspects related to clinical radiation therapy is treatment planning. This medical physics procedure is performed before radiotherapy treatments, and they consist in dose simulations prior to irradiation of the tissues. Several codes and programmes based on Monte Carlo calculation have been developed through the years to simulate radiation passage through biological tissues (e.g. Geant4, EGS, MCNP and PENELOPE: [38]). Early in the development of Au NPs as radiosensitizers, simulation studies have been conducted to identify the most critical parameters enabling dose enhancement. Quickly after the inception of the concept of Au NPs as radiosensitizers, it became necessary to evaluate the respective impact of size, concentration and distribution of Au NPs in vivo, when they are irradiated by fluxes of photons. Dosimetric studies were critical in identifying the more interesting combinations of parameters to be tested in experimental radiotherapy studies involving Au NPs.

The first simulation studies were based on calculating in a relatively conventional manner, the ratio between the mass energy attenuation coefficients of gold and tissue at the macroscopic scale (e.g. tumour scale). Dose enhancement factors were calculated as a quantitative value that exhibits the effect of Au NPs presence in tissue. Dose enhancement factors correspond to the ratio of the dose deposited in a tumour volume when Au NPs are present over the dose when no Au NPs are in the tissue. It was highlighted that this macroscopic approach had some drawbacks and did not represent accurately the dose enhancement phenomenon observed thorough experimental in vitro studies. A new dosimetric approach called microdosimetry was developed to bridge the experimental results and the dose enhancement deposition created by the generation of secondary electrons at the nanoscale (cellular scale). Through these studies, it was confirmed that low-energy photon sources such as brachytherapy seeds (e.g. ^{103}Pd and ^{125}I) offered a higher dose enhancement, as well as the benefit to use smaller Au NPs to reduce the amount of auto absorption of secondary electrons directly in the Au NPs [91, 111, 112]. These simulations offered new insights about dose deposition at the nanometric scale and confirmed the strong potential for radiooncology, of irradiating Au NPs with low-energy photons such as low dose-rate brachytherapy seeds.

6.4 Radioactive Au NPs

Historically, radioactive Au NPs have been first introduced to oncology procedures in the mid-1950s in the form of colloidal Au NPs [23, 61, 62]. However, it was highlighted that heterogeneous distribution of the solution inside the tumour influenced the efficacy of the treatment, as well as their quick excretion from the body. The size variability and the poor uniformity of Au NPs forced to abandon this strategy until recent advances in nanotechnology and materials characterization techniques (e.g. transmission electron microscopy), which helped to develop new synthesis routes that offer very controlled sizes and shapes. These advances allowed to revisit the concept of radioactive Au NPs for potential interstitial treatments.

Radioactive gold (^{198}Au; $t_{1/2}$ =2.7 days) is a high-energy beta-emitter (E_{moy} =312 keV) that also emits high-energy photons (E_{moy} =412 keV). It has been revisited recently in several preclinical studies involving tumour injections in animal models (Fig. 12b, c) [17, 94, 110, 168]. Beta-emitters have been investigated in internal therapy for many years, since they can create dense damages to the DNA. However, the high-energy photons emissions of ^{198}Au are not perfectly well suited for clinical applications. More recently, low-energy beta-emitters have been used to radiolabel Au NPs (e.g. ^{177}Lu) [29, 30, 201]. This radiolabeling is performed via chelate of linker molecules attached at the surface of the Au NPs. Au NPs radiolabeled through chelates can lead to the cleavage of the radioisotope in vivo. Recently, a new approach based on the low-energy photon emitter ^{103}Pd commonly used in brachytherapy has emerged in the form of core–shell Pd:Au NPs. Core–shell Au NPs incorporating radioisotope ^{103}Pd (20–23 keV photons) have demonstrated a strong potential in delaying tumour growth (Fig. 12) [50, 110, 170]. Moreover, the combination of low-energy photon and gold offers the best conditions to benefit dose enhancement via the radiosensitization effect [111].

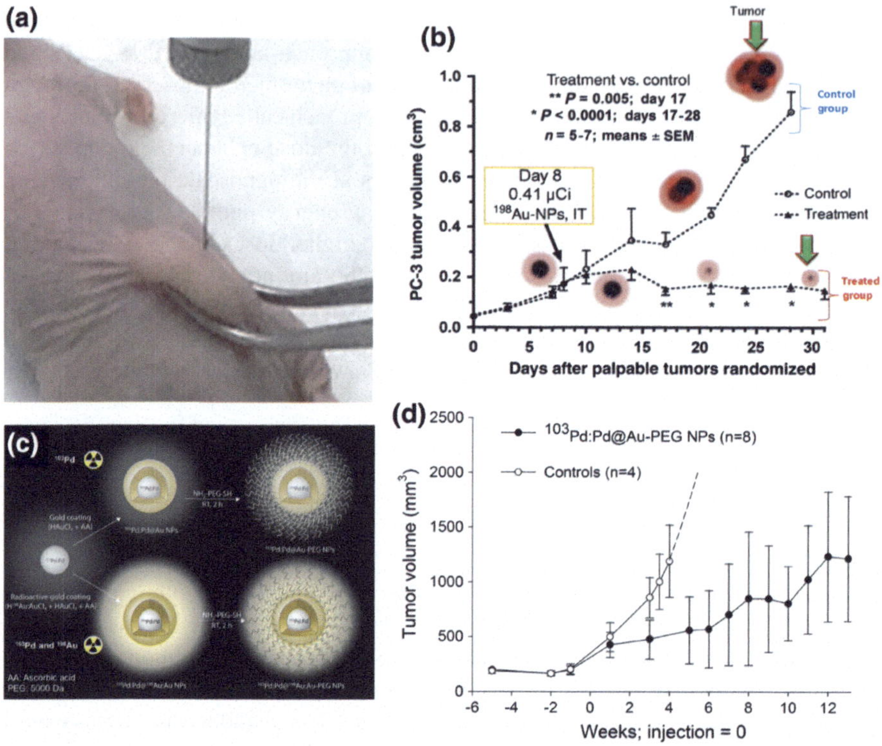

Fig. 12 **a** Intratumoural (i.t.) administration of Au NPs in xenograft tumours (murine model; *source* The authors). **b** Tumour growth curves in a prostate cancer model injected with radioactive ^{198}Au NPs: these results figure among the first in the recent years to have revealed the potential of radioactive Au NPs in oncology (reprinted from [168], with permission from the National Academy of Sciences of the United States of America). **c** Schematic representation of ^{103}Pd:Pd@Au-PEG NPs (single radioisotope) and ^{103}Pd:Pd@^{198}Au:Au-PEG NPs (dual radioisotopes, represented by the enhanced corona) synthesized by a direct reduction route forming core–shell particles (reprinted from [110], with permission from Wiley). **d** Tumour volume follow-up of prostate cancer, after injection of ~1 mCi of ^{103}Pd:Pd@Au-PEG NPs (reprinted from [110] with permission from Wiley)

6.5 Current Status of Au NPs as Radiosensitizers for Clinical Applications

Each of the experimental studies mentioned above showed interesting outcomes in terms of tumour growth delay. However, it is still very difficult to compare the efficacy of each one due to the various conditions employed. Every strategy has advantages and drawbacks that need to be balanced in the perspective of clinical implementation. The impact of key parameters such as the radiation source, the physico-chemical properties of Au NPs (e.g. size, concentration and functionalization), the dose or the administration route must be evaluated systematically to enable comparisons between the efficacy of each one of the Au NPs formulations.

For Au NPs to be translated into clinical procedures, a certain number of challenges must be overcome. First, the long-term retention of Au NPs into the tumours, and their excretion pathways, must be comprehensively investigated. Long-term biocompatibility issues, as well as the biodistribution and clearance routes, must be thoroughly studied. In addition to this, treatment planning and dose measurements tools currently used in medical physics must be adapted to the specific reality of Au NPs. Recent studies have clearly highlighted the importance of understanding Au NPs diffusion in tissues and the impact of this diffusion on treatment efficacy and on the dose given to the surrounding organs [21, 68, 109, 171]. It is also important to develop simulation tools for the diffusion of Au NPs that could be added to existing dose planification platforms. In conclusion, it is evident that more experiments are needed before health agencies approve Au NPs procedures in radiation therapy. Continued interdisciplinary research efforts are key to develop optimal radiosensitization, 'nanobrachytherapy' products and dose calculation tools that could make a wide impact in the future of oncology practice.

7 Biodistribution, Clearance and in Vivo Toxicity

7.1 Biodistribution and Clearance Routes of Gold Nanoparticles

The size, shape, surface chemistry, charge and concentration of Au NPs are determinant factors for their successful application in biomedical imaging and in radiotherapy [74, 76, 162]. These parameters influence the biodistribution, the accumulation, the diffusion and the uptake of the NPs by cells and tumours [55, 74, 77, 95, 162]. Therefore, the physico-chemical properties must be optimized according to the final application, so that the highest image quality or therapeutic efficiency could be achieved, while using the minimum concentration of Au NPs possible. Comprehensive reviews on the toxicity and biodistribution of Au NPs have been written by Dykman and Khlebstov [55, 95].

7.2 The Intravascular Injection Route

NPs can be administered through different routes and the intravenous (i.v.) has been the most common procedure reported in preclinical studies. It is not invasive, it provides a quick distribution of the NPs through the body [90], and it also opens possibilities for targeted CT imaging [33, 56, 114, 149, 150, 154, 195]. Upon their i.v. administration, Au NPs diffuse in the blood pool [77]. PEG is commonly used as a coating for Au NPs. This polymer efficiently stabilizes the particles by providing steric hindrance and surface charge modulation. The adsorption of opsonins at the surface of NPs, usually occurring very fast as soon as the nanoparticles are injected into the blood, is thereby delayed. As a consequence, it takes longer time for the mononuclear phagocytic system (MPS) cells to recognize the NPs [28]. After some hours in circulation, they end up accumulating in the liver and in the spleen [28, 71, 77, 148, 164]. Depending on surface properties and size, Au NPs can also significantly accumulate in the lungs [18] and in the kidneys [28, 71, 164]. To a lower extent, Au NPs can also be detected in the lymph nodes, in the small and large intestines, in the heart, in the skeletal muscle and in the brain [28, 164].

The effect of size on the blood clearance of NPs is clearly revealed by comparing the works performed by Hainfeld et al. [71] and Cai et al. [28]. In the first study, the authors used Au NPs with an estimated HD of 5.8 nm, while on the second work, NPs with an HD of 38 nm were used. As a consequence, 5 h after the injection, Hainfeld et al. observed that 77.5% of the injected dose (i.d.) had been eliminated by the kidneys, while Cai et al. reported that 72 h after the injection, only 6% of Au i.d. had been eliminated in the faeces. Also, Hirn et al. studied the biodistribution of negatively charged Au NPs with different core diameters (e.g. 1.4, 5, 18, 80 and 200 nm). The authors observed that the NPs of 5, 18, 80 and 200 nm were quickly eliminated from the bloodstream and accumulated mainly in the liver (91.9–96.9% of the i.d.). The 1.4 nm diameter NPs were also cleared relatively fast, and there was lower retention in the liver (e.g. 51.3%). The spleen was the second organ with the highest accumulation of Au NPs; however, the percentage of retained NPs, about 2%, was similar for all the particle sizes studied [77].

As mentioned in Sect. 5, the long blood half-life of Au NPs systems, and in particular the larger PEGylated ones, is a consequence of the inability of kidneys to excrete them. It is also due to the delayed uptake of the NPs by the macrophages residing in the organs of the MPS, such as the liver and the spleen [77, 148]. NPs with an HD smaller than 6 nm are easily excreted by the kidneys because the size is under the renal filtration threshold [148]. NPs with sizes ranging from 6 to 10 nm can still be eliminated by the kidneys, but usually, a certain retention in other organs such as liver and spleen also occurs. It is generally assumed that particles with positively charged surfaces are eliminated faster than neutral or negatively charged surfaces [117]. Finally, if the Au NPs are not eliminated relatively quickly by the kidneys, they accumulate in different organs and are assumed to be slowly eliminated by the biliary system [148].

When the nanoparticles are designed to remain in the blood for several minutes or even hours, they can reach and accumulate in tumours by the enhanced, permeability and retention (EPR) effect. In fact, tumours that grow fast present an irregular and leaky vasculature [4, 117]. For the NPs to accumulate in the tumour through the EPR effect, their size (i.e. their hydrodynamic diameter—HD), their surface charge, their hindrance (e.g. by PEGylation) and their shape must be carefully tuned to maximize the blood circulation times. The heterogeneity of blood vessels around and inside the tumours must be taken into account. Usually, the periphery of tumours is well vascularized, and therefore, the NPs can more easily accumulate in this area. However, the core of the tumour is generally less vascularized, and this effect must be taken into account when evaluating the potential of nanoparticles injected i.v. for oncology procedures [192].

Lately, there has been much debate on the efficiency, and therefore on the relevance of developing new therapeutics based on the EPR effect. In fact, an analysis of 232 studies reporting on the i.v. injection of NPs concluded that the median of NPs percentage accumulating in the tumours was 0.7% of the injected dose. Active targeting, in which biomolecules are attached at the surface of NPs to enable their binding to surface antigens, only raised the efficiency of delivery to 0.9% i.d. [192]. These observations suggest that approximately 99% of the overall doses of NPs described in the scientific literature where i.v. injections were performed in cancer xenograft animal models, accumulate non-specifically in organs, where they can possibly raise negative side effects by long-term exposition [77, 192]. Finally, the inefficient delivery of NPs to tumours would necessarily be translated into higher manufacturing costs per therapeutic dose [192].

7.3 The Intratumoral Injection Route

The percentage of Au NPs reaching tumours after i.v. injections in small-animal models have been reportedly low [192], and for this reason, alternative administration routes have been investigated. A few studies have reported on the intraperitoneal injection [32], whereas direct intratumoral injections (i.t.) are increasingly performed in the context of brachytherapy (see Fig. 12a). This appears as a promising way to improve the NP delivery efficiency in the tumour while minimizing the accumulation of Au NPs in non-specific organs [29, 33, 94, 110, 166].

The i.t. injection of Au NPs as radiosensitizers for the treatment of prostate cancer tumours, results in much lower accumulations of the product in non-specific organs (liver, spleen), compared with the i.v. route [94, 110, 166]. Chattopadhyay et al. clearly demonstrated the advantages of this approach by comparing the biodistribution of trastuzumab-^{111}In-Au NPs injected by either the i.v. or the i.t. route in mice bearing HER-2 positive breast tumours. Quantification of organ uptake at 48 h after the injections demonstrated that i.t. administration yielded 24 times higher concentration of trastuzumab-^{111}In-Au NPs in the tumour (29.6% i.d./g vs. 1.2% i.d./g for i.v. injection). In addition, this administration route lead to 10 times lower accumulation of the NPs in the spleen (1.8% i.d./g vs. 19.2% i.d./g for i.v. injection), and about 1.7 of the concentration in the liver (1.6% i.d./g vs. 2.7% i.d./g for i.v. injection) [34].

Shi et al. also compared the biodistribution of Au NPs injected either i.t. or i.v. In this work, tiopronin-coated Au NPs (core diameter $= 2.77 \pm 0.69$ nm) were administered to mice bearing an HCT116 tumour (e.g. colorectal tumour) [166]. The results revealed that 8 h after i.t. or i.v. injection, tumour uptake was 94 times higher in the animals injected i.t. (496 ± 106 μg Au/g) compared with i.v.-injected ones (5.27 ± 1.17 μg Au/g).

Another work demonstrating the advantages of i.t. injection was performed by Laprise-Pelletier et al. [110]. In this study, 4 μL of radioactive Pd@Au NPs (e.g. ^{103}Pd:Pd@Au-PEG, $HD_{number\text{-}w.} = 36.4$ nm) suspended in alginate gel was injected directly into prostate tumours (PC3). The alginate gel was used to maintain the particles in the tumour, because this gel polymerizes in the presence of Ca^{2+} ions, which are naturally available in the tissues. Biodistribution data was collected after harvesting the tumour (e.g. 27–33 or 88–90 d after i.t. injection), the results showed that most of the injected NPs (e.g.>75% of the total activity) were trapped in the tumour. Amounts of NPs were found in the liver (e.g. approximately 16% of the total activity) and in the spleen (e.g.≈3% of the total activity). Weekly inspections performed after i.t. injections, using a Geiger counter, indicated that NPs reach the liver and spleen within the first 2 weeks after the injection, and this suggests that improvements to the binding of Au NPs in the tumours must be made at the injection step. TEM observations demonstrated that ultra-small NPs (<5 nm) were the most likely to escape the tumour matrix in the first hours following i.t. injection [110].

7.4 In Vivo Toxicity Studies

To transfer Au NPs in clinical applications, either as contrast or as radiosensitizing agents, it is necessary to undertake comprehensive studies on their potential side effects should they remain in the body for several months [168]. The impact of long-term exposure of organs to various amounts of Au NPs, is studied through in vivo toxicity assays. Each new compound has unique physico-chemical properties, and these have a specific impact on the interaction with the biologic system.

Several research groups developing Au NP-based radiosensitizers have conducted in vivo toxicity assays (Table 1). Among the most commonly conducted tests figure (1) body weight monitoring; (2) organ histology (e.g. heart, liver, spleen, lung, kidney and reproductive organs); (3) complete blood count (CBC); and (4) liver and kidney function tests. A CBC examination often includes the quantification of white blood cells (WBC), red blood cell (RBC), hematocrit (HCT), mean corpuscular volume (MCV), haemoglobin (HGB), platelets (PLT), mean corpuscular haemoglobin (MCH) and mean corpuscular haemoglobin concentration (MCHC). On the other hand, liver and kidney function tests investigate the concentration of alanine aminotransferase (ALT), aspartate aminotransferase (AST), total protein (TP), albumin (ALB), alkaline phosphatase (ALP), gamma-glutamyl transferase (GGT), blood urea nitrogen (BUN), creatinine (CREA), globulin (GLOB) and total bilirubin (TB) [33, 203, 204].

Most of the studies summarized in Table 1 concluded that Au NPs do not induce significant systemic toxicity. Only two studies performed with Au NPs with an HD larger than the renal filtration threshold and which were injected intraperitoneally, detected liver damage [203, 204]. This was concluded after histologic observation of liver samples and due to alteration of the AST and RBC values for BSA-coated Au NPs [203], and because of AST and ALT abnormal results for PEG-coated Au NPs

Table 1 Summary of systemic in vivo toxicity investigations for the development of radiosensitizing Au NPs

Refs.	Study duration	Adm. route	Au NP: core size/coating	Dose ([Au])	Animal model	In vivo toxicity tests	Evidence of systemic toxicity
[203]	20 days	i.p.	2 nm/GSH or BSA	10 mg/kg	Mice	– Histology – CBC – ALT, AST, TP, ALB, BUN, CREA, GLOB, TB	– Liver tissue damage, increased AST and decreases RBC for BSA-coated Au NPs
[204]	24 days	i.p.	4.8, 12.1, 27.3 and 46.6 nm/PEG	4 mg/kg	Mice	– Histology – ALT, AST, ALB, GLOB	– Liver tissue damage, increased AST and ALT
[33]	118 days	i.t.	30 nm/PEG and Trastuzumab	0.8 mg	Mice	– Body weight – CBC – ALT, CREA	–
[15]	1 month	i.t.	12–15 nm/gum arabic	0.325 mg	Dogs	– Body weight – CBC – ALT, ALP, CREA, BUN	–
[201]	15 days	i.t.	30 nm/Panitumumab, PEG and DOTA-^{177}Lu	6×10^{11} particles	Mice	– Body weight – CBC – ALT, CREA	–
[31]	30 days	i.t.	12–18 nm/gum arabic	–	Mice	– CBC	–
[29]	70 days	i.t.	30 nm/Trastuzumab, PEG and DTPA-^{111}In	0.7 mg	Mice	– Body weight – CBC – ALT, CREA	–
[30]	16 days	i.t.	30 nm/Trastuzumab, PEG and DOTA-^{177}Lu	0.15 mg	Mice	– Body weight – CBC – ALT, CREA	–
[72]	14 days	i.v.	1.9 nm/–	0.8 g/kg	Mice	– CBC – ALT, AST, TP, ALP, ALB, GGT, TB, phosphorus CREA, BUN	–
[36]	20 days	i.v.	~18 nm/BSA	250 μl, 1.3 mg/mL	Mice	– Body weight – Histology	–
[133]	35 days	i.t.	~120 nm/^{103}Pd	40 μl, 2.03×10^{10} NPs/mL	Mice	– CBC – ALT, AST, BUN, CREA	–

[204]. Data from recent systemic in vivo toxicity studies conducted in the context of radiosensitizing with Au NPs are listed in Table 1. These studies point to the low systemic effects of Au NPs at the concentration required for radiosensitization, and in particular if the NPs are administered intratumorally.

However, longer and more standardized studies must be performed to identify the potential impacts in the long term, of long retentions of Au NPs in the liver and in other organs. In addition, assays to detect the effects of oxidative stress in the liver (e.g. lipid peroxidation, protein carbonylation and inflammatory markers such as TNF-α, IL-1ß, IL-6 and IL-10) [118] and abnormal gene expression should also be performed to clarify the interaction between the retained NPs and the organs [203]. Effects of long-term exposure to nanoparticles must be collected in the critical organs, in order to clearly and precisely identify the toxicity limits for guaranteeing a safe use of Au NPs in the human body.

8 Perspectives: Current Status of Au NPs in Clinical Trials

The clinical use of NPs is still scarce, and most of the approved systems are made of protein–drug conjugates or liposomes [49, 163]. Regarding inorganic nanoparticles, only iron-based ones have received approval so far [12]. The approved compounds are used for the treatment of cancer, for fungal infections, for macular degeneration, or as imaging agents, vaccines and anaesthetics [12].

Several Au NP systems have been investigated in clinical studies for cancer treatment. A list of current and past trials can be found on the U.S. National Library of Medicine website (www.clinicaltrials.gov), as well as in several recent reviews in [11]. Examples include tumour necrosis factor-bound colloidal gold (CYT-6091, NCT00436410 and NCT00356980), which has been evaluated in Phase 0 and I studies for treatment of primary, advanced and metastatic solid tumours [107, 115, 135]. PEGylated silica-gold nanoshells (AuroLase®) have also been tested for photothermal therapy in head and neck tumours (NCT00848042). Technical and biological challenges associated with these Au NPs systems hindered their clinical approval. In particular, relatively weak tumour targeting values were found following systemic administration [11]. To date, no clinical study has evaluated Au NPs as radiosensitization agents for cancer treatment; however, the fact that other nanoparticle systems (e.g. gadolinium and hafnium oxide-based) are being tested as radiosensitizers in Phase I, II and III oncology studies [45], opens the door to more clinical investigations in the field, including with Au NPs.

To reach clinical trials, more comprehensive studies must be performed at the preclinical stage on a per-compound basis. These studies must include (1) a thorough physico-chemical characterization of Au NPs before and after exposure to biological media; (2) a reproducible demonstration of the therapeutic efficiency of Au NPs in vivo and of an improved contrast enhancement; (3) a detailed evaluation of systemic toxicity; (4) a comprehensive assessment of pro-inflammatory and immunity response for each one of these compounds; and (5) a clear demonstration that these new compounds can be systematically produced with the utmost level of precision under good manufacturing practices.

9 Conclusion

As demonstrated in various preclinical studies, gold nanoparticles (Au NPs) have emerged as a promising platform for imaging and therapy. In this chapter, we introduced the main concepts underlying the potential of Au NPs as imaging agents for CT, as well as radiosensitizers for cancer therapy. The main synthesis routes for the production and functionalization of Au NPs were presented, followed by a detailed description of the mechanisms of interaction occurring between high-energy photons and Au NPs. The main developments in the field of Au NPs as contrasts agents for CT (X-ray imaging), as well as Au NPs radiosensitizers for radiotherapy (medical physics in oncology), were also described. More clinical trials will be necessary in the near future in order to circumscribe precisely the types of cancer as well as their grades that could be best treated with such technologies. Prior to this, the biodistribution and clearance data, the toxicity thresholds, the pro-inflammatory and the immune response characteristics of each new compound will have to be revealed through comprehensive in vitro and preclinical studies. In resume, Au NPs could provide more efficient imaging agents, as well as radiosensitizing products making possible the use of much lower doses of ionizing radiation in radiation oncology treatments.

Acknowledgements Mr. Mahmoud Omar and Mrs. Beatriz Ribeiro Nogueira are gratefully acknowledged for their work on the syntheses related to Fig. 1.

References

1. Abel B, Buck U, Sobolewski AL, Domcke W (2011) On the nature and signatures of the solvated electron in water. Phys Chem Chem Phys 14:22–34. https://doi.org/10.1039/C1CP21803D
2. Akolkar R, Sankaran RM (2013) Charge transfer processes at the interface between plasmas and liquids. J Vac Sci Technol, A 31:050811. https://doi.org/10.1116/1.4810786
3. Al Zaki A, Joh D, Cheng Z, De Barros ALB, Kao G, Dorsey J, Tsourkas A (2014) Gold-loaded polymeric micelles for computed tomography-guided radiation therapy treatment and radiosensitization. ACS Nano 8:104–112. https://doi.org/10.1021/nn405701q
4. Albanese A, Tang PS, Chan WCW (2012) The effect of nanoparticle size, shape, and surface chemistry on biological systems. Annu Rev Biomed Eng 14:1–16. https://doi.org/10.1146/annurev-bioeng-071811-150124
5. Alkilany AM, Nagaria PK, Hexel CR, Shaw TJ, Murphy CJ, Wyatt MD (2009) Cellular uptake and cytotoxicity of gold nanorods: molecular origin of cytotoxicity and surface effects. Small 5:701–708. https://doi.org/10.1002/smll.200801546
6. Allijn IE et al (2013) Gold nanocrystal labeling allows low-density lipoprotein imaging from the subcellular to macroscopic level. ACS Nano 7:9761–9770. https://doi.org/10.1021/nn403258w
7. Alric C et al (2013) The biodistribution of gold nanoparticles designed for renal clearance. Nanoscale 5:5930–5939. https://doi.org/10.1039/c3nr00012e
8. Amendola V, Meneghetti M (2009) Laser ablation synthesis in solution and size manipulation of noble metal nanoparticles. Phys Chem Chem Phys 11:3805. https://doi.org/10.1039/b900654k

9. Amendola V, Meneghetti M (2013) What controls the composition and the structure of nanomaterials generated by laser ablation in liquid solution? Phys Chem Chem Phys 15:3027. https://doi.org/10.1039/c2cp42895d
10. Amendola V, Polizzi S, Meneghetti M (2006) Laser ablation synthesis of gold nanoparticles in organic solvents. J Phys Chem B 110:7232–7237. https://doi.org/10.1021/jp0605092
11. Anselmo A, Mitragotri S (2015) A review of clinical translation of inorganic nanoparticles. AAPS J 17:1041–1054. https://doi.org/10.1208/s12248-015-9780-2
12. Anselmo AC, Mitragotri S (2016) Nanoparticles in the clinic. Bioeng Transl Med 1:10–29. https://doi.org/10.1038/520609a
13. Arvizo RR, Bhattacharyya S, Kudgus RA, Giri K, Bhattacharya R, Mukherjee P (2012) Intrinsic therapeutic applications of noble metal nanoparticles: past, present and future. Chem Soc Rev 41:2943–2970. https://doi.org/10.1039/c2cs15355f
14. Attix FH (1986) Introduction to radiological physics and radiation dosimetry. Wiley, New York
15. Axiak-Bechtel SM et al (2014) Gum arabic-coated radioactive gold nanoparticles cause no short-term local or systemic toxicity in the clinically relevant canine model of prostate cancer. IJN 9:5001–5011. https://doi.org/10.2147/ijn.s67333
16. Bednar N, Matović J, Stojanović G (2013) Properties of surface dielectric barrier discharge plasma generator for fabrication of nanomaterials. J Electrostat 71:1068–1075. https://doi.org/10.1016/j.elstat.2013.10.010
17. Black KCL et al (2014) Radioactive 198Au-doped nanostructures with different shapes for in vivo analyses of their biodistribution tumor uptake, and intratumoral distribution. ACS Nano 8:4385–4394
18. Boote E et al (2010) Gold nanoparticle contrast in a phantom and juvenile swine. Acad Radiol 17:410–417. https://doi.org/10.1016/j.acra.2010.01.006
19. Bouchard M et al (2015) Rapid nucleation of iron oxide nanoclusters in aqueous solution by plasma electrochemistry. Langmuir 31:7633–7643. https://doi.org/10.1021/acs.langmuir.5b01235
20. Bouchard M, Laprise-Pelletier M, Turgeon S, Fortin M-A (2017) Efficient and rapid synthesis of radioactive gold nanoparticles by dielectric barrier discharge. Part Part Syst Charact 34:1600231. https://doi.org/10.1002/ppsc.201600231
21. Brivio D, Nguyen PL, Sajo E, Ngwa W, Zygmanski P (2017) A Monte Carlo study of I-125 prostate brachytherapy with gold nanoparticles: dose enhancement with simultaneous rectal dose sparing via radiation shielding. Phys Med Biol 62:1935–1948
22. Brust M, Walker M, Bethell D, Schiffrin DJ, Whyman R (1994) Synthesis of thiol-derivatised gold nanoparticles in a two-phase liquid–liquid system. J Chem Soc, Chem Commun:801–802. https://doi.org/10.1039/c39940000801
23. Bulkley GJ, O'Conor VJ (1959) Treatment of carcinoma of the prostate by interstitial irradiation with radioactive gold. Experimental and clinical studies. Trans Am Assoc Genito-Urin Surg 51:126–133
24. Bushberg JT, Seibert JA, Leidholdt EM, Boone JM (2011) The essential physics of medical imaging. Wolters Kluwer Health
25. Butterworth KT, McMahon SJ, Currell FJ, Prise KM (2012) Physical basis and biological mechanisms of gold nanoparticle radiosensitization. Nanoscale 4:4830–4838. https://doi.org/10.1039/c2nr31227a
26. Butterworth KT, McMahon SJ, Taggart LE, Prise KM (2013) Radiosensitization by gold nanoparticles: effective at megavoltage energies and potential role of oxidative stress. Transl Cancer Res 2
27. Butterworth KT et al (2016) Preclinical evaluation of gold-DTDTPA nanoparticles as theranostic agents in prostate cancer radiotherapy. Nanomedicine (Lond) 11:2035–2047. https://doi.org/10.2217/nnm-2016-0062
28. Cai QY et al (2007) Colloidal gold nanoparticles as a blood-pool contrast agent for X-ray computed tomography in mice. Invest Radiol 42:797–806. https://doi.org/10.1097/RLI.0b013e31811ecdcd

29. Cai Z, Chattopadhyay N, Yang K, Kwon YL, Yook S, Pignol J-P, Reilly RM (2016) 111In-labeled trastuzumab-modified gold nanoparticles are cytotoxic in vitro to HER2-positive breast cancer cells and arrest tumor growth in vivo in athymic mice after intratumoral injection. Nucl Med Biol 43:818–826. https://doi.org/10.1016/j.nucmedbio.2016.08.009
30. Cai Z, Yook S, Lu Y, Bergstrom D, Winnik MA, Pignol J-P, Reilly RM (2017) Local radiation treatment of HER2-positive breast cancer using trastuzumab-modified gold nanoparticles labeled with ^{177}Lu. Pharm Res 34:579–590. https://doi.org/10.1007/s11095-016-2082-2
31. Chanda N et al (2010a) Radioactive gold nanoparticles in cancer therapy: therapeutic efficacy studies of GA-198AuNP nanoconstruct in prostate tumor-bearing mice. Nanomedicine 6:201–209. https://doi.org/10.1016/j.nano.2009.11.001
32. Chanda N et al (2010b) Bombesin functionalized gold nanoparticles show in vitro and in vivo cancer receptor specificity. Proc Natl Acad Sci 107:8760–8765. https://doi.org/10.1073/pnas.1002143107
33. Chattopadhyay N et al (2012) Role of antibody-mediated tumor targeting and route of administration in nanoparticle tumor accumulation in vivo. Mol Pharm 9:2168–2179. https://doi.org/10.1021/mp300016p
34. Chattopadhyay N, Cai Z, Kwon YL, Lechtman E, Pignol J-P, Reilly RM (2013) Molecularly targeted gold nanoparticles enhance the radiation response of breast cancer cells and tumor xenografts to X-radiation. Breast Cancer Res Treat 137:81–91. https://doi.org/10.1007/s10549-012-2338-4
35. Chauhan VP, Stylianopoulos T, Boucher Y, Jain RK (2011) Delivery of molecular and nanoscale medicine to tumors: transport barriers and strategies. Annu Rev Chem Biomol Eng 2:281–298. https://doi.org/10.1146/annurev-chembioeng-061010-114300
36. Chen N, Yang W, Bao Y, Xu H, Qin S, Tu Y (2015) BSA capped Au nanoparticle as an efficient sensitizer for glioblastoma tumor radiation therapy. RSC Adv 5:40514–40520. https://doi.org/10.1039/c5ra04013b
37. Cheng NN, Starkewolf Z, Davidson RA, Sharmah A, Lee C, Lien J, Guo T (2012) Chemical enhancement by nanomaterials under X-ray irradiation. J Am Chem Soc 134:1950–1953. https://doi.org/10.1021/ja210239k
38. Chetty I et al (2007) Report of the AAPM Task Group No. 105: issues associated with clinical implementation of Monte Carlo-based photon and electron external beam treatment planning. Med Phys 34:4818–4853
39. Chithrani DB et al (2010) Gold nanoparticles as radiation sensitizers in cancer therapy. Radiat Res 173:719–728. https://doi.org/10.1667/RR1984.1
40. Choi HS, Liu W, Liu F, Nasr K, Misra P, Bawendi MG, Frangioni JV (2010) Design considerations for tumour-targeted nanoparticles. Nat Nanotechnol 5:42–47. https://doi.org/10.1038/nnano.2009.314
41. Cobley CM, Chen JY, Cho EC, Wang LV, Xia YN (2011) Gold nanostructures: a class of multifunctional materials for biomedical applications. Chem Soc Rev 40:44–56. https://doi.org/10.1039/b821763g
42. Cormode DP et al (2008) Nanocrystal core high-density lipoproteins: a multimodality contrast agent platform. Nano Lett 8:3715–3723. https://doi.org/10.1021/nl801958b
43. Cormode DP et al (2010) Atherosclerotic plaque composition: analysis with multicolor CT and targeted gold nanoparticles. Radiology 256:774–782. https://doi.org/10.1148/radiol.10092473
44. Coulter JA, Hyland WB, Nicol J, Currell FJ (2013) Radiosensitising nanoparticles as novel cancer therapeutics—pipe dream or realistic prospect? Clin Oncol 25:593–603
45. Cui L, Her S, Borst GR, Bristow RG, Jaffray DA, Allen C (2017) Radiosensitization by gold nanoparticles: will they ever make it to the clinic? Radiother Oncol 124:344–356. https://doi.org/10.1016/j.radonc.2017.07.007
46. Cui L et al (2014) Hypoxia and cellular localization influence the radiosensitizing effect of gold nanoparticles (AuNPs) in breast cancer cells. Radiat Res 182:475–488. https://doi.org/10.1667/RR13642.1

47. Cui L et al (2017) Significant radiation enhancement effects by gold nanoparticles in combination with cisplatin in triple negative breast cancer cells and tumor xenografts. Radiat Res 187:147–160
48. Daniel MC, Astruc D (2004) Gold nanoparticles: assembly, supramolecular chemistry, quantum-size-related properties, and applications toward biology, catalysis, and nanotechnology. Chem Rev 104:293–346. https://doi.org/10.1021/cr030698+
49. Dawidczyk CM, Kim C, Park JH, Russell LM, Lee KH, Pomper MG, Searson PC (2014) State-of-the-art in design rules for drug delivery platforms: lessons learned from FDA-approved nanomedicines. J Controlled Release 187:133–144. https://doi.org/10.1016/j.jconrel.2014.05.036
50. Djoumessi D, Laprise-Pelletier M, Chevallier P, Lagueux J, Cote MF, Fortin MA (2015) Rapid, one-pot procedure to synthesise 103Pd:Pd@Au nanoparticles en route for radiosensitisation and radiotherapeutic applications. J Mater Chem B 3:2192–2205. https://doi.org/10.1039/c4tb01663g
51. Doane TL, Burda C (2012) The unique role of nanoparticles in nanomedicine: imaging, drug delivery and therapy. Chem Soc Rev 41:2885–2911. https://doi.org/10.1039/c2cs15260f
52. Dreaden EC, Alkilany AM, Huang X, Murphy CJ, El-Sayed MA (2012) The golden age: gold nanoparticles for biomedicine. Chem Soc Rev 41:2740–2779. https://doi.org/10.1039/c1cs15237h
53. Dreaden EC, Mackey MA, Huang X, Kang B, El-Sayed MA (2011) Beating cancer in multiple ways using nanogold. Chem Soc Rev 40:3391. https://doi.org/10.1039/c0cs00180e
54. Duong Q, Tan Y, Corey J, Anz S, Sun P (2015) Mechanism of the transfer of $AuCl4^-$ and TOA^+ ions across the liquid/liquid interface. J Phys Chem C 119:10365–10369. https://doi.org/10.1021/jp512909t
55. Dykman L, Khlebtsov N (2012) Gold nanoparticles in biomedical applications: recent advances and perspectives. Chem Soc Rev 41:2256–2282. https://doi.org/10.1039/c1cs15166e
56. Eck W, Nicholson AI, Zentgraf H, Semmler W, Bartling S (2010) Anti-CD4-targeted gold nanoparticles induce specific contrast enhancement of peripheral lymph nodes in X-ray computed tomography of live mice. Nano Lett 10:2318–2322. https://doi.org/10.1021/nl101019s
57. Eustis S, El-Sayed MA (2006) Why gold nanoparticles are more precious than pretty gold: noble metal surface plasmon resonance and its enhancement of the radiative and nonradiative properties of nanocrystals of different shapes. Chem Soc Rev 35:209–217. https://doi.org/10.1039/b514191e
58. Faraday M (1857) X. The Bakerian Lecture.—Experimental relations of gold (and other metals) to light. Philos Trans R Soc Lond 147:145–181
59. Ferrari M (2005) Cancer nanotechnology: opportunities and challenges. Nat Rev Cancer 5:161–171. https://doi.org/10.1038/nrc1566
60. Fievet F, Lagier JP, Blin B, Beaudoin B, Figlarz M (1989) Homogeneous and heterogeneous nucleations in the polyol process for the preparation of micron and submicron size metal particles. Solid State Ionics 32–33(Part 1):198–205. https://doi.org/10.1016/0167-2738(89)90222-1
61. Flocks RH, Kerr HD, Elkins HB, Culp D (1952) Treatment of carcinoma of the prostate by interstitial radiation with radio-active gold (Au 198): a preliminary report. J Urol 68:510–522
62. Flocks RH, Kerr HD, Elkins HB, Culp DA (1954) The treatment of carcinoma of the prostate by interstitial radiation with radioactive gold (Au198): a follow-up report. J Urol 71:628–633
63. Fojtik A, Henglein A (1993) Laser ablation of films and suspended particles in a solvent: formation of cluster and colloid solutions. Ber Bunsenges Phys Chem:252–254
64. Fratoddi I, Venditti I, Cametti C, Russo MV (2015) How toxic are gold nanoparticles? The state-of-the-art. Nano Res 8:1771–1799. https://doi.org/10.1007/s12274-014-0697-3
65. Frens G (1973) Controlled nucleation for the regulation of the particle size in monodisperse gold suspensions. Nature 241:20–22. https://doi.org/10.1038/10.1038/physci241020a0
66. Furuya K, Hirowatari Y, Ishioka T, Harata A (2007) Protective agent-free preparation of gold nanoplates and nanorods in aqueous $HAuCl_4$ solutions using gas–liquid interface discharge. Chem Lett 36:1088–1089. https://doi.org/10.1246/cl.2007.1088

67. Galper MW et al (2012) Effect of computed tomography scanning parameters on gold nanoparticle and iodine contrast. Invest Radiol 47:475–481. https://doi.org/10.1097/RLI.0b013e3182562ab9
68. Ghorbani M, Bakhshabadi M, Golshan A, Knaup C (2013) Dose enhancement by various nanoparticles in prostate brachytherapy. Australas Phys Eng Sci Med 36:431–440
69. Gore JC, Yankeelov TE, Peterson TE, Avison MJ (2009) Molecular imaging without radiopharmaceuticals? J Nucl Med 50:999–1007. https://doi.org/10.2967/jnumed.108.059576
70. Hainfeld J, Smilowitz H, O'Connor M, Dilmanian F, Slatkin D, Oconnor M (2013) Gold nanoparticle imaging and radiotherapy of brain tumors in mice. Nanomed: Nanotechnol Biol Med 8:1601–1609
71. Hainfeld JF, Slatkin DN, Focella TM, Smilowitz HM (2006) Gold nanoparticles: a new X-ray contrast agent. Br J Radiol 79:248–253. https://doi.org/10.1259/bjr/13169882
72. Hainfeld JF, Slatkin DN, Smilowitz HM (2004) The use of gold nanoparticles to enhance radiotherapy in mice. Phys Med Biol 49:N309–N315. https://doi.org/10.1088/0031-9155/49/18/n03
73. Hall EJ, Giaccia AJ (2006) Radiobiology for the radiologist. Lippincott Williams & Wilkins
74. Haume K et al (2016) Gold nanoparticles for cancer radiotherapy: a review. Cancer Nanotechnol 7:8. https://doi.org/10.1186/s12645-016-0021-x
75. Hazkani I et al (2017) Can molecular profiling enhance radiotherapy? Impact of personalized targeted gold nanoparticles on radiosensitivity and imaging of adenoid cystic carcinoma. Theranostics 7:3962–3971. https://doi.org/10.7150/thno.19615
76. Her S, Jaffray DA, Allen C (2017) Gold nanoparticles for applications in cancer radiotherapy: mechanisms and recent advancements. Adv Drug Deliv Rev 109:84–101. https://doi.org/10.1016/j.addr.2015.12.012
77. Hirn S et al (2011) Particle size-dependent and surface charge-dependent biodistribution of gold nanoparticles after intravenous administration. Eur J Pharm Biopharm 77:407–416. https://doi.org/10.1016/j.ejpb.2010.12.029
78. Hu M et al (2006) Gold nanostructures: engineering their plasmonic properties for biomedical applications. Chem Soc Rev 35:1084. https://doi.org/10.1039/b517615h
79. Huang X, Li Y, Zhong X (2014) Effect of experimental conditions on size control of Au nanoparticles synthesized by atmospheric microplasma electrochemistry. Nanoscale Res Lett 9:572. https://doi.org/10.1186/1556-276X-9-572
80. Hubbell JH, Seltzer SM. Tables of X-Ray mass attenuation coefficients and mass energy-absorption coefficients from 1 keV to 20 MeV for elements Z = 1 to 92 and 48 additional substances of dosimetric interest, May 1996 edn
81. Itina TE (2011) On nanoparticle formation by laser ablation in liquids. J Phys Chem C 115:5044–5048. https://doi.org/10.1021/jp1090944
82. Jackson PA, Rahman WNWA, Wong CJ, Ackerly T, Geso M (2010) Potential dependent superiority of gold nanoparticles in comparison to iodinated contrast agents. Eur J Radiol 75:104–109. https://doi.org/10.1016/j.ejrad.2009.03.057
83. Jain S et al (2011) Cell-specific radiosensitization by gold nanoparticles at megavoltage radiation energies. Int J Radiat Oncol Biol Phys 79:531–539. https://doi.org/10.1016/j.ijrobp.2010.08.044
84. Jain S, Hirst DG, O'Sullivan JM (2012) Gold nanoparticles as novel agents for cancer therapy. Br J Radiol 85:101–113. https://doi.org/10.1259/bjr/59448833
85. Jakhmola A (2012) Inorganic nanoparticles based contrast agents for X-ray computed tomography. Adv Healthc Mater 1:413–431
86. Jana NR, Gearheart L, Murphy CJ (2001) Evidence for seed-mediated nucleation in the chemical reduction of gold salts to gold nanoparticles. Chem Mater 13:2313–2322. https://doi.org/10.1021/cm000662n
87. Jeong S-Y, Park S-J, Yoon SM, Jung J, Woo HN, Yi SL (2009) Systemic delivery and preclinical evaluation of Au nanoparticle containing beta-lapachone for radiosensitization. J Controlled Release 139. https://doi.org/10.1016/j.jconrel.2009.07.007

88. Jeynes JCG, Merchant MJ, Spindler A, Wera AC, Kirkby KJ (2014) Investigation of gold nanoparticle radiosensitization mechanisms using a free radical scavenger and protons of different energies. Phys Med Biol 59:6431
89. Johns HE, Cunningham JR (1983) The physics of radiology. Charles C. Thomas
90. Johnston HJ, Hutchison G, Christensen FM, Peters S, Hankin S, Stone V (2010) A review of the in vivo and in vitro toxicity of silver and gold particulates: particle attributes and biological mechanisms responsible for the observed toxicity. Crit Rev Toxicol 40:328–346. https://doi.org/10.3109/10408440903453074
91. Jones BL, Krishnan S, Cho SH (2010) Estimation of microscopic dose enhancement factor around gold nanoparticles by Monte Carlo calculations. Med Phys 37:3809–3816. https://doi.org/10.1118/1.3455703
92. Kachelriess M (2006) Clinical X-ray computed tomography. In: Schlegel W, Bortfeld T, Grosu A-L (eds) New technologies in radiation oncology. Medical radiology. Springer, Berlin, pp 41–80. https://doi.org/10.1007/3-540-29999-8_7
93. Kattumuri V (2007) Gum arabic as a phytochemical construct for the stabilization of gold nanoparticles: in vivo pharmacokinetics and X-ray-contrast-imaging studies. Small 3:333–341
94. Khan MK et al (2008) Fabrication of $\{^{198}Au^{0}\}$ radioactive composite nanodevices and their use for nanobrachytherapy. Nanomed Nanotechnol Biol Med 4:57–69. https://doi.org/10.1016/j.nano.2007.11.005
95. Khlebtsov N, Dykman L (2011) Biodistribution and toxicity of engineered gold nanoparticles: a review of in vitro and in vivo studies. Chem Soc Rev 40:1647–1671
96. Kiessling F, Pichler BJ (2011) Small animal imaging: basics and practical guide. Springer, Berlin
97. Kim CK, Ghosh P, Rotello VM (2009) Multimodal drug delivery using gold nanoparticles. Nanoscale 1:61–67. https://doi.org/10.1039/b9nr00112c
98. Kim D, Park S, Lee JH, Jeong YY, Jon S (2007) Antibiofouling polymer-coated gold nanoparticles as a contrast agent for in vivo X-ray computed tomography imaging. J Am Chem Soc 129:7661–7665. https://doi.org/10.1021/ja071471p
99. Kim K et al (2016) Doxorubicin/gold-loaded core/shell nanoparticles for combination therapy to treat cancer through the enhanced tumor targeting. J Controlled Release 228:141–149. https://doi.org/10.1016/j.jconrel.2016.03.009
100. Kim ST, Chompoosor A, Yeh YC, Agasti SS, Solfiell DJ, Rotello VM (2012) Dendronized gold nanoparticles for siRNA delivery. Small 8:3253–3256. https://doi.org/10.1002/smll.201201141
101. Kircher MF, Willmann JK (2012) Molecular body imaging: MR imaging, CT, and US. Part I. Principles. Radiology 263:633–643. https://doi.org/10.1148/radiol.12102394
102. Kobayashi K, Usami N, Porcel E, Lacombe S, Le Sech C (2010) Enhancement of radiation effect by heavy elements. Mutat Res 704:123–131. https://doi.org/10.1016/j.mrrev.2010.01.002
103. Kojima C, Umeda Y, Ogawa M, Harada A, Magata Y, Kono K (2010) X-ray computed tomography contrast agents prepared by seeded growth of gold nanoparticles in PEGylated dendrimer. Nanotechnology 21:245104. https://doi.org/10.1088/0957-4484/21/24/245104
104. Koo IG, Lee MS, Shim JH, Ahn JH, Lee WM (2005) Platinum nanoparticles prepared by a plasma-chemical reduction method. J Mater Chem 15:4125–4128. https://doi.org/10.1039/B508420B
105. Koonce NA et al (2015) Combination of gold nanoparticle-conjugated tumor necrosis factor-α and radiation therapy results in a synergistic antitumor response in murine carcinoma models. Int J Radiat Oncol Biol Phys 93:588–596. https://doi.org/10.1016/j.ijrobp.2015.07.2275
106. Krause W (1999) Delivery of diagnostic agents in computed tomography. Adv Drug Deliv Rev 37:159–173. https://doi.org/10.1016/S0169-409X(98)00105-7
107. Kumar A, Zhang X, Liang X-J (2013) Gold nanoparticles: emerging paradigm for targeted drug delivery system. Biotechnol Adv 31:593–606. https://doi.org/10.1016/j.biotechadv.2012.10.002

108. Kuwahata H, Mikami I (2014) Generation of nitric acid and nitrous acid in distilled water irradiated with atmospheric-pressure plasma jet. e-J Surf Sci Nanotechnol 12:410–413. https://doi.org/10.1380/ejssnt.2014.410
109. Lai P, Lechtman E, Mashouf S, Pignol J-P, Reilly RM (2016) Depot system for controlled release of gold nanoparticles with precise intratumoral placement by permanent brachytherapy seed implantation (PSI) techniques. Int J Pharm 515:729–739. https://doi.org/10.1016/j.ijpharm.2016.11.001
110. Laprise-Pelletier M, Lagueux J, Côté M-F, Lagrange T, Fortin M-A (2017) Low-dose prostate cancer brachytherapy with radioactive palladium-gold nanoparticles. Adv Healthc Mater 6:1601120. https://doi.org/10.1002/adhm.201601120
111. Lechtman E, Chattopadhyay N, Cai Z, Mashouf S, Reilly R, Pignol JP (2011) Implications on clinical scenario of gold nanoparticle radiosensitization in regards to photon energy, nanoparticle size, concentration and location. Phys Med Biol 56:4631–4647. https://doi.org/10.1088/0031-9155/56/15/001
112. Leung MK, Chow JC, Chithrani BD, Lee MJ, Oms B, Jaffray DA (2011) Irradiation of gold nanoparticles by x-rays: Monte Carlo simulation of dose enhancements and the spatial properties of the secondary electrons production. Med Phys 38:624–631
113. Lewinski N (2008) Cytotoxicity of nanoparticles. Small 4:26–49
114. Li X, Anton N, Zuber G, Vandamme T (2014) Contrast agents for preclinical targeted X-ray imaging. Adv Drug Deliv Rev 76:116–133. https://doi.org/10.1016/j.addr.2014.07.013
115. Libutti SK et al (2010) Phase I and pharmacokinetic studies of CYT-6091, a novel PEGylated colloidal gold-rhTNF nanomedicine. Clin Cancer Res: Off J Am Assoc Can Res 16:6139–6149. https://doi.org/10.1158/1078-0432.CCR-10-0978
116. Liu R, Wang Y, Yuan Q, An D, Li J, Gao X (2014) The Au clusters induce tumor cell apoptosis via specifically targeting thioredoxin reductase 1 (TrxR1) and suppressing its activity. Chem Commun (Camb) 50:10687–10690. https://doi.org/10.1039/c4cc03320e
117. Longmire M, Choyke PL, Kobayashi H (2008) Clearance properties of nano-sized particles and molecules as imaging agents: consideration and caveats. Nanomedicine 3. https://doi.org/10.2217/17435889.3.5.703
118. Lopez-Chaves C, Soto-Alvaredo J, Montes-Bayon M, Bettmer J, Llopis J, Sanchez-Gonzalez C (2017) Gold nanoparticles: distribution, bioaccumulation and toxicity. In vitro and in vivo studies. Nanomedicine. https://doi.org/10.1016/j.nano.2017.08.011
119. Love JC, Estroff LA, Kriebel JK, Nuzzo RG, Whitesides GM (2005) Self-assembled monolayers of thiolates on metals as a form of nanotechnology. Chem Rev 105:1103–1170. https://doi.org/10.1021/cr0300789
120. Lu F, Doane TL, Zhu JJ, Burda C (2012) Gold nanoparticles for diagnostic sensing and therapy. Inorg Chim Acta 393:142–153. https://doi.org/10.1016/j.ica.2012.05.038
121. Lusic H, Grinstaff MW (2013) X-ray-computed tomography contrast agents. Chem Rev 113:1641–1666. https://doi.org/10.1021/cr200358s
122. Ma N et al (2016) Enhanced radiosensitization of gold nanospikes via hyperthermia in combined cancer radiation and photothermal therapy. ACS Appl Mater Interfaces 8:28480–28494. https://doi.org/10.1021/acsami.6b10132
123. Ma N et al (2017) Shape-dependent radiosensitization effect of gold nanostructures in cancer radiotherapy: comparison of gold nanoparticles, nanospikes, and nanorods. ACS Appl Mater Interfaces 9:13037–13048. https://doi.org/10.1021/acsami.7b01112
124. Mafune F, Kohno J, Takeda Y, Kondow T, Sawabe H (2001) Formation of gold nanoparticles by laser ablation in aqueous solution of surfactant. J Phys Chem B 105:5114–5120. https://doi.org/10.1021/jp0037091
125. Mariotti D, Patel J, Švrček V, Maguire P (2012) Plasma-liquid interactions at atmospheric pressure for nanomaterials synthesis and surface engineering. Plasma Processes Polym 9:1074–1085. https://doi.org/10.1002/ppap.201200007
126. Masood R et al (2012) Gold nanorod–sphingosine kinase siRNA nanocomplexes: a novel therapeutic tool for potent radiosensitization of head and neck cancer. Integr Biol 4:132–141

127. McKenna J, Patel J, Mitra S, Soin N, Švrček V, Maguire P, Mariotti D (2011) Synthesis and surface engineering of nanomaterials by atmospheric-pressure microplasmas. Eur Phys J Appl Phys 56. https://doi.org/10.1051/epjap/2011110203
128. McMahon SJ, Mendenhall MH, Jain S, Currell F (2008) Radiotherapy in the presence of contrast agents: a general figure of merit and its application to gold nanoparticles. Phys Med Biol 53:5635–5651. https://doi.org/10.1088/0031-9155/53/20/005
129. McQuaid HN et al (2016) Imaging and radiation effects of gold nanoparticles in tumour cells. Sci Rep 6:19442. https://doi.org/10.1038/srep19442
130. Merk V, Rehbock C, Becker F, Hagemann U, Nienhaus H, Barcikowski S (2014) In situ non-DLVO stabilization of surfactant-free, plasmonic gold nanoparticles: effect of Hofmeister's anions. Langmuir 30:4213–4222. https://doi.org/10.1021/la404556a
131. Mieszawska AJ, Mulder WJM, Fayad ZA, Cormode DP (2013) Multifunctional gold nanoparticles for diagnosis and therapy of disease. Mol Pharm 10:831–847. https://doi.org/10.1021/mp3005885
132. Misawa M, Takahashi J (2011) Generation of reactive oxygen species induced by gold nanoparticles under X-ray and UV Irradiations. Nanomed: Nanotechnol Biol Med 7:604–614. https://doi.org/10.1016/j.nano.2011.01.014
133. Moeendarbari S et al (2016) Theranostic nanoseeds for efficacious internal radiation therapy of unresectable solid tumors. Sci Rep 6:20614. https://doi.org/10.1038/srep20614
134. Moghimi SM, Hunter AC, Andresen TL (2012) Factors controlling nanoparticle pharmacokinetics: an integrated analysis and perspective. Annu Rev Pharmacol Toxicol 52:481–503. https://doi.org/10.1146/annurev-pharmtox-010611-134623
135. Movia D et al (2014) A safe-by-design approach to the development of gold nanoboxes as carriers for internalization into cancer cells. Biomaterials 35:2543–2557. https://doi.org/10.1016/j.biomaterials.2013.12.057
136. National Institute of Standards and Technology. http://www.nist.gov/pml/data/xraycoef. Accessed 17 Sept 2017
137. Nebuloni L, Kuhn GA, Muller R (2013) A comparative analysis of water-soluble and blood-pool contrast agents for in vivo vascular imaging with micro-CT. Acad Radiol 20:1247–1255. https://doi.org/10.1016/j.acra.2013.06.003
138. Neddersen J, Chumanov G, Cotton TM (1993) Laser ablation of metals: a new method for preparing sers active colloids. Appl Spectrosc 47:1959–1964. https://doi.org/10.1366/0003702934066460
139. Nel A, Xia T, Mädler L, Li N (2006) Toxic potential of materials at the nanolevel. Science 311:622
140. Neouze M-A, Schubert U (2008) Surface modification and functionalization of metal and metal oxide nanoparticles by organic ligands. Monatsh Chem 139:183–195. https://doi.org/10.1007/s00706-007-0775-2
141. O'Brien MN, Jones MR, Brown KA, Mirkin CA (2014) Universal noble metal nanoparticle seeds realized through iterative reductive growth and oxidative dissolution reactions. J Am Chem Soc 136:7603–7606. https://doi.org/10.1021/ja503509k
142. Otsuka H, Nagasaki Y, Kataoka K (2003) PEGylated nanoparticles for biological and pharmaceutical applications. Adv Drug Deliv Rev 55:403–419. https://doi.org/10.1016/s0169-409x(02)00226-0
143. Pan X, Cloutier P, Hunting D, Sanche L (2003) Dissociative electron attachment to DNA. Phys Rev Lett 90:208102. https://doi.org/10.1103/PhysRevLett.90.208102
144. Park J et al (2015) Multifunctional hollow gold nanoparticles designed for triple combination therapy and CT imaging. J Controlled Release 207:77–85. https://doi.org/10.1016/j.jconrel.2015.04.007
145. Peng C et al (2012) PEGylated dendrimer-entrapped gold nanoparticles for in vivo blood pool and tumor imaging by computed tomography. Biomaterials 33:1107–1119. https://doi.org/10.1016/j.biomaterials.2011.10.052
146. Perala SRK, Kumar S (2013) On the mechanism of metal nanoparticle synthesis in the Brust-Schiffrin method. Langmuir 29:9863–9873. https://doi.org/10.1021/la401604q

147. Perrault SD, Chan WCW (2009) Synthesis and surface modification of highly monodispersed, spherical gold nanoparticles of 50–200 nm. J Am Chem Soc 131:17042–17043. https://doi.org/10.1021/ja907069u
148. Perrault SD, Walkey C, Jennings T, Fischer HC, Chan WCW (2009) Mediating tumor targeting efficiency of nanoparticles through design. Nano Lett 9:1909–1915. https://doi.org/10.1021/nl900031y
149. Popovtzer A et al (2016) Actively targeted gold nanoparticles as novel radiosensitizer agents: an in vivo head and neck cancer model. Nanoscale 8:2678–2685. https://doi.org/10.1039/C5NR07496G
150. Popovtzer R, Agrawal A, Kotov NA, Popovtzer A, Balter J, Carey TE, Kopelman R (2008) Targeted gold nanoparticles enable molecular CT imaging of cancer. Nano Lett 8:4593–4596. https://doi.org/10.1021/nl8029114
151. Porcel E et al (2010) Platinum nanoparticles: a promising material for future cancer therapy? Nanotechnology 21:85103. https://doi.org/10.1088/0957-4484/21/8/085103
152. Reddy VR (2006) Gold nanoparticles: synthesis and applications. Synlett 2006:1791–1792. https://doi.org/10.1055/s-2006-944219
153. Rehbock C, Merk V, Gamrad L, Streubel R, Barcikowski S (2013) Size control of laser-fabricated surfactant-free gold nanoparticles with highly diluted electrolytes and their subsequent bioconjugation. Phys Chem Chem Phys 15:3057. https://doi.org/10.1039/c2cp42641b
154. Reuveni T, Motiei M, Romman Z, Popovtzer A, Popovtzer R (2011) Targeted gold nanoparticles enable molecular CT imaging of cancer: an in vivo study. Int J Nanomed 6:2859–2864. https://doi.org/10.2147/IJN.S25446
155. Richmonds C, Sankaran RM (2008) Plasma-liquid electrochemistry: rapid synthesis of colloidal metal nanoparticles by microplasma reduction of aqueous cations. Appl Phys Lett 93:131501–131503. https://doi.org/10.1063/1.2988283
156. Ritman EL (2007) Small-animal CT: its difference from, and impact on, clinical CT. Nucl Instrum Methods Phys Res, A 580:968–970. https://doi.org/10.1016/j.nima.2007.06.040
157. Rosa S, Connolly C, Schettino G, Butterworth KT, Prise KM (2017) Biological mechanisms of gold nanoparticle radiosensitization. Cancer Nanotechnol 8:2. https://doi.org/10.1186/s12645-017-0026-0
158. Rumbach P, Witzke M, Sankaran RM, Go DB (2013) Plasma-liquid interactions: separating electrolytic reactions from plasma/gas phase reactions. Proceedings of ESA Annual Meeting on Electrostatics
159. Saha K, Agasti SS, Kim C, Li X, Rotello VM (2012) Gold nanoparticles in chemical and biological sensing. Chem Rev 112:2739–2779. https://doi.org/10.1021/cr2001178
160. Sankaran RM (2011) Electrochemical cell including a plasma source and method of operating the electrochemical cell. US20110048960 A1, 3 Mar 2011
161. Scarboro SB, Followill DS, Howell RM, Kry SF (2011) Variations in photon energy spectra of a 6 MV beam and their impact on TLD response. Med Phys 38:2619–2628. https://doi.org/10.1118/1.3575419
162. Schuemann J et al (2016) Roadmap to clinical use of gold nanoparticles for radiation sensitization. Int J Radiat Oncol Biol Phys 94:189–205. https://doi.org/10.1016/j.ijrobp.2015.09.032
163. Searson PC (2014) Nanomedicines for cancer therapy: state-of-the-art and limitations to pre-clinical studies that hinder future developments:1–13. https://doi.org/10.3389/fchem.2014.00069/abstract
164. Semmler-Behnke M et al (2008) Biodistribution of 1.4- and 18-nm gold particles in rats. Small 4:2108–2111. https://doi.org/10.1002/smll.200800922
165. Seo D, Park JC, Song H (2006) Polyhedral gold nanocrystals with Oh symmetry: from octahedra to cubes. J Am Chem Soc 128:14863–14870. https://doi.org/10.1021/ja062892u
166. Shi M, Paquette B, Thippayamontri T, Gendron L, Guerin B, Sanche L (2016) Increased radiosensitivity of colorectal tumors with intra-tumoral injection of low dose of gold nanoparticles. Int J Nanomed 11:5323–5333. https://doi.org/10.2147/IJN.S97541
167. Shiba F (2013) Size control of monodisperse Au nanoparticles synthesized via a citrate reduction process associated with a pH-shifting procedure. CrystEngComm 15:8412–8415. https://doi.org/10.1039/C3CE41516C

168. Shukla R et al (2012) Laminin receptor specific therapeutic gold nanoparticles (^{198}AuNP-EGCg) show efficacy in treating prostate cancer. Proc Natl Acad Sci U S A 109:12426–12431. https://doi.org/10.1073/pnas.1121174109
169. Simao T et al (2016) Laser-synthesized ligand-free Au nanoparticles for contrast agent applications in computed tomography and magnetic resonance imaging. J Mater Chem B 4:6413–6427. https://doi.org/10.1039/C6TB01162D
170. Sina M et al (2016) Theranostic nanoseeds for efficacious internal radiation therapy of unresectable solid tumors. Sci Rep 6:1–9. https://doi.org/10.1038/srep20614
171. Sinha N et al (2015) Brachytherapy application with in situ dose painting administered by gold nanoparticle eluters. Int J Radiat Oncol Biol Phys 91:385–392. https://doi.org/10.1016/j.ijrobp.2014.10.001
172. Skotland T, Iversen TG, Sandvig K (2010) New metal-based nanoparticles for intravenous use: requirements for clinical success with focus on medical imaging. Nanomed: Nanotechnol Biol Med 6:730–737. https://doi.org/10.1016/j.nano.2010.05.002
173. Soenen SJ, Rivera-Gil P, Montenegro J-M, Parak WJ, De Smedt SC, Braeckmans K (2011) Cellular toxicity of inorganic nanoparticles: common aspects and guidelines for improved nanotoxicity evaluation. Nano Today 6:446–465. https://doi.org/10.1016/j.nantod.2011.08.001
174. Solomon SB, Silverman SG (2010) Imaging in interventional oncology. Radiology 257:624–640. https://doi.org/10.1148/radiol.10081490
175. Speck U (2008) Contrast agents: X-ray contrast agents and molecular imaging—a contradiction? In: Semmler W, Schwaiger M (eds) Molecular imaging I, vol 185/1. Handbook of experimental pharmacology. Springer, Berlin, pp 167–175. https://doi.org/10.1007/978-3-540-72718-7_8
176. Sperling RA, Rivera gil P, Zhang F, Zanella M, Parak WJ (2008) Biological applications of gold nanoparticles. Chem Soc Rev 37:1896–1908. https://doi.org/10.1039/b712170a
177. Su N, Dang Y, Liang G, Liu G (2015) Iodine-125-labeled cRGD-gold nanoparticles as tumor-targeted radiosensitizer and imaging agent. Nanoscale Res Lett 10:160. https://doi.org/10.1186/s11671-015-0864-9
178. Su S, Zuo XL, Pan D, Pei H, Wang LH, Fan CH, Huang W (2013) Design and applications of gold nanoparticle conjugates by exploiting biomolecule-gold nanoparticle interactions. Nanoscale 5:2589–2599. https://doi.org/10.1039/c3nr33870c
179. Sun Y, Xia Y (2002) Shape-controlled synthesis of gold and silver nanoparticles. Science 298:2176–2179. https://doi.org/10.1126/science.1077229
180. Sylvestre J-P, Poulin S, Kabashin AV, Sacher E, Meunier M, Luong JHT (2004) Surface chemistry of gold nanoparticles produced by laser ablation in aqueous media. J Phys Chem B 108:16864–16869. https://doi.org/10.1021/jp047134
181. Taggart LE, McMahon SJ, Currell FJ, Prise KM, Butterworth KT (2014) The role of mitochondrial function in gold nanoparticle mediated radiosensitisation. Cancer Nanotechnol 5:5. https://doi.org/10.1186/s12645-014-0005-7
182. Thakor AS, Jokerst J, Zavaleta C, Massoud TF, Gambhir SS (2011) Gold nanoparticles: a revival in precious metal administration to patients. Nano Lett 11:4029–4036. https://doi.org/10.1021/nl202559p
183. Torchilin VP, Frank-Kamenetsky MD, Wolf GL (1999) CT visualization of blood pool in rats by using long-circulating, iodine-containing micelles. Acad Radiol 6:61–65. https://doi.org/10.1016/s1076-6332(99)80063-4
184. Tsuji T, Iryo K, Nishimura Y, Tsuji M (2001) Preparation of metal colloids by a laser ablation technique in solution: influence of laser wavelength on the ablation efficiency (II). J Photochem Photobiol, A 145:201–207. https://doi.org/10.1016/s1010-6030(01)00583-4
185. Turkevich J, Stevenson PC, Hillier J (1951) A study of the nucleation and growth processes in the synthesis of colloidal gold. Discuss Faraday Soc 11:55–75. https://doi.org/10.1039/DF9511100055
186. Vincenzo A, Roberto P, Marco F, Onofrio MM, Maria Antonia I (2017) Surface plasmon resonance in gold nanoparticles: a review. J Phys: Condens Matter 29:203002

187. Wahab R et al (2014) Statistical analysis of gold nanoparticle-induced oxidative stress and apoptosis in myoblast (C2C12) cells. Colloids Surf, B Biointerfaces 123:664–672. https://doi.org/10.1016/j.colsurfb.2014.10.012
188. Wang R, Zuo S, Wu D, Zhang J, Zhu W, Becker KH, Fang J (2014) Microplasma-assisted synthesis of colloidal gold nanoparticles and their use in the detection of cardiac Troponin I (cTn-I). Plasma Processes Polym. https://doi.org/10.1002/ppap.201400127
189. Wang R, Zuo S, Zhu W, Wu S, Nian W, Zhang J, Fang J (2014) Microplasma-assisted growth of colloidal silver nanoparticles for enhanced antibacterial activity. Plasma Processes Polym 11:44–51. https://doi.org/10.1002/ppap.201300038
190. Webb JA, Bardhan R (2014) Emerging advances in nanomedicine with engineered gold nanostructures. Nanoscale 6:2502–2530. https://doi.org/10.1039/c3nr05112a
191. Weissleder R (2002) Scaling down imaging: molecular mapping of cancer in mice. Nat Rev Cancer 2:11–18. https://doi.org/10.1038/nrc701
192. Wilhelm S, Tavares AJ, Dai Q, Ohta S, Audet J, Dvorak HF, Chan WCW (2016) Analysis of nanoparticle delivery to tumours. Nat Rev Mater 1:16014. https://doi.org/10.1038/natrevmats.2016.14. https://www.nature.com/articles/natrevmats201614#supplementary-information
193. Wilson R et al (2009) Gold nanoparticle sensitize radiotherapy of prostate cancer cells by regulation of the cell cycle. Nanotechnology 20:375101
194. Woehrle GH, Brown LO, Hutchison JE (2005) Thiol-functionalized, 1.5-nm gold nanoparticles through ligand exchange reactions: scope and mechanism of ligand exchange. J Am Chem Soc 127:2172–2183. https://doi.org/10.1021/ja0457718
195. Wolfe T et al (2015) Targeted gold nanoparticles enhance sensitization of prostate tumors to megavoltage radiation therapy in vivo. Nanomed: Nanotechnol Biol Med 11:1277–1283. https://doi.org/10.1016/j.nano.2014.12.016
196. Xiao T et al (2013) Facile synthesis of acetylated dendrimer-entrapped gold nanoparticles with enhanced gold loading for CT imaging applications. J Mater Chem B 1:2773–2780. https://doi.org/10.1039/c3tb20399a
197. Yah CS (2013) The toxicity of Gold Nanoparticles in relation to their physiochemical properties. Biomed Res-India 24:400–413
198. Yang Y et al (2016) Tumor angiogenesis targeted radiosensitization therapy using gold nanoprobes guided by MRI/SPECT imaging. ACS Appl Mater Interfaces 8:1718–1732. https://doi.org/10.1021/acsami.5b09274
199. Yeh YC, Creran B, Rotello VM (2012) Gold nanoparticles: preparation, properties, and applications in bionanotechnology. Nanoscale 4:1871–1880. https://doi.org/10.1039/c1nr11188d
200. Yong KT, Swihart MT, Ding H, Prasad PN (2009) Preparation of gold nanoparticles and their applications in anisotropic nanoparticle synthesis and bioimaging. Plasmonics 4:79–93. https://doi.org/10.1007/s11468-009-9078-2
201. Yook S, Cai Z, Lu Y, Winnik MA, Pignol J-P, Reilly RM (2016) Intratumorally injected 177Lu-labeled gold nanoparticles: gold nanoseed brachytherapy with application for neoadjuvant treatment of locally advanced breast cancer. J Nucl Med 57:936–942. https://doi.org/10.2967/jnumed.115.168906
202. Zhang X-D et al (2015) Use of epidermal growth factor receptor antibody–gold cluster conjugates with good renal excretion in targeted cancer radiation treatment. J Mater Chem B: Mater Biol Med 3:4735–4741
203. Zhang X-D et al (2014) Enhanced tumor accumulation of sub-2 nm gold nanoclusters for cancer radiation therapy. Adv Healthc Mater 3:133–141. https://doi.org/10.1002/adhm.201300189
204. Zhang XD, Wu D, Shen X, Chen J, Sun YM, Liu PX, Liang XJ (2012) Size-dependent radiosensitization of PEG-coated gold nanoparticles for cancer radiation therapy. Biomaterials 33:6408–6419. https://doi.org/10.1016/j.biomaterials.2012.05.047
205. Zhao L, Jiang D, Cai Y, Ji X, Xie R, Yang W (2012) Tuning the size of gold nanoparticles in the citrate reduction by chloride ions. Nanoscale 4:5071–5076. https://doi.org/10.1039/C2NR30957B

206. Zhao N, Yang Z, Li B, Meng J, Shi Z, Li P, Fu S (2016) RGD-conjugated mesoporous silica-encapsulated gold nanorods enhance the sensitization of triple-negative breast cancer to megavoltage radiation therapy. Int J Nanomed 11:5595–5610. https://doi.org/10.2147/IJN.S104034
207. Zheng Q, Yang H, Wei J, Tong JL, Shu YQ (2013) The role and mechanisms of nanoparticles to enhance radiosensitivity in hepatocellular cell. Biomed Pharmacother 67:569–575. https://doi.org/10.1016/j.biopha.2013.04.003
208. Zheng Y, Cloutier P, Hunting DJ, Sanche L (2008) Radiosensitization by gold nanoparticles: comparison of DNA damage induced by low and high-energy electrons. J Biomed Nanotechnol 4:469–473. https://doi.org/10.1166/jbn.2008.012
209. Ziegler C, Eychmüller A (2011) Seeded growth synthesis of uniform gold nanoparticles with diameters of 15–300 nm. J Phys Chem C 115:4502–4506. https://doi.org/10.1021/jp1106982

Liposomes-Based Nanoparticles for Cancer Therapy and Bioimaging

Calvin Cheung and Wafa T. Al-Jamal

Abstract Liposome, one of the most well-established nanomedicines in cancer therapy and bioimaging, is a great delivery system with their flexibility and versatility. Liposomes resemble the biological cell membrane, adopting a lipid bilayer structure which provides protection and solubilisation of both hydrophilic and hydrophobic agents. A wide range of therapeutic drugs and imaging agents can, therefore, be encapsulated and delivered. Cationic liposomes, for example, are one of the popular choices of non-viral vector for the fast-growing field of gene therapy. Their physiochemical properties can be engineered and modified to suit specific applications simply by changing the lipid components and the corresponding ratio. This also expands the potential of having additional functionalities such as long-circulating, targeting and stimuli-responsiveness. In addition to delivering therapeutics and imaging agents, interactions between the lipids and the payloads can be beneficial for imaging enhancement. Stimuli-sensitive liposomes can be used along with diagnostic and therapeutics for image-guided drug delivery, providing real-time monitoring of the drug delivery process as well as spatiotemporal control over the release of drugs. Liposomes will be expected to be a promising delivery system and tool for personalised medicine.

Keywords Liposome · Drug delivery system · Gene delivery
Targeted delivery · Theranostic · Image-guided drug delivery

1 Introduction

Liposomes are a type of clinically well-established nanoparticle that have been commercially used to deliver cytotoxic drugs, antifungal drugs and vaccines. Some of the approved commercial liposome-based products are shown in Table 1. There are

C. Cheung · W. T. Al-Jamal (✉)
Queen's University Belfast, Belfast, Northern Ireland, UK
e-mail: w.al-jamal@qub.ac.uk

C. Cheung
e-mail: ccheung04@qub.ac.uk

G. Gonçalves and G. Tobias (eds.), *Nanooncology*, Nanomedicine and Nanotoxicology, https://doi.org/10.1007/978-3-319-89878-0_2

many more liposomal formulations undergoing clinical trials, which include new generations of liposomes as well as the delivery of genes and imaging agents [1–7]. Discovered by Alec Bangham in the 1960s, liposomes are simply formed from the self-assembly of lipids in aqueous media driven by their amphiphilic nature, resulting in spherical vesicles with an aqueous core enclosed by single or multiple concentric lipid bilayer [8]. The bilayer membrane structure of liposomes resembles the natural cell membranes, allowing them to be used as biological membrane model, for studying drug–membrane interactions [9, 10]. Liposomes offer exceptional biological performances, namely, biocompatibility, biodegradability, reduced toxicity and capacity for size and surface manipulations [11]. These properties comprise the outstanding profile that liposomes offer compared to other delivery systems.

1.1 Liposome Composition

The effectiveness of liposomes as a drug carrier for therapeutic and imaging agents is highly dependent on the physiochemical properties of liposomes, especially their membrane components, size and surface charge. Liposomes are mainly composed of amphiphilic phospholipid, which typically consist of a polar phosphate head group, and single or two non-polar fatty acid chains (Fig. 1). The chain length and the degree of unsaturation of the phospholipids greatly affect the phase transition (chain melting) temperature, which is closely related to the fluidity and integrity of the lipid bilayer [12, 13]. For example, liposomes prepared with unsaturated phospholipids showed higher blood clearance, and drug release rate, compared to that prepared with saturated phospholipids [14, 15]. Long-chain, saturated lipids also offer enhanced solubilisation capacity for hydrophobic agents attributed to the increased bilayer lipophilic area [16, 17]. Cholesterol is a non-bilayer forming lipid that can be incorporated into the bilayer membrane as a universal membrane sealer, minimising the bilayer permeability and enhancing the structural integrity [12]. Cholesterol can also increase the chemical stability of liposome membrane against degradation through peroxidation and hydrolysis [18].

The variation in the polar head leads to different type of lipids, classified into (a) zwitterionic phospholipids, including: phosphatidylcholine (PC) and phos-

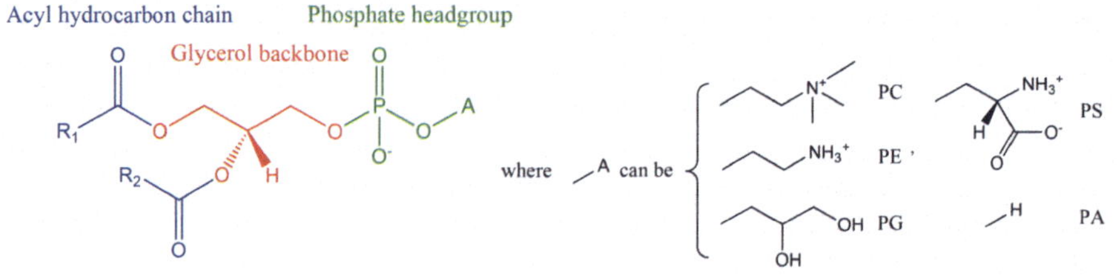

Fig. 1 General structure of phospholipids

Table 1 List of selected approved commercial liposome-based products (Data collected from U.S. Food and Drug Administration (https://www.fda.gov/) and European Medicines Agency (http://www.ema.europa.eu/)

Category	Brand name	Encapsulated drug	Liposome composition	Indications	Approval year
Cytotoxic	DaunoXome®	Daunorubicin	DSPC/Chol	Kaposi sarcoma	1996 (USA) (Discontinued)
	Doxil® Caelyx®	Doxorubicin	HSPC/Chol/DSPE-PEG$_{2000}$	Kaposi sarcoma, ovarian cancer, multiple myeloma	1995 (USA), 1996 (Europe)
	Lipo-dox®	Doxorubicin	DSPC/Chol/DSPE-PEG$_{2000}$	Kaposi sarcoma, ovarian cancer, breast cancer	2013 (USA)
	Marqibo®	Vincristine	SM/Chol	Acute lymphoblastic leukaemia	2012 (USA)
	Mepact®	Mifamurtide	POPC/DOPS	Osteosarcoma	2009 (Europe)
	Myocet®	Doxorubicin	EPC/Chol	Breast cancer	2000 (Europe)
	Onivyde®	Irinotecan	DSPC/Chol/DSPE-PEG$_{2000}$	Pancreatic adenocarcinoma	2011 (Europe), 2015 (USA)
	Vyxeos	Daunorubicin, cytarabine	DSPC/DSPG/Chol	Acute myeloid leukaemia	2017 (USA)
Pain management	DepoDur®	Morphine	DOPC/Chol/DPPG/Tricaprylin/Triolein	Pain management	2004 (USA) (Discontinued)
	Exparel®	Bupivacaine	DEPC/Chol/DPPG/Tricaprylin	Pain management	2011 (USA)

(continued)

Table 1 (continued)

Category	Brand name	Encapsulated drug	Liposome composition	Indications	Approval year
Others	AmBisome®	Amphotericin B	HSPC/DSPG/Chol	Systemic fungal infections, visceral leishmaniasis, cryptococcal meningitis	1990 (Europe), 1997 (USA)
	Depocyt®	Cytarabine	DOPC/Chol/Triolein/DPPG	Lymphomatous meningitis	1999 (USA)
	Visudyne®	Verteporfin	DMPC/EPG	Subfoveal choroidal neovascularisation	2000 (USA)

Chol cholesterol; *DEPC* 1,2-dierucoylphosphatidylcholine; *DMPC* 1,2-dimyristoyl-sn-glycero-3-phosphocholine; *DOPC* 1,2-dioleoyl-sn-glycero-3-phosphocholine; *DOPS* 1,2-dioleoyl-sn-glycero-3-phospho-L-serine; *DPPG* 1,2-dipalmitoyl-sn-glycero-3 phospho-rac-(1-glycerol); *DSPC* 1,2-distearoyl-sn-glycero-3-phosphocholine; *DSPE-PEG2000* 1,2-distearoyl-sn-glycero-3-phosphoethanolamine-*N*-[carboxy(polyethylene glycol)-2000]; *DSPG* 1,2-distearoyl-sn-glycero-3-phospho-(1′-rac-glycerol); *EPC* egg phosphatidylcholine; *EPG* egg phosphatidylglycerol; *HSPC* hydrogenated soy L-α-phosphatidylcholine; *POPC* 1-palmitoyl-2-oleoyl-sn-glycero-3-phosphocholine; *SM* sphingomyelin phatidylethanolamine (PE) or (b) anionic (negatively charged) phospholipids, including: phosphatidic acid (PA), phosphatidylserine (PS), phosphatidylglycerol (PG), phosphatidylinositol (PI) and cardiolipin (CL); and (c), cationic (positively charged) lipids, such as 1,2-dioleoyl-3-trimethylammonium-propane (DOTAP), 1,2-di-*O*-octadecenyl-3-trimethylammonium-propane (DOTMA) and 3β[*N*-(*N*′, *N*′-dimethylaminoethane)-carbamoyl] cholesterol (DC-Chol). Cationic lipids are not found naturally in biological membrane and are obtained exclusively through synthesis. They are extensively used for forming complexes with nucleic acids that enables their intracellular uptake and trafficking (see Sect. 2) [6, 19–21]. Figure 2 shows the chemical structures of different common lipids used in liposome preparation.

Different moieties can be synthetically anchored to these lipids and introduced to the bilayer to improve their physiochemical properties. A common example is the attachment of hydrophilic polymers as a steric stabiliser, also to reduce liposome recognition and clearance by immune system, such as poly(ethylene glycol)

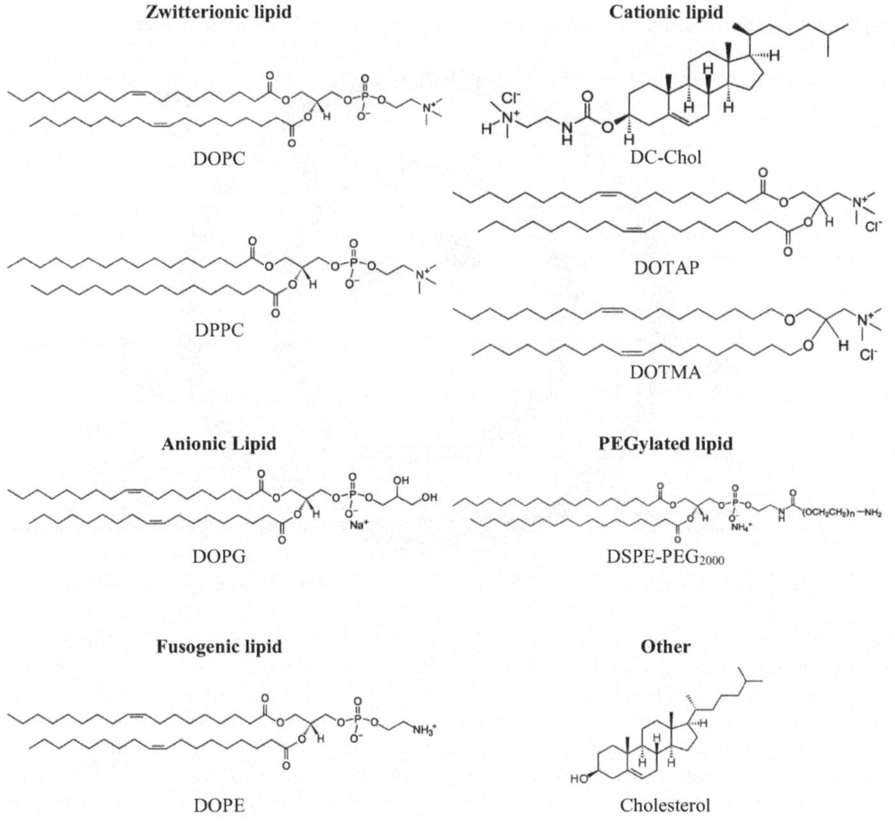

Fig. 2 Chemical structures of most common lipids used in liposome preparations

(PEG) [22], poly(vinyl alcohol) [23] and poly(*N*-vinylpyrrolidones) [24]. This type of liposomes is known as stealth liposomes or long-circulating liposomes.

The vesicle structure of liposomes inherently allows the entrapment of both lipophilic and hydrophilic cargos, in the lipid bilayer and the aqueous core, respectively, allowing solubilisation of a greater variety of drugs. When therapeutic is loaded into liposomes, it adopts the carrier's pharmacokinetics until it is released [1]. This leads to an increase in the therapeutic index, reduction in systemic toxicity [25–27], increase in drug stability [17, 25, 28] and/or the emergence of a sustained-release drug profile [1, 29, 30]. Moreover, protective hydrophilic polymer and targeting ligand can be anchored onto the liposomes to modulate their circulation time and targeting abilities, respectively [31]. In addition, liposomes responsive to external (e.g. magnetic field, light) [32, 33] or internal stimuli (e.g. pH, redox, enzymes) [34–36] that trigger their drug release at specific and controlled location and time have also been developed. Figure 3 shows different features of liposomes as drug delivery system.

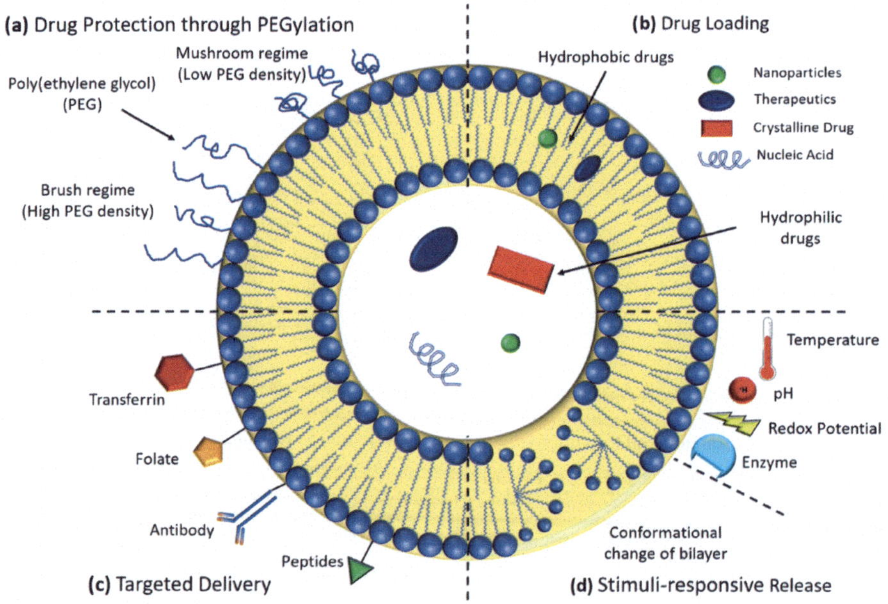

Fig. 3 Features of liposomes as drug delivery system. **a** Protection of drugs from degradation and immune clearance and extension of circulation time through surface modification with poly(ethylene glycol). **b** Entrapment of both hydrophilic and hydrophobic molecules into liposomes, as well as the adsorption onto the bilayer surface. Nanoparticles can also be incorporated into liposomes for imaging and therapeutic functions. **c** Conjugation of targeting moieties onto the surface of liposomes for targeted delivery. **d** Stimuli-responsive release of entrapped drugs using, e.g. lysolipid or cleavable linker, triggered by stimuli such as temperature, pH, redox potential and enzymatic activity

2 Cancer Therapy

A great amount of research effort has been invested in developing liposomal formulations to deliver a wide range of therapeutics, such as cytotoxins and nucleic acids. For both therapeutics, different strategies have been developed to promote therapeutics loading and release, resulting in formulations with promising anticancer activity.

2.1 Drug Delivery System

2.1.1 Drug Protection

Liposomes protect the encapsulated drug against degradation, inactivation, dilution. They also prevent the drugs from premature release and metabolism which causes systemic toxicity in vivo [1, 37]. The extended circulation time of liposomes provides a sustained-release profile of the drug [38] and increases the passive targeting of tumour through the enhanced permeability and retention (EPR) effect simultaneously [39]. These effects selectively increase the therapeutic efficacy of anticancer drugs, as higher amount of the drug reaches the tumour site.

For example, it was reported that PEGylated liposomal formulation of doxorubicin (DOX) showed comparable efficacy, versus its free form, when administered at lower doses and frequencies [26]. Moreover, the systemic toxicity, such as cardiotoxicity, bone marrow suppression, vomiting and hair loss, was lower compared to the free drug [26, 27]. Similar effects have been reported, with liposomal irinotecan (Onivyde®) and liposomal vincristine (Marqibo®), where both formulations demonstrated prolonged blood circulation, greater tumour growth inhibition, higher tumour accumulation and decreased systemic toxicity, compared to the free drugs [37, 39].

2.1.2 Drug Loading

The amphipathic nature of liposomes allows them to encapsulate both lipophilic and hydrophilic drugs, in the bilayer and in the aqueous core, respectively. The drugs can be loaded during the formation steps, known as *passive loading*, or loaded after the liposome formation, known as *active loading*.

The encapsulation efficiency for passive loading of hydrophilic drugs is generally low (<30%), and it is dependent on the solubility of the drug and the method of encapsulation, i.e. the ability of liposomes to trap aqueous medium containing the dissolved drugs during the vesicle formation [40]. In contrast, the encapsulation efficiency of hydrophobic drugs is usually high, as the drugs are dissolved together with the lipid in organic solvent during the vesicle formation. Encapsulation efficiency of 100% is achievable for hydrophobic drugs, although this is dependent on the drug–lipid interaction, and the solubility of the drugs in the liposome bilayer [40].

Undoubtedly, the ability of drugs to diffuse across lipid membranes will influence their loading and leakage. High molecular weight compounds, and polar solutes, such as glucose diffuse very slowly, while small molecules with neutral charges, such as water and ammonia can diffuse rapidly and freely across the membrane [41, 42]. Several drugs are lipophilic in nature and contain amines groups that can diffuse freely through the liposome bilayer in their unprotonated form [28]. The passive loading of these drugs results in rapid permeation, and undesirable leakage overtime. Therefore, active (remote) loading methods could be used as alternatives, to improve the loading capacity and efficiency of these drugs.

Doxil® is a liposomal formulation of doxorubicin, where the drug is actively loaded into liposomes, using a pH gradient method [43]. In active loading, a trans-membrane gradient is first established [28, 44–46]. As uncharged drugs (weak bases) permeate the membrane, they will be protonated at acidic pH, or form membrane-impermeable complexes with the Mn^{2+} or Ca^{2+} ions, depending on the gradient used. As protonated drugs or drug complexes decreases the free drug concentration inside the liposome, more uncharged free drugs will be further driven from the external medium into the liposome, increasing the amount of drugs encapsulated. In certain conditions, when the encapsulated (protonated or complex) drugs exceed their solubility in the aqueous core, they form precipitates, which results in high encapsulation efficiency, up to 100%, with minimum drug leakage [44].

Besides cytotoxic drugs, nanoparticles can be co-loaded to form liposome—nanoparticle hybrids to introduce multiple functionalities [47]. For examples, magnetic nanoparticles could be used to provide on-demand drug release, magnetic-induced hyperthermia and magnetic resonance imaging [48–51]. Similarly, the incorporation of gold nanoparticles into the liposome system can introduce therapeutic activity through photothermal effect [52–56], triggered drug release [57–59], beside imaging [60]. Other possible liposome hybrids include polymeric nanoparticles [61, 62], quantum dots [63, 64], copper [65] and other metal complexes [66]. The delivery of nucleic acids, also known as gene delivery will be discussed in detail in Sect. 2.2. The applications of liposomes encapsulating imaging agents for bioimaging will be discussed in Sect. 3.

2.1.3 Drug Release

Liposomes have been shown to offer great drug protection and reduced toxicity, however, this can also hamper drug release from liposomes at the target tissues [67]. An ideal drug delivery system should be able to not only protect the drugs from untimely release and causing systemic toxicity but also to distinguish between normal and cancer tissue, and release the drugs only when reaching the cancer tissue. Stimuli-responsive (or stimuli-sensitive) drug release strategy can be incorporated into liposomes to achieve on-demand drug release of drugs. Internal stimuli are attributed to the difference of internal environment between tumour and healthy tissues, such as decreased pH value, higher redox potential and overexpression of enzymes [68]. As internal stimuli-responsive systems exploit the pathological microenvironment of cancerous tissues, they are self-controlled and require no exotic tools to trigger such release. Stimuli can also be applied externally, which include temperature, magnetic field, ultrasound, light or electric pulses [69]. These stimuli are applied and controlled externally, thus can offer better spatial-, temporal- and dosage-precision and control; however, sophisticated tools need to be build up [69]. Examples of internal and external stimuli-responsive liposomal systems are shown in Tables 2 and 3.

For example, pH-sensitive liposomes can take advantage of the acidic environment in pathological tissues (tumours, inflamed and infected areas) as compared with normal tissues [80]. They could also utilise the acidic environment of endosomes to escape the hostile environment and avoid drug degradation [81, 82]. pH-sensitive liposome formulations are commonly composed of a fusogenic PE (e.g. DOPE) and ionisable lipids with compounds containing an acidic group, such as carboxylic group [80]. DOPE lipid has a smaller molecular area relative to the hydrophobic chains, and thus adopts a cone molecular shape and tends to form an inverted hexagonal phase. The carboxylic groups of ionisable lipids become ionised at physiological pH, which stabilises the fusogenic lipids and allows the formation of a stable bilayer structure

Table 2 Internal stimuli-responsive liposomal systems

Internal stimuli	Stimuli origin	Examples of responsive delivery system	Mechanism of action	References
Redox potential	Increased concentration of glutathione inside cancer cells	Thiolytically cleavable (disulphide conjugated) PEG–phospholipid conjugate	Thiolytic cleavage of grafted PEG and subsequent loss of bilayer stability of PE-containing liposomes	[35, 70]
pH	Decreased pH in pathological areas Acidic environment of endocytic vesicles	DOPE in combination with protonatable amphiphilic lipid, e.g. DOPE:CHEMS	Electrostatic repulsion of the amphiphilic molecules promotes the formation of bilayer structures at physiological pH and temperature. Protonation of the acidic group at low pH destabilises the bilayer leading to liposome collapse.	[34, 68]
Enzyme overexpression	Overexpressed enzymes in tumour tissues	sPLA2-cleavable lipid prodrug	sPLA2 hydrolysis of the lipid prodrug release the carried drug after a cyclization reaction	[36, 71]

CHEMS cholesteryl hemisuccinate; *DOPE* 1,2-dioleoyl-sn-glycero-3-phosphoethanolamine; *sPLA2* secretory phospholipase A2; *PEG* poly(ethylene glycol)

Table 3 External stimuli-responsive liposomal systems

External stimuli	Examples of responsive delivery system	Mechanism of action	References
Temperature	(i) LTSLs (ii) Leucine zipper peptide (iii) Gas-forming agents	(i) Stabilisation of pores, discs and other structural defects in liposomes bilayer by LTSLs (ii) Conformational change of the peptide forms pores the bilayer (iii) Thermal decomposition of gas-forming agents generates gas bubbles, disrupting the bilayer integrity	[72–75]
Light	(i) AuNP (ii) Bis-azobenzene PC	(i) Photothermal effect by surface plasmon resonance (ii) Cis-trans isomerisation promotes bulky cis isomer that interferes with bilayer packing resulting leakage	[54–56, 76]
Ultrasound	(i) TSLs (ii) Gas bubbles (e.g. PFCs)	(i) Thermal effect from high-intensity focused ultrasound (ii) Ultrasonic cavitation of encapsulated PFCs bubbles Low frequency ultrasound-induced structural defects of bilayer	[77, 78]
Magnetic Field	Superparamagnetic iron oxide nanoparticles	Magnetic nanoparticles induced hyperthermia by magnetic hysteresis	[32, 79]

AgNP silver nanoparticle; *AuNP* gold nanoparticle; *LTSL* lysolipid-containing thermosensitive liposomes; *PC* phosphatidylcholine; *PFC* perfluorocarbon; *TSL* thermosensitive liposome

[80]. Upon acidification, where these acidic groups become neutral (protonated), the stabilising effect disappears, leading to liposome destabilisation as the PE molecules revert into the inverted hexagonal phase. Such change causes transient alterations in the endosome structure, resulting in a cytoplasm drug release (Fig. 4). An alternative to the above strategy is to attach pH-sensitive polymer to the surface of the liposomes, which causes pH-sensitive destabilisation of the liposomes, and subsequent release of the payloads to the cytoplasm [83].

A second example of stimuli-responsive is temperature-sensitive liposomes which could be used in combination with hyperthermia. Hyperthermia can be induced locally using a magnetic field, laser, radiofrequency ablation and focused ultrasound, to increase drug bioavailability and reduce systemic side effects [74, 83–85]. In general, traditional temperature-sensitive liposomes are designed to release drugs at temperature above their phase transition, where the lipid hydrocarbon chains change from gel to liquid crystalline phase [86]. DPPC, with a transition temperature of 41 °C, slightly above physiological temperature while attainable by mild local hyperthermia, is therefore often included in temperature-sensitive liposome formulations. However, drug release from traditional formulations relying only on the changes in the phase transition of one lipid bilayer is considered relatively slow [87]. In order

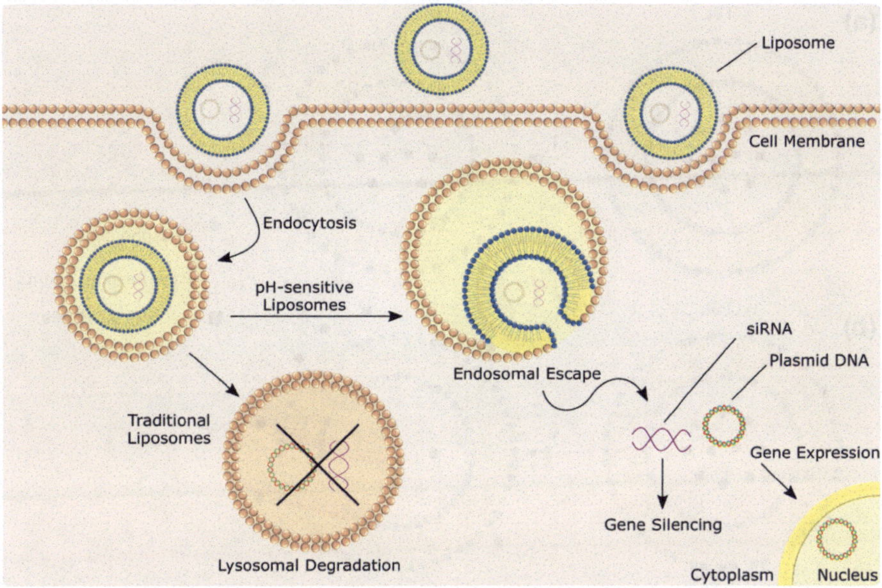

Fig. 4 pH-sensitive liposomes destabilise under acidic conditions of the endocytic pathway. Destabilisation or fusion of the liposomes results in endosomal escape and cytoplasmic delivery of the encapsulated therapeutics. Traditional liposomes unable to escape the endocytic pathway are degraded in the lysosomes, resulting in poor therapeutic effect

to improve the drug release rate, a permeabilising component can be introduced into the formulation, to induce thermally enhanced permeability to small molecules [87]. These components are usually pore-forming lysolipids, such as monopalmitoylphosphatidylcholine (MPPC) or monostearoylphosphatidylcholine (MSPC), and surfactant such as Brij® and Myrj® [88]. Above the phase transition, lysolipids promote the formation of pores and discs, that amplify the changes in the permeability of the bilayer resulting in burst release kinetics [87, 89]. These formulations are also known as lysolipid-containing thermosensitive liposomes or low temperature-sensitive liposomes (LTSL) (Fig. 5). In some other formulations, thermos-sensitivity is offered by incorporating thermosensitive agents, such as peptides (through conformational change) [72] and gas-forming agents [73, 75]. ThermoDox® is a LTSL formulations consisting of DPPC:MSPC:DSPE-PEG$_{2000}$ (86.5:9.7:3.8 molar ratio). It is arguably the most well-studied LTSL formulation, currently undergoing Phase III clinical trials for the treatment of primary liver cancer with standardised radiofrequency ablation [74, 87], and also for the treatment of breast cancer [90].

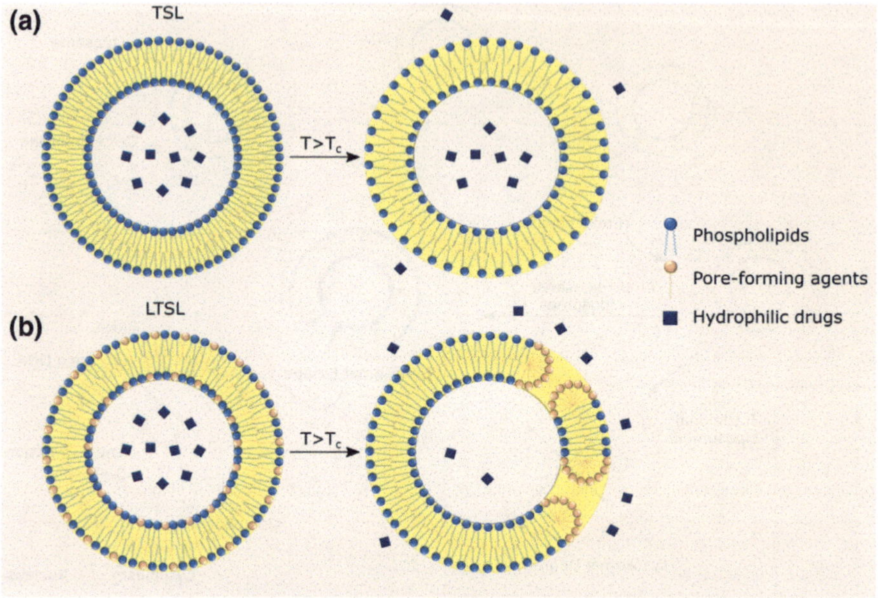

Fig. 5 **a** Traditional TSL releases drugs at temperature above their phase transition temperature, as their permeability increases. **b** LTSL contains pore-forming agents, which promote the formation of pores and discs, amplifying the change in permeability of the bilayer upon phase transition. This results in burst release of the encapsulated drugs

2.2 Liposomes in Gene Delivery

Treating cancers by replacing a lost or defective gene, known as *gene delivery*, has been seen as an alternative approach to traditional chemotherapy [91]. Research efforts have focused on designing effective carrier vectors that can protect nucleic acids, such as plasmid DNA, siRNA, mRNA and miRNA, from serum nucleases, and deliver these negatively charged molecules intracellularly [92].

Viral vectors, such as retrovirus, adenovirus or lentivirus have high gene transfer efficiency; however, they suffer from several drawbacks, such as non-specific expression of viral genes, immunogenicity and potential changes in the host genome structures [93, 94]. Besides the safety concerns, viral vectors have issues with scalability and cost, which has encouraged the development of non-viral vectors, such as liposomes [92].

Cationic liposomes can condense nucleic acids by forming complexes with their negatively charged phosphate groups, where the formed complexes are known as lipoplexes (Fig. 6). There are several advantages for using cationic liposomes as gene delivery systems [20]. Cationic liposomes are biodegradable in vivo, where endogenous enzymes can breakdown the lipid components of the liposomes. In addition, versatile liposome structure which enables lipid composition-dependent modula-

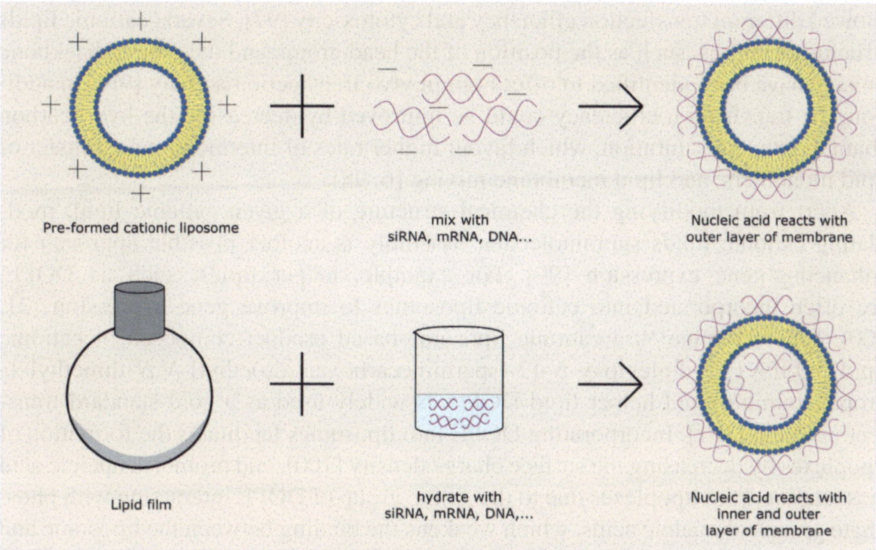

Fig. 6 Two protocols for the complexation of nucleic acids with cationic liposomes

tion of surface charge density for controlling the interaction forces of lipoplexes. Furthermore, surface modifications of liposomes are possible with the inclusion of PEGylated or functional lipids. Finally, co-delivery of therapeutic drugs has been achieved by encapsulating them into a single liposomal formulation [95, 96]. These versatile lipid-based gene vectors have overcome viral vectors shortcomings, where they exhibit reduced immunogenicity and ease of production; however, their poor gene transfer efficiency should be further enhanced [94].

2.2.1 Composition

Lipoplexes gene transfection efficiency is closely related to the physiochemical properties of the cationic liposomes, which is highly dependent on its composition [6, 97]. Cationic lipids, as the major structural component of liposomes, contribute greatly to the behaviour of cationic liposomes. The structure of cationic lipids can be divided into three components: a positively charged head group, linker backbone and hydrophobic anchor [6], which have affected the gene transfer efficiency in vitro and in vivo. For instance, multivalent cationic lipids have greater ability to condense and protect nucleic acid as compared to monovalent lipids, however, such strong electrostatic interactions could hinder nucleic acid release, resulting in poor gene transfection [6]. Moreover, these multivalent cationic lipids prefer an inverted cone molecular shape, which are more likely to form micelles that are less stable and more toxic than monovalent lipids [6]. Cationic lipids of similar head group structures, DOTMA and DOTAP, differed only by their backbone linker (see Fig. 2), and showed different transfection efficiency and cytotoxicity [97]. Several cationic lipids structural features, such as the position of the head group, and the type of backbone linkage have been identified to offer high in vivo transfection activity [98]. In addition, the transfection efficiency could be improved by decreasing the hydrocarbon chain length and saturation, which favour higher rates of intermembrane transfer of lipid monomers, and lipid membrane mixing [6, 98].

Apart from modifying the chemical structure of a given cationic lipid, modulating cationic lipids supramolecular assembly is another possible approach for enhancing gene expression [99]. For example, helper lipids, such as DOPE, are often incorporated into cationic liposomes to improve gene expression [21, 100]. Lipofectamine™, a cationic liposome-based product composed of cationic lipid DOSPA (2,3-dioleyloxy-*N*-[2-(sperminecarboxamido)ethyl]-*N*,*N*-dimethyl-1-propanaminium) and helper lipid DOPE, is widely used as a gold standard transfecting agent [101]. Incorporating DOPE into liposomes facilitates the formation of lipoplexes by decreasing the surface charge density [100], and promotes nucleic acid dissociation from lipoplexes due to the amine group of DOPE interacting with phosphate groups of nucleic acids, which weakens the binding between the liposome and the nucleic acid [102]. Furthermore, DOPE can undergo conformational changes and adopt an inverted hexagonal packing structure at low pH, which allows destabilisation of endosomal vesicles membrane and the subsequent release of lipoplexes to the cytoplasm [21]. PEG-derivatised lipid, cholesterol and its derivatives, e.g. DC-Chol, are other helper lipids that can also be used to modify liposome stability [103, 104], biological activity [6] and pharmacokinetics [104]. Table 4 summarises a wide range of liposomal formulations that have been used in gene delivery.

2.2.2 Mechanisms of Gene Delivery

The interactions of lipoplexes with mammalian cells can be summarised into lipoplexes cell binding and internalisation, followed by cytoplasmic and nuclear delivery of the nucleic acid.

Lipoplexes cell binding is primarily mediated by the electrostatic interactions between positively charged lipoplexes and negatively charged proteoglycans [112]. Following binding to the cell surface, lipoplexes are internalised mainly through endocytosis, via various size-dependent endocytic pathways [6, 113]. While internalisation through membrane fusion has also been recognised, it is not an efficient mean of gene delivery [113]. Following internalisation, the escape of genes from the endosomal compartment into the cytoplasm is essential to avoid degradation, and for their subsequent nuclear entry. As mentioned above, the presence of DOPE in the liposome can promote fusion with the endosomal membrane, allowing the release of the genes into the cytoplasm [6, 21]. While understanding the exact mechanism of nucleic acids trafficking into the nucleus in gene delivery is still limited, nuclear localisation sequence, or signal peptide (NLS) is known to mediate the translocation of particles up to 25 nm into the nucleus through the nuclear pore [114]. NLS has

Table 4 Liposomal formulations used in gene delivery

Lipid composition (% mole ratio)	Nucleic acid type	Gene target	Gene expression or silencing	References
DG:DOPE:Chol (60:20:20)	siRNA	Luciferase, MCL-1	Silencing	[105]
DOPE:Chol:TLO (33:33:33)	siRNA	MCL-1	Silencing	[95]
DOTAP:DOPE:C6 ceramide (50:40:10)	siRNA	STAT3	Silencing	[96]
DOTAP:DSPE:P (NIPAAm-*co*-DMAPAAm)-DOPE (30:65:5)	siRNA	GFP, luciferase	Silencing	[106]
DOPE:OH-Chol:MEL-A	siRNA	Bcl-2, luciferase	Silencing	[107]
Combinations of cationic lipid:helper lipid (50:50)	Plasmid DNA, mRNA, siRNA	GFP, luciferase LaminA/C	Expression Silencing	[99]
DOTMA:Chol:TPGS (49.5:49.5:1)	miRNA	MCL-1	Expression	[108]
PEG-OQLCS:FA-OQLCS:Chol (33:33:33)	Plasmid DNA	GFP	Expression	[61]
DOPE:OH-Chol	Plasmid DNA	Luciferase	Expression	[109]
DODAC:DOPE DODAC:DOPC (50:50)	Plasmid DNA	CAT	Expression	[102]
DOTAP:CDCA DOTAP:capsaicin DOTAP:PD98059 (2:1, mass ratio)	Plasmid DNA	FXR, TNF	Expression	[110]
DOTAP:DOPE:DPPC:Chol	Plasmid DNA	β-gal, GFP, p53	Expression	[111]

Bcl-2 B-cell lymphoma 2; *β-gal* β-galactosidase; *CAT* chloramphenicol acetyltransferase; *CDCA* chenodeoxycholic acid; *Chol* cholesterol; *DG* *N′-N″*-dioleylglutamide; *DODAC* *N-N*-dioleoyl-*N*,*N*-dimethylammonium chloride; *DOPC* 1,2-dioleoyl-sn-glycero-3-phosphocholine; *DOPE* 1,2-dioleoyl-sn-glycero-3-phosphoethanolamine; *DOTAP* 1,2-dioleoyl-3-trimethylammonium-propane; *DOTMA* 1,2-di-*O*-octadecenyl-3-trimethylammonium-propane; *DPPC* 1,2-dipalmitoyl-sn-glycero-3-phosphocholine; *FA* folic acid; *FXR* farnesoid X receptor; *GFP* green fluorescent protein; *MCL-1* myeloid cell leukaemia sequence 1; *MEL-A* mannosylerythritol lipid-A; *OH-Chol* cholesteryl-3β-carboxyamido ethylene-*N*-hydroxyethylamine; *OQLCS* octadecyl-quaternized lysine modified chitosan; *PD98059*, 2′-amino-3′-methoxyflavone; *P(NIPAAm-co-DMAPAAm)* poly(*N*-isopropylacrylamide–co-*N*,*N*′-dimethylaminopropylacrilamide); *STAT-3* signal transducer and activator of transcription 3; *TLO* trilysinoyl oleylamide; *TNF* tumour necrosis factor; *TPGS* D-α-tocopheryl polyethylene glycol 1000 succinate been demonstrated to increase the gene expression by simply anchoring NLS directly to the gene sequence [110, 114] or to the liposomes surface [115].

2.2.3 Limitations and Challenges

The applications of cationic liposome for gene delivery are still limited by drawbacks and limitations. Major technical challenges are the low efficiency of vector transport and gene transfection, poor stability, inducing immune responses, regulatory issues, commercialisation and cost [94].

A major limitation of gene delivery with cationic liposomes can be attributed to their cationic nature, where positive charges undergo non-specific interactions with negatively charged serum proteins and other blood components. This potentially disrupts the integrity of lipoplexes, or forms aggregates that are too large to be taken up by cells [20]. Cationic liposomes are also more likely to induce cell-mediated immune response, compared to neutral and anionic liposomes [116]. The straightforward strategy is to shield the surface charge of the liposomes through PEGylation, to reduce opsonisation and interactions with blood components [94]. However, PEGylation reduces cellular uptake and endosomal escape, which results in low gene expression or silencing activity [104]. Therefore, strategies to mitigate the drawbacks of PEGylation have been developed by incorporating targeting ligands, cleavable PEG linkers and stimuli-sensitive components into cationic liposomal formulations [104].

2.3 Targeted Delivery

Liposomal drug delivery system increases the specificity and lowers undesirable side effects of drugs. This can be achieved by selectively delivering the encapsulated drug to the tumour site with minimum cytotoxic accumulation in normal tissues. Fortunately, nanoparticles in the size range of 20–200 nm are able to accumulate at the tumour sites in higher concentration than in normal tissue for up to 50 times difference, due to the EPR effect [117, 118]. This phenomenon where nanoparticles extravasate the leaky blood vessels surrounding the tumour and accumulate through the EPR effect is known as the *passive targeting* [118].

The EPR effect is diffusional and slow in nature, therefore, it is essential to have liposomes with long blood circulation time to maximise their accumulation in tumour sites [29, 118]. Conventional liposomes are prone to rapid recognition and elimination from the bloodstream by the mononuclear phagocyte system (MPS), and end up in organs of the system, mainly the liver and spleen [119]. Unless the target of the drug delivery system is of the MPS, it should be avoided to maximise the accumulation in tumour sites located in other organs through the EPR effect. A strategy to avoid the MPS is to cover liposomes with a layer of hydrophilic polymer molecules, creating an extended hydration layer which sterically minimises the liposomes from interaction with the blood components and clearance by the MPS [29, 120–122]. This is achieved commonly by incorporating poly(ethylene glycol) (PEG) derivatised lipid, such as DSPE-PEG_{2000} (Fig. 3), into the liposome bilayer and this is known as PEGylation. These types of liposomes are known as long-circulating or stealth liposomes.

Active targeting could also be used to improve the specificity and the cellular uptake of the liposomes (Fig. 7). This could be achieved by attaching targeting ligands to the surface of the liposomes that recognise and bind to specific receptors that are overexpressed on cancer cells [118, 123].

Several studies have demonstrated enhanced uptake and efficacy of ligand-targeted liposomes in diseased tissue in comparison to non-targeted liposomes. These ligands have high binding affinity towards targets overexpressed in either the cancer cells or the tumour endothelium, as summarised in Table 5. A wide range of ligands have been conjugated to liposomes, such as small molecules, peptides, proteins and whole antibodies and their fragments. Viral surface glycoproteins can also be incorporated into the bilayer to form a special type of delivery system known as *virosomes*, combining the advantages of viral and non-viral vectors [93]. The enhanced therapeutic efficacy of actively targeted liposomes depends on multiple factors, such as targeting ligands affinity and surface density on the liposome, receptor density of the target, stability and biodistribution of the liposomes in vivo, mechanism of internalisation and intracellular delivery [118, 124–126].

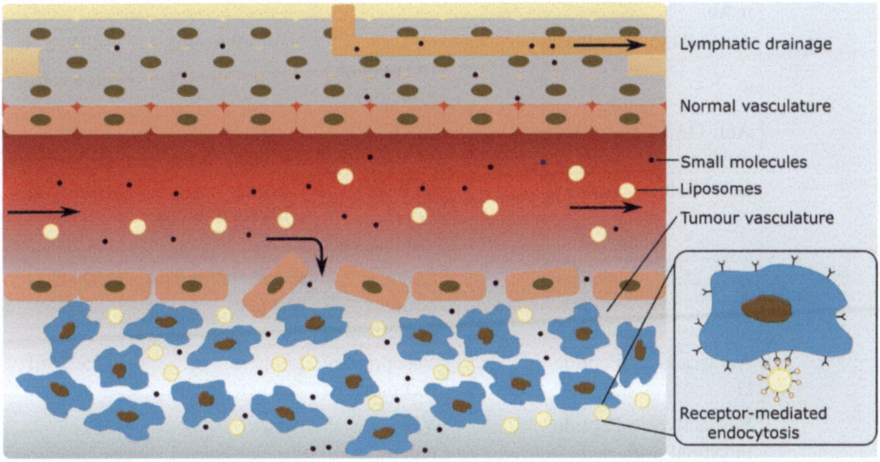

Fig. 7 Passive targeting of liposomes to tumours through the enhanced permeability and retention (EPR) effect. Angiogenic vessels in rapidly growing tumours are abnormally constructed with large vascular fenestrae and impaired lymphatic drainage. As a result, nanoparticles preferentially accumulate in the tumour interstitium

Table 5 Examples of actively targeted liposomes for tumour targeting

	Ligands	Targeting ligand type	Target	Cancer type	References
Cancer cell targeting	Transferrin	Glycoprotein	Transferrin receptor	Ovarian, glioma, colorectal, hepatoma	[127–130]
	Estrone	Steroid hormone	Estrogen receptor	Ovarian	[131]
	Folate	Small molecule	Folate receptor	Various	[132, 133]
	Anti-EGRF mAb	Antibody	EGRF receptor	Various	[134, 135]
	Anti-HER2 mAb	Antibody	HER-2 receptor	Various	[136–141]
	Anti-CD19 mAb, Fab' and scFv	Antibody	CD19	B-cell lymphoma	[142–145]
	Hyaluronan	Macromolecule	CD44	Various melanoma, lymphoma, breast, colorectal and lung)	[146]
	Anti-CD74 mAb	Antibody	CD74	B-cell lymphoma	[147]
	Anti-GD2	Antibody	GD2 (disialogan-glioside)	Neuroblastoma	[148]
	Anti-OA3 mAb	Antibody	OA3	Ovarian	[149]
	GAH Fab'	Antibody	Cell surface GAH binding site	Gastric, colorectal, breast	[150, 151]
	mAb 2C5	Antibody	Surface-bound nucleosome	Breast, colorectal, prostate	[152–154]
	Antagonist G	Peptide	Vasopressin receptor	Lung	[155, 156]
	Vasoactive intestinal peptide (VIP)	Peptide	VIP receptors	Breast	[157]
	sgc8 aptamer	Nucleic acid	Protein tyrosine kinase 7	Lymphoblastic leukaemia	[158]
	AS1411 aptamer	Nucleic acid	Nucleolin	Breast	[159]

(continued)

Table 5 (continued)

	Ligands	Targeting ligand type	Target	Cancer type	References
Tumoural endothelium targeting	Anti-VEGF mAb	Antibody	VEGF	Various (lung, breast, gastric, colorectal, ovarian)	[160]
	RGD peptide	Peptide	Integrins ($\alpha_v\beta_3$)	Melanoma, ovarian cancer	[161, 162]
	Anti-VCAM1	Antibody	VCAM-1	Endothelial cell	[163, 164]
	P_0 peptide	Peptide	ICAM-1	Melanoma	[165, 166]
	NGR peptide	Peptide	CD13	Neuroblastoma	[167]
	GPLPLR	Peptide	MT1-MMP	Endothelial cell	[168]
	Anti-MT1-MMP Fab'	Antibody	MT1-MMP	Fibrosarcoma	[169, 170]

CD cluster of differentiation; *EGRF* epidermal growth factor receptor; *Fab'* antigen-binding fragment; *GD2* disialoganglioside; *GPLPLR* Gly-Pro-Leu-Pro-Leu-Arg; *HER2* human epidermal growth factor receptor 2; *ICAM* intercellular adhesion molecule; *mAb* monoclonal antibody; *MT1-MMP* membrane type 1 matrix metalloproteinase; *NGR* Asn-Gly-Arg; *scFv* single-chain variable fragment; *VCAM* vascular cell adhesion molecule; *VEGF* vascular endothelial growth factor; *VIP* vasoactive intestinal peptide

3 Bioimaging

Besides therapeutics, liposomes have been used as delivery systems for medical diagnostics. Since imaging agents can be encapsulated into the aqueous core or the liposome bilayer through passive or active loading (see Sect. 2.1.2). Alternatively, they can be conjugated directly on the surface of the liposomes to avoid the issue of premature release [171]. Depending on the nature of the imaging agents, they can also be chelated onto the liposomal system by encapsulating the corresponding chelator in the core [172] or attaching the chelator to lipids of the liposome [173]. The approaches to load imaging agents into liposomes are illustrated in Fig. 8.

Various kinds of imaging agents have been encapsulated, conjugated or chelated onto the liposomes for bioimaging modalities, including radionuclides [173–177], nanoparticles and peptides [178]. These formulations have been used for magnetic resonance imaging (MRI), fluorescence imaging, nuclear medicine [single photon emission computed tomography (SPECT) and positron emission tomography/computed tomography (PET/CT)], multispectral optoacoustic tomography (MSOT) and ultrasound imaging (Table 6).

Liposomes functionalised with multimodality imaging agents have been reported by Li et al., which enable non-invasive MRI, near-infrared (NIR) fluorescent and

Table 6 Selected examples of liposomes as imaging agents and theranostics

Imaging Agent	Imaging modality	Therapeutics	Targeting ligands	References
^{64}Cu	PET/CT	–	–	[173]
^{64}Cu	PET/CT	^{177}Lu	–	[174]
^{18}F	PET/CT	siRNA	–	[175]
^{18}F, ^{111}In	PET/CT, SPECT	^{18}F	–	[176]
^{99}Tc, ^{64}Cu, IRDye 800CW, Gd	SPECT, PET/CT, Fluorescence, MRI	Doxorubicin	–	[177]
Gd, ^{111}In	MRI, SPECT	Doxorubicin	–	[179]
Gd	MRI	Doxorubicin	C3d peptide which binds to NCAM	[180]
Gd, Mn	MRI	Doxorubicin	–	[181]
SPION	MRI	Mitoxantrone	Gonadorelin which binds to LHRH receptor	[182]
SPION	MRI	–	–	[79, 183–187]
Gold nanoparticles	NIR Imaging, CT	–	–	[53, 188]
Gold nanorods	MSOT	–	–	[60]
ICG	MSOT	Doxorubicin	hCTM01 which binds to MUC-1 antigen	[189]
QDs (CdSe/ZnS)	Fluorescence	Camptothecin, irinotecan	–	[190]
QDs (CdSe/ZnS)	Fluorescence	Docetaxel	RGD sequence that binds to integrin receptors $\alpha_v\beta_3$ in the brain	[191]
Perfluoropentane	Ultrasound	Doxorubicin	–	[192]
GLuc, HPTS	Bioluminescence, Fluorescence	–	EC1 peptide targeting ErbB2	[178]

GLuc Gaussia luciferase; *HPTS* 8-hydroxypyrene-1,3,6-trisulfonic acid; *ICG* indocyanine green; *IGDD* image-guided drug delivery; *LHRH* luteinising hormone-releasing hormone; *MRI* magnetic resonance imaging; *MSOT* multispectral optoacoustic tomography; *NCAM* neural cell adhesion molecule; *NIR* near-infrared; *siRNA* small interfering ribonucleic acid; *SPECT* single photon emission computed tomography; *PET/CT* positron emission tomography/computed tomography

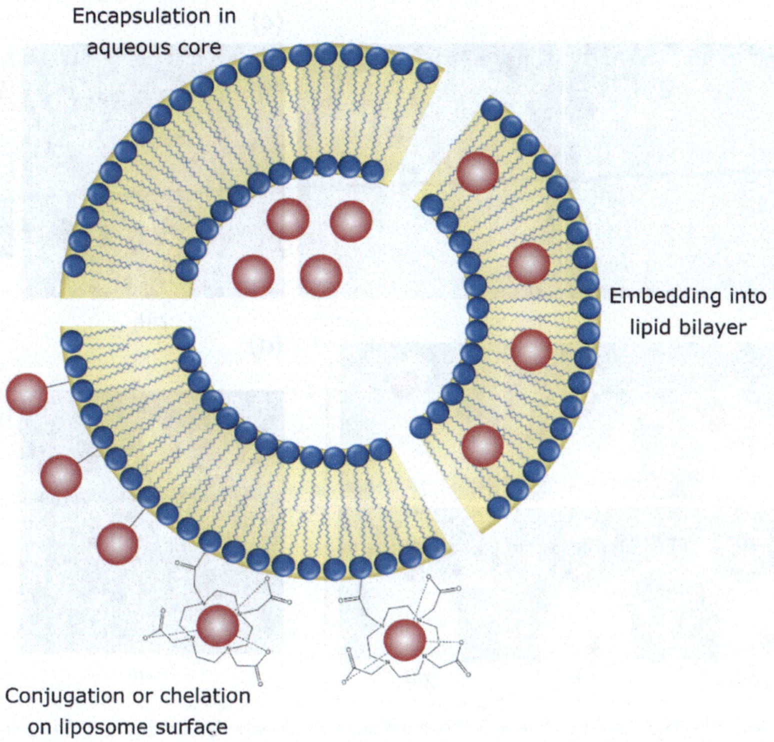

Fig. 8 Schematic diagram of three different approaches to incorporate imaging agents into liposomes. Hydrophilic imaging agents encapsulated in the aqueous core (top); hydrophobic imaging agents embedded into the lipid bilayer (right); and imaging agents conjugated or chelated to the liposome surface (bottom)

nuclear imaging of liposomal drug delivery as shown in Fig. 9, and therapy monitoring and prediction [177]. Liposomal formulations co-loaded with therapeutics, and imaging agents or diagnostics, are known as *theranostics*. For instance, porphysomes, a type of vesicle form from the self-assembly of phospholipid-porphyrin conjugates, where one hydrocarbon chain of the phospholipid is replaced by a porphyrin molecule, are capable of multimodal therapeutic and imaging modalities, as a result of their unique fluorescence self-quenching, photothermal and photoacoustic properties [193, 194]. Porphyrins could also act as metal chelators to chelate metal atom or radionuclides for MRI and nuclear medicine applications [195]. Liposomes containing phospholipid–porphyrin conjugates have demonstrated near-infrared light-induced permeabilisation property which is reversible, and with excellent spatial and temporal control.

As the encapsulated diagnostics adopt the liposome's pharmacokinetic, this can be useful for transporting imaging agents, such as radionuclide, to specific target sites, such as tumours, bone marrow [173, 196] and across the blood–brain barrier [197,

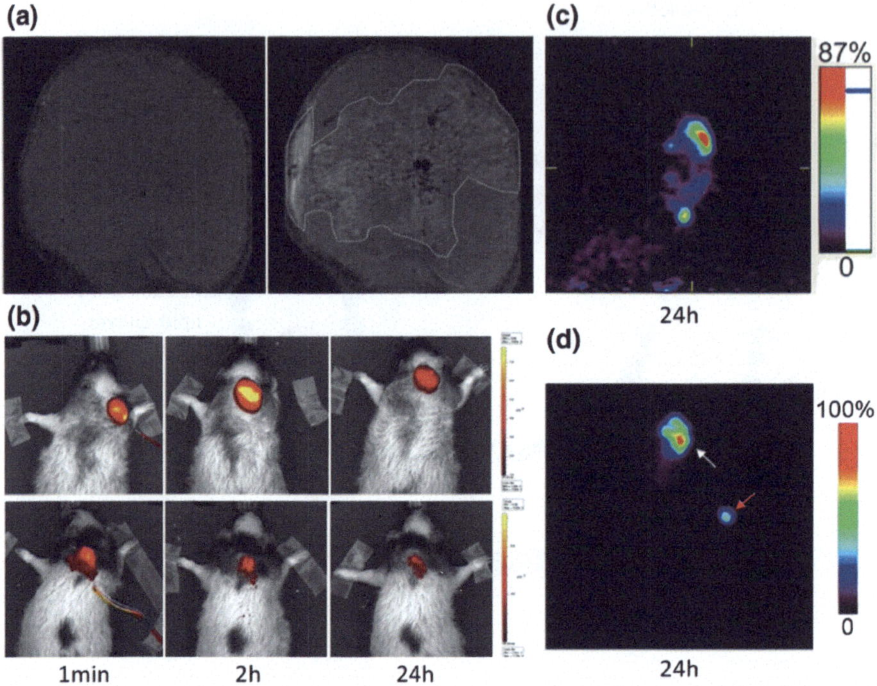

Fig. 9 Multimodal imaging of fluorescently labelled and radiolabelled Gd-liposomes in squamous cell carcinoma of head and neck (SCCHN) tumour xenografts in nude rats. **a** T1-weighted 7T MR images pre-(left) and post-(right) intratumoural injection. Dashed white line represents intratumoural distribution of liposomes, exhibiting higher T1 relaxivity for contrast-enhanced MR imaging. **b** NIR fluorescent images at 1 min, 2 and 24 h post intratumoural injection of IRDye-Gd-liposomes (upper panel) and free IRDye (lower panel). Free IRDye were cleared from the tumour as shown by decreasing fluorescent intensities across time. Fluorescent intensities of the liposome group were highest at 2 h post injection, demonstrating stable intratumoural retention and local diffusion of the liposomes. **c** Micro-PET sagittal images at 24 h post intratumoural infusion of ^{64}Cu-Gd-liposomes. The image showed the local retention and microdistribution of radioactivity in the tumour. **d** Planar gamma camera images at 24 h post intratumoural infusion of ^{99m}Tc-Gd-liposomes (white arrow: tumour; red arrow: reference standard outside the animal body). The image revealed a high level of Gd-liposome inside the tumour (Figure reproduced with permission from Ref. [177], © ACS 2012)

198] as shown in Fig. 10. PEGylated liposomes exhibit prolonged blood circulation and reduced clearance, thus preferentially accumulate at tumour tissues through the EPR effect [199], which results in higher signal-to-noise ratio between the tumour and the surrounding tissues (i.e. enhances sensitivity). Toxicity associated with the imaging agents can also be circumvented with the phospholipid coating [49, 200].

Apart from solely serving as a delivery device of imaging agents, liposomes may play a role in enhancing the imaging quality. For magnetic resonance imaging (MRI), paramagnetic MRI contrast agents such as gadolinium [179, 180, 202], manganese [181] and iron oxide [79, 182–187] have been encapsulated. The incorporation of large amount of Gd-containing lipids in the PEGylated liposomes results in higher longitudinal relaxivity r_1 compared with traditional Gd-DTPA; the degree of saturation of the lipids and the incorporation of liposomes were also found to affect the resultant relaxivity [202]. For iron oxide particles, degradation in endosomes reduced their magnetic resonance contrast, while the liposomal formulations induced intracellular clustering, which enhanced their T_2 contrast [187]. This intracellular clustering effect is important on the MRI detectability and the labelling efficiency of rapidly proliferating cells [187].

By exploiting the high encapsulation capacity of liposomes, fluorescence dye can be protected and self-quenched within the liposomes [203, 204]. These formulations demonstrate weak fluorescence intensity when the fluorescence dyes were encapsulated and exhibit stronger fluorescence intensity upon release and enabled prominent signal-to-noise ratio of tumour tissues. Some near-infrared imaging agents are theranostics, can be encapsulated into liposomes for both imaging and therapeutic. For example, gold nanoparticles CT imaging and photothermal therapy agent [53], and indocyanine green for optical imaging and photodynamic therapy [205]. Gold nanoparticles, in specific, can benefit from liposomal encapsulation as a better CT contrast can be obtained compared to free gold nanoparticles [53]. It was suggested that the presence of liposome enables a scattered accumulation of gold atoms within the lipid system that can provide a larger area to enhance X-ray attenuation [53].

3.1 Image-Guided Drug Delivery

The concept of image-guided drug delivery has been implemented using theranostic nanoparticles to provide real-time visualisation, evaluation, prediction and analysis of the drug delivery process [206]. This provides valuable information to predict the therapeutic outcomes in patients. Liposome is one of the drug delivery systems that have been developed for such application. For example, low temperature-sensitive liposomes (LTSL) containing a combined payload of MR contrast agents and doxorubicin is used for magnetic resonance image-guided drug delivery (IGDD) [179, 181, 207–209]. MRI signals depend on the interaction between the contrast agents with surrounding water to generate signals. This interaction varies substantially when the agents are encapsulated in and released out of liposomes, thereby such changes could be used as indicators of drug release. Mild local hyperthermia is preferentially established with high-intensity focused ultrasound (HIFU) to establish deep and local hyperthermia. MRI can be integrated with the HIFU unit (MR-HIFU) to provide temperature measurement for spatial and temperature feedback to maintain the desired temperature, and also allowing the diagnosis of the tumour for treatment [179, 210]. In MR-HIFU, the accumulation and target site distribution of the LTSL theranostic formulation is first visualised, then HIFU is applied to trigger drug release in tumour tissues, where LTSLs have accumulated with high spatiotemporal specificity. A similar approach was reported with LTSL and Gd-complexes and Mn-complexes [181, 210, 211]. Ultrasound-based IGDD with HIFU has been also

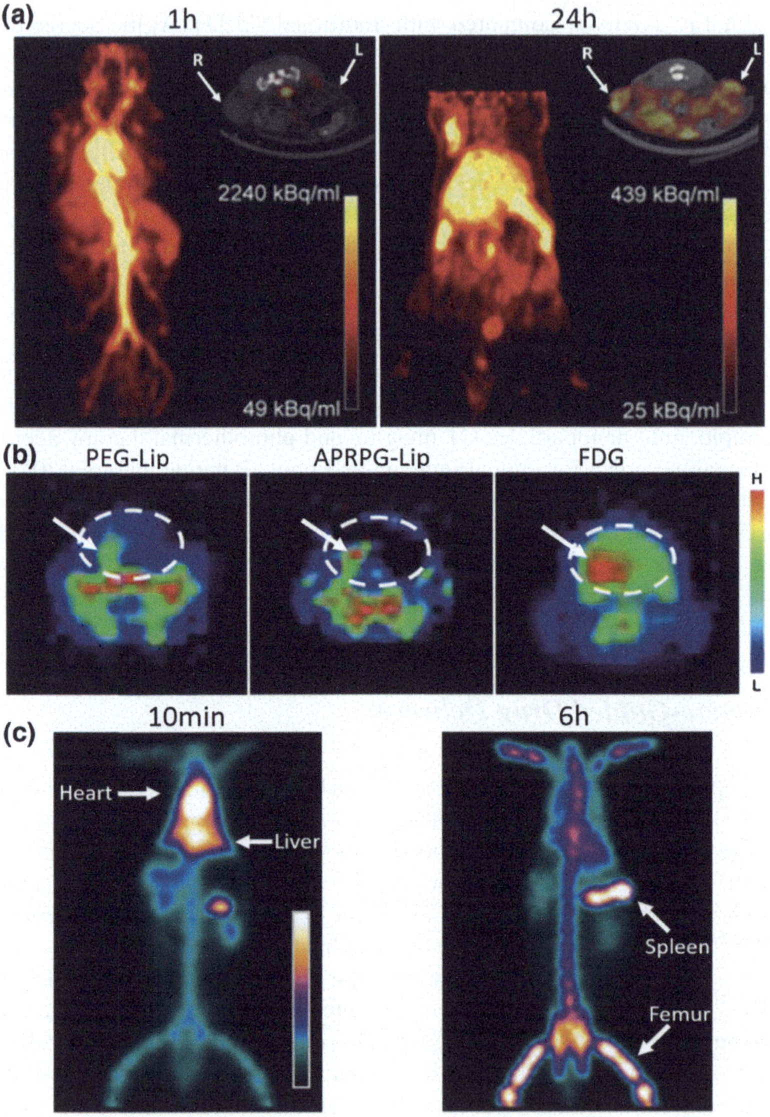

Fig. 10

developed, where perfluoropentane, an echogenic contrast agent, is encapsulated within LTSLs to create echogenic LTSLs [192]. These echogenic LTSLs support tumour imaging with ultrasound and trigger release through HIFU hyperthermia.

◂**Fig. 10** In vivo targeted delivery and bioimaging of radiolabelled liposomes. **a** PET/CT images of tumour accumulation of PEGylated ^{64}Cu-liposomes in human neuroendocrine tumours (NCI-H727) xenograft in nude mice at 1 and 24 h post intravenous administration (left: coronal whole body PET-image; right: Axial PET/CT-fusion image of tumours. The tumours on the left (L) and right (R) flanks are marked with white arrows). At 24 h after administration, the liposomes accumulated in the tumours and were visualised as radioactive hot spots distributed throughout the whole tumour area **b** PET images of brain tumours of glioma-bearing rats acquired at 1 h after intravenous injection with 1-[^{18}F]fluoro-3,6-dioxatetracosane-labelled PEG-modified liposomes (PEG-Lip) and Ala-Pro-Arg-Pro-Gly-PEG-modified liposomes (APRPG-Lip), and [2-^{18}F]-2-deoxyfluoro-d-glucose (FDG). White dotted ovals indicated the estimated brain position and the white arrows indicate the tumour mass. Both PEG-Lip and APRPG-Lip accumulated in the tumour region and imaged the tumour, with low background in other regions of the brain, as they failed to accumulate in normal brain tissue due to blood–brain barrier protection. The higher intratumoural localisation of APRPG-Lip compared to PEG-Lip could be explained by the angiogenic vessel-targeting property of the APRPG peptide. FDG also imaged tumoural accumulation, although the background signal in normal brain tissue was high. **c** Gamma camera image of rabbits acquired at 10 min and 6 h post intravenous infusion (lipids: 15 mg/kg b.w.). The images showed the injected radioactivity initially accumulated in the heart and liver, since both organs having large blood pool contributions. Over 60% of the infused dose deposited in the bone marrow after 6 h. (Figure reproduced with permission from: **a** Ref. [174], © 2016 Springer Berlin Heidelberg; **b** Ref. [201], © 2011 Elsevier; **c** Ref. [196], © 2007 Elsevier)

4 Conclusion

Liposomes have been successful in delivering a variety of therapeutics and imaging agents with hydrophilic or hydrophobic nature. They are biocompatible, biodegradable and exhibit reduced systemic toxicity and immunogenicity. PEGylated liposomes have been engineered to escape the immune system in vivo, which increases their tumour accumulation by utilising EPR effect. Targeting ligands and stimuli-sensitive components can also be incorporated into the same system to improve their tissue specificity of and selective drug release.

Cationic liposomes have been developed as a non-viral vector for gene delivery, since they have the capability to complex to negatively charged nucleic acids. Such interaction protects nucleic acids from degradation and enhances their intracellular uptake. The transfection efficiency of cationic liposomes depends on the liposomes surface charge, size and composition and the structure of the complex. Liposomes have been an alternative to viral gene vectors to co-deliver drugs and genes, however, their low gene transfer in vivo is considered the main hurdle to overcome.

Preparing imaging agent loaded liposomes is highly advantageous, since liposomes could protect diagnostics from aggregation in physiological conditions, prolong their blood residency and efficiently accumulate at tumour tissues. Liposomal encapsulation could also enhance the imaging quality and tissue specificity. Furthermore, liposomes can offer multiple imaging modalities, and theranostic applications, by loading different imaging agents and/or drugs in a single delivery system.

Finally, combining stimuli-responsive, imaging and therapeutic functionalities in liposomes, image-guided drug delivery can be achieved, providing real-time visualisation and feedback of the drug delivery process in vivo, which could help predicting the therapeutic outcome in patients.

References

1. Bozzuto G, Molinari A (2015) Liposomes as nanomedical devices. Int J Nanomed 10:975–999. https://doi.org/10.2147/IJN.S68861
2. Allen TM, Cullis PR (2013) Liposomal drug delivery systems: from concept to clinical applications. Adv Drug Deliv Rev 65:36–48. https://doi.org/10.1016/j.addr.2012.09.037
3. Wicki A, Witzigmann D, Balasubramanian V, Huwyler J (2015) Nanomedicine in cancer therapy: challenges, opportunities, and clinical applications. J Controlled Release 200:138–157. https://doi.org/10.1016/j.jconrel.2014.12.030
4. Sercombe L, Veerati T, Moheimani F et al (2015) Advances and challenges of liposome assisted drug delivery. Front Pharmacol 6:1–13. https://doi.org/10.3389/fphar.2015.00286
5. Chang H, Yeh M-K (2012) Clinical development of liposome-based drug: formulation, characterization, and therapeutic efficacy. Int J Nanomed 7:49–60. https://doi.org/10.2147/IJN.S26766
6. Simões S, Filipe A, Faneca H et al (2005) Cationic liposomes for gene delivery. Expert Opin Drug Deliv 2:237–254. https://doi.org/10.1517/17425247.2.2.237
7. Xing H, Hwang K, Lu Y (2016) Recent developments of liposomes as nanocarriers for theranostic applications. Theranostics 6:1336–1352. https://doi.org/10.7150/thno.15464
8. Bangham AD, Standish MM, Watkins JC (1965) Diffusion of univalent ions across the lamellae of swollen phospholipids. J Mol Biol 13:238–252. https://doi.org/10.1016/S0022-2836(65)80093-6
9. Peetla C, Stine A, Labhasetwar V (2009) Biophysical interactions with model lipid membranes: applications in drug discovery and drug delivery. Mol Pharm 6:1264–1276. https://doi.org/10.1021/mp9000662
10. Matos C, Moutinho C, Lobão P (2012) Liposomes as a model for the biological membrane: studies on daunorubicin bilayer interaction. J Membr Biol 245:69–75. https://doi.org/10.1007/s00232-011-9414-2
11. Al-Jamal WT, Kostarelos K (2011) Liposomes: from a clinically established drug delivery system to a nanoparticle platform for theranostic nanomedicine. Acc Chem Res 44:1094–1104. https://doi.org/10.1021/ar200105p
12. Ulrich AS (2002) Biophysical aspects of using liposomes as delivery vehicles. Biosci Rep 22:129–150. https://doi.org/10.1023/A:1020178304031
13. Li J, Wang X, Zhang T et al (2014) A review on phospholipids and their main applications in drug delivery systems. Asian J Pharm Sci 10:81–98. https://doi.org/10.1016/j.ajps.2014.09.004
14. Senior J, Gregoriadis G (1982) Is half-life of circulating liposomes determined by changes in their permeability? FEBS Lett 145:109–114. https://doi.org/10.1016/0014-5793(82)81216-7
15. Xu H, Ye F, Hu M et al (2014) Influence of phospholipid types and animal models on the accelerated blood clearance phenomenon of PEGylated liposomes upon repeated injection. Drug Deliv 7544:1–10. https://doi.org/10.3109/10717544.2014.885998
16. Ali MH, Moghaddam B, Kirby DJ et al (2013) The role of lipid geometry in designing liposomes for the solubilisation of poorly water soluble drugs. Int J Pharm 453:225–232. https://doi.org/10.1016/j.ijpharm.2012.06.056
17. Mohammed AR, Weston N, Coombes AGA et al (2004) Liposome formulation of poorly water soluble drugs: optimisation of drug loading and ESEM analysis of stability. Int J Pharm 285:23–34. https://doi.org/10.1016/j.ijpharm.2004.07.010

18. Samuni AM, Lipman A, Barenholz Y (2000) Damage to liposomal lipids: protection by antioxidants and cholesterol- mediated dehydration. Chem Phys Lipids 105:121–134. https://doi.org/10.1016/S0009-3084(99)00136-X
19. Barenholz Y (2001) Liposome application: problems and prospects. Curr Opin Colloid Interface Sci 6:66–77. https://doi.org/10.1016/S1359-0294(00)00090-X
20. Shim G, Kim M-G, Park JY, Oh Y-K (2013) Application of cationic liposomes for delivery of nucleic acids. Asian J Pharm Sci 8:72–80. https://doi.org/10.1016/j.ajps.2013.07.009
21. Balazs DA, Godbey W (2011) Liposomes for use in gene delivery. J Drug Deliv 2011:326497. https://doi.org/10.1155/2011/326497
22. Klibanov AL, Maruyama K, Torchilin VP, Huang L (1990) Amphipathic polyethyleneglycols effectively prolong the circulation time of liposomes. FEBS Lett 268:235–237. https://doi.org/10.1016/0014-5793(90)81016-H
23. Takeuchi H, Kojima H, Yamamoto H, Kawashima Y (2001) Evaluation of circulation profiles of liposomes coated with hydrophilic polymers having different molecular weights in rats. J Controlled Release 75:83–91. https://doi.org/10.1016/S0168-3659(01)00368-6
24. Torchilin VP, Levchenko TS, Whiteman KR et al (2001) Amphiphilic poly-N-vinylpyrrolidones: synthesis, properties and liposome surface modification. Biomaterials 22:3035–3044. https://doi.org/10.1016/S0142-9612(01)00050-3
25. Hamill RJ (2013) Amphotericin B formulations: a comparative review of efficacy and toxicity. Drugs 73:919–934. https://doi.org/10.1007/s40265-013-0069-4
26. O'Brien MER (2004) Reduced cardiotoxicity and comparable efficacy in a phase III trial of pegylated liposomal doxorubicin HCl (CAELYX™/Doxil®) versus conventional doxorubicin for first-line treatment of metastatic breast cancer. Ann Oncol 15:440–449. https://doi.org/10.1093/annonc/mdh097
27. Rafiyath SM, Rasul M, Lee B et al (2012) Comparison of safety and toxicity of liposomal doxorubicin vs. conventional anthracyclines: a meta-analysis. Exp Hematol Oncol 1:10. https://doi.org/10.1186/2162-3619-1-10
28. Gubernator J (2011) Active methods of drug loading into liposomes: recent strategies for stable drug entrapment and increased in vivo activity. Expert Opin Drug Deliv 8:565–580. https://doi.org/10.1517/17425247.2011.566552
29. Gabizon A, Goren D, Horowitz AT et al (1997) Long-circulating liposomes for drug delivery in cancer therapy: a review of biodistribution studies in tumor-bearing animals. Adv Drug Deliv Rev 24:337–344. https://doi.org/10.1016/S0169-409X(96)00476-0
30. Zhigaltsev IV, Maurer N, Akhong QF et al (2005) Liposome-encapsulated vincristine, vinblastine and vinorelbine: a comparative study of drug loading and retention. J Controlled Release 104:103–111. https://doi.org/10.1016/j.jconrel.2005.01.010
31. Torchilin VP (2005) Recent advances with liposomes as pharmaceutical carriers. Nat Rev Drug Discov 4:145–160. https://doi.org/10.1038/nrd1632
32. Kulshrestha P, Gogoi M, Bahadur D, Banerjee R (2012) In vitro application of paclitaxel loaded magnetoliposomes for combined chemotherapy and hyperthermia. Colloids Surf, B 96:1–7. https://doi.org/10.1016/j.colsurfb.2012.02.029
33. Leung SJ, Romanowski M (2012) Light-activated content release from liposomes. Theranostics 2:1020–1036. https://doi.org/10.7150/thno.4847
34. Simões S, Nuno Moreira J, Fonseca C et al (2004) On the formulation of pH-sensitive liposomes with long circulation times. Adv Drug Deliv Rev 56:947–965. https://doi.org/10.1016/j.addr.2003.10.038
35. Candiani G, Pezzoli D, Ciani L et al (2010) Bioreducible liposomes for gene delivery: from the formulation to the mechanism of action. PLoS ONE 5:1–8. https://doi.org/10.1371/journal.pone.0013430
36. Linderoth L, Peters GH, Madsen R, Andresen TL (2009) Drug delivery by an enzyme-mediated cyclization of a lipid prodrug with unique bilayer-formation properties. Angew Chemie Int Ed 48:1823–1826. https://doi.org/10.1002/anie.200805241
37. Passero FC, Grapsa D, Syrigos KN, Saif MW (2016) The safety and efficacy of Onivyde (irinotecan liposome injection) for the treatment of metastatic pancreatic cancer following

gemcitabine-based therapy. Expert Rev Anticancer Ther 16:697–703. https://doi.org/10.1080/14737140.2016.1192471
38. Mantripragada S (2002) A lipid based depot (DepoFoam® technology) for sustained release drug delivery. Prog Lipid Res 41:392–406. https://doi.org/10.1016/S0163-7827(02)00004-8
39. Silverman JA, Deitcher SR (2013) Marqibo® (vincristine sulfate liposome injection) improves the pharmacokinetics and pharmacodynamics of vincristine. Cancer Chemother Pharmacol 71:555–564. https://doi.org/10.1007/s00280-012-2042-4
40. Akbarzadeh A, Rezaei-Sadabady R, Davaran S et al (2013) Liposome: classification, preparation, and applications. Nanoscale Res Lett 8:102. https://doi.org/10.1186/1556-276X-8-102
41. New RR (1990) Liposomes: a practical approach. IRL Press at Oxford University Press
42. Antonenko YN, Pohl P, Denisov GA (1997) Permeation of ammonia across bilayer lipid membranes studied by ammonium ion selective microelectrodes. Biophys J 72:2187–2195. https://doi.org/10.1016/S0006-3495(97)78862-3
43. Barenholz Y (Chezy) (2012) Doxil®—The first FDA-approved nano-drug: lessons learned. J Controlled Release 160:117–134. https://doi.org/10.1016/j.jconrel.2012.03.020
44. Fritze A, Hens F, Kimpfler A et al (2006) Remote loading of doxorubicin into liposomes driven by a transmembrane phosphate gradient. Biochim Biophys Acta Biomembr 1758:1633–1640. https://doi.org/10.1016/j.bbamem.2006.05.028
45. Abraham SA, Edwards K, Karlsson G et al (2002) Formation of transition metal-doxorubicin complexes inside liposomes. Biochim Biophys Acta Biomembr 1565:41–54. https://doi.org/10.1016/S0005-2736(02)00507-2
46. Cheung BCL, Sun THT, Leenhouts JM, Cullis PR (1998) Loading of doxorubicin into liposomes by forming Mn^{2+}-drug complexes. Biochim Biophys Acta Biomembr 1414:205–216. https://doi.org/10.1016/S0005-2736(98)00168-0
47. Al-Jamal WT, Kostarelos K (2007) Liposome-nanoparticle hybrids for multimodal diagnostic and therapeutic applications. Nanomedicine (Lond) 2:85–98. https://doi.org/10.2217/17435889.2.1.85
48. Monnier CA, Burnand D, Rothen-Rutishauser B et al (2014) Magnetoliposomes: opportunities and challenges. Eur J Nanomed 6:201–215. https://doi.org/10.1515/ejnm-2014-0042
49. Fattahi H, Laurent S, Liu F et al (2011) Magnetoliposomes as multimodal contrast agents for molecular imaging and cancer nanotheragnostics. Nanomedicine (Lond) 6:529–544. https://doi.org/10.2217/nnm.11.14
50. Soenen SJ, Vande Velde G, Ketkar-Atre A et al (2011) Magnetoliposomes as magnetic resonance imaging contrast agents. Wiley Interdiscip Rev Nanomed Nanobiotechnol 3:197–211. https://doi.org/10.1002/wnan.122
51. Marie H, Plassat V, Lesieur S et al (2013) Magnetic-fluid-loaded liposomes for MR imaging and therapy of cancer. J Drug Deliv Sci Technol 23:25–37. https://doi.org/10.1016/S1773-2247(13)50004-9
52. Kasili PM, Dinh TV (2005) Liposome encapsulated gold nanoshells for nanophototherapy induced hyperthermia. Int J Nanotechnol 2:397. https://doi.org/10.1504/IJNT.2005.008076
53. Rengan AK, Jagtap M, De A et al (2014) Multifunctional gold coated thermo-sensitive liposomes for multimodal imaging and photo-thermal therapy of breast cancer cells. Nanoscale 6:916–923. https://doi.org/10.1039/c3nr04448c
54. Rengan AK, Bukhari AB, Pradhan A et al (2015) In vivo analysis of biodegradable liposome gold nanoparticles as efficient agents for photothermal therapy of cancer. Nano Lett 15:842–848. https://doi.org/10.1021/nl5045378
55. Liu Y, Zhang X, Liu Z, Wang L (2017) Gold nanoshell-based betulinic acid liposomes for synergistic chemo-photothermal therapy. Nanomed Nanotechnol Biol Med 13:1891–1900. https://doi.org/10.1016/j.nano.2017.03.012
56. Paasonen L, Laaksonen T, Johans C et al (2007) Gold nanoparticles enable selective light-induced contents release from liposomes. J Controlled Release 122:86–93. https://doi.org/10.1016/j.jconrel.2007.06.009
57. Volodkin DV, Skirtach AG, Möhwald H (2009) Near-IR remote release from assemblies of liposomes and nanoparticles. Angew Chemie Int Ed 48:1807–1809. https://doi.org/10.1002/anie.200805572

58. Wu G, Mikhailovsky A, Khant HA et al (2008) Remotely triggered liposomal release by near-infrared light absorption via hollow gold nanoshells. J Am Chem Soc 130:8175–8177. https://doi.org/10.1021/ja802656d
59. Pornpattananangkul D, Olson S, Aryal S et al (2011) Stimuli-responsive liposome fusion mediated by gold nanoparticles. ACS Nano 4:1935–1942. https://doi.org/10.1021/nn9018587
60. Lozano N, Al-Jamal WT, Taruttis A et al (2012) Liposome-gold nanorod hybrids for high-resolution visualization deep in tissues. J Am Chem Soc 134:13256–13258. https://doi.org/10.1021/ja304499q
61. Wang H, Zhao P, Su W et al (2010) PLGA/polymeric liposome for targeted drug and gene co-delivery. Biomaterials 31:8741–8748. https://doi.org/10.1016/j.biomaterials.2010.07.082
62. Zhang L, Granick S (2006) How to stabilize phospholipid liposomes (using nanoparticles). Nano Lett 6:694–698. https://doi.org/10.1021/nl052455y
63. Qu W, Zuo W, Li N et al (2017) Design of multifunctional liposome-quantum dot hybrid nanocarriers and their biomedical application. J Drug Target 2330:1–12. https://doi.org/10.1080/1061186X.2017.1323334
64. Al-Jamal WT, Al-Jamal KT, Tian B et al (2008) Lipid-quantum dot bilayer vesicles enhance tumor cell uptake and retention in vitro and in vivo. ACS Nano 2:408–418. https://doi.org/10.1021/nn700176a
65. Wang Y, Zeng S, Lin TM et al (2014) Evaluating the anticancer properties of liposomal copper in a nude xenograft mouse model of human prostate cancer: formulation, in vitro, in vivo, histology and tissue distribution studies. Pharm Res 31:3106–3119. https://doi.org/10.1007/s11095-014-1403-6
66. Lee J-H, Shin Y, Lee W et al (2016) General and programmable synthesis of hybrid liposome/metal nanoparticles. Sci Adv 2:e1601838. https://doi.org/10.1126/sciadv.1601838
67. Anderson M, Omri A (2004) The effect of different lipid components on the in vitro stability and release kinetics of liposome formulations. Drug Deliv 11:33–39. https://doi.org/10.1080/10717540490265243
68. Lee Y, Thompson DH (2017) Stimuli-responsive liposomes for drug delivery. Wiley Interdiscip Rev Nanomed Nanobiotechnol. https://doi.org/10.1002/wnan.1450
69. Mura S, Nicolas J, Couvreur P (2013) Stimuli-responsive nanocarriers for drug delivery. Nat Mater 12:991–1003. https://doi.org/10.1038/NMAT3776
70. Kirpotin D, Hong K, Mullah N et al (1996) Liposomes with detachable polymer coating: destabilization and fusion of dioleoylphosphatidylethanolamine vesicles triggered by cleavage of surface-grafted poly(ethylene glycol). FEBS Lett 388:115–118. https://doi.org/10.1016/0014-5793(96)00521-2
71. Murata K, Egami H, Kiyohara H et al (1993) Expression of group-II phospholipase A_2 in malignant and non-malignant human gastric mucosa. Br J Cancer 68:103–111. https://doi.org/10.1038/bjc.1993.294
72. Al-Ahmady ZS, Al-Jamal WT, Bossche JV et al (2012) Lipid-peptide vesicle nanoscale hybrids for triggered drug release by mild hyperthermia in vitro and in vivo. ACS Nano 6:9335–9346. https://doi.org/10.1021/nn302148p
73. Chen KJ, Liang HF, Chen HL et al (2013) A thermoresponsive bubble-generating liposomal system for triggering localized extracellular drug delivery. ACS Nano 7:438–446. https://doi.org/10.1021/nn304474j
74. Poon RT, Borys N (2011) Lyso-thermosensitive liposomal doxorubicin: an adjuvant to increase the cure rate of radiofrequency ablation in liver cancer. Future Oncol 7:937–945. https://doi.org/10.2217/fon.11.73
75. Lee JM, Kwag DS, Youn YS, Lee ES (2017) Gas-forming liposomes prepared using a liposomal magnetoporation method. Colloids Surf, B 155:209–214. https://doi.org/10.1016/j.colsurfb.2017.04.017
76. Bisby RH, Mead C, Morgan CG (2000) Wavelength-programmed solute release from photosensitive liposomes. Biochem Biophys Res Commun 276:169–173. https://doi.org/10.1006/bbrc.2000.3456

77. Schroeder A, Kost J, Barenholz Y (2009) Ultrasound, liposomes, and drug delivery: principles for using ultrasound to control the release of drugs from liposomes. Chem Phys Lipids 162:1–16. https://doi.org/10.1016/j.chemphyslip.2009.08.003
78. Meng L, Deng Z, Niu L et al (2015) A disposable microfluidic device for controlled drug release from thermal-sensitive liposomes by high intensity focused ultrasound. Theranostics 5:1203–1213. https://doi.org/10.7150/thno.12295
79. Amstad E, Kohlbrecher J, Müller E et al (2011) Triggered release from liposomes through magnetic actuation of iron oxide nanoparticle containing membranes. Nano Lett 11:1664–1670. https://doi.org/10.1021/nl2001499
80. Karanth H, Murthy RSR (2007) pH-sensitive liposomes-principle and application in cancer therapy. J Pharm Pharmacol 59:469–483. https://doi.org/10.1211/jpp.59.4.0001
81. Straubinger RM, Düzgünes N, Papahadjopoulos D (1985) pH-sensitive liposomes mediate cytoplasmic delivery of encapsulated macromolecules. FEBS Lett 179:148–154. https://doi.org/10.1016/0014-5793(85)80210-6
82. Connor J, Yatvin MB, Huang L (1984) pH-sensitive liposomes: acid-induced liposome fusion. Proc Natl Acad Sci U S A 81:1715–1718
83. Deshpande PP, Biswas S, Torchilin VP (2013) Current trends in the use of liposomes for tumor targeting. Nanomedicine (Lond) 8:1509–1528. https://doi.org/10.2217/nnm.13.118
84. Torchilin V (2009) Multifunctional and stimuli-sensitive pharmaceutical nanocarriers. Eur J Pharm Biopharm 71:431–444. https://doi.org/10.1016/j.ejpb.2008.09.026
85. Chatterjee DK, Diagaradjane P, Krishnan S (2011) Nanoparticle-mediated hyperthermia in cancer therapy. Ther Deliv 2:1001–1014. https://doi.org/10.4155/tde.11.72
86. Ganta S, Devalapally H, Shahiwala A, Amiji M (2008) A review of stimuli-responsive nanocarriers for drug and gene delivery. J Controlled Release 126:187–204. https://doi.org/10.1016/j.jconrel.2007.12.017
87. Needham D, Park J, Wright AM, Tong J (2013) Materials characterization of the low temperature sensitive liposome (LTSL): effects of the lipid composition (lysolipid and DSPE-PEG2000) on the thermal transition and release of doxorubicin. Faraday Discuss 161:515–534. https://doi.org/10.1039/c2fd20111a
88. Tagami T, Ernsting MJ, Li SD (2011) Optimization of a novel and improved thermosensitive liposome formulated with DPPC and a Brij surfactant using a robust in vitro system. J Controlled Release 154:290–297. https://doi.org/10.1016/j.jconrel.2011.05.020
89. Ickenstein LM, Arfvidsson MC, Needham D et al (2003) Disc formation in cholesterol-free liposomes during phase transition. Biochim Biophys Acta Biomembr 1614:135–138. https://doi.org/10.1016/S0005-2736(03)00196-2
90. Zagar TM, Vujaskovic Z, Formenti S et al (2014) Two phase I dose-escalation/pharmacokinetics studies of low temperature liposomal doxorubicin (LTLD) and mild local hyperthermia in heavily pretreated patients with local regionally recurrent breast cancer. Int J Hyperth 30:285–294. https://doi.org/10.3109/02656736.2014.936049
91. Yang ZR, Wang HF, Zhao J et al (2007) Recent developments in the use of adenoviruses and immunotoxins in cancer gene therapy. Cancer Gene Ther 14:599–615. https://doi.org/10.1038/sj.cgt.7701054
92. Mintzer MA, Simanek EE (2009) Nonviral vectors for gene delivery. Chem Rev 109:259–302. https://doi.org/10.1021/cr800409e
93. Liu H, Tu Z, Feng F et al (2015) Virosome, a hybrid vehicle for efficient and safe drug delivery and its emerging application in cancer treatment. Acta Pharm 65:105–116. https://doi.org/10.1515/acph-2015-0019
94. Ramamoorth M (2015) Non viral vectors in gene therapy—an overview. J Clin Diagn Res 9:1–6. https://doi.org/10.7860/JCDR/2015/10443.5394
95. Shim G, Han SE, Yu YH et al (2011) Trilysinoyl oleylamide-based cationic liposomes for systemic co-delivery of siRNA and an anticancer drug. J Controlled Release 155:60–66. https://doi.org/10.1016/j.jconrel.2010.10.017
96. Jose A, Labala S, Venuganti VVK (2016) Co-delivery of curcumin and STAT3 siRNA using deformable cationic liposomes to treat skin cancer. J Drug Target. https://doi.org/10.1080/1061186X.2016.1258567

97. Song YK, Liu F, Chu S, Liu D (1997) Characterization of cationic liposome-mediated gene transfer in vivo by intravenous administration. Hum Gene Ther 8:1585–1594. https://doi.org/10.1089/hum.1997.8.13-1585
98. Ren T, Song YK, Zhang G, Liu D (2000) Structural basis of DOTMA for its high intravenous transfection activity in mouse. Gene Ther 7:764–768. https://doi.org/10.1038/sj.gt.3301153
99. Colombani T, Peuziat P, Dallet L et al (2017) Self-assembling complexes between binary mixtures of lipids with different linkers and nucleic acids promote universal mRNA, DNA and siRNA delivery. J Controlled Release 249:131–142. https://doi.org/10.1016/j.jconrel.2017.01.041
100. Zuidam NJ, Barenholz Y (1998) Electrostatic and structural properties of complexes involving plasmid DNA and cationic lipids commonly used for gene delivery. Biochim Biophys Acta Biomembr 1368:115–128. https://doi.org/10.1016/S0005-2736(97)00187-9
101. Dalby B, Cates S, Harris A et al (2004) Advanced transfection with Lipofectamine 2000 reagent: primary neurons, siRNA, and high-throughput applications. Methods 33:95–103. https://doi.org/10.1016/j.ymeth.2003.11.023
102. Harvie P, Wong FM, Bally MB (1998) Characterization of lipid DNA interactions. I. Destabilization of bound lipids and DNA dissociation. Biophys J 75:1040–1051. https://doi.org/10.1016/S0006-3495(98)77593-9
103. Nakamura T, Noma Y, Sakurai Y, Harashima H (2017) Modifying cationic liposomes with cholesteryl-PEG prevents their aggregation in human urine and enhances cellular uptake by bladder cancer cells. Biol Pharm Bull 40:234–237. https://doi.org/10.1248/bpb.b16-00770
104. Hatakeyama H, Akita H, Harashima H (2013) The polyethyleneglycol dilemma: advantage and disadvantage of PEGylation of liposomes for systemic genes and nucleic acids delivery to tumors. Biol Pharm Bull 36:892–899. https://doi.org/10.1248/bpb.b13-00059
105. Kang SH, Cho HJ, Shim G et al (2011) Cationic liposomal co-delivery of small interfering RNA and a MEK inhibitor for enhanced anticancer efficacy. Pharm Res 28:3069–3078. https://doi.org/10.1007/s11095-011-0569-4
106. Wang J, Ayano E, Maitani Y, Kanazawa H (2017) Enhanced cellular uptake and gene silencing activity of siRNA using temperature-responsive polymer-modified liposome. Int J Pharm 523:217–228. https://doi.org/10.1016/j.ijpharm.2017.03.035
107. Inoh Y, Furuno T, Hirashima N et al (2011) Rapid delivery of small interfering RNA by biosurfactant MEL-A-containing liposomes. Biochem Biophys Res Commun 414:635–640. https://doi.org/10.1016/j.bbrc.2011.09.147
108. Wu Y, Crawford M, Yu B et al (2011) MicroRNA delivery by cationic lipoplexes for lung cancer therapy. Mol Pharm 8:1381–1389. https://doi.org/10.1021/mp2002076
109. Inoh Y, Nagai M, Matsushita K et al (2017) Gene transfection efficiency into dendritic cells is influenced by the size of cationic liposomes/DNA complexes. Eur J Pharm Sci 102:230–236. https://doi.org/10.1016/j.ejps.2017.03.023
110. Liu F, Conwell CC, Yuan X et al (2007) Novel nonviral vectors target cellular signaling pathways: regulated gene expression and reduced toxicity. J Pharmacol Exp Ther 321:777. https://doi.org/10.1124/jpet.106.118117.lipid
111. Jinturkar KA, Anish C, Kumar MK et al (2012) Liposomal formulations of Etoposide and Docetaxel for p53 mediated enhanced cytotoxicity in lung cancer cell lines. Biomaterials 33:2492–2507. https://doi.org/10.1016/j.biomaterials.2011.11.067
112. Mislick KA, Baldeschwieler JD (1996) Evidence for the role of proteoglycans in cation-mediated gene transfer. Proc Natl Acad Sci U S A 93:12349–12354. https://doi.org/10.1073/pnas.93.22.12349
113. Chesnoy S, Huang L (2000) Structure and function of lipid-DNA complexes for gene delivery. Annu Rev Biophys Biomol Struct 29:27–47. https://doi.org/10.1146/annurev.biophys.29.1.27
114. Zanta MA, Belguise-Valladier P, Behr JP (1999) Gene delivery: a single nuclear localization signal peptide is sufficient to carry DNA to the cell nucleus. Proc Natl Acad Sci U S A 96:91–96. https://doi.org/10.1073/pnas.96.1.91
115. Aronsohn AI, Hughes JA (1998) Nuclear localization signal peptides enhance cationic liposome-mediated gene therapy. J Drug Target 5:163–169. https://doi.org/10.3109/10611869808995871

116. Nakanishi T, Kunisawa J, Hayashi A et al (1999) Positively charged liposome functions as an efficient immunoadjuvant in inducing cell-mediated immune response to soluble proteins. J Controlled Release 61:233–240. https://doi.org/10.1016/S0168-3659(99)00097-8
117. Iyer AK, Khaled G, Fang J, Maeda H (2006) Exploiting the enhanced permeability and retention effect for tumor targeting. Drug Discov Today 11:812–818. https://doi.org/10.1016/j.drudis.2006.07.005
118. Danhier F, Feron O, Préat V (2010) To exploit the tumor microenvironment: Passive and active tumor targeting of nanocarriers for anti-cancer drug delivery. J Controlled Release 148:135–146. https://doi.org/10.1016/j.jconrel.2010.08.027
119. Gabizon A, Chisin R, Amselem S et al (1991) Pharmacokinetic and imaging studies in patients receiving a formulation of liposome-associated adriamycin. Br J Cancer 64:1125–1132. https://doi.org/10.1038/bjc.1991.476
120. Immordino ML, Dosio F, Cattel L (2006) Stealth liposomes: Review of the basic science, rationale, and clinical applications, existing and potential. Int J Nanomed 1:297–315. https://doi.org/10.1023/A:1020134521778
121. Ishida T, Harashima H, Kiwada H (2001) Interactions of liposomes with cells in vitro and in vivo: opsonins and receptors. Curr Drug Metab 2:397–409. https://doi.org/10.2174/1389200013338306
122. Tirosh O, Kohen R, Katzhendler J et al (1997) Novel synthetic phospholipid protects lipid bilayers against oxidation damage: role of hydration layer and bound water. J Chem Soc Perkin Trans 2:383–390. https://doi.org/10.1039/A601955B
123. Torchilin VP (2010) Passive and active drug targeting: drug delivery to tumors as an example. Handb Exp Pharmacol 197:3–53. https://doi.org/10.1007/978-3-642-00477-3_1
124. Yingchoncharoen P, Kalinowski DS, Richardson DR (2016) Lipid-based drug delivery systems in cancer therapy: what is available and what is yet to come. Pharmacol Rev 68:701–787. https://doi.org/10.1124/pr.115.012070
125. Noble GT, Stefanick JF, Ashley JD et al (2014) Ligand-targeted liposome design: challenges and fundamental considerations. Trends Biotechnol 32:32–45. https://doi.org/10.1016/j.tibtech.2013.09.007
126. Sawant RR, Torchilin VP (2012) Challenges in development of targeted liposomal therapeutics. AAPS J 14:303–315. https://doi.org/10.1208/s12248-012-9330-0
127. Krieger ML, Eckstein N, Schneider V et al (2010) Overcoming cisplatin resistance of ovarian cancer cells by targeted liposomes in vitro. Int J Pharm 389:10–17. https://doi.org/10.1016/j.ijpharm.2009.12.061
128. Ying X, Wen H, Lu WL et al (2010) Dual-targeting daunorubicin liposomes improve the therapeutic efficacy of brain glioma in animals. J Controlled Release 141:183–192. https://doi.org/10.1016/j.jconrel.2009.09.020
129. Suzuki R, Takizawa T, Kuwata Y et al (2008) Effective anti-tumor activity of oxaliplatin encapsulated in transferrin-PEG-liposome. Int J Pharm 346:143–150. https://doi.org/10.1016/j.ijpharm.2007.06.010
130. Li X, Ding L, Xu Y et al (2009) Targeted delivery of doxorubicin using stealth liposomes modified with transferrin. Int J Pharm 373:116–123. https://doi.org/10.1016/j.ijpharm.2009.01.023
131. Paliwal SR, Paliwal R, Mishra N et al (2010) A novel cancer targeting approach based on estrone anchored stealth liposome for site-specific breast cancer therapy. Curr Cancer Drug Targets 10:343–353. https://doi.org/10.2174/156800910791190210
132. Gabizon A, Horowitz AT, Goren D et al (2003) In vivo fate of folate-targeted polyethylene-glycol liposomes in tumor-bearing mice. Clin Cancer Res 9:6551–6559
133. Yamada A, Taniguchi Y, Kawano K et al (2008) Design of folate-linked liposomal doxorubicin to its antitumor effect in mice. Clin Cancer Res 14:8161–8168. https://doi.org/10.1158/1078-0432.CCR-08-0159
134. Beuttler J, Rothdiener M, Müller D et al (2009) Targeting of epidermal growth factor receptor (EGFR)-expressing tumor cells with sterically stabilized affibody liposomes (SAL). Bioconjug Chem 20:1201–1208. https://doi.org/10.1021/bc900061v

135. Mamot C, Drummond DC, Noble CO et al (2005) Epidermal growth factor receptor-targeted immunoliposomes significantly enhance the efficacy of multiple anticancer drugs in vivo. Cancer Res 65:11631–11638. https://doi.org/10.1158/0008-5472.CAN-05-1093
136. Kirpotin DB, Drummond DC, Shao Y et al (2006) Antibody targeting of long-circulating lipidic nanoparticles does not increase tumor localization but does increase internalization in animal models. Cancer Res 66:6732–6740. https://doi.org/10.1158/0008-5472.CAN-05-4199
137. Park JW, Hong K, Kirpotin DB et al (2002) Anti-HER2 immunoliposomes: enhanced efficacy attributable to targeted delivery. Clin Cancer Res 8:1172–1181
138. Puri A, Kramer-Marek G, Campbell-Massa R et al (2008) HER2-specific affibody-conjugated thermosensitive liposomes (Affisomes) for improved delivery of anticancer agents. J Liposome Res 18:293–307. https://doi.org/10.1080/08982100802457377
139. Alexis F, Basto P, Levy-Nissenbaum E et al (2008) HER-2-targeted nanoparticle-affibody bioconjugates for cancer therapy. ChemMedChem 3:1839–1843. https://doi.org/10.1002/cmdc.200800122
140. Kirpotin D, Park JW, Hong K et al (1997) Sterically stabilized anti-HER2 immunoliposomes: design and targeting to human breast cancer cells in vitro. Biochemistry 36:66–75. https://doi.org/10.1021/bi962148u
141. Yang T, Choi MK, De Cui F et al (2007) Antitumor effect of paclitaxel-loaded PEGylated immunoliposomes against human breast cancer cells. Pharm Res 24:2402–2411. https://doi.org/10.1007/s11095-007-9425-y
142. Allen TM, Mumbengegwi DR, Charrois GJR (2005) Anti-CD19-targeted liposomal doxorubicin improves the therapeutic efficacy in murine B-cell lymphoma and ameliorates the toxicity of liposomes with varying drug release rates. Clin Cancer Res 11:3567–3573. https://doi.org/10.1158/1078-0432.CCR-04-2517
143. Cheng WWK, Das D, Suresh M, Allen TM (2007) Expression and purification of two anti-CD19 single chain Fv fragments for targeting of liposomes to CD19-expressing cells. Biochim Biophys Acta Biomembr 1768:21–29. https://doi.org/10.1016/j.bbamem.2006.09.004
144. Cheng WWK, Allen TM (2008) Targeted delivery of anti-CD19 liposomal doxorubicin in B-cell lymphoma: a comparison of whole monoclonal antibody, Fab′ fragments and single chain Fv. J Controlled Release 126:50–58. https://doi.org/10.1016/j.jconrel.2007.11.005
145. Sapra P, Moase EH, Ma J, Allen TM (2004) Improved therapeutic responses in a xenograft model of human B lymphoma (Namalwa) for liposomal vincristine versus liposomal doxorubicin targeted via anti-CD19 IgG2a or Fab′ fragments. Clin Cancer Res 10:1100–1111. https://doi.org/10.1158/1078-0432.CCR-03-0041
146. Eliaz RE, Nir S, Marty C, Szoka FC (2004) Determination and modeling of kinetics of cancer cell killing by doxorubicin and doxorubicin encapsulated in targeted liposomes. Cancer Res 64:711–718. https://doi.org/10.1158/0008-5472.CAN-03-0654
147. Lundberg BB (2007) Cellular association and cytotoxicity of doxorubicin-loaded immunoliposomes targeted via Fab′ fragments of an anti-CD74 antibody. Digestion 7544:171–175. https://doi.org/10.1080/10717540601036831
148. Pastorino F, Brignole C, Marimpietri D et al (2003) Doxorubicin-loaded Fab′ fragments of anti-disialoganglioside immunoliposomes selectively inhibit the growth and dissemination of human neuroblastoma in nude mice. Cancer Res 63:86–92
149. Vingerhoeds MH, Steerenberg PA, Hendriks JJ et al (1996) Immunoliposome-mediated targeting of doxorubicin to human ovarian carcinoma in vitro and in vivo. Br J Cancer 74:1023–1029. https://doi.org/10.1038/bjc.1996.484
150. Matsumura Y, Gotoh M, Muro K et al (2004) Phase I and pharmacokinetic study of MCC-465, a doxorubicin (DXR) encapsulated in PEG immunoliposome, in patients with metastatic stomach cancer. Ann Oncol 15:517–525. https://doi.org/10.1093/annonc/mdh092
151. Hosokawa S, Tagawa T, Niki H et al (2003) Efficacy of immunoliposomes on cancer models in a cell-surface-antigen-density-dependent manner. Br J Cancer 89:1545–1551. https://doi.org/10.1038/sj.bjc.6601341

152. Elbayoumi TA, Torchilin VP (2009) Tumor-targeted nanomedicines: enhanced antitumor efficacy in vivo of doxorubicin-loaded, long-circulating liposomes modified with cancer-specific monoclonal antibody. Clin Cancer Res 15:1973–1980. https://doi.org/10.1158/1078-0432.CCR-08-2392
153. Elbayoumi TA, Torchilin VP (2008) Tumor-specific antibody-mediated targeted delivery of Doxil® reduces the manifestation of auricular erythema side effect in mice. Int J Pharm 357:272–279. https://doi.org/10.1016/j.ijpharm.2008.01.041
154. Zhu L, Kate P, Torchilin VP (2012) Matrix metalloprotease 2-responsive multifunctional liposomal nanocarrier for enhanced tumor targeting. ACS Nano 6:3491–3498. https://doi.org/10.1021/nn300524f
155. Moreira JN, Gaspar R, Allen TM (2001) Targeting Stealth liposomes in a murine model of human small cell lung cancer. Biochim Biophys Acta 1515:167–176. https://doi.org/10.1016/s0005-2736(01)00411-4
156. Moreira JNN, Hansen CB, Gaspar R, Allen TM (2001) A growth factor antagonist as a targeting agent for sterically stabilized liposomes in human small cell lung cancer. Biochim Biophys Acta 1514:303–317. https://doi.org/10.1016/S0005-2736(01)00386-8
157. Dagar S, Sekosan M, Lee BS et al (2001) VIP receptors as molecular targets of breast cancer: implications for targeted imaging and drug delivery. J Controlled Release 74:129–134. https://doi.org/10.1016/S0168-3659(01)00326-1
158. Kang H, O'Donoghue MB, Liu H, Tan W (2010) A liposome-based nanostructure for aptamer directed delivery. Chem Commun (Camb) 46:249–251. https://doi.org/10.1039/b916911c
159. Xing H, Tang L, Yang X et al (2013) Selective delivery of an anticancer drug with aptamer-functionalized liposomes to breast cancer cells in vitro and in vivo. J Mater Chem B Mater Biol Med 1:5288–5297. https://doi.org/10.1039/C3TB20412J
160. Shi C, Cao H, He W et al (2015) Novel drug delivery liposomes targeted with a fully human anti-VEGF165 monoclonal antibody show superior antitumor efficacy in vivo. Biomed Pharmacother 73:48–57. https://doi.org/10.1016/j.biopha.2015.05.008
161. Zhao H, Wang J-C, Sun Q-S et al (2009) RGD-based strategies for improving antitumor activity of paclitaxel-loaded liposomes in nude mice xenografted with human ovarian cancer. J Drug Target 17:10–18. https://doi.org/10.1080/10611860802368966
162. Xiong XB, Huang Y, Lu WL et al (2005) Intracellular delivery of doxorubicin with RGD-modified sterically stabilized liposomes for an improved antitumor efficacy: in vitro and in vivo. J Pharm Sci 94:1782–1793. https://doi.org/10.1002/jps.20397
163. Voinea M, Manduteanu I, Dragomir E et al (2005) Immunoliposomes directed toward VCAM-1 interact specifically with activated endothelial cells - A potential tool for specific drug delivery. Pharm Res 22:1906–1917. https://doi.org/10.1007/s11095-005-7247-3
164. Gosk S, Moos T, Gottstein C, Bendas G (2008) VCAM-1 directed immunoliposomes selectively target tumor vasculature in vivo. Biochim Biophys Acta Biomembr 1778:854–863. https://doi.org/10.1016/j.bbamem.2007.12.021
165. Jaafari MR, Foldvari M (1999) P0 protein mediated targeting of liposomes to melanoma cells with high level of ICAM-1 expression. J Drug Target 7:101–112. https://doi.org/10.3109/10611869909085495
166. Jaafari MR, Foldvari M (2002) Targeting of liposomes to melanoma cells with high levels of ICAM-1 expression through adhesive peptides from immunoglobulin domains. J Pharm Sci 91:396–404. https://doi.org/10.1002/jps.10062
167. Pastorino F, Brignole C, Marimpietri D et al (2003) Vascular damage and anti-angiogenic effects of tumor vessel-targeted liposomal chemotherapy. Cancer Res 63:7400–7409
168. Kondo M, Asai T, Katanasaka Y et al (2004) Anti-neovascular therapy by liposomal drug targeted to membrane type-1 matrix metalloproteinase. Int J Cancer 108:301–306. https://doi.org/10.1002/ijc.11526
169. Atobe K, Ishida T, Ishida E et al (2007) In vitro efficacy of a sterically stabilized immunoliposomes targeted to membrane type 1 matrix metalloproteinase (MT1-MMP). Biol Pharm Bull 30:972–978. https://doi.org/10.1248/bpb.30.972

170. Hatakeyama H, Akita H, Ishida E et al (2007) Tumor targeting of doxorubicin by anti-MT1-MMP antibody-modified PEG liposomes. Int J Pharm 342:194–200. https://doi.org/10.1016/j.ijpharm.2007.04.037
171. Dearling JLJ, Packard AB (2017) Molecular imaging in nanomedicine—a developmental tool and a clinical necessity. J Controlled Release 261:23–30. https://doi.org/10.1016/j.jconrel.2017.06.011
172. Lee H, Zheng J, Gaddy D et al (2015) A gradient-loadable 64Cu-chelator for quantifying tumor deposition kinetics of nanoliposomal therapeutics by positron emission tomography. Nanomed Nanotechnol Biol Med 11:155–165. https://doi.org/10.1016/j.nano.2014.08.011
173. Lee SG, Gangangari K, Kalidindi TM et al (2016) Copper-64 labeled liposomes for imaging bone marrow. Nucl Med Biol 43:781–787. https://doi.org/10.1016/j.nucmedbio.2016.08.011
174. Petersen AL, Henriksen JR, Binderup T et al (2016) In vivo evaluation of PEGylated 64Cu-liposomes with theranostic and radiotherapeutic potential using micro PET/CT. Eur J Nucl Med Mol Imaging 43:941–952. https://doi.org/10.1007/s00259-015-3272-6
175. Asai T (2012) Positron emission tomography (PET) imaging of small interfering RNA (siRNA) delivery in drug development. Yakugaku Zasshi 132:1159–1163. https://doi.org/10.1002/clc.22060
176. Lamichhane N, Dewkar GK, Sundaresan G et al (2017) [18F]-fluorinated carboplatin and [111in]-liposome for image-guided drug delivery. Int J Mol Sci. https://doi.org/10.3390/ijms18051079
177. Li S, Goins B, Zhang L, Bao A (2012) Novel multifunctional theranostic liposome drug delivery system: construction, characterization, and multimodality MR, near-infrared fluorescent, and nuclear imaging. Bioconjug Chem 23:1322–1332. https://doi.org/10.1021/bc300175d
178. Han XJ, Wei YF, Wan YY et al (2014) Development of a novel liposomal nanodelivery system for bioluminescence imaging and targeted drug delivery in ErbB2-overexpressing metastatic ovarian carcinoma. Int J Mol Med 34:1225–1232. https://doi.org/10.3892/ijmm.2014.1922
179. De Smet M, Langereis S, Van Den Bosch S et al (2013) SPECT/CT imaging of temperature-sensitive liposomes for MR-image guided drug delivery with high intensity focused ultrasound. J Controlled Release 169:82–90. https://doi.org/10.1016/j.jconrel.2013.04.005
180. Grange C, Geninatti-Crich S, Esposito G et al (2010) Combined delivery and magnetic resonance imaging of neural cell adhesion molecule-targeted doxorubicin-containing liposomes in experimentally induced Kaposi's sarcoma. Cancer Res 70:2180–2190. https://doi.org/10.1158/0008-5472.CAN-09-2821
181. Yeo SY, de Smet M, Langereis S et al (2014) Temperature-sensitive paramagnetic liposomes for image-guided drug delivery: Mn^{2+} versus [Gd(HPDO3A)(H_2O)]. Biochim Biophys Acta Biomembr 1838:2807–2816. https://doi.org/10.1016/j.bbamem.2014.07.019
182. He Y, Zhang L, Song C, Zhu D (2014) Design of multifunctional magnetic iron oxide nanoparticles/mitoxantrone-loaded liposomes for both magnetic resonance imaging and targeted cancer therapy. Int J Nanomed 4055. https://doi.org/10.2147/ijn.s61880
183. Garnier B, Tan S, Miraux S et al (2012) Optimized synthesis of 100 nm diameter magnetoliposomes with high content of maghemite particles and high MRI effect. Contrast Media Mol Imaging 7:231–239. https://doi.org/10.1002/cmmi.487
184. Martínez-González R, Estelrich J, Busquets M (2016) Liposomes loaded with hydrophobic iron oxide nanoparticles: suitable T2 contrast agents for MRI. Int J Mol Sci 17:1209. https://doi.org/10.3390/ijms17081209
185. Shirmardi Shaghasemi B, Virk MM, Reimhult E (2017) Optimization of magneto-thermally controlled release kinetics by tuning of magnetoliposome composition and structure. Sci Rep 7:7474. https://doi.org/10.1038/s41598-017-06980-9
186. Skouras A, Mourtas S, Markoutsa E et al (2011) Magnetoliposomes with high USPIO entrapping efficiency, stability and magnetic properties. Nanomed Nanotechnol Biol Med 7:572–579. https://doi.org/10.1016/j.nano.2011.06.010
187. Soenen SJH, Himmelreich U, Nuytten N et al (2010) Intracellular nanoparticle coating stability determines nanoparticle diagnostics efficacy and cell functionality. Small 6:2136–2145. https://doi.org/10.1002/smll.201000763

188. Chithrani BD, Dunne M, Stewart J et al (2010) Delivery of smaller gold nanoparticles by liposomal incorporation. In: Proceedings of the 2010 International Conference on Nanoscience and Nanotechnology, ICONN 2010, pp 241–243. https://doi.org/10.1109/iconn.2010.6045194
189. Lozano N, Al-Ahmady ZS, Beziere NS et al (2015) Monoclonal antibody-targeted PEGylated liposome-ICG encapsulating doxorubicin as a potential theranostic agent. Int J Pharm 482:2–10. https://doi.org/10.1016/j.ijpharm.2014.10.045
190. Wen CJ, Sung CT, Aljuffali IA et al (2013) Nanocomposite liposomes containing quantum dots and anticancer drugs for bioimaging and therapeutic delivery: a comparison of cationic, PEGylated and deformable liposomes. Nanotechnology 24:325101. https://doi.org/10.1088/0957-4484/24/32/325101
191. Sonali Singh RP, Sharma G et al (2016) RGD-TPGS decorated theranostic liposomes for brain targeted delivery. Colloids Surf, B 147:129–141. https://doi.org/10.1016/j.colsurfb.2016.07.058
192. Maples D, McLean K, Sahoo K et al (2015) Synthesis and characterisation of ultrasound imageable heat-sensitive liposomes for HIFU therapy. Int J Hyperth 31:674–685. https://doi.org/10.3109/02656736.2015.1057622
193. Lovell JF, Jin CS, Huynh E et al (2011) Porphysome nanovesicles generated by porphyrin bilayers for use as multimodal biophotonic contrast agents. Nat Mater 10:324–332. https://doi.org/10.1038/nmat2986
194. Miranda D, Carter K, Luo D et al (2017) Multifunctional liposomes for image-guided intratumoral chemo-phototherapy. Adv Healthc Mater 1700253:1–9. https://doi.org/10.1002/adhm.201700253
195. Huynh E, Zheng G (2014) Porphysome nanotechnology: a paradigm shift in lipid-based supramolecular structures. Nano Today 9:212–222. https://doi.org/10.1016/j.nantod.2014.04.012
196. Sou K, Goins B, Takeoka S et al (2007) Selective uptake of surface-modified phospholipid vesicles by bone marrow macrophages in vivo. Biomaterials 28:2655–2666. https://doi.org/10.1016/j.biomaterials.2007.01.041
197. Vieira DB, Gamarra LF (2016) Getting into the brain: liposome-based strategies for effective drug delivery across the blood–brain barrier. Int J Nanomed 11:5381–5414. https://doi.org/10.2147/IJN.S117210
198. Mehta A, Ghaghada K, Mukundan S (2016) Molecular imaging of brain tumors using liposomal contrast agents and nanoparticles. Magn Reson Imaging Clin N Am 24:751–763. https://doi.org/10.1016/j.mric.2016.06.004
199. Hansen AE, Petersen AL, Henriksen JR et al (2015) Positron emission tomography based elucidation of the enhanced permeability and retention effect in dogs with cancer using copper-64 liposomes. ACS Nano 9:6985–6995. https://doi.org/10.1021/acsnano.5b01324
200. Wen CJ, Zhang LW, Al-Suwayeh SA et al (2012) Theranostic liposomes loaded with quantum dots and apomorphine for brain targeting and bioimaging. Int J Nanomed 7:1599–1611. https://doi.org/10.2147/IJN.S29369
201. Oku N, Yamashita M, Katayama Y et al (2011) PET imaging of brain cancer with positron emitter-labeled liposomes. Int J Pharm 403:170–177. https://doi.org/10.1016/j.ijpharm.2010.10.001
202. Strijkers GJ, Mulder WJM, Van Heeswijk RB et al (2005) Relaxivity of liposomal paramagnetic MRI contrast agents. Magn Reson Mater Phys Biol Med 18:186–192. https://doi.org/10.1007/s10334-005-0111-y
203. Tansi FL, Rüger R, Rabenhold M et al (2015) Fluorescence-quenching of a liposomal-encapsulated near-infrared fluorophore as a tool for in vivo optical imaging. JoVE (Journal Vis Exp) e52136. https://doi.org/10.3791/52136
204. Wu B, Wan B, Lu S-T et al (2017) Near-infrared light-triggered theranostics for tumor-specific enhanced multimodal imaging and photothermal therapy. Int J Nanomed 12:4467–4478. https://doi.org/10.2147/IJN.S137835

205. Shemesh CS, Moshkelani D, Zhang H (2015) Thermosensitive liposome formulated indocyanine green for near-infrared triggered photodynamic therapy: in vivo evaluation for triple-negative breast cancer. Pharm Res 32:1604–1614. https://doi.org/10.1007/s11095-014-1560-7
206. Lammers T, Kiessling F, Hennink WE, Storm G (2010) Nanotheranostics and image-guided drug delivery: current concepts and future directions. Mol Pharm 7:1899–1912. https://doi.org/10.1021/mp100228v
207. Staruch RM, Hynynen K, Chopra R (2015) Hyperthermia-mediated doxorubicin release from thermosensitive liposomes using MR-HIFU: therapeutic effect in rabbit Vx2 tumours. Int J Hyperth 31:118–133. https://doi.org/10.3109/02656736.2014.992483
208. Ranjan A, Jacobs GC, Woods DL et al (2012) Image-guided drug delivery with magnetic resonance guided high intensity focused ultrasound and temperature sensitive liposomes in a rabbit Vx2 tumor model. J Controlled Release 158:487–494. https://doi.org/10.1016/j.jconrel.2011.12.011
209. Kneepkens E, Fernandes A, Nicolay K, Grüll H (2016) Iron(III)-based magnetic resonance-imageable liposomal T1 contrast agent for monitoring temperature-induced image-guided drug delivery. Invest Radiol 51:735–745. https://doi.org/10.1097/RLI.0000000000000297
210. De Smet M, Heijman E, Langereis S et al (2011) Magnetic resonance imaging of high intensity focused ultrasound mediated drug delivery from temperature-sensitive liposomes: an in vivo proof-of-concept study. J Controlled Release 150:102–110. https://doi.org/10.1016/j.jconrel.2010.10.036
211. Viglianti BL, Ponce AM, Michelich CR et al (2006) Chemodosimetry of in vivo tumor liposomal drug concentration using MRI. Magn Reson Med 56:1011–1018. https://doi.org/10.1002/mrm.21032

Quantum Dots for Cancer Therapy and Bioimaging

Fu-Gen Wu, Xiaodong Zhang, Xiaokai Chen, Wei Sun, Yan-Wen Bao, Xian-Wu Hua, Ge Gao and Hao-Ran Jia

Abstract Quantum dots (QDs) usually refer to very small nanoparticles of only few nanometers in size. The optical and electronic properties of QDs differ from those of larger particles. QDs will emit light of specific frequencies if electricity or light is applied to them, and these frequencies can be precisely tuned by changing the dots' size, shape, and material, giving rise to many applications. In this chapter, apart from the most common QDs, e.g., the cadmium (Cd)-containing semiconductor QDs, other types of QDs, including silver chalcogenide quantum dots, carbon quantum dots, silicon quantum dots, black phosphorus quantum dots, germanium quantum dots, and polymer dots are also introduced with an emphasis on their cancer therapy and imaging applications.

Keywords Quantum dots · Fluorescence imaging · Cell imaging · Cancer theranostics · Anticancer

1 Semiconductor Quantum Dots

Semiconductor quantum dots are one of the most important QDs, whose size and shape can be precisely controlled by the duration, temperature, and ligand molecules used in the synthesis [1]. In comparison with conventional molecular fluorophores, the emission wavelength of QDs can be tuned by varying their size and composition. Due to the narrow emission and broad excitation spectra, QDs perform multicolor imaging with minimal spectral overlap and have the possibility to excite all colors of QDs simultaneously with a single light source. Furthermore, QDs display excellent photostability over molecular fluorophores, so that long-term imaging can be achieved without artifacts from photobleaching. Because of these unique optical properties, QDs are of wide interest and have emerged as a strong competitor as fluorescent probes for biomedical imaging and diagnostics applications [2, 3]. In

F.-G. Wu (✉) · X. Zhang · X. Chen · W. Sun · Y.-W. Bao · X.-W. Hua · G. Gao · H.-R. Jia
State Key Laboratory of Bioelectronics, School of Biological Science and Medical Engineering, Southeast University, 2 Sipailou Road, Nanjing 210096, People's Republic of China
e-mail: wufg@seu.edu.cn

G. Gonçalves and G. Tobias (eds.), *Nanooncology*, Nanomedicine and Nanotoxicology, https://doi.org/10.1007/978-3-319-89878-0_3

particular, as one type of multifunctional materials, QDs exhibit specific advantages in tumor imaging and tumor therapy due to the large surface area which enables them to be conjugated with different agents including imaging substances, targeting molecules, and therapeutic agents. Herein, we focus on QDs or QDs-based nanomaterials for tumor imaging and therapy in recent years.

1.1 QDs for Tumor Imaging

Cancer remains one of the leading causes of death in the world. Diagnostic tumor imaging has gained a major role in the management of tumor therapy with qualitative and quantitative analyses of the biological processes of tumors by monitoring changes of tumor cells at tissue, cellular, or subcellular levels. There are various kinds of imaging techniques such as magnetic resonance imaging (MRI), computed tomography (CT), positron emission tomography (PET), and fluorescence imaging. These techniques and the combined use of them help to provide accurate information of tumors from a variety of aspects. For this section, we overview the recent progress of visible- and near-infrared (NIR)-emitting QDs-based fluorescence imaging and multi-model imaging of tumor.

1.1.1 QDs-Based Fluorescence Imaging

Fluorescence imaging is commonly used as one of the most potent tools for cancer diagnosis by tumor-targeted imaging from cells and tissues of living animals. On the other hand, functionalized QDs which are modified with biomolecules possess many advantages over conventional organic fluorophores, such as high photoluminescence efficiency, great photostability, size-dependent emission wavelength, and sharp emission profile. In addition to their good biocompatibility and low toxicity, QDs emitting in the visible and NIR ranges have been widely applied in fluorescence imaging in vitro and in vivo as novel fluorescent probes [4–11].

Visible Fluorescence Imaging

Fluorescence imaging using light in the visible wavelength regime (400–700 nm) is a useful technology for cancer diagnosis due to its fast feedback as well as relatively good spatial resolution. Visible-emitting QDs have gained considerable attention in the last decade for in vitro and in vivo tumor imaging [12–20]. Zhang et al. offered a strategy to synthesize DNA-functionalized Zn^{2+}-doped CdTe QDs (DNA-QDs) through a facile one-pot hydrothermal route [13]. The as-prepared QDs exhibit high quantum yield (up to 80.5%), excellent photostability, and low toxicity. Moreover, DNA has been designed as an aptamer specific for mucin 1 overexpressed in many cancer cells including lung adenocarcinoma, and the aptamer-QDs are designed for the first time for application in active tumor-targeted imaging in vitro and in vivo (Fig. 1a). Furthermore, an intelligent “on-off” or “off-on” switch system for tumor-targeted fluorescence imaging without unnecessary background signal is urgently desired. Zhang et al. reported a novel fluorescence turn-on probe for targeted imaging of folate receptor (FR)-overexpressed cancer cells based on the self-assembly of folic acid (FA) and polyethyleneimine (PEI)-coated CdS/ZnS QDs (PEI-CdS/ZnS QDs) [14]. The primary fluorescence of PEI-CdS/ZnS QDs turns off first upon the electrostatic adsorption of FA onto PEI-CdS/ZnS QDs based on electron transfer to produce negligible fluorescence background. The presence of FR expressed on the surface of cancer cells then makes FA desorb from PEI-CdS/ZnS QDs due to the specific and high affinity of FA for FR. As a result, the primary fluorescence of PEI-CdS/ZnS QDs adhering to the cells turns on due to the inhibition of electron transfer (Fig. 1b). Besides tumor-targeted imaging, recognition of a specific cancer cell type among various cell types is also essential for cancer diagnosis and targeted cancer therapy. In a recent study, Wang et al. presented a pattern recognition of cells via multiplexed imaging with three types of monosaccharide imprinted QDs [18]. Aberrant expression of glycan structures on the cell surface is a universal hallmark of cancer cells. For instance, sialic acid (SA) and fucose (Fuc) are overexpressed on most cancer cells, while mannose (Man) is overexpressed on certain cancer cells, such as liver cancer cells. Therefore, the combination of multiple monosaccharides can be effective for specific cancer recognition. As shown in Fig. 1c, the three-dimensional (3D) plot intuitively shows the spatial distributions of these cell lines in the coordinate system. The results disclose the similarities and differences of different cell lines, allowing for not only the recognition of cancer cells from normal cells but also the recognition of specific cancer cells. Thus, the study paves a solid ground for the design and preparation of novel cancer cell-targeting reagents and nanoprobes.

On the other hand, QDs-based cancer molecular imaging has emerged as an important technique for cancer detection, personalized treatment, drug development, and imaging-guided surgery. Over the past decade, various types of cancer molecules, including nucleic acids [6, 8], proteins [12, 19], cell-surface receptors [17], and antigen [21] have been conjugated onto QDs to identify and target cancer cells. Notably, protein-targeting strategy plays a significant role and has attracted much attention. To improve protein labeling efficiency, Wichner et al. reported compact aqueous CdSe/CdS QDs with superior single-molecule optical property [19]. These QDs are able to label SNAP-tagged proteins ~10-fold more efficiently than existing SNAP ligands. Furthermore, the QDs show that 99% of time is spent in the fluorescence on-state, ~fourfold higher quantum efficiency than standard CdSe/ZnS QDs. Figure 1d shows that these bright QDs can track the stepping movement of a kinesin motor in vitro, and the improved labeling efficiency enables the tracking of single kinesins in live cells. Additionally, labeling protein biomarkers by multicolor QD probes is an effective method for investigating tumor heterogeneity and complexity. For example, Liu et al. utilized QD-based spectral imaging for high-throughput digital mapping of molecular, cellular, and glandular variations on surgical prostate cancer specimens [12]. They detected and identified a single malignant tumor cell from the complex microenvironments of radical prostatectomy and needle biopsy

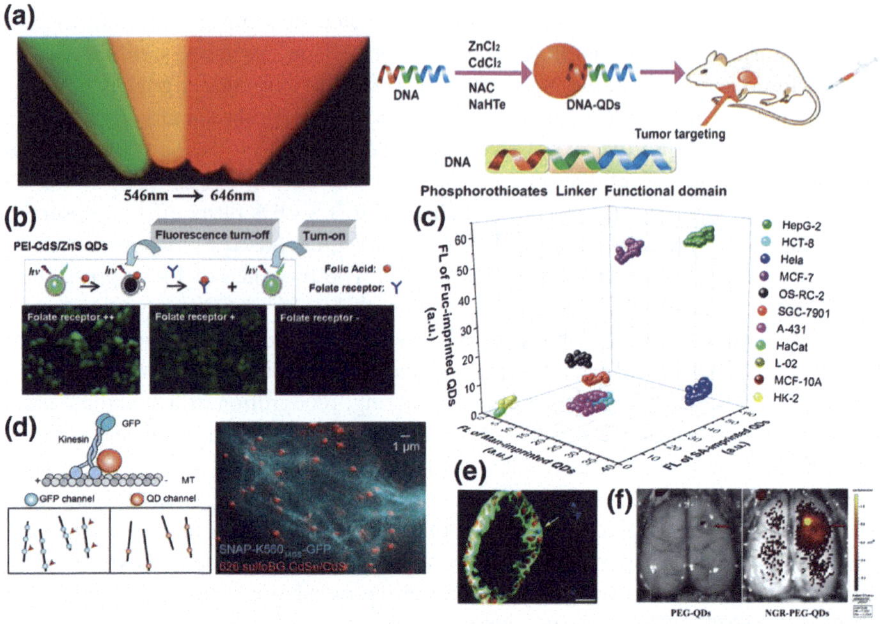

Fig. 1 **a** Photograph of aqueous solution of DNA-QDs under UV condition, and schematic illustration of the synthesis and tumor-targeted fluorescence imaging of DNA-QDs. Reprinted with permission from Ref. [13]. Copyright 2013 American Chemical Society. **b** Schematic illustration of FA-assembled PEI-CdS/ZnS QDs as fluorescence turn-on probe for imaging FR overexpressed cancer cells. Reprinted with permission from Ref. [14]. Copyright 2013 American Chemical Society. **c** Spatial distribution of the 11 cell lines constructed by multiplexed imaging via simultaneously staining with Man-, SA-, and Fuc-imprinted QDs. Reprinted with permission from Ref. [18]. Copyright 2017 American Chemical Society. **d** Schematic in vitro microscopy and live-cell imaging of kinesin labeled with seven Gly-Ser repeats (SNAP-K560$_{14GS}$-GFP) and red sulfoBG QDs. Reprinted with permission from Ref. [19]. Copyright 2017 American Chemical Society. **e** Identification of single malignant tumor cells in a predominantly benign prostate gland by QD multiplexed staining of four protein biomarkers (E-cadherin, green; CK HMW, white; p63, red; and AMACR, blue). Reprinted with permission from Ref. [12]. Copyright 2010 American Chemical Society. **f** Fluorescence imaging of the tumors and normal brains using PEG-QDs or NGR-PEG-QDs after the tail vein injection. Reprinted with permission from Ref. [20]. Copyright 2017 Elsevier

tissue specimens using a panel of just four protein biomarkers (E-cadherin, high-molecular-weight cytokeratin, p63, and α-methylacyl CoA racemase) (Fig. 1e). The multiplexed QD mapping provides correlated molecular and morphological information that is not available from traditional tissue staining and molecular profiling methods. For the in vivo tumor fluorescence imaging, delivery of imaging agents to brain glioma is challenging because the blood–brain barrier (BBB) functions as a physiological checkpoint guarding the central nervous system from circulating large molecules [22]. Thus, BBB limits drug delivery to brain parenchyma, attenuating the diagnosis and therapy effect of brain tumors. Fortunately, QDs offer great promises for crossing the BBB and reaching brain parenchyma based on their ultrasmall sizes, contributing to the development of theranostic nanoprobes for various neurological disorders. In a very recent study, Huang et al. synthesized a novel nanoprobe by conjugating biotinylated asparagines-glycine-arginine (NGR) peptides to avidin-PEG-coated QDs [20]. These QDs can cross the BBB and target CD13-overexpressing glioma and tumor vasculature in vitro and in vivo, contributing to the fluorescence imaging of the brain malignancy (Fig. 1f). This nanotechnology highlights a novel prospect for the molecular diagnosis and image-guided neurosurgery of glioma.

NIR Fluorescence Imaging

NIR-emitting QDs (NIR QDs), allowing lower tissue absorption and scattering, lower undesirable NIR autofluorescence, and deeper penetration depth, have recently been explored as highly promising imaging probes. These high-quality NIR-emitting QDs are especially useful in cellular labeling, deep-tissue imaging, and tumor targeting. In particular, two optimal wavelength ranges of 700–950 and 1000–1350 nm, known as the first and second biological windows (I-BW and II-BW), respectively, have been identified. In the I-BW (700–950 nm), the light is minimally absorbed by tissue components as compared to visible light, resulting in greater penetration and thus deeper imaging. At even longer wavelengths of the II-BW (1000–1350 nm), a greater reduction in the scattering cross section leads to a further improvement in the detection depth and resolution. In this section, we will summarize the recent advancements of the NIR QDs for in vitro and in vivo tumor imaging.

To date, a variety of QDs emitting within I-BW (NIR-I QDs) with well-controlled structure and multifunctional properties as novel biolabeling agents have been developed [4, 23–26]. For instance, Liu et al. synthesized *N*-acetyl-L-cysteine (NAC)-capped CdHgTe/CdS core/shell QDs. The QDs have NIR-I fluorescence (Fig. 2A, a) and are successfully applied for in vivo tumor imaging of nude mice by passive targeting (Fig. 2A, b), indicating that these highly fluorescent probes can be very effective in long-term diagnostics and therapy in in vivo observation [26]. Miyashita et al. developed a new immunohistochemical (IHC) technique with NIR-I QD-conjugated trastuzumab using single-particle imaging to quantitatively measure the HER2 expression level (Fig. 2B) [25]. Moreover, they precisely calculated the number of QD-conjugated trastuzumab particles binding specifically to a cancer cell as the IHC-QD score. The use of IHC-QD score is believed as a predictive factor for trastuzumab therapy.

Compared to the widely used NIR-I QDs, the QDs emitting in the II-BW (NIR-II QDs) can realize in vivo imaging with much higher signal-to-noise (S/N) ratio and spatial resolution. Thus, the development of an effective aqueous synthetic route to high-quality and biocompatible NIR-II emitting QDs is highly appealing. Many efforts have been devoted to developing NIR-II QDs for in vivo fluorescence imaging [9, 11, 27–29]. Ren et al. reported water-dispersible PbS/CdS/ZnS core/shell/shell QDs emitting NIR-II fluorescence (Fig. 2C, a and b) with excellent colloidal stability and photostability [28]. These QDs are injected into mice for tumor imaging in small animals and show high brightness even at quite low concentrations, under

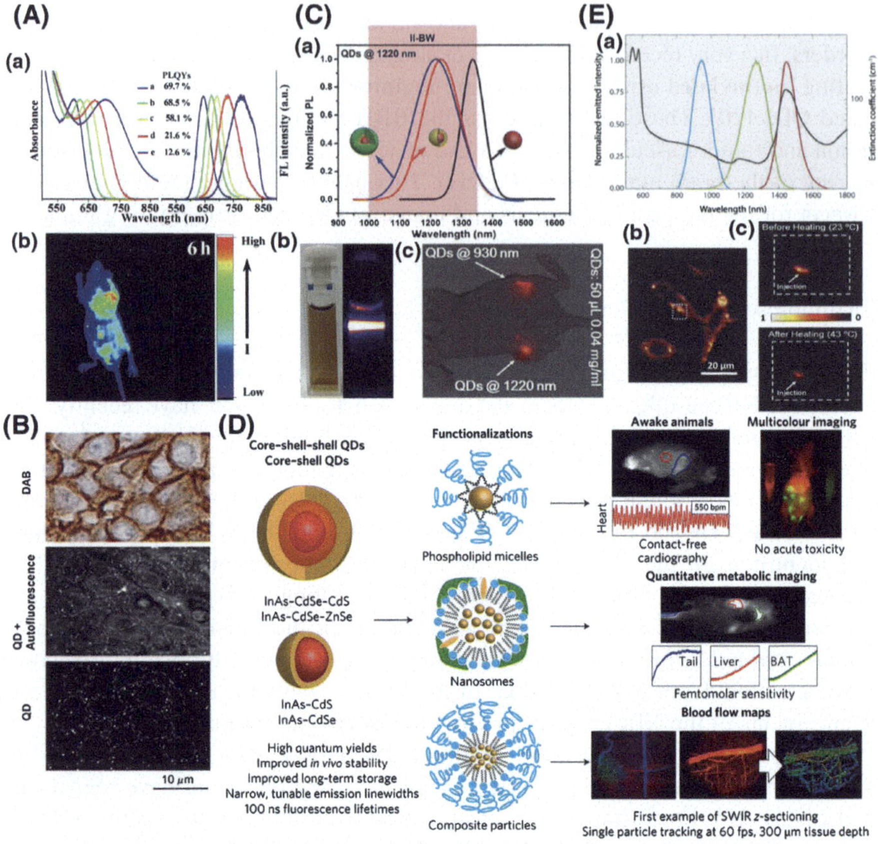

Fig. 2 **A** UV-PL spectra (a) and in vivo tumor targeting (b) of the NIR CdHgTe/CdS/ZnS QDs. Reprinted with permission from Ref. [26]. Copyright 2017 Royal Society of Chemistry. **B** Images of IHC stained with DAB-labeled HER2 antibody (top row) or QD-conjugated trastuzumab (middle and bottom rows). The middle row shows the images observed with 695–740 nm band-pass filter. The bottom row shows the images subtracting autofluorescence in tissues from the images of middle row to visualize the fluorescence of QDs (bright spots) only. Reprinted with permission from Ref. [25]. Copyright 2016 Wiley. **C** (a) PL spectra of initial PbS QDs, PbS/CdS QDs, and final PbS/CdS/ZnS QDs emitting at 1220 nm. (b) Optical (left) and luminescence (right) images of 930 nm emitting PbS/CdS/ZnS core/shell/shell QDs. (c) In vivo optical + fluorescence images of the mouse after injection with the two differently sized QDs solutions emitting in the first and second biological windows, respectively. Reprinted with permission from Ref. [28]. Copyright 2017 Wiley. **D** A schematic overview of the synthesis of core–shell and core–shell–shell short-wave infrared (SWIR) QDs and the subsequent functionalization for next-generation imaging applications. Reprinted with permission from Ref. [29]. Copyright 2017 Nature Publishing Group. **E** (a) Three different emission spectra from different PbS/CdS/ZnS QDs (blue, QDs emitting at 930 nm; green, QDs emitting at 1270 nm; red, QDs emitting at 1450 nm). The black line shows the extinction coefficient (comprising both absorption and scattering contributions) of biological tissues for the spectral range under consideration. (b) Fluorescence image of four HeLa cells incubated with a solution of 1270 nm emitting QDs with the excitation of 808 nm. (c) Sub-tissue fluorescence images taken at two different temperatures from the chicken breast previously injected with QDs emitting at 1270 nm. Reprinted with permission from Ref. [27]. Copyright 2015 Wiley

significantly reduced NIR laser density excitations, with very short signal integration time and deep injection (Fig. 2C, c). Bruns et al. introduced a class of short-wavelength infrared region (SWIR: 1000–2000 nm)-emissive indium-arsenide-based QDs that are readily modifiable for various imaging applications [29]. These QDs exhibit narrow and size-tunable emission and a dramatically higher emission quantum yield than previously described SWIR probes. Then, they quantify, in mice, the metabolic turnover rates of lipoproteins in several organs simultaneously and in real time as well as heartbeat and breathing rates in awake and unrestrained animals, and generate detailed 3D quantitative flow maps of the mouse brain vasculature (Fig. 2D). Benayas et al. used water-dispersible core/shell/shell PbS/CdS/ZnS QDs as NIR imaging probes fabricated through a rapid, cost-effective microwave-assisted cation exchange procedure [27]. The emission wavelength of the QDs probe is within the second biological window (1000–1350 nm) (Fig. 2E, a). The in vitro (Fig. 2E, b) and ex vivo (Fig. 2E, c) experiments prove that the QDs are capable of high-resolution thermal sensing in the physiological temperature range. Together with their intense fluorescence, these PbS/CdS/ZnS QDs represent multifunctional probes both for in vitro and in vivo applications in biomedicine.

1.1.2 QDs-Based Multimodal Imaging

With the outstanding optical properties of QDs, the QDs-based fluorescence imaging techniques show great potential in extracting detailed biomedical information with high imaging sensitivity and low-cost imaging facilities in comparison to clinically used MRI, CT, and PET methods. However, the poor tissue penetration restricts the better biomedical research and related clinical applications of light-emitting QDs as fluorescence imaging probes. Therefore, the combination of QDs-based fluorescence imaging and other imaging techniques within a single nanoplatform has emerged as an effective approach to collect reliable biomedical information, thus improving the efficiency and sensitivity of clinical imaging diagnostics. In recent years, multimodal imaging with the combination of fluorescence imaging and other imaging techniques such as dark-field imaging, PET, CT, photoacoustic (PA), and MRI offers revolutionary imaging tools for biomedical applications and has ignited intense research interest worldwide [30–34].

Recently, the integration of MRI agents with QDs has been designed as a useful imaging modality pair for more accurate biomedical detections [30, 35, 36]. MRI is one of the most powerful medical diagnosis tools, which can visualize the anatomical structure of the body and extract physiological information with high spatial resolution and soft tissue contrast [37]. The MRI/fluorescence imaging technology enables significant improvement in diagnostic accuracy and therapeutic strategy, in comparison with standalone imaging. For example, Lai et al. presented In_2S_3/ZnS core/shell QDs co-doped with Ag^+ and Mn^{2+} (referred to as AgMn:In_2S_3/ZnS) [36]. Ag^+ can alter the optical properties of the host QDs, whereas the spin magnetic moment ($S = 5/2$) of Mn^{2+} efficiently induces the longitudinal relaxation of water protons. This is the first report of the aqueous synthesis of color-tunable AgMn:In_2S_3/ZnS core/shell

QDs with magnetic properties. The obtained QDs have a satisfactory quantum yield (45%), high longitudinal relaxivity (6.84 mM^{-1} s^{-1}), and robust photostability. As seen by confocal microscopy and MRI, AgMn:In_2S_3/ZnS conjugated to hyaluronic acid (referred to as AgMn:In_2S_3/ZnS@HA) can efficiently and specifically target to cancer cells. Moreover, AgMn:In_2S_3/ZnS@HA shows negligible cytotoxicity in vitro and in vivo, rendering it a promising diagnostic probe for dual-modal imaging in clinical applications (Fig. 3a).

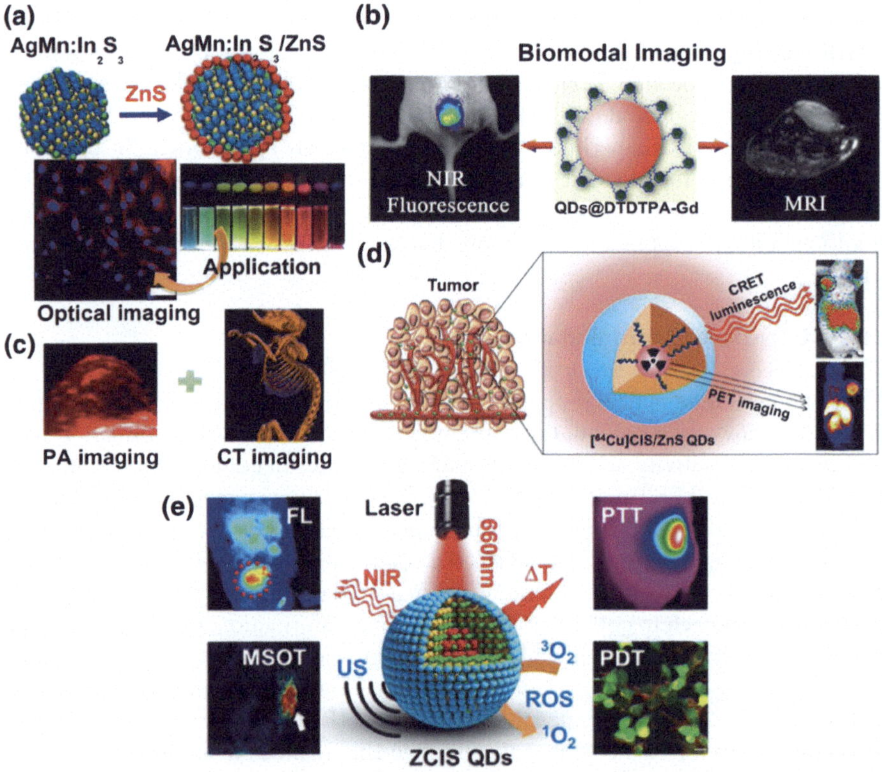

Fig. 3 **a** Synthesis of AgMn:In_2S_3/ZnS core/shell QDs and their application in optical imaging and MRI. Reprinted with permission from Ref. [36]. Copyright 2017 Elsevier. **b** Schematic illustration, in vivo NIR fluorescence images, and T_1-weighted MRI of QDs@DTDTPA-Gd NPs. Reprinted with permission from Ref. [34]. Copyright 2017 American Chemical Society. **c** In vivo PA and CT images of 4T1 tumors on mice after intravenously injected with MoS_2 QD@polyaniline nanohybrids. Reprinted with permission from Ref. [33]. Copyright 2016 American Chemical Society. **d** Illustration of the intrinsically radioactive [^{64}Cu]CIS/ZnS QDs for PET/Cerenkov resonance energy transfer (CRET) luminescence imaging. Reprinted with permission from Ref. [32]. Copyright 2015 American Chemical Society. **e** Illustration of ZCIS QDs used as theranostic nanomedicines with intrinsic fluorescence/MSOT imaging and PTT/PDT therapy abilities. Reprinted with permission from Ref. [40]. Copyright American Chemical Society

Apart from the visible light-emitting QDs, the NIR-emitting QDs have gained more attention when integrating with MRI agents since they can minimize light absorption and diffusion and therefore maximize imaging depth [34, 38]. Yang et al. developed a bimodal contrast nanoagent by chelating gadolinium ions to 2-[bis[2-[carboxymethyl-[2-oxo-2-(2-sulfanylethyl-amino)ethyl]amino]ethyl]amino]acetic acid (DTDTPA)-modified $CuInS_2/ZnS$ QDs [34]. The longitudinal relaxivity of the resulted QDs@DTDTPA-Gd NPs is calculated to be 9.91 $mM^{-1}\ s^{-1}$, which is 2.5 times as high as that of the clinically approved Gd-DTPA (3.9 $mM^{-1}\ s^{-1}$). In addition, the in vivo imaging experiments show that QDs@DTDTPA-Gd NPs can enhance both NIR fluorescence and T_1-weighted MRI of tumor tissue through passive targeting accumulation. Moreover, the high colloidal and fluorescence stabilities and good biocompatibility ensure the potential use of QDs@DTDTPA-Gd NPs as an efficient nanoagent to integrate the extremely high sensitivity of fluorescence imaging to the high resolution of MRI (Fig. 3b).

CT and PA imaging, with higher density resolution and deeper tissue penetration, are widely used as diagnostic tools in biomedical applications. Wang et al. introduced a versatile nanomaterial based on MoS_2 QD@polyaniline nanohybrids, which exhibit not only fluorescence imaging of cancer but also enhanced PA imaging and CT signal in vivo (Fig. 3c) [33]. This versatile nanohybrid shows good potential to facilitate the multimodal imaging for a better imaging-guided tumor therapy.

Besides, PET is also a powerful biomedical imaging technique widely used for diagnostic applications in clinical oncology with the advantages of high sensitivity and quantitative accuracy. Attaching positron-emitting radioisotopes onto QDs has also attracted intense interest in both preclinical research and clinical applications. As a representative example, Guo et al. described a straightforward synthesis of intrinsically radioactive [^{64}Cu]CuInS/ZnS QDs by directly incorporating ^{64}Cu into CuInS/ZnS nanostructure with $^{64}CuCl_2$ as a synthetic precursor (Fig. 3d) [32]. The [^{64}Cu]CuInS/ZnS QDs are demonstrated to have an excellent radiochemical stability with less than 3% free ^{64}Cu detected even after exposure to serum containing EDTA (5 mM) for 24 h. The PEGylated radioactive QDs show high tumor uptake (10.8% ID/g) in a U87MG mouse xenograft model. Overall, these [^{64}Cu]CuInS/ZnS QDs are successfully applied as an efficient PET/self-illuminating luminescence in vivo imaging agent.

On the other hand, multispectral optoacoustic tomography (MSOT) is also an indispensable imaging technique that overcomes the optical diffusion limitation by integrating the spectral selectivity of molecular excitation with the high resolution of ultrasound detection based on the PA effect [39]. Therefore, it is highly desirable to achieve theranostic nanomedicines with intrinsic fluorescence/MSOT dual-modal imaging ability. Lv et al. exploited CuInS/ZnS quantum dots (ZCIS QDs) as "all-in-one" versatile nanomedicines that possess intrinsic fluorescence/PA imaging for synergistic photothermal therapy (PTT)/photodynamic therapy (PDT) (Fig. 3e) [40]. The ZCIS QDs allow noninvasively monitoring tumor site localization profiles and thus hold great potential as precision theranostic nanomedicines.

Other multimodal technologies including fluorescence/CT and fluorescence/plasmic imaging based on QDs have also acquired wide attention [31, 41].

These nanomaterials have highly integrated multifunctions. For example, combining CT contrast agents and QDs into one nanoplatform provides more complementary and accurate information about the anatomical structure, as well as high-resolution and sensitive imaging capability at both tissue and cell levels. Besides, the integration of the fluorescence and plasmic imaging is potentially powerful in precise identification of subcellular targets by co-localization of fluorescence and light scattering (dark field) imaging.

1.2 QDs in Cancer Therapy

QDs are luminescent nanocrystals with rich surface chemistry and unique optical properties, which make them useful as probes or carriers for drug delivery in cancer therapy. Besides, due to their unique properties, QDs can also produce reactive oxygen species (ROS) or generate heat under irradiation to kill cancer cells. In this part, we summarize the recent advancements of QDs in cancer therapy.

1.2.1 QDs as Photosensitizers

PDT involves the administration of photosensitizers (PSs) followed by local illumination of the lesion using light of a specific wavelength to activate the PS. A series of photochemical reactions triggered by the PS can lead to the death of cancerous or bacterial cells [42]. Due to their unique optical properties, QDs can act as PSs and kill cancer cells under light irradiation. Dong et al. synthesized photoluminescent MoS_2 QDs with superior singlet oxygen ($^1O_2$) production ability exceeding the commercial photosensitizer PpIX, which has great potential for PDT [43]. He et al. demonstrated that CdSe/ZnS QDs with illumination can cause ultrastructural changes in pancreatic cancer cells, such as organelle degeneration and chromatin condensation and aggregation at the periphery of the nucleus due to the ROS generation [44].

Although QDs can be PS candidates, their photosensitizing efficacy is still not satisfactory for clinical applications. Thus, a new type of photosensitizer consisting of CdTe QDs with good photosensitizing efficacy, excellent water dispersibility, and stability was reported by Sun et al. [45]. Different from most of the previous reports, the as-prepared QDs do not inhibit the growth of normal cells in the experimental concentration range, but can act as a photosensitizer to specifically and remarkably inhibit the proliferation of human hepatoma cells. Mechanistic studies reveal that the QDs can be specifically internalized by hepatoma cells, considerably induce the generation of intracellular ROS under light illumination, and significantly induce the necrosis of hepatoma cells. This work provides an inspiration for the direct application of QDs as a new type of photosensitizer to treat human hepatoma through PDT.

1.2.2 QDs as Photothermal Agents

Besides being as PSs, QDs can also act as photothermal agents in photothermal therapy (PTT). The QDs which have strong absorbance in NIR region can produce heat for selectively killing/disrupting cancer cells. Wang et al. prepared MoS_2 QD@polyaniline nanohybrids, and realized successful PTT of cancer [33]. Yong et al. demonstrated that WS_2 QDs with small size (3 nm) possess not only significant X-ray CT/PA imaging signal enhancement but also remarkable PTT/radiotherapy (RT) synergistic effect for tumor treatment [46]. Chu et al. reported that CdTe and CdSe QDs can rapidly convert light energy into heat upon 671-nm laser irradiation [47]. The growth of mouse melanoma tumors injected with silica-coated CdTe QDs is significantly inhibited after laser irradiation.

1.2.3 QDs as PSs and Photothermal Agents

As stated above, QDs can be PSs and photothermal agents simultaneously. The combination therapy can be more effective than a single treatment. Lv et al. synthesized CuInS/ZnS quantum dots (ZCIS QDs) for synergistic PTT/PDT therapy [40]. Under a single 660 nm laser irradiation, the ZCIS QDs have simultaneous photothermal and photodynamic effects, resulting in high therapeutic efficacy against tumors. Ding et al. realized PA imaging-guided PTT/PDT in a single material MoO_{3-x} QDs [48]. Due to their strong NIR harvesting ability, MoO_{3-x} QDs can convert incident light into hyperthermia and sensitize the formation of singlet oxygen synchronously as evidenced by the in vitro assay.

1.2.4 QDs in Drug Delivery Systems

QDs represent a versatile platform for designing and engineering drug delivery systems. QDs facilitate the in-depth studies on the interactions between nanocarriers and biological systems through real-time monitoring of biodistribution, intracellular uptake, drug release, and long-term nanocarrier fate. At the same time, the compact size and compatibility with a variety of surface modification strategies of QDs enable the substitution of any NP core with a QD in single-NP drug delivery systems, or the incorporation of QD tags within larger multicomponent vehicles.

QDs as Carriers in PDT

In PDT, the excellent tissue-penetrating ability of an external excitation light and the good match between the wavelength of laser emission and the absorption wavelength of each PS are the two main factors that affect the generation of ROS and cell killing behavior. In an attempt made by Hsu et al., Renilla luciferase-immobilized QDs-655 (QD-RLuc8) was used for bioluminescence resonance energy transfer (BRET)-

mediated PDT to solve the abovementioned problems [49]. The bioluminescent QD-RLuc8 conjugate exhibits self-illumination at 655 nm after coelenterazine addition, which can activate the photosensitizer Foscan-loaded micelles for PDT. This nanotechnology-based PDT possesses several clinical benefits, such as overcoming light penetration issues and treating deeper lesions that are intractable by PDT alone. Tsay et al. developed peptide-coated QD-photosensitizer conjugates using covalent conjugation strategies [50]. Rose bengal and chlorin e6, which generate singlet oxygen in high yield, are covalently attached to phytochelatin-related peptides. The photosensitizer–peptide conjugates are subsequently used to coat green- and red-emitting CdSe/CdS/ZnS nanocrystals. Generation of singlet oxygen can be achieved via indirect excitation through Förster resonance energy transfer (FRET) from the QDs to PSs, or by direct excitation of the PSs. In the latter case, by using two color excitations, the conjugate can be simultaneously used for fluorescence imaging and singlet oxygen generation. Singlet oxygen quantum yields as high as 0.31 were achieved using the 532-nm excitation wavelength. Similarly, Martynenko et al. prepared ZnSe/ZnS QDs and chlorin e6 complexes [51]. These complexes have shown ~50% intracomplex FRET from QDs to chlorin e6. The PDT test shows that the complexes had a twofold enhancement of the cancer cell photodynamic destruction as compared to free chlorin e6 molecules. They believe the enhanced PDT effect is attributed to two factors: the efficient QD–chlorin e6 photoexcitation energy transfer and the enhanced cellular uptake of the photosensitizer in the presence of ZnSe/ZnS QDs.

Owing to the two-photon excitation (TPE) property of QDs, enhanced photodynamic therapeutic efficacy through combining QDs with PSs under TPE was achieved by several research groups [52–54]. The singlet oxygen generation under TPE in the QD–PS systems is much higher than that in the free PSs. These studies underline the potential of QD-combined PSs for TPE PDT.

QDs as Carriers in Chemotherapy

Although chemotherapy is widely used in cancer treatment, it is usually ineffective due to the low cellular uptake and low tumor-targeting efficiency of the chemotherapeutics used. To solve these problems, Zhou et al. prepared 3-mercapitalpropionic acid (MPA)-capped CdTe QDs (MPA-CdTe QDs) to facilitate the interaction of the anticancer agent daunorubicin (DNR) with leukemia cells and kill drug-resistant leukemia K562/A02 cell lines [55]. In another study, Ye et al. synthesized ZnO QDs with polymer shells, onto which Gd^{3+} ions and the anticancer drug doxorubicin (DOX) were adsorbed form a new kind of multifunctional ZnO-Gd-DOX nanoplatform [56]. The as-prepared nanoplatforms are pH-sensitive and can release DOX to cancer cells in vitro and to mouse tumors in vivo, and have better specificity and lower toxicity than free DOX, and even better therapeutic efficacy than an FDA-approved commercial DOX-loading drug DOX-Liposome Injection (DOXIL, NDA#050718). The ZnO-Gd-DOX nanoplatforms exhibit strong red fluorescence, which is beneficial to the fluorescence imaging on live mice. Further, the nanoplatforms possess a high longitudinal relaxivity of 52.5 mM^{-1} s^{-1} at 0.55 T, which is superior to many other Gd^{3+}-based nanoparticles. Thus, both fluorescence imaging and MRI can be applied simultaneously on the tumor-bearing mice along with drug delivery.

QDs as Labels in PTT

In PTT, a photothermal agent can heat and kill abnormal cells or tissues under light irradiation. However, this strategy usually has low selectivity. Thus, an imaging agent is urgently needed to "see" the tumor site in PTT. Due to the excellent photoluminescence property, QDs are commonly used as an imaging agent in PTT. Xia et al. developed multifunctional NPs by incorporating gold nanorods (GNRs) and CdSe/ZnS QDs into silica [41]. Cell imaging experiments reveal that the NPs exhibit strong X-ray attenuation for X-ray CT imaging and strong fluorescence for fluorescence imaging. Nair et al. developed a hybrid nanosystem based on QDs and single-wall carbon nanotubes (SWCNTs) which is found to be useful not only in imaging applications but also in selective cancer cell destruction [57]. Besides, a novel, multifunctional, and low-toxic QD-reduced graphene oxide (rGO) nanocomposite was designed to serve as an imaging agent in the visible light region and a photothermal agent in the NIR region by Hu et al. [58]. Since the photothermal effect of the irradiated rGO can cause not only cell killing but also the degradation of the QDs, the QDs also serve as an optical indicator for the heat dosage and therapeutic progress.

QDs as Labels in Other Treatments

Cell-derived microparticles (MPs) have been recently recognized as critical intercellular information conveyors. However, a further understanding of their biological behaviors and potential applications has been hampered by the limitations of current labeling techniques. Chen et al. proposed a universal donor-cell-assisted membrane biotinylation strategy for labeling MPs by utilizing the natural membrane phospholipid exchange of their donor cells [59]. This innovative strategy conveniently leads to the specific, efficient, reproducible, and biocompatible QD labeling of MPs, thereby reliably conferring the valuable traceability of MPs. By further loading with small interfering RNA (siRNA), QD-labeled MPs that have inherent cell-targeting and biomolecule-conveying ability were successfully employed for combined bioimaging and tumor-targeted therapy. Kim et al. designed and synthesized immunomodulatory hybrid nanoconjugates (HNCs) based on polymer nanocomposites containing QDs (as imaging tracers) conjugated with CpG oligodeoxynucleotides (ODNs) (as a TLR9 ligand) and STAT3 siRNAs (as an immunosuppressive gene silencer) [60]. These HNCs can efficiently target immune cells, induce TLR activation, and silence immunosuppressive genes. Simultaneous delivery of STAT3 siRNAs and CpG ODNs to the tumor microenvironment causes the inhibition of STAT3 along with the activation of dendritic cells (DCs) by CpG ODNs, and their antitumor effects are found to be synergistic. By using NIR-emitting QDs, the migration of DCs to lymph nodes was tracked by real-time NIR fluorescence imaging.

To be able to label a gene and monitor its migration are key important approaches for the clinical application of cancer suicide gene therapy. One of the most promising suicide genes—herpes simplex virus thymidine kinase (HSV-TK) gene was successfully linked with CdTe/CdS core/shell QDs [61]. From confocal microscopy, it is demonstrated that plasmid TK intracellular trafficking can be traced via monitoring the luminescence of the QDs up to 96 h after transfection of QDs-TK conjugates into HeLa cells. 3-(4,5-Dimethylthiazol-2-yl)-2,5-diphenyltetrazolium bromide (MTT) assay results show that the QDs-TK conjugates have a high cytotoxicity after adding ganciclovir (GCV) into HeLa cells, whereas the QDs exert no detectable deleterious effects on the cellular processes. These results indicate that the QDs-based labeling technique is suitable for monitoring TK gene delivery and anticancer activity.

1.3 Limitations and Future Perspectives

QDs as one type of multifunctional materials, have shown promising advantages in tumor imaging and therapy due to their specific physicochemical properties. Using the rich surface functionalization chemistry of QDs, targeting biomolecules and drug formulations can be integrated with QDs for traceable drug delivery and therapy in vitro and in vivo. Many studies have demonstrated that the incorporation of drug formulations with QDs do not compromise the drug efficacy. More importantly, the QD-involving nanostructures are able to serve as an excellent platform for the development of a new generation of traceable drug delivery strategies for real-time monitoring of the drug biodistribution in vitro and in vivo.

Because of toxicity concerns, heavy-metal-based QDs might not be the best candidate for in vivo drug delivery and therapy. Thus, many researchers are currently synthesizing cadmium-free QDs for in vivo applications. However, for in vitro-based drug studies, heavy-metal-based QDs will still be used, since toxicity is not a key concern. Another potential concern for the use of QDs in delivery and therapy is the overall QD size. In general, it is preferable to minimize the overall size of QDs for in vivo applications to reduce their accumulation in the reticuloendothelial system. Further, passivation of the QD surface with a long-lasting and robust polymer coating is essential to prevent the breakdown of QDs in the biological environment that gives rise to their toxicity. Some reports have suggested that capping the QD core with a higher bandgap semiconductor or biomolecule can minimize the toxicity of the QDs. However, it is worth noting that each additional step toward functionalizing the QDs will contribute to their final hydrodynamic size and could directly or indirectly affect their biodistribution. If these issues could be solved, we envision that QDs will become one type of promising material for real-time tumor-targeted imaging and therapy in the future.

2 Silver Chalcogenide Quantum Dots

Silver chalcogenide QDs (SCQDs) composed of silver and a chalcogen (S, Se, or Te) are another type of newly emerged semiconductor QDs. Their narrow and controllable bandgap and extremely low solubility in aqueous solutions endow SCQDs with outstanding photoluminescence properties like NIR emission, adjustable emission peaks, high photostability, and good biocompatibility. These properties make SCQDs competitive candidates in bioimaging, detection, photocatalysis, thermal-electric applications, and QDs-sensitized solar cells. Currently, the synthetic methods and applications of SCQDs have been systemically reviewed by Gui et al. [62] In this part of the chapter, we will focus on utilizing SCQDs in bioimaging and cancer therapy. Subsequently, we are going to introduce SCQDs in three parts: Ag_2S QDs, Ag_2Se QDs, and Ag_2Te QDs. Considering that Ag_2Te QDs are seldom reported in the literature, our discussion will be mainly concentrated on Ag_2S QDs and Ag_2Se QDs.

The applications of traditional QDs (most commonly CdSe@ZnS QDs) in in vivo bioimaging are limited because these QDs usually have fluorescence emissions in the regime of visible spectrum (400–700 nm) in which the autofluorescence of living tissues makes the results ambiguous. Subsequently, NIR-emissive QDs emerge as a new group of luminescent probes benefiting from their deeper tissue penetration and less skin scattering. Compared with other heavy-metal-containing NIR QDs including PbS, PbSe, and CdHgTe QDs, SCQDs may be more biocompatible and more suitable for biomedical applications.

2.1 *Silver Sulfide (Ag_2S) QDs*

Although the synthesis of Ag_2S nanocrystals has been intensively explored in the past 20 years, there were no reports about the photoluminescence of this type of nanomaterials prior to 2010. The discovery of the NIR emission property of SCQDs has attracted great interest from the researchers in the world because it makes the non-invasive imaging in deep tissues possible. In addition, their extremely low aqueous solubility ensures that a minimum amount of silver ions is released to the environment, ensuring their safety in biomedical applications. Besides Ag_2S QDs, there are a number of NIR-II emitting fluorophores, such as single-walled carbon nanotubes (SWCNTs), organic dyes, rare-earth-doped nanoparticles, and conjugated copolymers [63]. However, SWCNTs have relatively low photoluminescence quantum yields (PLQYs) and their safety is under discussion. Organic dyes are prone to be photobleached easily. In contrast, Ag_2S QDs are bright, biocompatible, and resistant to photobleaching. Concerning that the utilization of Ag_2S QDs in bioimaging and cancer therapy has been mainly reported in the recent 5 years, we will review the work from 2012 to 2017.

In 2012, the application of Ag_2S QDs for bioimaging was demonstrated by several research groups [64–67]. For example, Jiang et al. synthesized water-dispersible Ag_2S QDs using 3-mercaptopropionic acid (3-MPA) as capping agent in ethylene glycol medium [64]. The photoluminescence of the as-prepared Ag_2S QDs can be tuned from 510 to 1221 nm and the 910 nm emitting sample with QY = 2.1% was employed for in vivo imaging (Fig. 4a–d). Under either subcutaneous or celiac injection occasion, the Ag_2S QDs display bright fluorescence, which is significantly distinct from the autofluorescence of body. Zhang et al. synthesized NIR-II emitting Ag_2S QDs modified by dihydrolipoic acid (denoted as DHLA-Ag_2S QDs) [65]. They realized in vitro cell imaging with high selectivity by conjugating DHLA-Ag_2S QDs with specific molecules (Erbitux protein for targeting to epidermal growth factor receptor (EGRF), and cyclic arginine-glycine-aspartic acid (RGD) peptide for targeting to $\alpha_v\beta_3$ integrin overexpressed on the surface of several types of cancer cells). In another study, Hong et al. conjugated the DHLA-Ag_2S QDs with six-armed polyethylene glycol (6PEG) to yield 6PEG-Ag_2S QDs (QY = 15.5%) [67]. Due to its high brightness and proper size, 6PEG-Ag_2S QDs were used for fast tumor detection guided by enhanced permeability and retention (EPR) effect (Fig. 5, a–f). The biodistribution and pharmacokinetics results (Fig. 5, g–i) show that >10% ID/g 6PEG-Ag_2S QDs accumulate in the tumor site and 6PEG-Ag_2S QDs can be metabolized by biliary excretion.

In 2013, the Ag_2S QDs were further functionalized and their long-term potential toxicity was systematically evaluated [68, 69]. Wang et al. reported a method of preparing ultrasmall bovine serum albumin (BSA)-stabilized Ag_2S QDs with tunable NIR fluorescence and vascular endothelial growth factor (VEGF) antibody (antiVEGF) was additionally conjugated to enhance the in vivo U-87 MG human glioblastoma tumor-targeting efficiency on mice [68]. However, the QY of the as-prepared BSA-Ag_2S QDs was determined to be only 1.8%. In another example, Zhang et al. provided a detailed study of long-term in vivo biodistribution and examined the potential toxicity of PEGylated Ag_2S QDs [69]. The results showed that the PEGylated Ag_2S QDs are mainly accumulated in liver and spleen after intravenous injection and mostly eliminated by fecal excretion. Negligible toxicity is observed with an intravenous injection dose of as high as 30 mg/kg during the 2-month-long experiments, which is evidenced by the blood biochemical, hematological, and histological analyses. Besides tumor-targeting imaging, Tan et al. also used Ag_2S QDs as NIR fluorescent probes in nitric oxide delivery system [70, 71].

In view of the excellent imaging properties of Ag_2S QDs with ultrahigh spatial (30 μm) and temporal resolution (<50 ms), they were applied in more circumstances in 2014 [72–74]. Li et al. used PEGylated Ag_2S QDs to develop a noninvasive approach of visualizing lymphatic drainage and vascular networks and tracking angiogenesis mediated by a minuscule tumor in vivo [72]. Chen et al. further modified the PEGylated Ag_2S QDs with TAT peptide to form TAT-Ag_2S QDs, which can efficiently label human mesenchymal stem cells (hMSCs) [73]. Applying the TAT-Ag_2S QDs, the dynamic monitoring of the transplanted hMSCs in vivo can be realized with a temporal resolution of 100 ms. Further, Chen et al. conjugated 3-MPA-capped Ag_2S QDs with doxorubicin (DOX, a commercial drug for cancer

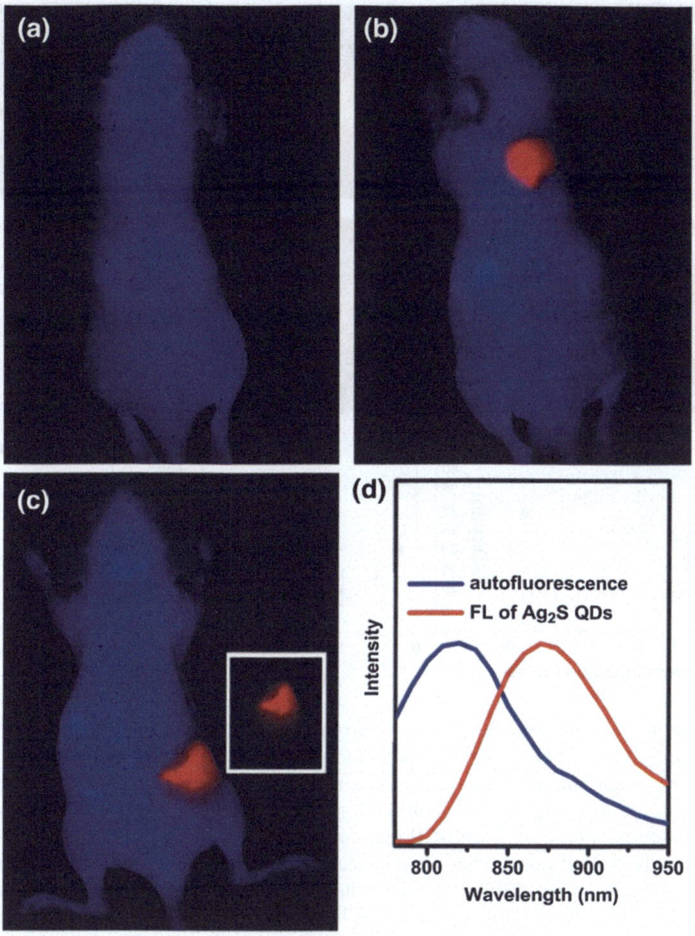

Fig. 4 In vivo NIR fluorescence imaging (pseudocoloured image) of nude mice. Control experiment (**a**), with subcutaneous injection (**b**) and with celiac injection (**c**) of Ag_2S QDs emitting at 910 nm; Unmixed image of Ag_2S QDs fluorescence signal (**c**, inset); The corresponding emission spectra of the autofluorescence and QDs fluorescence of mice with celiac injection (**d**). (In images **a**–**d**, the blue corresponds to the mice autofluorescence and the red corresponds to QDs fluorescence.) Reprinted with permission from Ref. [64]. Copyright 2012 Elsevier

chemotherapy) and cyclic RGD peptide as a multifunctional nanoplatform to achieve NIR imaging-guided tumor-targeting therapy [74]. This work implies that Ag_2S QDs can not only be used as NIR fluorescent probes but also have potential applications in cancer therapy.

Because of their remarkable NIR fluorescence property and proper size for tumor targeting by EPR effect, Ag_2S QDs are increasingly regarded as one of the ideal nanoplatforms for detecting tumors, tracking biological process, and monitoring drug biodistribution in 2015 [75–81]. Other modalities such as CT and MRI were

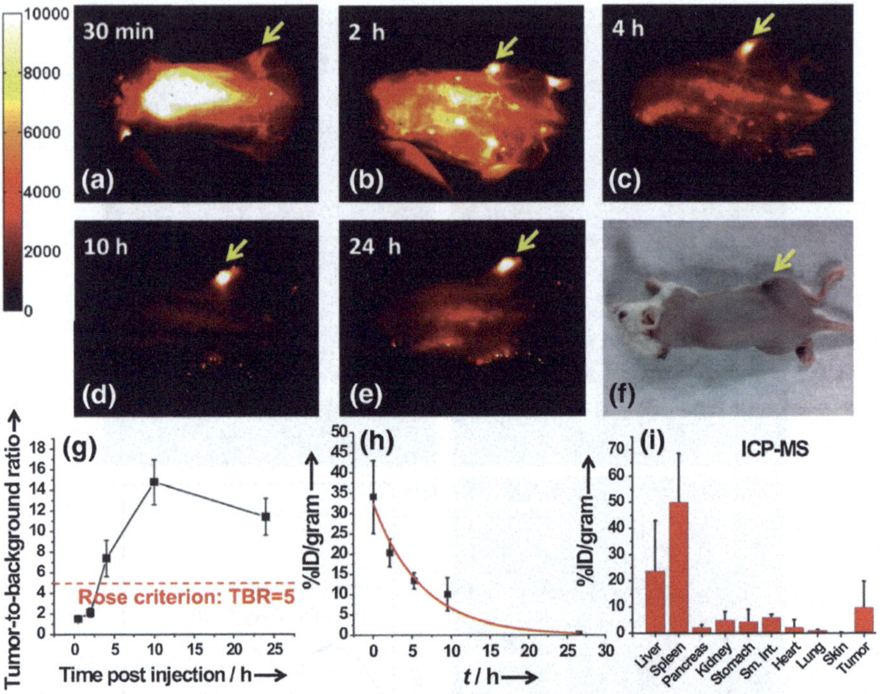

Fig. 5 NIR-II fluorescence imaging of a xenograft 4T1 tumor with high uptake of 6PEG-Ag_2S QDs. **a–e** Time course of NIR-II fluorescence images of the same mouse injected with 6PEG-Ag_2S QDs. **f** A white-light optical image of the same mouse at 24 h postinjection (p.i.). The tumor mass was visibly darkened owing to high uptake of 6PEG-Ag_2S QDs. **g** The tumor-to-background ratio (TBR) plotted as a function of time p.i. for NIR-II images, indicating a 100% certainty of the tumor being identified from the NIR-II image after 4 h p.i., according to the Rose criterion. **h** A representative plot of the %ID/gram of the 6PEG-Ag_2S QDs in the blood versus time after tail vein injection, as determined by the NIR-II fluorescence of the blood samples. A first-order exponential fits the data points with a half-life of circulation for the 6PEG-Ag_2S QDs of 4.37 ± 0.75 h. **i** Quantitative biodistribution of 6PEG-Ag_2S QDs in various organs and the tumor 72 h after injection based on inductively coupled plasma mass spectrometry (ICP-MS). Reprinted with permission from Ref. [67]. Copyright 2012 Wiley

employed together with NIR fluorescence imaging by combining Ag_2S QDs with iodinated oil and Gd^{3+}, respectively [77, 78]. Since 2016, multimodal imaging has become popular for the reason that each individual imaging method has its own disadvantages. Wu et al. reported the synthesis of ICG@PEG-Ag_2S QDs using an FDA-approved NIR dye indocyanine green (ICG) [82]. The as-prepared ICG@PEG-Ag_2S QDs exhibit dual-modal imaging ability (fluorescence and PA imaging) of atherosclerosis. Furthermore, after precisely targeting atherosclerosis, the photothermal effect of ICG can be employed to treat the disease. Zhang et al. prepared Ag_2S@BSA-$DTPA^{Gd}$ via a biomineralization method to realize fluorescence imaging and MRI of tiny tumors [83]. Inspired by the strong absorption of Ag_2S QDs in NIR regime, Gao et al. directly utilized these QDs to achieve cancer PTT [84]. They used aptamer to further functionalize the Ag_2S QDs to increase the tumor-targeting efficiency. The results show that Ag_2S QDs are not only NIR fluorescent probes but also excellent photothermal agents. Subsequently, in 2017, Yang et al. reported the synthesis of size-dependent Ag_2S QDs in hollow human serum albumin (HSA) nanocages for dual-modal imaging (fluorescence and PA imaging) and PTT [85]. The as-prepared Ag_2S QDs show various advantages including photobleaching resistance, preferable endocytosis, effective tumor accumulation, and in vivo body elimination. Also in 2017, Li et al. developed an Ald/DOX@Ag_2S nanoplatform for bone tumor chemotherapy and osteolysis inhibition [86]. In their design, alendronate (Ald) which is conjugated to the surface of Ag_2S QDs plays a role of osteolysis inhibition and pain alleviation. Meanwhile, DOX as the antitumor drug is encapsulated around Ag_2S QDs by hydrophobic interaction. The key component Ag_2S QDs is responsible for prolonging drug circulation time and visualizing the drug delivery process.

2.2 Silver Selenide (Ag_2Se) QDs

Selenium is in the same group with sulfur, so Ag_2Se QDs are akin to Ag_2S QDs in many properties. Ag_2Se QDs were first used in bioimaging in 2012 [87]. As shown in Fig. 6, the Ag_2Se QDs were injected into the abdominal cavity of a nude mouse and the emitting light successfully penetrated the tissues even on the back side. In 2013, C_{18}-PMH-PEG-Ag_2Se QDs with bright photoluminescence centered at 1300 nm, excellent water-dispersity, great colloidal stability and photostability, and good biocompatibility were employed for deep imaging of organs and vascular structures in vivo [88]. Compared with intravenously injected ICG, the C_{18}-PMH-PEG-Ag_2Se QDs show a much clearer image of liver which is buried deep in the body and can visualize the branched blood vessels even as cramped as 123 μm. Although Ag_2Se QDs are thought to be safe, there has been no detailed investigation about their in vivo behavior and toxicity until 2016 [89]. As reported by Tang et al., PEG-Ag_2Se QDs are quickly cleared from the circulation system in mice with a half-life of 0.4 h. Additionally, the PEG-Ag_2Se QDs are mainly amassed in liver and spleen and are converted to Ag and Se in a week. Similar to Ag_2S QDs, Ag_2Se QDs can also be used in multimodal imaging-guided theranostics. Zhao et al. functionalized Ag_2Se QDs with Mn^{2+} to form Ag_2Se@Mn QDs, which integrated the remarkable NIR fluorescence of Ag_2Se QDs and MRI ability of Mn^{2+} [90]. Further, the Ag_2Se@Mn QDs were loaded into circulating MPs (CMPs) freshly purified from the peripheral blood of oral squamous cell carcinoma (OSCC) patients via electroporation, and the thus-obtained Ag_2Se@Mn QD-labeled CMPs were developed as dual-modally traceable and actively tumor-targeted nanoplatform for cancer theranostics [91]. A similar design was reported by Zhu et al., in which cetuximab (a clinical drug for tumor therapy) was conjugated to Ag_2Se QDs to achieve simultaneous imaging and therapy of orthotopic tongue cancer [92].

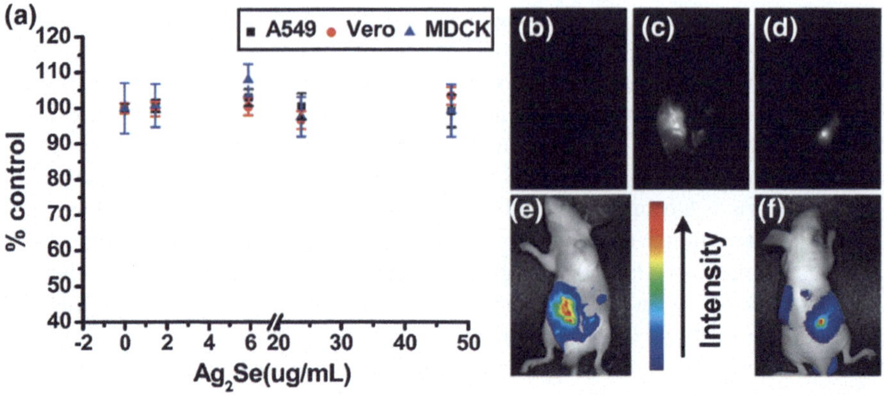

Fig. 6 MTT assay and NIR images of a living nude mouse after injection of Ag_2Se QDs: **a** MTT assay on A549, Vero, and MDCK cells exposed to Ag_2Se QDs at different concentrations from 0 to 47.4 μg/mL for 24 h; **b** fluorescence image of the nude mouse; **c** fluorescence imaging from the abdominal cavity of the nude mouse with Ag_2Se QDs injected into the abdominal cavity; **d** fluorescence imaging on the back side of the nude mouse with Ag_2Se QDs injected into the abdominal cavity; **e**, **f** merged images of the bright-field and the threshold false color of (**c**) and (**d**), respectively. Reprinted with permission from Ref. [87]. Copyright 2012 American Chemical Society

2.3 Silver Telluride (Ag_2Te) QDs

In comparison with Ag_2S QDs and Ag_2Se QDs, Ag_2Te QDs seem to be enigmatic. Because of the high electron mobility and low thermal conductivity of Ag_2Te, it is usually used for thermoelectric applications rather than biomedicine. The photoluminescence property of Ag_2Te QDs was first reported by Yarema et al. [93]. They synthesized 3.2 nm Ag_2Te QDs with the emission peak centered at 1300 nm. Chen et al. prepared luminescent Ag_2Te QDs via a cation exchange method [94]. The as-synthesized Ag_2Te QDs possess photoluminescence ranging from 900 to 1300 nm and the PLQY was determined to be 2.1%. By further growth of ZnS, the PLQY of Ag_2Te/ZnS core/shell QDs can be increased to 5.6%. Neither the Ag_2Te QDs nor the Ag_2Te/ZnS core/shell QDs at concentrations ranging from 50 to 200 nM show cytotoxicity after incubation with HeLa cells for 24 h. Nevertheless, their relatively low PLQYs limit their further imaging applications. In 2015, Yang et al. employed a multivalent polymer poly(maleic anhydride)-graft-cysteamine (PMAC) as both stabilizer and capping agent to fabricate Ag_2Te QDs in aqueous solution [95]. The as-prepared Ag_2Te QDs with adjustable NIR fluorescence peaks ranging from 995 to 1068 nm have relatively high PLQYs of 13.1–15.9%. They also show excellent photostability and aqueous stability as illustrated in Fig. 7b–f. Until now, the bioimaging and cancer therapy applications of Ag_2Te QDs are still lacking. Considering that Ag_2Te has even lower solubility than Ag_2S and Ag_2Se but similar NIR-II photoluminescence properties as compared with Ag_2S and Ag_2Se, we assume that Ag_2Te QDs are probably

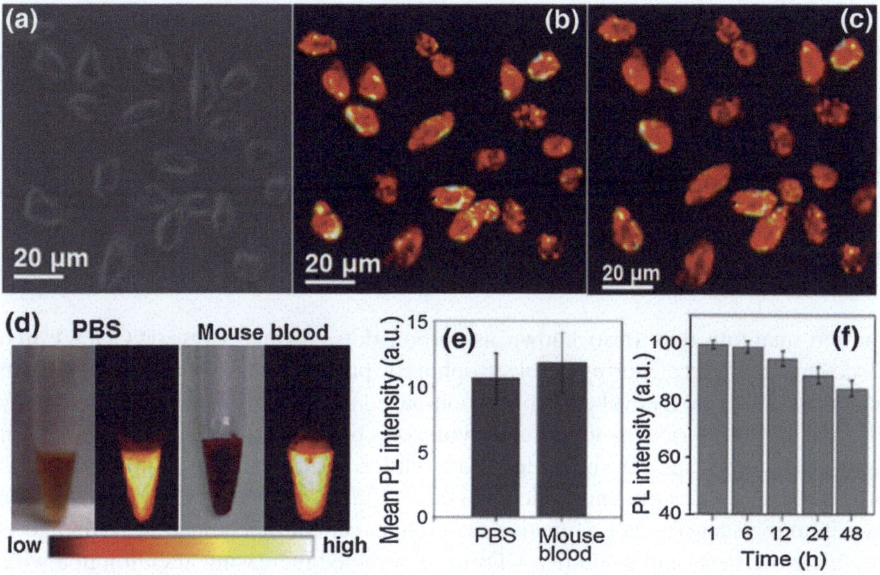

Fig. 7 **a** Bright-field images of L929 cells, **b** NIR-II images acquired immediately upon irradiation (808 nm), and **c** acquired after irradiation for 2 h of L929 cells. **a–c** The cells were incubated with Ag_2Te QDs (0.1 mg/mL) for 1 h. **d** Bright-field and NIR-II images of Ag_2Te QDs suspended in PBS (1 mM, pH = 7.4) or 100-fold diluted mouse whole blood. **e** Mean PL intensities of the two samples determined based on the NIR-II images. **f** PL stability of Ag_2Te QDs in mouse whole blood within 48 h. Reprinted with permission from Ref. [95]. Copyright 2015 Elsevier

an alternative in bioimaging. Furthermore, the intrinsic heavy atom Te may endow the Ag_2Te QDs with satisfying performance as a CT contrast agent.

2.4 *Limitations and Future Perspectives*

SCQDs possess tunable, bright NIR photoluminescence and have been extensively reported as fluorescent probes in vitro and in vivo. Their proper sizes and functional surfaces suggest their promising cancer theranostic applications. Compared with Cd/Pb/Hg-based NIR QDs, there are no concerns on SCQDs about toxicity caused by release of heavy metal ions. In recent years, the synthetic methods of SCQDs have been greatly explored and the various SCQDs have been developed into multimodal imaging probes and drug delivery platforms. However, there are still some problems which hamper the further applications of SCQDs in bioimaging. For example, the small size of SCQDs endows them with large surface area to be functionalized but also means more defects on the surface which decrease the PLQY of SCQDs. For Cd-based QDs, this issue can be solved by wrapping the QDs with ZnS to form a core–shell structure. Nevertheless, the materials that match the lattice of SCQD cores and form a shell on SCQD cores have been seldom reported. That is why the PLQY of Cd-based QDs can achieve even 60%, while it is difficult to find a kind of SCQDs with PLQY above 20%. If this problem can be solved, the PLQY and photostability of SCQDs could be further improved. We believe that SCQDs are a competitive class of promising theranostic agents in the future.

3 Carbon Quantum Dots

Carbon quantum dots (also known as carbon dots, C-dots, CDs, or CQDs), first discovered by Xu et al. during the electrophoretic purification of single-walled carbon nanotubes [96], are a novel class of carbon-based nanomaterials which are typically discrete, quasispherical nanoparticles, with sizes below 10 nm [97, 98]. Due to their fantastic features, such as simple low-cost synthesis and scalability, superior optical properties (e.g., tunable and wide emissions, high photostability, and two-photon excited fluorescence), facile functionalization, excellent biocompatibility, and good chemical inertness and solubility, CDs have attracted increasing attention in a wide range of applications including bioimaging, sensing, catalysis, photoelectric devices, and theranostics [99–104]. Synthetic approaches for CDs are generally classified into two categories—top-down and bottom-up [105]. Herein, we focus on the recent progress of CDs in cancer cell imaging and cancer therapy.

3.1 CDs for Cancer Bioimaging

CDs usually show the excitation wavelength-dependent emission property, which is mainly attributed to the surface state of CDs [106, 107]. A surface defect-based luminescence mechanism has been suggested for CDs [108, 109]. In addition, the use of surface passivating agents to provide uniform photoluminescence trapping sites on the CD surface and the introduction of electron-donating heteroatoms as dopants help to tune the photoluminescence properties of CDs [98]. The emission spectra of CDs are usually broad, ranging from deep ultraviolet to visible, or even extended to NIR. Jiang et al. prepared red, green, and blue emissive CDs using three kinds of phenylenediamine through a solvothermal method [110]. Ding et al. reported the first one-pot syntheses of full-color light-emitting CDs, and then collected the CDs exhibiting excitation-independent luminescence from blue to red via column chromatography following the hydrothermal treatment [111].

Bioimaging is one of the most important applications of CDs because of their unique optical properties and low cytotoxicity. Compared to the conventional semiconductor QDs, which usually contain cadmium or other heavy metals, CDs comprise nontoxic elements, ensuring their good biocompatibility [108, 112]. The first preliminary assessment of the bioimaging potential of CDs was presented by Sun et al., who observed cellular uptake of CDs by Caco-2 cells by confocal microscopy

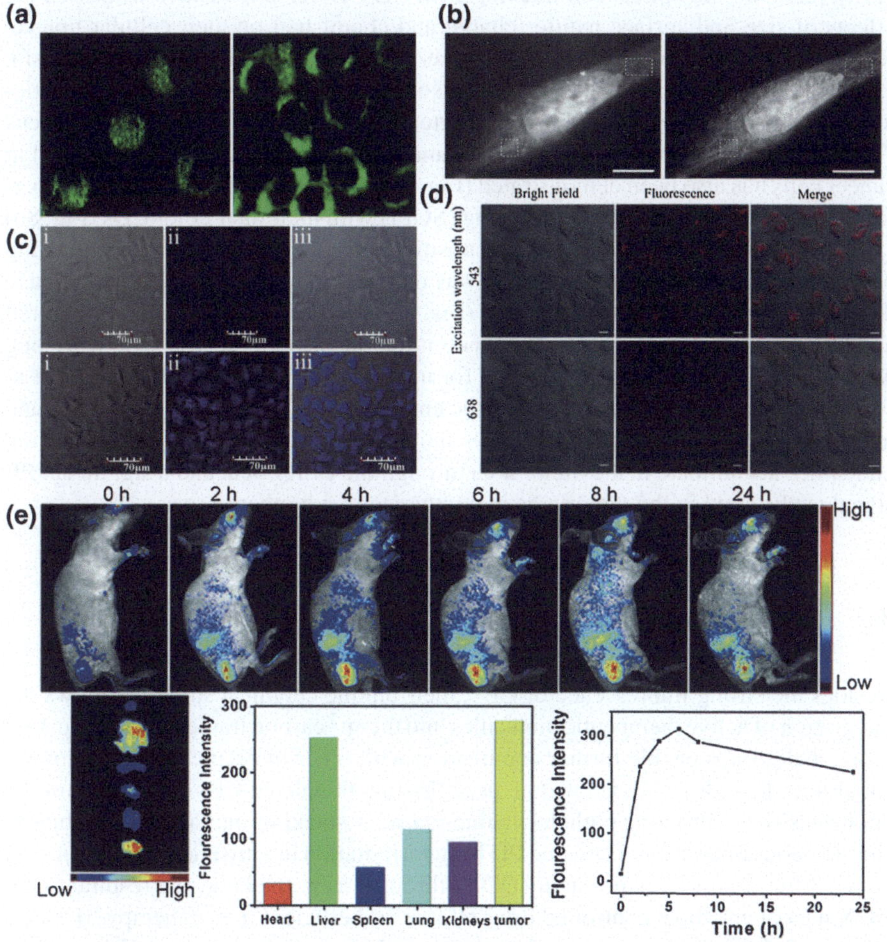

Fig. 8 **a** Representative two-photon luminescence images (800 nm excitation) of MCF-7 cells with internalized CDs. Reprinted with permission from Ref. [119]. Copyright 2007 American Chemical Society. **b** Fluorescence wide-field image and second-order SOFI image of a Saos-2 cell, respectively. Reprinted with permission from Ref. [120]. Copyright 2016 American Chemical Society. **c** Confocal images of 293T cells (top) and HeLa cells (bottom) incubated with CDs for 4 h. (i) bright-field, (ii) confocal fluorescence, and (iii) overlay images. Reprinted with permission from Ref. [121]. Copyright 2016 Elsevier. **d** Confocal images of the CDs in HeLa cells with bright field and excitation at 543 and 638 nm, respectively. Reprinted with permission from Ref. [123]. Copyright 2015 Wiley. **e** In vivo and ex vivo FL imaging. Top: Real-time in vivo FL images after intravenous injection of CDs in nude mice at different time points. Bottom: Ex vivo FL images (left) and average FL intensities (middle) of main organs and tumor tissue at 24 h postinjection, and average FL intensities of the tumor area at different time points postinjection (right). Reprinted with permission from Ref. [123]. Copyright 2015 Wiley

[113], realizing the optical cell tracking with CDs. Diverse CDs have witnessed the effects of size and surface nature (charge and chemistry) on their cellular uptake. Recent studies demonstrated that CDs are mostly internalized in the cytoplasm, especially in endosomes/lysosomes, but also in mitochondria or endoplasmic reticulum [114–116]. Very few CDs were reported to stain the cell membrane or nucleus [117, 118]. The two-photon fluorescence imaging capability of CDs internalized in cancer cells has also been demonstrated (Fig. 8a) [119]. Chizhik et al. realized super-resolution optical fluctuation bioimaging (SOFI) with their dual-color CDs (Fig. 8b) [120]. On the other hand, one kind of positively charged CDs fabricated by using polyethyleneimine and folic acid (FA) was reported to be able to selectively image the folate receptor-positive cancer cells (Fig. 8c) [121]. Furthermore, since deep-red and NIR light exhibits deeper tissue penetration, the development of CDs with long wavelength emissions is highly desired for in vivo imaging [122]. Ge et al. successfully utilized red fluorescent CDs with the emission peak at ~640 nm for in vitro and in vivo imaging (Fig. 8d, e) [123]. CDs intravenously injected into tumor-bearing nude mice accumulate in the tumor area through the EPR effect and a significant FL signal is observed in the tumor area in comparison with other tissues.

3.2 CDs for Cancer Therapy

Besides the strong fluorescence of CDs, their unique chemical structure allows the integration of active therapeutic molecules into the sp^2 carbon frame, and their surface functional groups enable further conjugation with other molecules such as biological affinity ligands [124]. Choi et al. used FA-functionalized CDs as carriers for the photosensitizer (PS) zinc phthalocyanine via π–π stacking interactions, leading to simultaneous imaging and targeted PDT after irradiation in vitro and in vivo (Fig. 9A) [124]. After loading doxorubicin (DOX) through π–π stacking, the resultant CD-DOX nanoagent show controlled drug release and efficient tumor therapy. Besides, Gong et al. fabricated innovative phosphorus and nitrogen dual-doped hollow carbon dots (PNHCDs) with negative surface charge and proved that PNHCDs (as the carriers) have strong electrostatic and hydrogen bonding interactions with the DOX drug molecules [125]. Apart from the physical interactions between CDs and drugs, the covalent bonding is also an important choice for loading drugs on CDs. In this regard, Yang et al. prepared CDs–DOX complexes using DOX-SH and amine-functionalized CDs via a coupling linker, *N*-hydroxysuccinimide ester-poly(ethylene glycol)$_{10}$-maleimide, ensuring a slower and prolonged DOX accumulation in the nucleus, which results in enhanced anti-tumor efficacy and less side effects compared to free DOX (Fig. 9B) [126]. Hua et al. utilized the CDs with intrinsic mitochondrial targeting ability to deliver PSs to mitochondria after covalent conjugation, and achieved significantly enhanced PDT efficacy (Fig. 9C) [127]. Zheng et al. developed a novel theranostic nanomedicine via the condensation reaction between the amine groups on the CD's surface and the carboxyl group of Pt(IV) complex (Oxa(IV)–COOH, a chemotherapeutic prodrug), which may hold great potential for cancer diagnostics

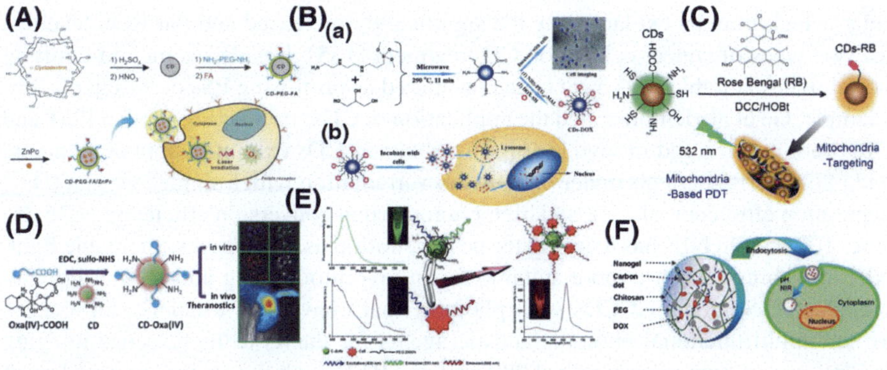

Fig. 9 **A** Schematic illustration of the preparation of carbon nanodots from α-cyclodextrin and targeted PDT with folic acid-functionalized carbon nanodots loaded with zinc phthalocyanine (CD-PEG-FA/ZnPc). Reprinted with permission from Ref. [124]. Copyright 2014 Wiley. **B** Schematic illustration of the synthesis of CDs and their applications in cell imaging and anticancer drug delivery (a). The interaction of CDs–DOX with cancer cells (b). Reprinted with permission from Ref. [126]. Copyright 2017 Royal Society of Chemistry. **C** Carbon quantum dots with intrinsic mitochondrial targeting ability for enhanced mitochondria-based PDT. Reprinted with permission from Ref. [127]. Copyright 2017 Royal Society of Chemistry. **D** Synthetic scheme for CD-Oxa and its applications in bioimaging and theranostics. Reprinted with permission from Ref. [128]. Copyright 2014 Wiley. **E** FRET process between CDs and Ce6. Reprinted with permission from Ref. [129]. Copyright 2012 Wiley. **F** PEG-chitosan@CDs hybrid nanogels for pH/NIR dual-responsive drug release and two-photon fluorescence cellular imaging. Reprinted with permission from Ref. [130]. Copyright 2015 Wiley

and therapy (Fig. 9D) [128]. Huang et al. designed novel multifunctional chlorin e6-conjugated CDs (CDs-Ce6, via amide condensation) as a light-triggerable theranostics agent for simultaneous enhanced-photosensitizer fluorescence detection (PFD) and PDT by FRET mechanism (Fig. 9E) [129]. The CDs as the carriers to load PSs or dyes can improve the stability and solubility of PSs or dyes in aqueous/biological media, extend blood circulation time, and enhance biocompatibility, thus making CDs-Ce6 a good candidate with excellent imaging and tumor-homing ability for FL imaging-guided PDT treatment. Furthermore, many pH-, redox-, enzyme-, or light-responsive CDs-involving complexes have been developed for imaging-guided drug delivery with enhanced therapeutic efficiency [130–133]. For instance, Wang et al. designed the biocompatible PEG-chitosan@CDs hybrid nanogels to realize pH and NIR light dual-responsive drug release and combined chemo-photothermal treatment (Fig. 9F) [130].

Besides being used as drug carriers, CDs can also be used as nanomedicines. In 2011, Christensen et al. found that CDs can be used as PSs to generate ROS under the irradiation of blue light in vitro [134]. However, the shallower tissue penetration of blue light limits the applications of the above CDs in vivo. In 2013, Hsu et al. validated that CDs without laser have a remarkable inhibitory effect on the growth of MCF-7 and MDA-MB-231 cancer cells, with lower toxicity to the MCF-10A normal cells, which may be explained by the significantly increased cellular ROS levels in the two types of cancer cells upon CD treatment [135]. Besides, some red-emitting or red light-excitable CDs have been considered as promising PSs or PT agents. For example, Ge et al. demonstrated the application of CDs for imaging-guided PDT and PTT both in vitro and in vivo [136]. The obtained CDs exhibit dual photodynamic and photothermal effects under 635 nm laser irradiation with a singlet oxygen ($^1O_2$) generation efficiency of 27% and high photothermal conversion efficiency of 36.2% (Fig. 10a). Since NIR has a deeper tissue penetration as compared with visible light, NIR-absorbing CDs are more suitable for in vivo bioimaging and PTT. Lan et al. prepared S, Se-codoped CDs with photothermal conversion efficiency of ~58.2% as new multifunctional phototheranostic agents for the TPE fluorescence imaging (with the excitation wavelength at 880 nm) and PTT (with the excitation wavelength at 635 nm) of cancer cells [137]. Another kind of CDs (named CyCD) showing strong

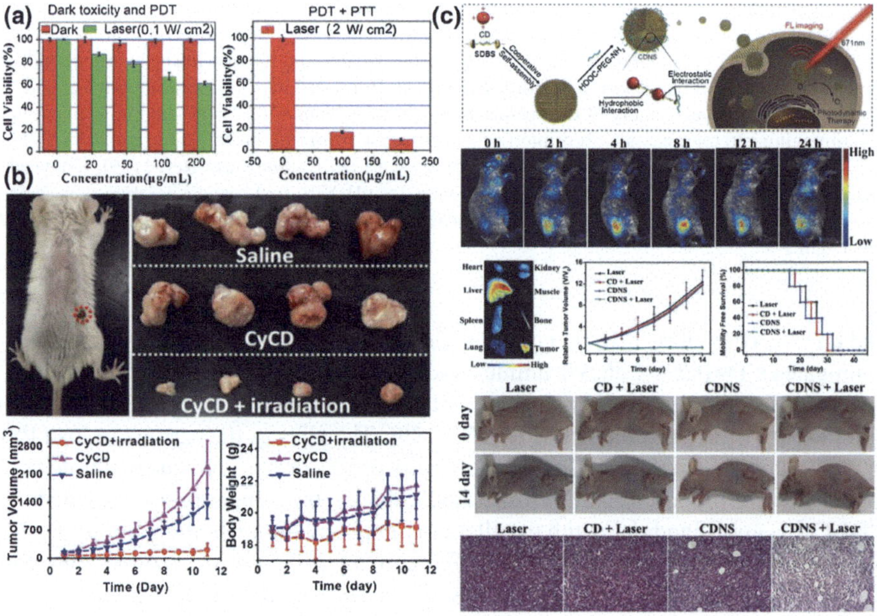

Fig. 10 **a** In vitro PDT/PTT. Relative viability of B16-F0 skin cancer cells incubated with various concentrations of CDs under dark or irradiation by 635 nm laser at power densities of 0.1 W cm^{-2} (left, PDT) or 2 W cm^{-2} (right, PDT/PTT) for 10 min. Reprinted with permission from Ref. [136]. Copyright 2015 Wiley. **b** In vivo PTT. Top: Photographs of a CT26 tumor-bearing mouse after PTT (the tumor is circled with a red dotted line) and the excised tumors on the 11th day. Bottom: The tumor volume and body weight of the CT26 cancer bearing mice after different treatments (saline, CyCD, and CyCD + irradiation). Reprinted with permission from Ref. [138]. Copyright 2016 American Chemical Society. **c** Top: Schematic representations of the fabrication of CD nanosphere (CDNS) from individual CDs and amphipathic SDBS and the application as a nanophotosensitizer for simultaneous NIR-light-responsive FL imaging and PDT of cancer. Bottom: In vivo FL imaging and PDT efficacy of CDNS. Reprinted with permission from Ref. [139]. Copyright 2017 Wiley

absorption and NIR emission within the range from 600 to 900 nm was proved to be an ideal theragnostic agent for NIR fluorescence imaging and photothermal therapy in vitro and in vivo (Fig. 10b) [138]. Jia et al. prepared CD nanospheres via the self-assembly of individual CDs and sodium dodecyl benzene sulfonate (SDBS) to achieve NIR-light-responsive FL imaging and PDT of cancer (Fig. 10c) [139].

4 Silicon Quantum Dots

Silicon or silicon-containing nanomaterials, a series of important nanomaterials with attractive properties including huge surface-to-volume ratios, favorable biocompatibility, improved multifunctionality, and excellent electronic/mechanical properties [140, 141], have been developed for various applications ranging from electronics to biology. Various silicon nanomaterials have been developed, such as silicon nanorods [142], silicon noanowires [143], and silicon nanodots [144]. Among these silicon nanomaterials, silicon nanodots or silicon quantum dots (SiQDs) are especially suitable for bioimaging and cancer therapy due to their ultrasmall size, bright fluorescence, and good biocompatibility.

4.1 Silicon Quantum Dots for Bioimaging Applications

The room-temperature synthesis of water-soluble silicon quantum dots that exhibit strong blue photoluminescence (PL) has been reported by Warner et al. [145]. The ease of synthesis and optical properties make the SiQDs excellent candidates for the imaging of cancer cells, as demonstrated in HeLa cells. For the applications in biological imaging and diagnosis, SiQDs must remain luminescent and be stably dispersed in biological fluids with a wide pH range of pH and a high salt concentration. Erogbogbo et al. reported the preparation of highly stable aqueous suspensions of SiQDs using phospholipid micelles, where the optical properties of SiQDs are well retained [146]. The micelle-encapsulated SiQDs were used as luminescent labels for pancreatic cancer cells in vitro (Fig. 11), thereby highlighting their potential as a nontoxic optical probe for biomedical diagnostics.

To overcome the shortcomings of severe photobleaching and cytotoxicity associated with the traditional dyes and the fluorescent II/VI QDs, He et al. reported a type of silicon-based nanospheres with the merits of excellent water-dispersibility, strong photoluminescence, and robust photostability, which can be used for cellular imaging [147]. To establish their utility as cellular probes, HEK293T human kidney cells were chosen to be labeled by the as-prepared SiQDs. As shown in Fig. 12, the confocal images indicate that the PL of the SiQDs-labeled HEK293T cells is intense and can be directly observed under excitation at different wavelengths. Besides, robust anti-photobleaching of the obtained SiQDs was verified by comparing with the fluorescein isothiocyanate (FITC) dye and fluorescent II/VI QDs.

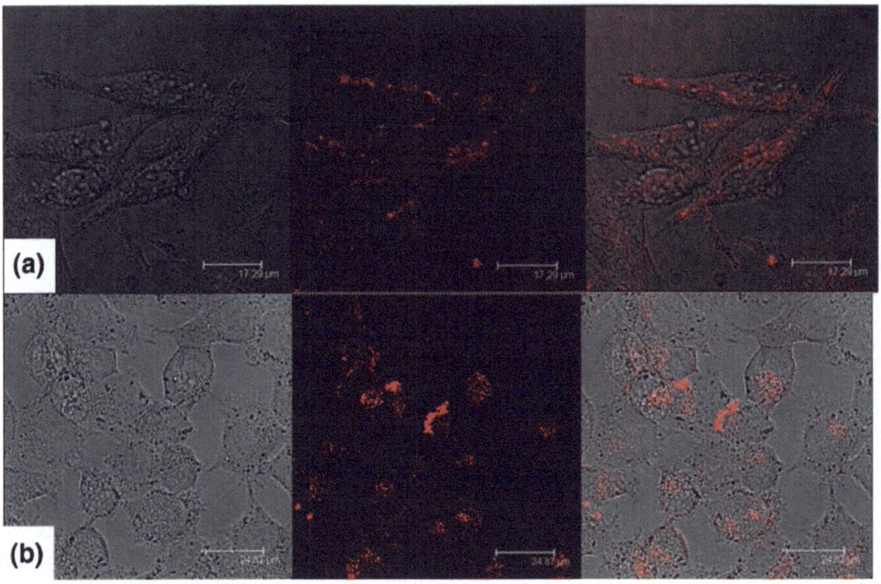

Fig. 11 Confocal microscopic visualization of live pancreatic cancer cells treated with **a** amine-terminated micelle-encapsulated SiQDs, **b** transferrin-conjugated micelle-encapsulated SiQDs. Reprinted with permission from Ref. [146]. Copyright 2008 American Chemical Society

A novel kind of oxidized silicon nanospheres (*O*-SiNSs) prepared via thermal oxidation of the precursor SiNSs was reported by He et al. [148]. The *O*-SiNSs possess the properties of excellent aqueous dispersibility, high PLQY of 25%, wide pH stability, superior photostability, and favorable biocompatibility. Moreover, the *O*-SiNSs are conjugated with antibody (abbreviated as *O*-SiNSs/antibody bioconjugates), which are successfully applied in immunofluorescence cell imaging. Moreover, simultaneous detection of two biological targets and removal of background autofluorescence can also be realized by antibody-conjugated silicon quantum dot nanoparticles and organic dyes, which was reported by Tu et al. [149]. To realize selective cancer imaging using SiQDs, Erogbogbo et al. reported the development of SiQDs modified with folate and antimesothelin [150]. The successful targeted cancer cell imaging will be highly beneficial for targeted cancer diagnosis and therapy.

4.2 Silicon Quantum Dots for Drug Delivery and Cancer Therapy

Owing to the wide and growing applications of fluorescence in biomedicine and bioengineering fields, fluorescence imaging-guided cancer therapy has attracted great interest. Moreover, due to the low cytotoxicity of silicon materials, SiQDs are good

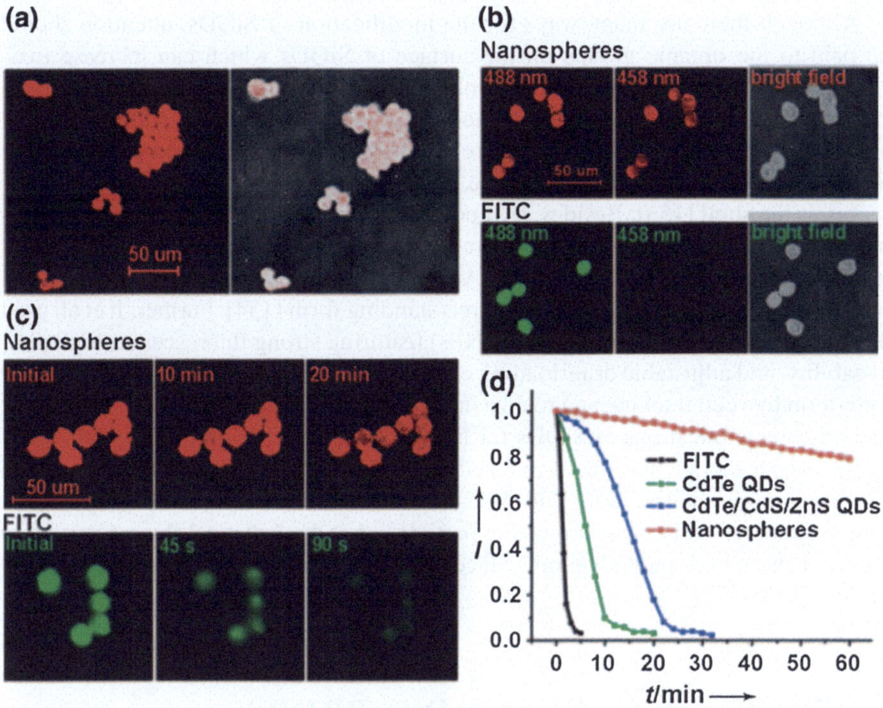

Fig. 12 **a** Fluorescence microscopy images of HEK2-293T cells labeled with the as-prepared SiQDs. **b** Comparison of fluorescence signals of HEK293T cells imaging with the nanospheres (top) and FITC (bottom). **c** Temporal evolution of fluorescence of the HEK293T cells labeled with the as-prepared nanospheres (top) and FITC (bottom). **d** Photostability comparison of fluorescent II/VI QDs and the obtained SiQDs. Reprinted with permission from Ref. [147]. Copyright 2009 Wiley

carriers for delivering drugs into cells. SiQDs for delivering siRNA into tumor cells have been realized by Klein et al [151]. The internalization of SiQDs was found to occur via endocytosis, which was observed by transmission electron microscopy (TEM) and confocal microscopy. Moreover, the SiQD-siRNA complexes can significantly reduce the transporter efficiencies for the P-glycoprotein substrate Rhodamine [123].

Although there are many ways for the modification of SiQDs, attention should be paid to the organic groups on the surface of SiQDs which can increase toxicity. Ruizendaal et al. reported that amine-functionalized SiQDs exhibit cytotoxicity, whereas carboxylic acid-terminated analogues do not [152]. Multifunctional nanocarriers with a core–shell structure for drug delivery have been developed by Xu et al., in which SiQDs serve as the core and a water-soluble block copolymer serves as the shell [153]. Besides, as reported by Wang et al., poly(ethylene glycol)-block-polylactide (PEG-PLA) NPs were also used to encapsulate fluorescent SiQDs to deliver the anticancer drug quercetin, which suppress human hepatoma HepG2 cell proliferation more effectively than the free-standing form [154]. Further, Ji et al. used the ultrasmall silicon nanoparticles (SiNPs) featuring strong fluorescence, high photostability, and adjustable drug-loading capacity to load the anticancer drug DOX for long-term live cell tracking and realize in vivo cancer treatment (Fig. 13) [155]. Very recently, the applications of SiQDs for fluorescence imaging-guided high photodynamic cancer therapy have been realized by Liu et al., who selected phthalocyanine (Pc) as a representative drug and prepared SiQD-based composite nanoparticles. The as-prepared Si/Pc nanocomposite particles emit dual channel fluorescence signals and show high photodynamic cancer therapeutic efficiencies both in vitro and in vivo [156].

5 Black Phosphorus Quantum Dots (BP QDs)

As a new member of two-dimensional (2D) layered materials, black phosphorus (BP) has attracted considerable research interest and exhibits many potential applications in various areas such as nanoelectronics, optoelectronics, bioimaging, and phototherapy [157–160]. In the BP family, zero-dimensional BP quantum dots (BP QDs, the ultrasmall BP nanosheets) have been successfully synthesized through chemical methods recently. In 2015, BP QDs were synthesized for the first time by Zhang's group through a facile top-down method [161]. Besides, the approaches of solvothermal synthesis and ultrasonication were reported by Xu et al. and Gao et al. to prepare BP QDs, respectively [162, 163]. Compared with traditional semiconductor QDs, BP QDs show superior biocompatibility due to their in vivo stability and their final nontoxic degradation products (including phosphate and phosphonate) [164–166]. Consequently, BP QDs may be suitable for various biomedical applications, such as bioimaging, drug delivery, and cancer therapy (especially photothermal therapy). Sun et al. reported a new type of BP QDs (synthesized by a simple liquid exfoliation technique), which possesses excellent NIR photothermal properties (with a large extinction coefficient of 14.8 Lg^{-1} cm^{-1} at 808 nm and a photothermal conversion efficiency of 28.4%), good biocompatibility, as well as high PTT efficiency, suggesting great potential of the BP QDs in PTT applications [160]. Besides, Shao et al. synthesized biodegradable BP-based nanospheres using BP QDs and poly(lactic-*co*-glycolic acid) (PLGA), which have improved stability, good biocompatibility, excellent tumor targeting

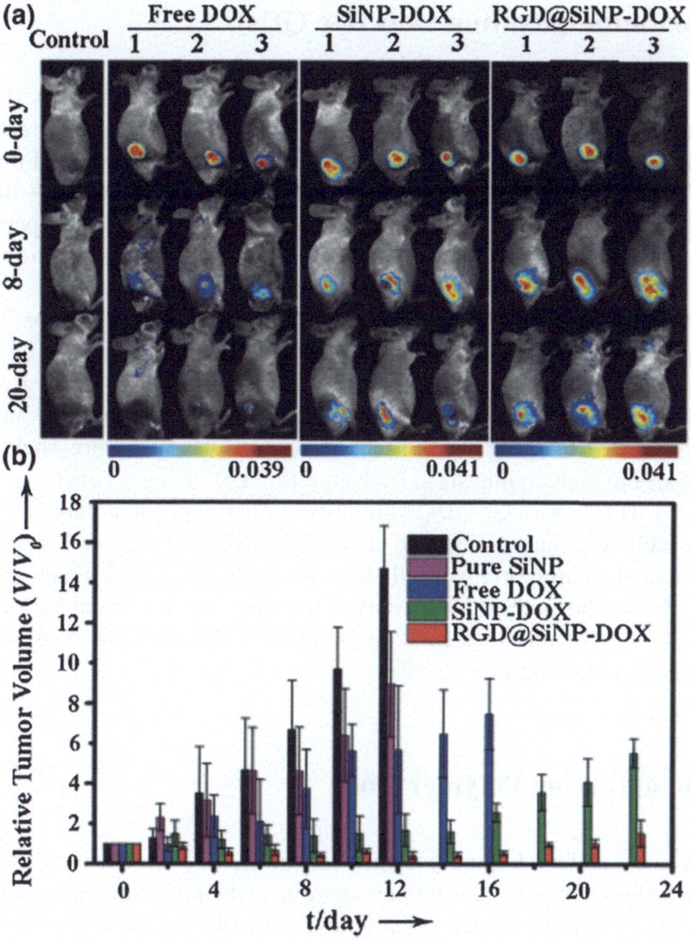

Fig. 13 **a** In vivo fluorescence images of mice at different time points after intratumoral injection of physiological saline (control), free DOX, SiNP-DOX, and RGD@SiNP-DOX. **b** Inhibition of tumor growth at different times points after intratumoral injection of physiological saline (control), pure SiNP, free DOX, SiNP-DOX, and RGD@SiNP-DOX. Reprinted with permission from Ref. [155]. Copyright 2015 Wiley

ability, and high PTT efficiency [167]. For cancer imaging application, BP QDs with excellent fluorescence properties were also obtained by ultrasonication-assisted solution method, and were successfully utilized in HeLa cell imaging [164].

6 Germanium Quantum Dots (Ge QDs)

Group IV materials (C, Si, and Ge), especially the Ge quantum dots (Ge QDs), which were synthesized for the first time in 1982 [168, 169], have attracted much attention and hold great potential for the applications in the biomedical field [169]. Compared with II–VI (e.g., CdX, X=S, Se, Te), III–V (e.g., GaAs, InP, InAs), and IV–VI (e.g., PbX, X=S, Se) QDs, Ge QDs possess various superior properties, such as valuable semiconducting and special optical properties, ultralow cytotoxicity, and electrochemical stability [169–171].

Based on their excellent fluorescence properties and good biocompatibility, Ge QDs may have excellent performance in the applications of cancer imaging. Li et al. developed blue-emitting pH-sensitive Ge QDs via a facile and green aqueous solution-based route and employed the Ge QDs to monitor the lysosome pH via cancer fluorescence imaging [172]. Meanwhile, Ge QDs with core–shell structure and size of ~3 nm were synthesized by Karatutlu et al. using a bench-top colloidal method [173]. This type of Ge QDs exhibits excitation-dependent fluorescence property with excellent photostability and superior biocompatibility and these superior properties render them a promising fluorescent probe for cancer imaging. In summary, considering the undesirable toxicity of traditional heavy-metal-based QDs and poor photostability of various organic dyes, the Ge QDs may hold great promise for cancer imaging and therapy.

7 Semiconducting Polymer Dots

In recent years, semiconducting polymer nanoparticles have emerged as attractive fluorescent probes in biomedical applications due to their outstanding optical characteristics [174]. Generally, these nanoparticles refer to nanomaterials consisting of π-conjugated polymers which in their pristine state are wide-band-gap semiconductors. These semiconducting polymers possess a direct band gap, resulting in an efficient absorption or emission at the band edge. According to the semiconductor band theory, an electron is excited from the highest occupied energy band (the π band) to the lowest unoccupied energy band (the π* band), which leaves a hole in the π band. The recombination of the excited electron with the hole generates a fluorescent photon. By tuning the π–π* band gaps, different semiconducting polymers can emit fluorescence of various wavelengths [174]. Therefore, semiconducting polymer-based fluorescent nanomaterials possess tunable electrical and optical properties.

Polymer dots (Pdots), a small subset of semiconducting polymer nanoparticles, possess a particle size (typically smaller than 20–30 nm) comparable to that of conventional QDs. To distinguish Pdots from conventional dye-loaded latex spheres or nanoparticles which possess only a small fraction of semiconducting polymers, Pdots are required to contain semiconducting polymers with a volume or weight fraction higher than 50%, preferably 80–90% [174]. Besides, Pdots should also contain a hydrophobic polymer interior which is essential to their colloidal stability, packing density of fluorophores, and fluorescence brightness. To date, a number of researches have demonstrated that Pdots exhibit extraordinary fluorescence brightness, fast emission rate, excellent photostability, and nonblinking feature [175–178]. These advantageous optical properties make them well-suited for applications in light-emitting devices [179]. In addition, the past decade has also witnessed the rapid development of Pdots in a wide range of biological applications, including cellular imaging [177, 178, 180–183], biological detection (e.g., pH [184], temperature [185], oxygen [186], and blood glucose level [187]), and high-resolution single-particle tracking [188]. Particularly, Pdots have attracted considerable interest in cancer diagnosis and therapeutics, due to their proper size, extraordinary brightness, and minimal cytotoxicity.

7.1 Cancer Cell-Specific Imaging and in Vivo Tumor Imaging

One of the most important applications of Pdots is the fluorescence imaging for cancer therapy. Benefiting from their flexible polymer matrix, Pdots can be functionalized via surface modification to realize cancer cell-specific fluorescence imaging [177, 189, 190]. For example, Zhang et al. prepared streptavidin (SA)-conjugated Pdots and successfully labeled the cancer cell-surface marker HER2 in human breast cancer cells through the specific recognition between biotin and streptavidin (Fig. 14) [189]. Besides, Geng et al. embedded Pdots into silica nanoparticles for targeted cellular imaging of HER2-overexpressed SK-BR-3 breast cancer cells [190]. Apart from cellular imaging, Pdots were also utilized for in vivo tumor targeting [191, 192]. Wu et al. developed a type of highly emissive Pdots (approximately 15 times brighter than the commercial QDs) modified with a tumor-specific peptide ligand and demonstrated their specific targeting to malignant brain tumors [192]. In view of the advantage that NIR light has a large penetration depth in tissue and causes minimal cellular autofluorescence, Pdots with NIR emission are highly desirable for bioimaging. To solve this challenge, Wu et al. synthesized squaraine-based Pdots with large Stokes shifts and narrow-band emissions in the NIR region and employed the Pdots to label EpCAM receptors on the surface of human breast cancer MCF-7 cells [193].

7.2 Pdots-Based Cancer Therapy

Apart from bioimaging applications, Pdots are also widely used in designing nanomedicines for cancer treatment, especially PDT. Recently, nanoparticle-based photosensitizing agents have established their great potential in PDT as a result of

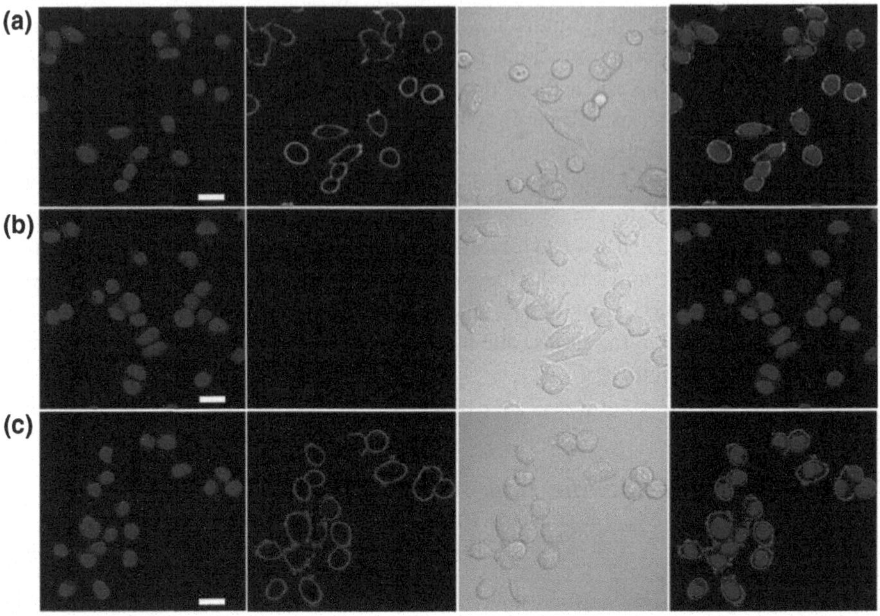

Fig. 14 Fluorescence images of SK-BR-3 breast cancer cells treated with Pdot–SA probes. **a** Positive labeling of cancer cells using PFBT-C2-SA probe. **b** Negative labeling of cancer cells in the absence of EDC in the bioconjugation step. **c** Positive labeling of cancer cells using red-emitting PFTBT/PFBT-C2-SA probe. Reprinted with permission from Ref. [189]. Copyright 2012 American Chemical Society

the enhanced selectivity toward cancerous tissues and improved therapeutic efficacy [194, 195]. However, many PDT nanoagents are limited by low absorptivity and less efficient energy transfer. Owing to the light-harvesting and energy transfer properties of semiconducting polymers, Pdots provide new opportunities to overcome these limitations. One strategy is to entrap photosensitizers into Pdots where the polymer can efficiently absorb and transfer energy to the photosensitizer, thereby enabling enhanced $^1O_2$ generation [196–199]. For example, Li et al. developed an energy transfer-mediated Pdot platform doped with a molecular photosensitizer tetraphenylporphyrin (TPP) for in vitro and in vivo PDT studies. In this system, the highly fluorescent Pdots are completely quenched, realizing an energy transfer efficiency of nearly 100% and $^1O_2$ generation quantum yield of ~50% [197]. Similarly, Zhang and co-workers reported folic acid and horseradish peroxidase (HRP)-bifunctionalized Pdots incorporated with meta-tetra(hydroxyphenyl)-chlorin (m-THPC) for targeted PDT and cancer cell imaging [198]. On the other hand, considering the relatively large triplet energy values of most semiconducting polymers (typically exceed 0.98 eV), the energy from the excited state of Pdots can be directly delivered to the ground state of molecular oxygen, resulting in the generation of $^1O_2$. Therefore, Pdots can serve as not only imaging agents but also photosensitizers for enhanced PDT. Shi et al.

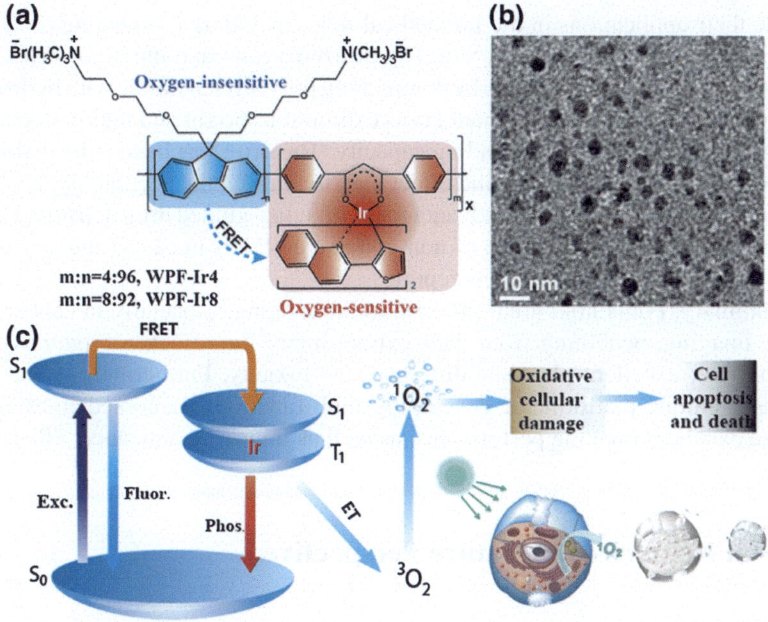

Fig. 15 **a** Chemical structures of phosphorescent conjugated Pdots with the Ir(III) complexes. **b** High-resolution transmission electron microscopy (HR-TEM) image of WPF-Ir4 in aqueous solution. **c** Schematic illustration of the oxygen sensing and PDT. Reprinted with permission from Ref. [200]. Copyright 2014 Wiley

have successfully synthesized a series of ultrasmall phosphorescent Pdots composed of phosphorescent Ir(III) complexes and fluorescent fluorene units [200]. As shown in Fig. 15, the Pdots can transfer the energy from excited state upon irradiation to molecular oxygen for ratiometric oxygen sensing as well as photodynamic cancer therapy.

7.3 Non-conjugated Pdots and Their Biomedical Applications

Recently, non-conjugated Pdots which are different from traditional conjugated Pdots have emerged as a new type of fluorescent materials. Non-conjugated Pdots contain no typical fluorophore groups but only sub-fluorophores (such as C=O, C=N, and N=O), and thereby are not supposed to possess strong photoluminescence. Nonetheless, the photoluminescence of these sub-fluorophores can be significantly enhanced via polymerization and crosslinking, hydrothermal treatment, self-assembly, or physical immobilization, which is called crosslink-enhanced emission (CEE) effect [201]. Such CEE effect endows non-conjugated Pdots with outstanding photoluminescence property. Besides, these Pdots also possess good stability and low toxicity, which promote their applications in the biomedical field, including bioimaging [202] and drug delivery [203]. For example, Sun et al. utilized a general route to construct multifunctional non-conjugated Pdots by conjugating polyethyleneimine with hydrophobic polylactide [203]. The as-formed Pdots exhibit ultrabright and multicolorful fluorescence with excellent drug-loading capacity. The drug (paclitaxel)-loaded Pdots show not only improved therapeutic effect but also substantial accumulation around tumor, demonstrating their great advantage for imaging-guided drug delivery. Unfortunately, the potential application of non-conjugated Pdots in cancer therapy is still largely unexplored despite of a few reports.

In summary, Pdots hold great promise in bioimaging, especially in cancer cell-specific imaging, benefiting from their extraordinary fluorescence brightness, fast emission rate, excellent photostability, and low toxicity. Furthermore, Pdots also serve as a unique platform for developing multifunctional nanomedicines which integrate excellent imaging performance as well as improved anticancer effect.

8 Final Remarks and Future Perspective

In this chapter, we have presented a comprehensive overview of recent advancements in the area of photoluminescent QDs for cancer imaging and therapy. These quantum dots can be mainly classified into metal-containing or metal-free groups. The most common QDs, the semiconducting, metal-containing QDs, have been extensively investigated due to their superior optical properties. To date, several new types of QDs, especially the metal-free QDs (such as carbon, silicon, phosphorus, germanium, and organic molecule-based QDs) have attracted increasing interest from researchers in the field of cancer theranostics.

There are several issues that must be considered if these QDs are to be used for cancer imaging and therapy. First, the QDs should have suitable fluorescence emission property and good aqueous dispersibility and stability. Second, these QDs should have good biocompatibility, if they are used as fluorescent probes or drug carriers. Third, they can interact adequately with cells or can reach tumor tissues via passive targeting or active targeting. All these three issues are related with the structures and properties of the QDs. To design and synthesize QDs with suitable core materials, sizes, and surface chemistries are still the main challenges in the field of cancer theranostics. The development of highly effective, low-cost, and eco-friendly synthetic approaches and the elaborate control of the chemical, optical, and biological properties of these QDs are highly desired for future applications of QDs in the biomedical field. On the other hand, since a deep understanding of the interactions between QDs and cancer cells or tumor microenvironments will largely affect the imaging and therapeutic outcomes, developing new imaging techniques or analytical methods to monitor and decipher the details of the interaction processes will promote the clinical translation of these QDs.

References

1. Michalet X, Pinaud FF, Bentolila LA, Tsay JM, Doose S, Li JJ, Sundaresan G, Wu AM, Gambhir SS, Weiss S (2005) Quantum dots for live cells, in vivo imaging, and diagnostics. Science 307:538–544
2. Vu TQ, Lam WY, Hatch EW, Lidke DS (2015) Quantum dots for quantitative imaging: from single molecules to tissue. Cell Tissue Res 360:71–86
3. Zhou J, Yang Y, Zhang CY (2015) Toward biocompatible semiconductor quantum dots: from biosynthesis and bioconjugation to biomedical application. Chem Rev 115:11669–11717
4. Roy M, Niu CJ, Chen YH, McVeigh PZ, Shuhendler AJ, Leung MK, Mariampillai A, DaCosta RS, Wilson BC (2012) Estimation of minimum doses for optimized quantum dot contrast-enhanced vascular imaging in vivo. Small 8:1780–1792
5. Zhou RH, Li M, Wang SL, Wu P, Wu L, Hou XD (2014) Low-toxic Mn-doped ZnSe/ZnS quantum dots conjugated with nano-hydroxyapatite for cell imaging. Nanoscale 6:14319–14325
6. Wei W, He XW, Ma N (2014) DNA-templated assembly of a heterobivalent quantum dot nanoprobe for extra- and intracellular dual-targeting and imaging of live cancer cells. Angew Chem In Ed 53:5573–5577
7. Li ZS, Xu W, Wang YT, Shah BR, Zhang CL, Chen YJ, Li Y, Li B (2015) Quantum dots loaded nanogels for low cytotoxicity, pH-sensitive fluorescence, cell imaging and drug delivery. Carbohydr Polym 121:477–485
8. Lee J, Kang HJ, Jang H, Lee YJ, Lee YS, Ali BA, Al-Khedhairy AA, Kim S (2015) Simultaneous imaging of two different cancer biomarkers using aptamer-conjugated quantum dots. Sensors 15:8595–8604
9. Sasaki A, Tsukasaki Y, Komatsuzaki A, Sakata T, Yasuda H, Jin T (2015) Recombinant protein (EGFP-protein G)-coated PbS quantum dots for in vitro and in vivo dual fluorescence (visible and second-NIR) imaging of breast tumor. Nanoscale 7:5115–5119
10. Sureshkumar S, Jothimani B, Sridhar TM, Venkatachalapathy B (2016) Synthesis and characterization of gadolinium doped ZnSe quantum dots for fluorescence imaging of cancer cells. RSC Adv 6:16081–16086
11. Kong YF, Chen J, Fang HW, Heath G, Wo Y, Wang WL, Li YX, Guo Y, Evans SD, Chen SY, Zhou DJ (2016) Highly fluorescent ribonuclease-A-encapsulated lead sulfide quantum dots for ultrasensitive fluorescence in vivo imaging in the second near-infrared window. Chem Mater 28:3041–3050
12. Liu J, Lau SK, Varma VA, Moffitt RA, Caldwell M, Liu T, Young AN, Petros JA, Osunkoya AO, Krogstad T, Leyland-Jones B, Wang MD, Nie SM (2010) Molecular mapping of tumor heterogeneity on clinical tissue specimens with multiplexed quantum dots. ACS Nano 4:2755–2765
13. Zhang CL, Ji XH, Zhang Y, Zhou GH, Ke XL, Wang HZ, Tinnefeld P, He ZK (2013) One-pot synthesized aptamer-functionalized CdTe: Zn^{2+} quantum dots for tumor-targeted fluorescence imaging in vitro and in vivo. Anal Chem 85:5843–5849
14. Zhang Y, Liu JM, Yan XP (2013) Self-assembly of folate onto polyethyleneimine-coated CdS/ZnS quantum dots for targeted turn-on fluorescence imaging of folate receptor overexpressed cancer cells. Anal Chem 85:228–234
15. Wang HN, Sun HF, Wei H, Xi P, Nie SM, Ren QS (2014) Biocompatible hyaluronic acid polymer-coated quantum dots for CD44(+) cancer cell-targeted imaging. J Nanopart Res 16:2621
16. Mohammad F, Bwatanglang IB, Yusof NA, Abdullah J, Hussein MZ, Alitheen NB, Abu N (2016) Folic acid targeted Mn: ZnS quantum dots for theranostic applications of cancer cell imaging and therapy. Int J Nanomed 11:413–428
17. Li Z, He XW, Luo X, Wang L, Ma N (2016) DNA-programmed quantum dot polymerization for ultrasensitive molecular imaging of cancer cells. Anal Chem 88:9355–9358
18. Wang SS, Wen YR, Wang YJ, Ma YY, Liu Z (2017) Pattern recognition of cells via multiplexed imaging with monosaccharide-imprinted quantum dots. Anal Chem 89:5646–5652

19. Wichner SM, Mann VR, Powers AS, Segal MA, Mir M, Bandaria JN, DeWitt MA, Darzacq X, Yildiz A, Cohen BE (2017) Covalent protein labeling and improved single-molecule optical properties of aqueous CdSe/CdS quantum dots. ACS Nano 11:6773–6781
20. Huang N, Cheng S, Zhang X, Tian Q, Pi JL, Tang J, Huang Q, Wang F, Chen J, Xie ZY, Xu ZY, Chen WF, Zheng HZ, Cheng Y (2017) Efficacy of NGR peptide-modified PEGylated quantum dots for crossing the blood–brain barrier and targeted fluorescence imaging of glioma and tumor vasculature. Nanomedicine 13:83–93
21. Au GHT, Mejias L, Swami VK, Brooks AD, Shih WY, Shih WH (2014) Quantitative assessment of Tn antigen in breast tissue micro-arrays using CdSe aqueous quantum dots. Biomaterials 35:2971–2980
22. Karamanos Y, Pottiez G (2016) Proteomics and the blood-brain barrier: how recent findings help drug development. Expert Rev Proteomics 13:251–258
23. Foda MF, Huang L, Shao F, Han HY (2014) Biocompatible and highly luminescent near-infrared $CuInS_2/ZnS$ quantum dots embedded silica beads for cancer cell imaging. ACS Appl Mater Interfaces 6:2011–2017
24. He L, Li L, Wang W, Abdel-Halim ES, Zhang J, Zhu JJ (2016) Highly luminescent and biocompatible near-infrared core-shell CdSeTe/CdS/C quantum dots for probe labeling tumor cells. Talanta 146:209–215
25. Miyashita M, Gonda K, Tada H, Watanabe M, Kitamura N, Kamei T, Sasano H, Ishida T, Ohuchi N (2016) Quantitative diagnosis of HER2 protein expressing breast cancer by single-particle quantum dot imaging. Cancer Med 5:2813–2824
26. Liu XJ, Zhou PJ, Zhan HJ, Liu HY, Zhang JW, Zhao YN (2017) Synthesis and characterization of near-infrared-emitting CdHgTe/CdS/ZnS quantum dots capped by *N*-acetyl-*L*-cysteine for in vitro and in vivo imaging. RSC Adv 7:29998–30007
27. Benayas A, Ren FQ, Carrasco E, Marzal V, del Rosal B, Gonfa BA, Juarranz A, Sanz-Rodríguez F, Jaque D, García-Solé J, Verone F, Ma DL (2015) PbS/CdS/ZnS quantum dots: a multifunctional platform for in vivo near-infrared low-dose fluorescence imaging. Adv Funct Mater 25:6650–6659
28. Ren FQ, del Rosal B, An SY, Yang F, Carrasco E, Benayas A, Oh JK, Jaque D, de la Fuente ÁJ, Verone F, Ma DL (2017) Development and investigation of ultrastable PbS/CdS/ZnS quantum dots for near-infrared tumor imaging. Part Part Syst Charact 34:1600242
29. Bruns O, Bischof T, Harris D, Franke D, Shi Y, Riedemann L, Bartelt A, Jaworski F, Carr J, Rowlands C, Wilson M, Chen O, Wei H, Hwang G, Montana D, Coropceanu I, Achorn O, Kloepper J, Heeren J, So P, Fukumura D, Jensen K, Jain R, Bawendi M (2017) Next-generation in vivo optical imaging with short-wave infrared quantum dots. Nature Biomed Eng 1:0056
30. Ding K, Jing LH, Liu CY, Hou Y, Gao MY (2014) Magnetically engineered Cd-free quantum dots as dual-modality probes for fluorescence/magnetic resonance imaging of tumors. Biomaterials 35:1608–1617
31. Wu Q, Chen L, Huang L, Wang J, Liu JW, Hu C, Han HY (2015) Quantum dots decorated gold nanorod as fluorescent-plasmonic dual-modal contrasts agent for cancer imaging. Biosens Bioelectron 74:16–23
32. Guo WS, Sun XL, Jacobson O, Yan XF, Min K, Srivatsan A, Niu G, Kiesewetter DO, Chang J, Chang XY (2015) Intrinsically radioactive [^{64}Cu] CuInS/ZnS quantum dots for PET and optical imaging: improved radiochemical stability and controllable cerenkov luminescence. ACS Nano 9:488–495
33. Wang JP, Tan XX, Pang XJ, Liu L, Tan FP, Li N (2016) MoS_2 quantum dot@polyaniline inorganic-organic nanohybrids for in vivo dual-modal imaging guided synergistic photothermal/radiation therapy. ACS Appl Mater Interfaces 8:24331–24338
34. Yang YB, Lin L, Jing LJ, Yue XL, Dai ZF (2017) $CuInS_2/ZnS$ quantum dots conjugating Gd(III) chelates for near infrared fluorescence and magnetic resonance bimodal imaging. ACS Appl Mater Interfaces 9:23450–23457
35. Lin BB, Yao XZ, Zhu YH, Shen JH, Yang XL, Li CZ (2014) Multifunctional gadolinium-labeled silica-coated core/shell quantum dots for magnetic resonance and fluorescence imaging of cancer cells. RSC Adv 4:20641–20648

36. Lai PY, Huang CC, Chou TH, Ou KL, Chang JY (2017) Aqueous synthesis of Ag and Mn co-doped In_2S_3/ZnS quantum dots with tunable emission for dual-modal targeted imaging. Acta Biomater 50:522–533
37. Jing LH, Ding K, Kershaw SV, Kempson IM, Rogach AL, Gao MY (2014) Magnetically engineered semiconductor quantum dots as multimodal imaging probes. Adv Mater 26:6367–6386
38. Sitbon G, Bouccara S, Tasso M, Francois A, Bezdetnaya L, Marchal F, Beaumont M, Pons T (2014) Multimodal Mn-doped I–III–VI quantum dots for near infrared fluorescence and magnetic resonance imaging: from synthesis to in vivo application. Nanoscale 6:9264–9272
39. Pu KY, Shuhendler AJ, Jokerst JV, Mei JG, Gambhir SS, Bao ZN, Rao JH (2014) Semiconducting polymer nanoparticles as photoacoustic molecular imaging probes in living mice. Nat Nanotechnol 9:233–239
40. Lv GX, Guo WS, Zhang W, Zhang TB, Li SY, Chen SZ, Eltahan AS, Wang DL, Wang YQ, Zhang JC, Wang PC, Chang J, Liang XJ (2016) Near-infrared emission CuInS/ZnS quantum dots: all-in-one theranostic nanomedicines with intrinsic fluorescence/photoacoustic imaging for tumor phototherapy. ACS Nano 10:9637–9645
41. Xia HX, Yang XQ, Song JT, Chen J, Zhang MZ, Yan DM, Zhang L, Qin MY, Bai LY, Zhao YD, Ma ZY (2014) Folic acid conjugated silica-coated gold nanorods and quantum dots for dual-modality CT and fluorescence imaging and photothermal therapy. J Mater Chem B 2:1945–1953
42. Zhang XD, Xia LY, Chen XK, Chen Z, Wu FG (2017) Hydrogel-based phototherapy for fighting cancer and bacterial infection. Sci China Mater 60:487–503
43. Dong HF, Tang SS, Hao YS, Yu HZ, Dai WH, Zhao GF, Cao Y, Lu HT, Zhang XJ, Ju HX (2016) Fluorescent MoS_2 quantum dots: ultrasonic preparation, up-conversion and down-conversion bioimaging, and photodynamic therapy. ACS Appl Mater Interfaces 8:3107–3114
44. He SJ, Cao J, Li YS, Yang JC, Zhou M, Qu CY, Zhang Y, Shen F, Chen Y, Li MM, Xu LM (2016) CdSe/ZnS quantum dots induce photodynamic effects and cytotoxicity in pancreatic cancer cells. World J Gastroenterol 22:5012–5022
45. Sun JJ, Guo YM, Zhu L, Yang L, Shi WK, Wang K, Zhang H (2017) Photodynamic therapy of human hepatoma using semiconductor quantum dots as sole photosensitizer. Part Part Syst Charact 34:1600413
46. Yong Y, Cheng XJ, Bao T, Zu M, Yan L, Yin WY, Ge CC, Wang DL, Gu ZJ, Zhao YL (2015) Tungsten sulfide quantum dots as multifunctional nanotheranostics for in vivo dual-modal image-guided photothermal/radiotherapy synergistic therapy. ACS Nano 9:12451–12463
47. Chu MQ, Pan XJ, Zhang D, Wu Q, Peng JL, Hai WX (2012) The therapeutic efficacy of CdTe and CdSe quantum dots for photothermal cancer therapy. Biomaterials 33:7071–7083
48. Ding DD, Guo W, Guo CS, Sun JZ, Zheng NN, Wang F, Yan M, Liu SQ (2017) MoO_{3-x} quantum dots for photoacoustic imaging guided photothermal/photodynamic cancer treatment. Nanoscale 9:2020–2029
49. Hsu CY, Chen CW, Yu HP, Lin YF, Lai PS (2013) Bioluminescence resonance energy transfer using luciferase-immobilized quantum dots for self-illuminated photodynamic therapy. Biomaterials 34:1204–1212
50. Tsay JM, Trzoss M, Shi LX, Kong XX, Selke M, Jung ME, Weiss S (2007) Singlet oxygen production by peptide-coated quantum dot-photosensitizer conjugates. J Am Chem Soc 129:6865–6871
51. Martynenko IV, KuznetsovaVA, Orlova AO, Kanaev PA, Maslov VG, Loudon A, Zaharov V, Parfenov P, Gun'ko YK, Baranov AV, Fedorov AV (2015) Chlorin e6–ZnSe/ZnS quantum dots based system as reagent for photodynamic therapy. Nanotechnology 26:055102
52. Qi ZD, Li DW, Jiang P, Jiang FL, Li YS, Liu Y, Wong WK, Cheah KW (2011) Biocompatible CdSe quantum dot-based photosensitizer under two-photon excitation for photodynamic therapy. J Mater Chem 21:2455–2458
53. Fowley C, Nomikou N, McHale AP, McCarron PA, McCaughan B, Callan JF (2012) Water soluble quantum dots as hydrophilic carriers and two-photon excited energy donors in photodynamic therapy. J Mater Chem 22:6456–6462

54. Chou KL, Won N, Kwag J, Kim S, Chen JY (2013) Femto-second laser beam with a low power density achieved a two-photon photodynamic cancer therapy with quantum dots. J Mater Chem B 1:4584–4592
55. Zhou YY, Shi LX, Li QN, Jiang H, Lv G, Zhao J, Wu CH, Selke M, Wang XM (2010) Imaging and inhibition of multi-drug resistance in cancer cells via specific association with negatively charged CdTe quantum dots. Biomaterials 31:4958–4963
56. Ye DX, Ma YY, Zhao W, Cao HM, Kong JL, Xiong HM, Möhwald H (2016) ZnO-based nanoplatforms for labeling and treatment of mouse tumors without detectable toxic side effects. ACS Nano 10:4294–4300
57. Nair LV, Nagaoka Y, Maekawa T, Sakthikumar D, Jayasre RS (2014) Quantum dot tailored to single wall carbon nanotubes: a multifunctional hybrid nanoconstruct for cellular imaging and targeted photothermal therapy. Small 10:2771–2775
58. Hu SH, Chen YW, Hung WT, Chen IW, Chen SY (2012) Quantum-dot-tagged reduced graphene oxide nanocomposites for bright fluorescence bioimaging and photothermal therapy monitored in situ. Adv Mater 24:1748–1754
59. Chen G, Zhu JY, Zhang ZL, Zhang W, Ren JG, Wu M, Hong ZY, Lv C, Pang DW, Zhao YF (2015) Transformation of cell-derived microparticles into quantum-dot labeled nanovectors for antitumor siRNA delivery. Angew Chem Int Ed 54:1036–1040
60. Kim JH, Noh YW, Heo MB, Cho MY, Lim YT (2012) Multifunctional hybrid nanoconjugates for efficient in vivo delivery of immunomodulating oligonucleotides and enhanced antitumor immunity. Angew Chem Int Ed 51:9670–9673
61. Shao D, Zeng QH, Fan Z, Li J, Zhang M, Zhang YL, Li O, Chen L, Kong XG, Zhang H (2012) Monitoring HSV-TK/ganciclovir cancer suicide gene therapy using CdTe/CdS core/shell quantum dots. Biomaterials 33:4336–4344
62. Gui RJ, Jin H, Wang ZH, Tan LJ (2015) Recent advances in synthetic methods and applications of colloidal silver chalcogenide quantum dots. Coordin Chem Rev 296:91–124
63. Hong GS, Diao S, Antaris AL, Dai HJ (2015) Carbon nanomaterials for biological imaging and nanomedicinal therapy. Chem Rev 115:10816–10906
64. Jiang P, Zhu CN, Zhang ZL, Tian ZQ, Pang DW (2012) Water-soluble Ag_2S quantum dots for near-infrared fluorescence imaging in vivo. Biomaterials 33:5130–5135
65. Zhang Y, Hong GS, Zhang YJ, Chen GC, Li F, Dai HJ, Wang QB (2012) Ag_2S quantum dot: a bright and biocompatible fluorescent nanoprobe in the second near-infrared window. ACS Nano 6:3695–3702
66. Wang CX, Wang Y, Xu L, Zhang D, Liu MX, Li XW, Sun HC, Lin Q, Yang B (2012) Facile aqueous-phase synthesis of biocompatible and fluorescent Ag_2S nanoclusters for bioimaging: tunable photoluminescence from red to near infrared. Small 8:3137–3142
67. Hong GS, Robinson JT, Zhang YJ, Diao S, Antaris AL, Wang QB, Dai HJ (2012) In vivo fluorescence imaging with Ag_2S quantum dots in the second near-infrared region. Angew Chem Int Ed 51:9818–9821
68. Wang Y, Yan XP (2013) Fabrication of vascular endothelial growth factor antibody bioconjugated ultrasmall near-infrared fluorescent Ag_2S quantum dots for targeted cancer imaging in vivo. Chem Commun 49:3324–3326
69. Zhang Y, Zhang YJ, Hong GS, He W, Zhou K, Yang K, Li F, Chen GC, Liu Z, Dai HJ, Wang QB (2013) Biodistribution, pharmacokinetics and toxicology of Ag_2S near-infrared quantum dots in mice. Biomaterials 34:3639–3646
70. Tan LJ, Wan AJ, Li HL (2013) Conjugating S-nitrosothiols with glutathiose stabilized silver sulfide quantum dots for controlled nitric oxide release and near-infrared fluorescence imaging. ACS Appl Mater Interfaces 5:11163–11171
71. Tan LJ, Wan AJ, Li HL (2013) Ag_2S quantum dots conjugated chitosan nanospheres toward light-triggered nitric oxide release and near-infrared fluorescence imaging. Langmuir 29:15032–15042
72. Li CY, Zhang YJ, Wang M, Zhang Y, Chen GC, Li L, Wu DM, Wang QB (2014) In vivo real-time visualization of tissue blood flow and angiogenesis using Ag_2S quantum dots in the NIR-II window. Biomaterials 35:393–400

73. Chen GC, Tian F, Zhang Y, Zhang YJ, Li CY, Wang QB (2014) Tracking of transplanted human mesenchymal stem cells in living mice using near-infrared Ag_2S quantum dots. Adv Funct Mater 24:2481–2488
74. Chen HY, Li BW, Zhang M, Sun K, Wang YR, Peng KR, Ao MD, Guo YR, Gu YQ (2014) Characterization of tumor-targeting Ag_2S quantum dots for cancer imaging and therapy in vivo. Nanoscale 6:12580–12590
75. Hu F, Li CY, Zhang YJ, Wang M, Wu DM, Wang QB (2015) Real-time in vivo visualization of tumor therapy by a near-infrared-II Ag_2S quantum dot-based theranostic nanoplatform. Nano Res. 8:1637–1647
76. Tang R, Xue JP, Xu BG, Shen DW, Sudlow GP, Achilefu S (2015) Tunable ultrasmall visible-to-extended near-infrared emitting silver sulfide quantum dots for integrin-targeted cancer imaging. ACS Nano 9:220–230
77. Qin MY, Yang XQ, Wang K, Zhang XS, Song JT, Yao MH, Yan DM, Liu B, Zhao YD (2015) In vivo cancer targeting and fluorescence-CT dual-mode imaging with nanoprobes based on silver sulfide quantum dots and iodinated oil. Nanoscale 7:19484–19492
78. Li CY, Cao LM, Zhang YJ, Yi PW, Wang M, Tan B, Deng ZW, Wu DM, Wang QB (2015) Pre-operative detection and intraoperative visualization of brain tumors for more precise surgery: a new dual-modality MRI and NIR nanoprobe. Small 11:4517–4525
79. Chen GC, Tian F, Li CY, Zhang YJ, Weng Z, Zhang Y, Peng R, Wang QB (2015) In vivo real-time visualization of mesenchymal stem cells tropism for cutaneous regeneration using NIR-II fluorescence imaging. Biomaterials 53:265–273
80. Li CY, Li F, Zhang YJ, Zhang WJ, Zhang XE, Wang QB (2015) Real-time monitoring surface chemistry-dependent in vivo behaviors of protein nanocages via encapsulating an NIR-II Ag_2S quantum dot. ACS Nano 9:12255–12263
81. Huang S, Peng S, Li Y, Cui J, Chen H, Wang L (2015) Development of NIR-II fluorescence image-guided and pH-responsive nanocapsules for cocktail drug delivery. Nano Res 8:1932–1943
82. Wu CX, Zhang YJ, Li Z, Li CY, Wang QB (2016) A novel photoacoustic nanoprobe of ICG@PEG-Ag_2S for atherosclerosis targeting and imaging in vivo. Nanoscale 8:12531–12539
83. Zhang J, Hao GY, Yao CF, Yu JN, Wang J, Yang WT, Hu CH, Zhang BB (2016) Albumin-mediated biomineralization of paramagnetic NIR Ag_2S QDs for tiny tumor bimodal targeted imaging in vivo. ACS Appl Mater Interfaces 8:16612–16621
84. Gao JW, Wu CL, Deng D, Wu P, Cai CX (2016) Direct synthesis of water-soluble aptamer-Ag_2S quantum dots at ambient temperature for specific imaging and photothermal therapy of cancer. Adv Healthcare Mater 5:2437–2449
85. Yang T, Tang YA, Liu L, Lv XY, Wang QL, Ke HT, Deng YB, Yang H, Yang XL, Liu G, Zhao YL, Chen HB (2017) Size-dependent Ag_2S nanodots for second near-infrared fluorescence/photoacoustics imaging and simultaneous photothermal therapy. ACS Nano 11:1848–1857
86. Li CY, Zhang YJ, Chen GC, Hu F, Zhao K, Wang QB (2017) Engineered multifunctional nanomedicine for simultaneous stereotactic chemotherapy and inhibited osteolysis in an orthotopic model of bone metastasis. Adv Mater 29:1605754
87. Gu YP, Cui R, Zhang ZL, Xie ZX, Pang DW (2012) Ultrasmall near-infrared Ag_2Se quantum dots with tunable fluorescence for in vivo imaging. J Am Chem Soc 134:79–82
88. Dong BH, Li CY, Chen GC, Zhang YJ, Zhang Y, Deng MJ, Wang QB (2013) Facile synthesis of highly photoluminescent Ag_2Se quantum dots as a new fluorescent probe in the second near-infrared window for in vivo imaging. Chem Mater 25:2503–2509
89. Tang H, Yang ST, Yang YF, Ke DM, Liu JH, Chen X, Wang H, Liu Y (2016) Blood clearance, distribution, transformation, excretion, and toxicity of near-infrared quantum dots Ag_2Se in mice. ACS Appl Mater Interfaces 8:17859–17869
90. Zhao JY, Chen G, Gu YP, Cui R, Zhang ZL, Yu ZL, Tang B, Zhao YF, Pang DW (2016) Ultrasmall magnetically engineered Ag_2Se quantum dots for instant efficient labeling and whole-body high-resolution multimodal real-time tracking of cell-derived microvesicles. J Am Chem Soc 138:1893–1903

91. Yu ZL, Zhang W, Zhao JY, Zhong WQ, Ren JG, Wu M, Zhang ZL, Pang DW, Zhao YF, Chen G (2017) Development of a dual-modally traceable nanoplatform for cancer theranostics using natural circulating cell-derived microparticles in oral cancer patients. Adv Funct Mater 27:1703482
92. Zhu CN, Chen G, Tian ZQ, Wang W, Zhong WQ, Li Z, Zhang ZL, Pang DW (2017) Near-infrared fluorescent Ag_2Se-cetuximab nanoprobes for targeted imaging and therapy of cancer. Small 13:1602309
93. Yarema M, Pichler S, Sytnyk M, Seyrkammer R, Lechner RT, Fritz-Popovski G, Jarzab D, Szendrei K, Resel R, Korovyanko O, Loi MA, Paris O, Hesser G, Heiss W (2011) Infrared emitting and photoconducting colloidal silver chalcogenide nanocrystal quantum dots from a silylamide-promoted synthesis. ACS Nano 5:3758–3765
94. Chen C, He XW, Gao L, Ma N (2013) Cation exchange-based facile aqueous synthesis of small, stable, and nontoxic near-infrared Ag_2Te/ZnS core/shell quantum dots emitting in the second biological window. ACS Appl Mater Interfaces 5:1149–1155
95. Yang M, Gui RJ, Jin H, Wang ZH, Zhang FF, Xia JF, Bi S, Xia YZ (2015) Ag_2Te quantum dots with compact surface coatings of multivalent polymers: ambient one-pot aqueous synthesis and the second near-infrared bioimaging. Colloids Surf B: Biointerfaces 126:115–120
96. Xu XY, Ray R, Gu YL, Ploehn HJ, Gearheart L, Raker K, Scrivens WA (2004) Electrophoretic analysis and purification of fluorescent single-walled carbon nanotube fragments. J Am Chem Soc 126:12736–12737
97. Baker SN, Baker GA (2010) Luminescent carbon nanodots: emergent nanolights. Angew Chem Int Ed 49:6726–6744
98. Hill S, Galan MC (2017) Fluorescent carbon dots from mono- and polysaccharides: synthesis, properties and applications. Beilstein J Org Chem 13:675–693
99. Hua XW, Bao YW, Wang HY, Chen Z, Wu FG (2017) Bacteria-derived fluorescent carbon dots for microbial live/dead differentiation. Nanoscale 9:2150–2161
100. Yang JJ, Zhang XD, Ma YH, Gao G, Chen XK, Jia HR, Li YH, Chen Z, Wu FG (2016) Carbon dot-based platform for simultaneous bacterial distinguishment and antibacterial applications. ACS Appl Mater Interfaces 8:32170–32181
101. Zheng XT, Ananthanarayanan A, Luo KQ, Chen P (2015) Glowing graphene quantum dots and carbon dots: properties, syntheses, and biological applications. Small 11:1620–1636
102. Lim SY, Shen W, Gao ZQ (2015) Carbon quantum dots and their applications. Chem Soc Rev 44:362–381
103. Du Y, Guo SJ (2016) Chemically doped fluorescent carbon and graphene quantum dots for bioimaging, sensor, catalytic and photoelectronic applications. Nanoscale 8:2532–2543
104. Wang R, Lu KQ, Tang ZR, Xu YJ (2017) Recent progress in carbon quantum dots: synthesis, properties and applications in photocatalysis. J Mater Chem A 5:3717–3734
105. Zuo PL, Lu XH, Sun ZG, Guo YH, He H (2016) A review on syntheses, properties, characterization and bioanalytical applications of fluorescent carbon dots. Microchim Acta 183:519–542
106. Wang JL, Qiu JJ (2016) A review of carbon dots in biological applications. J Mater Sci 51:4728–4738
107. Namdari P, Negahdari B, Eatemadi A (2017) Synthesis, properties and biomedical applications of carbon-based quantum dots: an updated review. Biomed Pharmacother 87:209–222
108. Jaleel JA, Pramod K (2018) Artful and multifaceted applications of carbon dot in biomedicine. J Control Release 269:302–321
109. Yuan FL, Li SH, Fan ZT, Meng XY, Fan LZ, Yang SH (2016) Shining carbon dots: synthesis and biomedical and optoelectronic applications. Nano Today 11:565–586
110. Jiang K, Sun S, Zhang L, Lu Y, Wu AG, Cai CZ, Lin HW (2015) Red, green, and blue luminescence by carbon dots: full-color emission tuning and multicolor cellular imaging. Angew Chem Int Ed 54:5360–5363
111. Ding H, Yu SB, Wei JS, Xiong HM (2016) Full-color light-emitting carbon dots with a surface-state-controlled luminescence mechanism. ACS Nano 10:484–491
112. Shi HT, Wei JF, Qiang L, Chen X, Meng XW (2014) Fluorescent carbon dots for bioimaging and biosensing applications. J Biomed Nanotechnol 10:2677–2699

113. Sun YP, Zhou B, Lin Y, Wang W, Fernando KAS, Pathak P, Meziani MJ, Harruff BA, Wang X, Wang HF, Luo PG, Yang H, Kose ME, Chen BL, Veca LM, Xie SY (2006) Quantum-Sized carbon dots for bright and colorful photoluminescence. J Am Chem Soc 128:7756–7757
114. Hola K, Zhang Y, Wang Y, Giannelis EP, Zboril R, Rogach AL (2014) Carbon dots—emerging light emitters for bioimaging, cancer therapy and optoelectronics. Nano Today 9:590–603
115. Luo PG, Yang F, Yang ST, Sonkar SK, Yang LJ, Broglie JJ, Liu Y, Sun YP (2014) Carbon-based quantum dots for fluorescence imaging of cells and tissues. RSC Adv 4:10791–10807
116. Gao G, Jiang YW, Jia HR, Yang JJ, Wu FG (2018) On-off-on fluorescent nanosensor for Fe^{3+} detection and cancer/normal cell differentiation via silicon-doped carbon quantum dots. Carbon 134:232–243
117. Zhang Z, Hao JH, Zhang J, Zhang BL, Tang JL (2012) Protein as the source for synthesizing fluorescent carbon dots by a one-pot hydrothermal route. RSC Adv 2:8599–8601
118. Hua XW, Bao YW, Wu FG (2018) Fluorescent carbon quantum dots with intrinsic nucleolus-targeting capability for nucleolus imaging and enhanced cytosolic and nuclear drug delivery. ACS Appl Mater Interfaces 10:10664–10677
119. Cao L, Wang X, Meziani MJ, Lu FS, Wang HF, Luo PG, Lin Y, Harruff BA, Veca LM, Murray D, Xie SY, Sun YP (2007) Carbon dots for multiphoton bioimaging. J Am Chem Soc 129:11318–11319
120. Chizhik AM, Stein S, Dekaliuk MO, Battle C, Li WX, Huss A, Platen M, Schaap IAT, Gregor I, Demchenko AP, Schmidt CF, Enderlein J, Chizhik AI (2016) Super-resolution optical fluctuation bio-imaging with dual-color carbon nanodots. Nano Lett 16:237–242
121. Yang XD, Wang Y, Shen XR, Su CY, Yang JH, Piao MJ, Jia F, Gao GH, Zhang L, Lin Q (2017) One-step synthesis of photoluminescent carbon dots with excitation-independent emission for selective bioimaging and gene delivery. J Colloid Interface Sci 492:1–7
122. Lu SY, Shi LZ, Liu JJ, Zhu SJ, Chen AM, Jin MX, Yang B (2017) Near-infrared photoluminescent polymer-carbon nanodots with two-photon fluorescence. Adv Mater 29:1603443
123. Ge JC, Jia QY, Liu WM, Guo L, Liu QY, Lan MH, Zhang HY, Meng XM, Wang PF (2015) Red-emissive carbon dots for fluorescent, photoacoustic, and thermal theranostics in living mice. Adv Mater 27:4169–4177
124. Choi Y, Kim S, Choi MH, Ryoo SR, Park J, Min DH, Kim BS (2014) Highly biocompatible carbon nanodots for simultaneous bioimaging and targeted photodynamic therapy in vitro and in vivo. Adv Funct Mater 24:5781–5789
125. Gong XJ, Zhang QY, Gao YF, Shuang SM, Choi MMF, Dong C (2016) Phosphorus and nitrogen dual-doped hollow carbon dot as a nanocarrier for doxorubicin delivery and biological imaging. ACS Appl Mater Interfaces 8:11288–11297
126. Yang JJ, Gao G, Zhang XD, Ma YH, Jia HR, Jiang YW, Wang ZF, Wu FG (2017) Ultra-small and photostable nanotheranostic agents based on carbon quantum dots passivated with polyamine-containing organosilane molecules. Nanoscale 9:15441–15452
127. Hua XW, Bao YW, Chen Z, Wu FG (2017) Carbon quantum dots with intrinsic mitochondrial targeting ability for mitochondria-based theranostics. Nanoscale 9:10948–10960
128. Zheng M, Liu S, Li J, Qu D, Zhao HF, Guan XG, Hu XL, Xie ZG, Jing XB, Sun ZC (2014) Integrating oxaliplatin with highly luminescent carbon dots: an unprecedented theranostic agent for personalized medicine. Adv Mater 26:3554–3560
129. Huang P, Lin J, Wang XS, Wang Z, Zhang CL, He M, Wang K, Chen F, Li ZM, Shen GX, Cui DX, Chen XY (2012) Light-triggered theranostics based on photosensitizer-conjugated carbon dots for simultaneous enhanced-fluorescence imaging and photodynamic therapy. Adv Mater 24:5104–5110
130. Wang H, Di J, Sun YB, Fu JP, Wei ZY, Matsui H, Alonso A. del C, Zhou SQ (2015) Biocompatible PEG-Chitosan@Carbon dots hybrid nanogels for two-photon fluorescence imaging, near-infrared light/pH dual-responsive drug carrier, and synergistic therapy. Adv Funct Mater 25:5537–5547
131. Feng T, Ai XZ, An GH, Yang PP, Zhao YL (2016) Charge-convertible carbon dots for imaging-guided drug delivery with enhanced in vivo cancer therapeutic efficiency. ACS Nano 10:4410–4420

132. Feng T, Ai XZ, Ong HM, Zhao YL (2016) Dual-responsive carbon dots for tumor extracellular microenvironment triggered targeting and enhanced anticancer drug delivery. ACS Appl Mater Interfaces 8:18732–18740
133. Zhao QF, Wang SY, Yang Y, Li X, Di DH, Zhang CG, Jiang TY, Wang SL (2017) Hyaluronic acid and carbon dots-gated hollow mesoporous silica for redox and enzyme-triggered targeted drug delivery and bioimaging. Mater Sci Eng, C 78:475–484
134. Christensen IL, Sun YP, Juzenas P (2011) Carbon dots as antioxidants and prooxidants. J Biomed Nanotechnol 7:667–676
135. Hsu PC, Chen PC, Ou CM, Chang HY, Chang HT (2013) Extremely high inhibition activity of photoluminescent carbon nanodots toward cancer cells. J Mater Chem B 1:1774–1781
136. Ge JC, Jia QY, Liu WM, Lan MH, Zhou BJ, Guo L, Zhou HY, Zhang HY, Wang Y, Gu Y, Meng XM, Wang PF (2016) Carbon dots with intrinsic theranostic properties for bioimaging, red-light-triggered photodynamic/photothermal simultaneous therapy in vitro and in vivo. Adv Healthcare Mater 5:665–675
137. Lan MH, Zhao SJ, Zhang ZY, Yan L, Guo L, Niu GL, Zhang JF, Zhao JF, Zhang HY, Wang PF, Zhu GY, Lee CS, Zhang WJ (2017) Two-photon-excited near-infrared emissive carbon dots as multifunctional agents for fluorescence imaging and photothermal therapy. Nano Res 10:3113–3123
138. Zheng M, Li Y, Liu S, Wang WQ, Xie ZG, Jing XB (2016) One-pot to synthesize multifunctional carbon dots for near infrared fluorescence imaging and photothermal cancer therapy. ACS Appl Mater Interfaces 8:23533–23541
139. Jia QY, Ge JC, Liu WM, Guo L, Zheng XL, Chen SQ, Chen MX, Liu S, Zhang LP, Wang MQ, Zhang HY, Wang PF (2017) Self-Assembled carbon dot nanosphere: a robust, near-infrared light-responsive, and vein injectable photosensitizer. Adv Healthcare Mater 6:1601419
140. Allen JE, Hemesath ER, Perea DE, Lensch-falk JL, Li ZY, Yin F, Gass MH, Wang P, Bleloch AL, Palmer RE, Lauhon LJ (2008) High-resolution detection of au catalyst atoms in si nanowires. Nat Nanotechnol 3:168–173
141. Grom GF, Lockwood DJ, McCaffrey JP, Labbé HJ, Fauchet PM, White B Jr, Diener J, Kovalev D, Koch F, Tsybeskov L (2000) Ordering and self-organization in nanocrystalline silicon. Nature 407:358–361
142. Fan JG, Tang XJ, Zhao YP (2004) Water contact angles of vertically aligned Si nanorod arrays. Nanotechnology 15:501–504
143. Schmidt V, Wittemann JV, Senz S, Gösele U (2009) Silicon nanowires: a review on aspects of their growth and their electrical properties. Adv Mater 21:2681–2702
144. Chen XK, Zhang XD, Xia LY, Wang HY, Chen Z, Wu FG (2018) One-step synthesis of ultrasmall and ultrabright organosilica nanodots with 100% photoluminescence quantum yield: long-term lysosome imaging in living, fixed, and permeabilized cells. Nano Lett 18:1159−1167.
145. Warner JH, Hoshino A, Yamamoto K, Tilley RD (2005) Water-soluble photoluminescent silicon quantum dots. Angew Chem Int Ed 44:4550–4554
146. Erogbogbo F, Yong KT, Roy I, Xu GX, Prasad PN, Swihart MT (2008) Biocompatible luminescent silicon quantum dots for imaging of cancer cells. ACS Nano 2:873–878
147. He Y, Kang ZH, Li QS, Tsang CHA, Fan CH, Lee ST (2009) Ultrastable, highly fluorescent, and water-dispersed silicon-based nanospheres as cellular probes. Angew Chem Int Ed 48:128–132
148. He Y, Su YY, Yang XB, Kang ZH, Xu TT, Zhang RQ, Fan CH, Lee ST (2009) Photo and pH stable, highly-luminescent silicon nanospheres and their bioconjugates for immunofluorescent cell imaging. J Am Chem Soc 131:4434–4438
149. Tu CC, Awasthi K, Chen KP, Lin CH, Hamada M, Ohta N, Li YK (2017) Time-gated imaging on live cancer cells using silicon quantum dot nanoparticles with long-lived fluorescence. ACS Photonics 4:1306–1315
150. Erogbogbo F, Tien CA, Chang CW, Yong KT, Law WC, Ding H, Roy I, Swihart MT, Prasad PN (2011) Bioconjugation of luminescent silicon quantum dots for selective uptake by cancer cells. Bioconjugate Chem 22:1081–1088

151. Klein S, Zolk O, Fromm MF, Schrödl F, Neuhuber W, Kryschi C (2009) Functionalized silicon quantum dots tailored for targeted sirna delivery. Biochem Biophys Res Commun 387:164–168
152. Ruizendaal L, Bhattacharjee S, Pournazari K, Rosso-Vasic M, de Haan LHJ, Alink GM (2009) Synthesis and cytotoxicity of silicon nanoparticles with covalently attached organic monolayers. Nanotoxicology 3:339–347
153. Xu ZG, Wang DD, Guan M, Liu XY, Yang YJ, Wei DF (2012) Photoluminescent silicon nanocrystal-based multifunctional carrier for ph-regulated drug delivery. ACS Appl Mater Interfaces 4:3424–3431
154. Wang Q, Bao YP, Ahire J, Chao YM (2013) Co-encapsulation of biodegradable nanoparticles with silicon quantum dots and quercetin for monitored delivery. Adv Healthcare Mater 2:459–466
155. Ji XY, Peng F, Zhong YL, Su YY, Jiang XX, Song CX, Yang L, Chu BB, Lee ST, He Y (2015) Highly fluorescent, photostable, and ultrasmall silicon drug nanocarriers for long-term tumor cell tracking and in-vivo cancer therapy. Adv Mater 27:1029–1034
156. Liu JJ, Chang Q, Bao MM, Yuan B, Yang K, Ma YQ (2017) Silicon quantum dots delivered phthalocyanine for fluorescence guided photodynamic therapy of tumor. Chin Phys B 26:098102
157. Qiao JS, Kong XH, Hu ZX, Yang F, Ji W (2014) High-mobility transport anisotropy and linear dichroism in few-layer black phosphorus. Nat Commun 5:4475
158. Wu JX, Mao NN, Xie LM, Xu H, Zhang J (2015) Identifying the crystalline orientation of black phosphorus using angle-resolved polarized raman spectroscopy. Angew Chem Int Ed 54:2366–2369
159. Wang H, Yang XZ, Shao W, Chen SC, Xie JF, Zhang XD, Wang J, Xie Y (2015) Ultrathin black phosphorus nanosheets for efficient singlet oxygen generation. J Am Chem Soc 137:11376–11382
160. Sun ZB, Xie HH, Tang SY, Yu XF, Guo ZN, Shao JD, Zhang H, Huang H, Wang HY, Chu PK (2015) Ultrasmall black phosphorus quantum dots: synthesis and use as photothermal agents. Angew Chem Int Ed 54:11526–11530
161. Zhang X, Xie HM, Liu ZD, Tan CL, Luo ZM, Li H, Lin JD, Sun LQ, Chen W, Xu ZC, Xie LH, Huang W, Zhang H (2015) Black phosphorus quantum dots. Angew Chem Int Ed 54:3653–3657
162. Xu YH, Wang ZT, Guo ZN, Huang H, Xiao QL, Zhang H, Yu XF (2016) Solvothermal synthesis and ultrafast photonics of black phosphorus quantum dots. Adv. Opt. Mater. 4:1223–1229
163. Gao LF, Xu JY, Zhu ZY, Hu CX, Zhang L, Wang Q, Zhang HL (2016) Small molecule-assisted fabrication of black phosphorus quantum dots with a broadband nonlinear optical response. Nanoscale 8:15132–15136
164. Lee HU, Park SY, Lee SC, Choi S, Seo S, Kim H, Won J, Choi K, Kang KS, Park HG, Kim HS, An HR, Jeong KH, Lee YC, Lee J (2016) Black phosphorus (BP) nanodots for potential biomedical applications. Small 12:214–219
165. Ling X, Wang H, Huang SX, Xia FN, Dresselhaus MS (2015) The renaissance of black phosphorus. Proc Natl Acad Sci U S A 112:4523–4530
166. Gu W, Yan YH, Pei XY, Zhang CL, Ding CP, Xian YZ (2017) Fluorescent black phosphorus quantum dots as label-free sensing probes for evaluation of acetylcholinesterase activity. Sens Actuator B-Chem 250:601–607
167. Shao JD, Xie HH, Huang H, Li ZB, Sun ZB, Xu YH, Xiao QL, Yu XF, Zhao YT, Zhang H, Wang HY, Chu PK (2016) Biodegradable black phosphorus-based nanospheres for in vivo photothermal cancer therapy. Nat Commun 7:12967
168. Hayashi S, Ito M, Kanamori H (1982) Raman study of gas-evaporated germanium microcrystals. Solid State Commun 44:75–79
169. Carolan D (2017) Recent advances in germanium nanocrystals: synthesis, optical properties and applications. Prog Mater Sci 90:128–158
170. Reiss P, Carrière M, Lincheneau C, Vaure L, Tamang S (2016) Synthesis of semiconductor nanocrystals, focusing on nontoxic and earth-abundant materials. Chem Rev 116:10731–10819

171. Lambert TN, Andrews NL, Gerung H, Boyle TJ, Oliver JM, Wilson BS, Han SM (2007) Water-soluble germanium(0) nanocrystals: cell recognition and near-infrared photothermal conversion properties. Small 3:691–699
172. Li F, Wang J, Sun SQ, Wang H, Tang ZY, Nie GJ (2015) Facile synthesis of ph-sensitive germanium nanocrystals with high quantum yield for intracellular acidic compartment imaging. Small 11:1954–1961
173. Karatutlu A, Song MY, Wheeler AP, Ersoy O, Little WR, Zhang YP, Puech P, Boi FS, Luklinska Z, Sapelkin AV (2015) Synthesis and structure of free-standing germanium quantum dots and their application in live cell imaging. RSC Adv 5:20566–20573
174. Wu CF, Chiu DT (2013) Highly fluorescent semiconducting polymer dots for biology and medicine. Angew Chem Int Ed 52:3086–3109
175. Wu CF, Szymanski C, Cain Z, McNeill J (2007) Conjugated polymer dots for multiphoton fluorescence imaging. J Am Chem Soc 129:12904–12905
176. Yu JB, Wu CF, Sahu SP, Fernando LP, Szymanski C, McNeill J (2009) Nanoscale 3D tracking with conjugated polymer nanoparticles. J Am Chem Soc 131:18410–18414
177. Wu CF, Schneider T, Zeigler M, Yu JB, Schiro PG, Burnham DR, McNeill JD, Chiu DT (2010) Bioconjugation of ultrabright semiconducting polymer dots for specific cellular targeting. J Am Chem Soc 132:15410–15417
178. Wu CF, Jin YH, Schneider T, Burnham DR, Smith PB, Chiu DT (2010) Ultrabright and bioorthogonal labeling of cellular targets using semiconducting polymer dots and click chemistry. Angew Chem Int Ed 49:9436–9440
179. Sun C, Zhang Y, Sun K, Reckmeier C, Zhang TQ, Zhang XY, Zhao J, Wu CF, Yu WW, Rogach AL (2015) Combination of carbon dot and polymer dot phosphors for white light-emitting diodes. Nanoscale 7:12045–12050
180. Wu CF, Bull B, Szymanski C, Christensen K, McNeill J (2008) Multicolor conjugated polymer dots for biological fluorescence imaging. ACS Nano 2:2415–2423
181. Ye FM, Wu CF, Jin YH, Wang M, Chan YH, Yu JB, Sun W, Hayden S, Chiu DT (2012) A compact and highly fluorescent orange-emitting polymer dot for specific subcellular imaging. Chem Commun 48:1778–1780
182. Li Q, Zhang JN, Sun W, Yu JB, Wu CF, Qin WP, Chiu DT (2014) Europium-complex-grafted polymer dots for amplified quenching and cellular imaging applications. Langmuir 30:8607–8614
183. Chan YH, Ye FM, Gallina ME, Zhang XJ, Jin YH, Wu IC, Chiu DT (2012) Hybrid semiconducting polymer dot–quantum dot with narrowband emission, near-infrared fluorescence, and high brightness. J Am Chem Soc 134:7309–7312
184. Chan YH, Wu CF, Ye FM, Jin YH, Smith PB, Chiu DT (2011) Development of ultrabright semiconducting polymer dots for ratiometric ph sensing. Anal Chem 83:1448–1455
185. Ye FM, Wu CF, Jin YH, Chan YH, Zhang XJ, Chiu DT (2011) Ratiometric temperature sensing with semiconducting polymer dots. J Am Chem Soc 133:8146–8149
186. Wu CF, Bull B, Christensen K, McNeill J (2009) Ratiometric single-nanoparticle oxygen sensors for biological imaging. Angew Chem Int Ed 48:2741–2745
187. Sun K, Tang Y, Li Q, Yin SY, Qin WP, Yu JB, Chiu DT, Liu YB, Yuan Z, Zhang XJ, Wu CF (2016) In vivo dynamic monitoring of small molecules with implantable polymer-dot transducer. ACS Nano 10:6769–6781
188. Yu JB, Wu CF, Tian ZY, McNeill J (2012) Tracking of single charge carriers in a conjugated polymer nanoparticle. Nano Lett 12:1300–1306
189. Zhang XJ, Yu JB, Wu CF, Jin YH, Rong Y, Ye FM, Chiu DT (2012) Importance of having low-density functional groups for generating high-performance semiconducting polymer dots. ACS Nano 6:5429–5439
190. Geng JL, Liu J, Liang J, Shi HB, Liu B (2013) A general approach to prepare conjugated polymer dot embedded silica nanoparticles with a SiO_2@CP@SiO_2 structure for targeted HER2-positive cellular imaging. Nanoscale 5:8593–8601
191. Chang KW, Liu ZH, Fang XF, Chen HB, Men XJ, Yuan Y, Sun K, Zhang XJ, Yuan Z, Wu CF (2017) Enhanced phototherapy by nanoparticle-enzyme via generation and photolysis of hydrogen peroxide. Nano Lett 17:4323–4329

192. Wu CF, Hansen SJ, Hou Q, Yu JB, Zeigler M, Jin YH, Burnham DR, McNeill JD, Olson JM, Chiu DT (2011) Design of highly emissive polymer dot bioconjugates for in vivo tumor targeting. Angew Chem Int Ed 50:3430–3434
193. Wu IC, Yu JB, Ye FM, Rong Y, Gallina ME, Fujimoto BS, Zhang Y, Chan YH, Sun W, Zhou XH, Wu CF, Chiu DT (2015) Squaraine-based polymer dots with narrow, bright near-infrared fluorescence for biological applications. J Am Chem Soc 137:173–178
194. Jia HR, Jiang YW, Zhu YX, Li YH, Wang HY, Han XF, Yu ZW, Gu N, Liu PD, Chen Z, Wu FG (2017) Plasma membrane activatable polymeric nanotheranostics with self-enhanced light-triggered photosensitizer cellular influx for photodynamic cancer therapy. J Control Release 255:231–241
195. Zhu YX, Jia HR, Chen Z, Wu FG (2017) Photosensitizer (PS)/polyhedral oligomeric silsesquioxane (poss)-crosslinked nanohybrids for enhanced imaging-guided photodynamic cancer therapy. Nanoscale 9:12874–12884
196. Tang Y, Chen HB, Chang KW, Liu ZH, Wang Y, Qu SN, Xu H, Wu CF (2017) Photo-cross-linkable polymer dots with stable sensitizer loading and amplified singlet oxygen generation for photodynamic therapy. ACS Appl Mater Interfaces 9:3419–3431
197. Li SY, Chang KW, Sun K, Tang Y, Cui N, Wang Y, Qin WP, Xu H, Wu CF (2016) Amplified singlet oxygen generation in semiconductor polymer dots for photodynamic cancer therapy. ACS Appl Mater Interfaces 8:3624–3634
198. Zhang YR, Pang L, Ma C, Tu Q, Zhang R, Saeed E, Mahmoud AE, Wang JY (2014) Small molecule-initiated light-activated semiconducting polymer dots: an integrated nanoplatform for targeted photodynamic therapy and imaging of cancer cells. Anal Chem 86:3092–3099
199. Chang KW, Tang Y, Fang XF, Yin SY, Xu H, Wu CF (2016) Incorporation of porphyrin to π-conjugated backbone for polymer-dot-sensitized photodynamic therapy. Biomacromol 17:2128–2136
200. Shi HF, Ma X, Zhao Q, Liu B, Qu QY, An ZF, Zhao YL, Huang W (2014) Ultrasmall phosphorescent polymer dots for ratiometric oxygen sensing and photodynamic cancer therapy. Adv Funct Mater 24:4823–4830
201. Zhu SJ, Song YB, Shao JR, Zhao XH, Yang B (2015) Non-conjugated polymer dots with crosslink-enhanced emission in the absence of fluorophore units. Angew Chem Int Ed 54:14626–14637
202. Zhu SJ, Wang L, Zhou N, Zhao XH, Song YB, Maharjan S, Zhang JH, Lu LJ, Wang HY, Yang B (2014) The crosslink enhanced emission (CEE) in non-conjugated polymer dots: from the photoluminescence mechanism to the cellular uptake mechanism and internalization. Chem Commun 50:13845–13848
203. Sun Y, Cao WP, Li SL, Jin SB, Hu KL, Hu LM, Huang YY, Gao XY, Wu Y, Liang XJ (2013) Ultrabright and multicolorful fluorescence of amphiphilic polyethyleneimine polymer dots for efficiently combined imaging and therapy. Sci Rep 3:3036

Polymeric Nanoparticles for Cancer Therapy and Bioimaging

Eva Espinosa-Cano, Raquel Palao-Suay, María Rosa Aguilar, Blanca Vázquez and Julio San Román

Abstract Polymeric nanoparticles have demonstrated to be good candidates as vehicles of drugs or molecules for combined treatment and diagnosis of cancer. In comparison with inorganic nanoparticulated systems, they present remarkable benefits in terms of stability, biocompatibility, biodegradability, tailorability and low cost. Polymeric nanoparticles can be design to passively or actively accumulate in tumor sites by controlling their hydrodynamic properties or functionalizing their surface with targeting molecules. Moreover, polymers responding to particular tumor microenvironment conditions like reduced pH, high levels of reactive oxygen species or overexpressed enzymes, can be used to trigger a controlled drug delivery, a contrast agent exposure, or to enhance the therapeutic effect of a theranostic system. This chapter focuses on the most recent advances in this field by discussing in depth examples of nanoparticles that, exploiting all these strategies, can be visualized with one or more imaging techniques: optical imaging, MRI, US, PA, PET, SPECT or CT; and present therapeutic effect (i.e. chemotherapy, gene therapy, photothermal or photodynamic therapy) due to the presence of active moieties.

Keywords Passive targeting · Active targeting · pH-responsive polymers · ROS-responsive polymers · enzyme-responsive polymers · External stimuli

---

E. Espinosa-Cano · R. Palao-Suay · M. R. Aguilar (✉) · B. Vázquez · J. S. Román
Group of Biomaterials, Institute of Polymer Science and Technology, ICTP-CSIC, C/Juan de La Cierva, 3, 28006 Madrid, Spain
e-mail: mraguilar@ictp.csic.es

E. Espinosa-Cano · R. Palao-Suay · M. R. Aguilar · B. Vázquez · J. S. Román
Networking Biomedical Research Centre in Bioengineering, Biomaterials and Nanomedicine, CIBER-BBN, Av. Monforte de Lemos, 3-5. Pabellón 11. Planta 0, 28029 Madrid, Spain

G. Gonçalves and G. Tobias (eds.), *Nanooncology*, Nanomedicine and Nanotoxicology, https://doi.org/10.1007/978-3-319-89878-0_4

1 Introduction

Polymer chemistry allows the design of on-demand macromolecules that combine multiple bioactive elements into a single system. The great versatility of the synthetic processes, such as the reversible addition-fragmentation chain transfer (RAFT) polymerization or the atom transfer radical polymerization (ATRP), let a precise control of the molecular architecture and functionality of the polymers. The polymer chemist can design and synthesize macromolecules with an appropriate monomer distribution forming amphiphilic macromolecules that self-assemble in appropriate conditions developing polymeric nanoparticles (NPs). The NPs can serve as vehicles of drugs or molecules for the treatment and diagnosis of cancer. Moreover, the surface of these NPs can also be decorated with multiple targeting moieties in order to have a more selective and more active theranostic system with reduced undesired side effects of the treatment. In addition, polymeric systems present remarkable benefits in terms of stability, biocompatibility, biodegradability, tailorability, and low cost [1].

This chapter focuses on polymer-based nanoparticulated systems for theranostics of cancer highlighting the most relevant advances achieved in this field during the last decade. To set the basis for the design of these polymeric nanocarriers, it is important to identify the biomarkers of the tumor microenvironment and the main differences in healthy tissue. One important hallmark of cancer tissue is the increased permeability of the blood vessels due to an accelerated and uncontrolled angiogenesis. The fenestrated leaky vessels present endothelial gaps between 10 and 100 times larger in tumor models than in normal tissue (200–1000 nm and 10 nm, respectively) [2]. The leakiness together with the lack of lymphatic drainage favors the passive accumulation of high molecular weight ($\geq$40 kDa), long-circulating macromolecules into the tumor site, a phenomenon known as enhanced permeability and retention (EPR) effect [3]. PEGylation is the main strategy used to increase the circulation time of organic and inorganic nanoparticulated theranostic systems favoring their passive accumulation into tumor sites and enhancing both imaging and therapy [1]. The carriers are designed to present an external layer of poly(ethylene glycol) (PEG), a highly hydrophilic polymer, that avoids opsonization and so an early recognition and elimination by the reticuloendothelial system (RES).

Therefore, cancer theranostic systems should present sizes in the nanoscale but larger than 10 nm. Moreover, it has been described that particles smaller than 100 nm are rapidly eliminated from the circulation by the RES presenting a low rate of extravasation into permeable tissues, and those particles larger than 200 nm have difficulties to be endocyted by the cells. Therefore, the optimal size to take advantage of the EPR effect is between 100 and 200 nm.

Accumulation of nanocarriers in the tumor site can also be achieved by active targeting by the incorporation of aptamers, antibodies, proteins, peptides, nucleic acids, or small molecules that specifically interact with receptors or surface ligands at the region of interest [2, 4].

The particular tumor microenvironment conditions can be used to trigger a controlled drug delivery, a contrast agent (CA) exposure, or to enhance the therapeutic effect of a theranostic system using smart polymers (i.e., polymers that change their physicochemical characteristics in response to specific external stimuli) [5]. The most commonly used are pH-responsive polymers that undergo changes in response to the acidic pH of the tumor sites (between pH 4.5 and 6.0) [6–9], ROS-responsive polymers especially disulfide-based structures [10], and enzyme-responsive polymers that can be cleaved by cancer overexpressed enzymes like matrix metalloproteinases (MMPs) or cathepsins [11].

Optical imaging, magnetic resonance imaging (MRI), ultrasound (US), photoacoustic (PA) imaging, positron emission tomography (PET), single-photon emission computed tomography (SPECT), and X-ray computed tomography (CT) are the most commonly used techniques for cancer diagnosis. This chapter focuses on the polymer-based nanoparticulated theranostic systems that can be visualized with one or more of these imaging techniques and present therapeutic effect due to the presence of active moieties to perform chemotherapy, gene therapy, photothermal, or photodynamic therapy.

2 Optical Imaging

Optical imaging is one of the most widely used imaging techniques for cancer detection, providing great information related to the tumor anatomical structure as well as the tumor metabolism and biochemistry. Among the optical imaging technologies, near-infrared (NIR) fluorescence has been extensively explored owing to the extremely low absorption and autofluorescence of biological tissues in the wavelength range from 700 to 1000 nm. These characteristics significantly enhance the image sensitivity, minimizing background interference, and increasing tissue depth penetration [12].

Currently, most common NIR contrast molecules are inorganic compounds such as different noble metals, quantum dots, and oxide nanoparticles (NPs), among others. However, all these different alternatives are characterized by a nondegradable nature, nonspecific biodistribution, and long-term toxicity. As a consequence, organic NIR contrast molecules have gained special attention in the recent years. NIR fluorescent probes, such as cyanine compounds [13–15], are small organic molecules with excellent optical properties (high molar absorptivity, good photostability, and strong fluorescent emission in the NIR region). These molecules absorb NIR wavelengths to reach an excited singlet state being part of this energy dissipated in the form of fluorescence, part transformed into heat, and part producing free radicals or reactive oxygen species (ROS). Depending on the efficiency of these processes, NIR dyes can be used as promising agents not only for imaging purposes but also for photothermal (PTT) [16] and photodynamic (PDT) [17] therapy.

Indocyanine green (ICG) is a U.S. Food and Drug Administration (FDA)-approved dye that effectively fluoresces in the region of 650–850 nm, also producing heat upon NIR irradiation. Other remarkable cyanines are pentamethine and heptamethine derivatives (such as Cy5, Cy5.5, and Cy7, among others) that exhibit high molar absorption coefficient and fluorescence quantum yield. However, these molecules can be easily degraded in aqueous medium and are characterized by a short circulation time in physiological conditions. To palliate these drawbacks, a wide range of cyanines have been incorporated into different polymeric formulations with the aim to improve its bioavailability and stability [18]. For example, ICG has been encapsulated in the core of polymeric nanoparticles that also incorporated chemotherapeutics drugs as doxorubicin (DOX) in their inner core. Some interesting examples of these polymeric formulations are poly(ethylene glycol) and decylamine-grafted poly(L-aspartic acid) (mPEG-*b*-PAsp) [19], poly(lactic-*co*-glycolic acid) (PLGA)-lecithin-PEG NPs [20], and pH-sensitive terpolymer based on poly(ethylene glycol)-poly(ε-caprolactone)-poly(*N*-isopropylacrylamide) (PEG-PCL-PNIPA) [21]. Additionally, Cy5.5 was conjugated to mPEG-denatured bovine serum albumin (dBSA) NPs [22] and encapsulated into NPs based on amphiphilic block *N*-(2-hydroxypropyl) methacrylamide (HPMA) copolymer and gemcitabine conjugate [23]. Natural polymers have been also considered for these polymeric formulations. Particularly, ICG was physically entrapped and Cy7.5 covalently conjugated to nanoformulations derived from hyaluronic acid (HA) [24]. In all these examples, theranostic NPs demonstrated great potential in cancer imaging and chemo-photothermal synergized tumor ablation.

Actively targeted systems have also been extensively described in the last decade, incorporating active targeted molecules such as folic acid or monoclonal antibodies (mAbs) into cyanine-based polymeric NPs. For example, Lee et al. encapsulated ICG in the inner core of polyethylene glycol-coated poly(lactic-*co*-glycolic acid) NPs. These authors also successfully conjugated anti-HER2-mAbs on the surface of PEG-coated NPs. This strategy allowed to increase the circulation time in the bloodstream of these targeted NPs and to improve the anticancer activity in comparison to free ICG [25].

Smart polymers that exploit the acidic pH of the tumor microenvironment to facilitate the release chemotherapeutic drugs and cyanines have also been described in the literature. For instance, Hung et al. successfully synthesized polymeric NPs based on PLGA hydrophobic core where ICG and DOX were effectively encapsulated. These PLGA NPs were coated with pH-responsive *N*-acetyl histidine-modified D-α-tocopheryl polyethylene glycol 1000 succinate (NAcHis-TPGS), enhancing the selective release of encapsulated molecules into the acidic tumor area, owing to the protonation of the NAcHis moieties (see Fig. 1) [26]. Fu et al. conjugated folic acid to a triblock amphiphilic polymer that was synthesized by the combination of reversible addition-fragmentation chain transfer (RAFT) polymerization and ring opening polymerization using 2-(*N*-tert-butoxycarbonylamino)ethyl methacrylate (BEMA) and oligo(ethylene glycol)methacrylate (OEGMA). An amino-modified cyanine dye was conjugated to the polymer and DOX was successfully encapsulated. The folic acid targeted on the NP surface enhanced their endocytosis by tumor cells and the pH sensitivity of these NPs significantly favored the controlled release of DOX [27].

In order to enhance the photothermal treatment of tumor, the preparation of thermosensitive polymers has also been investigated. In this context, He et al. designed poly(ether amine)-based NPs with a phase transition temperature of 43 °C where

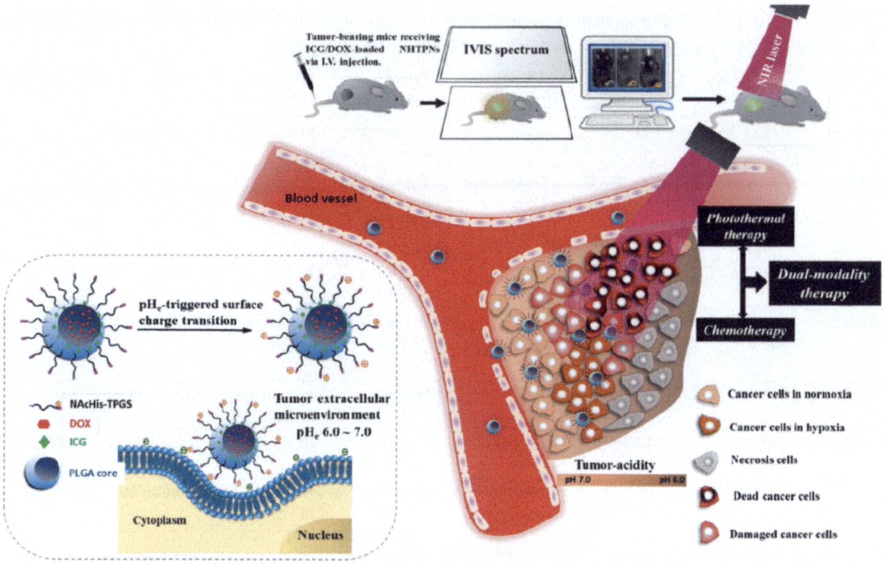

Fig. 1 Schematic diagram of uptake of theranostic NPs with pH sensitivity loaded with DOX and mixed with 131I-labeled hyaluronic acid (131I-HA) for photothermal therapy and chemotherapy. Reproduced with permission from [26]. Copyright Ivyspring International Publisher

ICG was successfully encapsulated in their inner core. Under NIR irradiation, the size of these nanoparticles increased as a result of their phase transition, favoring their accumulation in tumor tissues [28].

All these strategies to improve the effective administration of cyanines for cancer treatments have some limitations that hinder their transfer into clinical studies. In this sense, the specificity, affinity, and distribution of NIR-targeted NPs in cells and tissues are still unpredictable. For that reason, the research in this area is focused on the development of innovative strategies to overcome the limitations of most common NIR dyes.

Recently, novel NIR polymethine cyanine dyes have been synthesized, including a rigid carbocyclic ring in the middle of the polymethine linker. This structural modification significantly enhances the aqueous stability, fluorescence quantum yield, and photostability of these dyes. However, the high number of π-conjugated bonds into the chemical structure of these NIR dyes increases significantly their hydrophobicity, causing aggregation and poor water solubility. To mitigate these drawbacks, different NIR polymethine cyanines have been incorporated into polymeric self-assembled NPs [13, 14]. Some interesting examples of different researches described in the literature are recapitulated in Table 1. For example, IR820 and IR825 are two interesting examples of these studies. These molecules have a chemical structure similar to ICG. However, their chemical stability in vitro and in vivo is significantly improved in comparison to ICG. Fernandez et al. chemically conjugated IR825 to PEG-diamine,

Table 1 Summary of the most representative example of polymeric NPs that incorporates different NIR polymethine cyanines in their structure for theranostic purposes

Dye	Polymeric formulation	NP size (d.nm)	Therapeutic molecules	Type of therapy	Biological experiments	Type of cancer	References
IR-820	PEG-diamine	150	–	Diagnosis PTT	In vitro and in vivo	Ovarian and uterine	[29]
IR825	Poly(maleic anhydride-alt-1-octadecene)-PEG-amine	100-200	Gadolinium Ce6	Triple modal imaging PTT PDT	In vitro and in vivo	Breast cancer	[38]
IR-775/797	PEG-*b*-PCL	40	–	Diagnosis PDT PTT	In vitro and in vivo	Ovarian cancer	[30]
IR-780	Heparin–folic acid conjugate	80-120	–	Diagnosis PTT	In vitro and in vivo	Breast cancer	[36]
IR780	PEG_{2000}	100	–	Diagnosis PTT	In vitro and in vivo	Colon cancer	[35]
IR780	PEG-*b*-MTOS	110-160	IR-780	Dignosis, PTT and PDT	In vitro	Breast cancer	[37]
IR780	DSPE-PEG	100-130	FA conjugated	Diagnosis, smart responsive, PTT	In vitro and in vivo	Lung cancer	[39]

forming micelles with a promising potential for diagnosis by fluorescence imaging and PTT [29].

Currently, NIR heptamethine indocyanine has also gained special attention for the development of theranostic NPs for cancer treatment [14]. Very recently, Doung et al. designed biodegradable and biocompatible PEG-*b*-PCL polymeric NPs for the encapsulation of IR775 that is a non-soluble cyanine dye with photodynamic activity and optimal fluorescence for diagnosis. Authors demonstrated that IR775-NPs accumulated in the tumor after systemic administration, successfully delineated them with NIR fluorescence signal, and completely eradicated tumor areas after a single dose of combinatorial phototherapy. IR775-NPs presented higher efficacy than promising water-soluble heptamethine dye, IR783, and the FD- approved ICG (see Fig. 2) [30].

Additionally, some NIR heptamethine indocyanines have demonstrated preferential tumor accumulation without the conjugation of actively targeted ligands [14, 31]. In this sense, IR-780 dye is a lipophilic cation heptamethine dye with a rigid cyclohexenyl ring in the heptamethine chain and a central chlorine atom, which enhance photostability in comparison to other NIR nanoabsorbers [32]. Moreover, different studies have also demonstrated that this dye can preferentially accumulate into the mitochondria of tumor cells as a result of its lipophilic cationic nature and it shows high affinity for organic anion transporter peptides (OATPs), frequently overexpressed in tumor cells [33]. Nevertheless, the high hydrophobicity of this dye significantly hinders their real clinical application [34].

To overcome these limitations, IR-780 dye has been encapsulated and also conjugated into polymeric multifunctional micelles, as is summarized in Table 1. For example, Yuan et al. modified the structure of IR-780 using the central chlorine atom

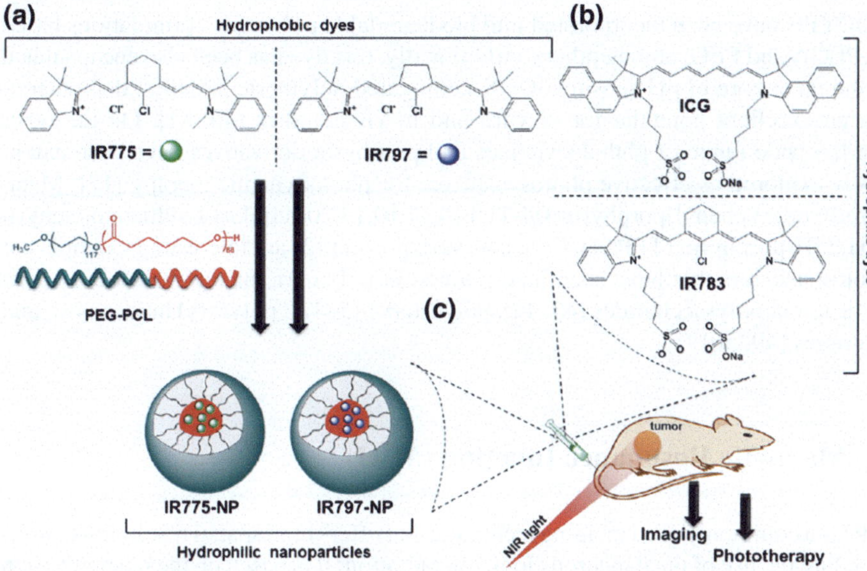

Fig. 2 **a** Schematic representation of PEG-*b*-PCL polymeric NPs where IR775 and IR797 were encapsulated. **b** The efficacy of these NPs was compared to analogous water-soluble dyes, ICG and I783. **c** IR775-NPs successfully eradicated tumor areas with high efficacy due to a phototherapeutic mechanism [30]. Copyright (2017), with permission from Elsevier

with the aim to conjugate a hydrophilic PEG_{2000} [35]. This conjugate was able to form micelles that significantly improved the water solubility of the dye, favoring its preferential tumor accumulation due to the EPR effect. Moreover, Yue et al. encapsulated IR780 into polymeric NPs with unsoluble folic acid hydrophobic core, heparin as hydrophilic shell, and water-soluble folic acid targeted into their surface. These micelles are promising nanoformulations for NIR fluorescence imaging and photothermal therapy in vitro and in vivo for the treatment of folate-overexpressed tumors [36]. Finally, Palao-Suay et al. explored the design of synthetic nanoassemblies based on amphiphilic block copolymers obtained by RAFT polymerization. Particularly, these authors conjugated IR780 dye to the hydrophilic segment of copolymer based on PEG and a methacrylic derivative of α-tocopheryl succinate (MTOS) that is a well-known anticancer compound. These copolymers were able to form micelles and to encapsulate additional free dye in their inner core. In vitro biological assays demonstrated an excellent phototoxicity of these fluorescence NPs against breast cancer cells that could be associated with photothermal and photodynamic events [37].

Finally, borondipyrromethane (BODYPI) and porphyrin derivatives are other interesting examples of NIR dyes with promising potential for the development of theranostic nanoplatforms for cancer treatment in clinic. On the one hand, BODIPY dyes exhibit high quantum yield and excellent thermal and photochemical stability. However, their fluorescence emission from yellow to deep-red emission is relatively short with low extinction coefficients. In spite of these drawbacks, different

BODYPIs have been incorporated into biodegradable polymeric formulations based on PLGA and PEG, among others. Additionally, this dye has been also encapsulated in the inner core of pH-sensitive DOX-conjugated polymeric micelles, demonstrating an excellent potential for in vitro and in vivo studies [40, 41]. On the other hand, a wide range of phthalocyanines and porphyrin derivatives have been extensively explored as effective photosensitizers for photodynamic therapy [17]. Meta-tetra(hydroxyphenyl)porphyrin (*m*-THPP), 5,10,15,20-tetrakis(4-sulfanatophenyl)-21H,23H-porphyrin (TPPS4), Ce6, and Verteporfin [42] are interesting examples of photosensitizers that have been incorporated in polymeric nanoformulations based on different polysaccharides [43, 44], polyesters [45–47], polyacrylamides [48], and pluronics [49, 50].

3 Magnetic Resonance Imaging (MRI)

MRI is a commonly used imaging technique that offers high spatial resolution (1 mm) avoiding the use of the dangerous ionizing radiation. It is based on the magnetization properties of atomic nuclei [1]. Briefly, an external magnetic field is applied to align the magnetic moments of hydrogen atoms in tissues and this alignment is then disrupted by the application of an external radio frequency. As the magnetic moments of protons return to their ground state through various relaxation processes, they emit a radiofrequency signal that is detected and transformed into an image [51]. The contrast is achieved, thanks to the different relaxation times (i.e., time taken for the magnetic moments to return to their ground state) of each tissue's protons [4].

However, the application of this technique for cancer diagnosis is limited due to the low sensitivity or poor contrast [4]. Therefore, it is essential the use of MRI contrast agents (CAs) that have the capacity to alter relaxation times of protons in different tissues by their interaction with the external magnetic field [1].

There are two main types of CAs: (a) T_1-weighed CAs, like Gd^{3+}, Mn^{2+}, or Cu^{2+}-based compounds, with positive contrast enhancement (i.e., white signal) by increasing the longitudinal relaxation time (T_1) of the surrounding protons; and (b) T_2-weighed CAs, mainly superparamagnetic iron oxide nanoparticles (SPIONs), with negative contrast enhancement (i.e., black signal) by increasing the transverse relaxation time (T_2) of the surrounding protons.

Low molecular weight complexes of these metals do not provide accurate tumor MRI imaging because of their low molecular relaxivity together with their low selectivity to tumor tissue as they are quickly excreted by the organism having no time to accumulate at disease sites. Moreover, the need of high doses to provide improved images of tumors significantly increases the risk of systemic toxicity [4]. Therefore, the main challenges of MRI-based cancer diagnosis are tumor targeting and the improvement of T_1 and T_2 relaxivity to increase contrast and to minimize the dosage of MRI CAs [52]. The encapsulation or chelation of these CAs by polymeric nanoparticles has demonstrated great potential to overcome these drawbacks [51]. Apart from the already mentioned advantages that polymeric systems present with respect to those based on other materials, with an appropriate design, polymeric nanocarriers have shown to improve accumulation in tumor areas of MRI CAs by passive or active targeting while protecting CAs from degradation [1]. Besides, the joint encapsulation of the CAs and chemotherapeutics allows the minimization of the dose and, therefore, the reduction of the systemic toxicity [51].

Moreover, smart polymers responding to tumor-specific stimuli (i.e., acidic pH, overexpressed ROS, etc.) have successfully been used to specifically trigger a controlled delivery of drugs, genetic material, and/or activate CAs exposure at tumor sites enhancing theranostic capabilities [53, 54].

Here, it is summarized the most recent bibliography of polymer-based theranostic systems using MRI CAs. The main polymeric architectures found in the literature for theranostic purposes include nanogels [55, 56], nanocomplexes [57], micelles [58–60], and polymersomes [52, 61].

Regarding T_2-weighed CAs, the recent literature focuses on SPION-containing drug delivery systems for combined MRI imaging and chemotherapy [52], gene therapy [62], or cancer thermotherapy, thanks to their magnetically induced hyperthermia capability [63]. Although the negative contrast enhancement by T_2-weighed CAs is not always easy to interpret [4], SPIONs have emerged as a promising MRI imaging CA in the recent years as they present higher biocompatibility and provide an enhanced magnetic relaxivity when compared to gadolinium complexes, the most common T_1-weighed CAs [64].

The simplest designs are passively targeted nanocarriers based on biodegradable polymers like poly (ε-caprolactone) (PCL) or poly(lactic-*co*-glycolic acid) that accumulate into cancer tissue by EPR effect. For example, Schleich et al. [65] developed a paclitaxel (PTX)/SPION co-loaded PLGA-based micelles. Authors demonstrated that co-loaded NPs increased CT26 colon carcinoma cells death in vitro when compared to SPION-loaded or non-loaded PLGA NPs and they delayed tumor growth in an in vivo model of CT26-tumor-bearing mice. Moreover, they showed by relaxometry studies and phantom MRI superior properties with respect to a commercial T_2 contrast agent called Resovist®. Some years later, Qin et al. [64] proposed a new strategy consisting of the conjugation of SPIONs in the surface of DOX-loaded polymeric vesicles. The outer layer of PAA was decorated with SPIONs, and the chemotherapeutic drug (i.e., DOX) was loaded into the vesicle that presented a biodegradable membrane of PCL. The system was tested, in vitro and in vivo, as MRI CAs presenting high T_2-relaxivity and its in vitro anticancer activity were also demonstrated against HeLa cells. Yoon et al. [66] designed PTX/SPION co-loaded hydrotropic micelles based on an amphiphilic hydrotropic block copolymer, PEG-block-poly(2-[4-vinylbenzyloxy]-*N*,*N*-diethylnicotinamide(PEG-*b*-P(VBODENA)). The nanoparticles showed higher cytotoxicity against SCC7 squamous carcinoma cells in vitro in comparison to SPION-loaded nanoparticles but lower than PTX alone. The authors attributed this effect to the controlled release of PTX from the nanocarriers and they demonstrated that the co-loaded NPs led to a significantly higher long-term tumor volume reduction than the drug alone in an in vivo model of SC7-tumor-bearing mice.

In addition, passively targeted PLA-*b*-PEG-based micelles coated with PEI and chitosan have been successfully tested in vivo for gene therapy and enhanced MRI-combined cancer theranostics [62].

Actively targeted systems have also been extensively described in the last decade and mainly those incorporating folic acid (FA) [53, 54, 67] as it has been demonstrated a significant upregulation of folate receptor in cancer cells. As a recent example, Liu et al. [52] developed superparamagnetic polymersomes based on a folate-functionalized biodegradable diblock copolymer, folic acid-poly(L-glutamic acid)-*block*-poly(ε-caprolactone) [FA-PGA-b-PCL]. The SPIONs were in situ generated by chemical precipitation of Fe_3O_4 in the hydrophilic PGA corona and DOX was loaded into the inner core of the vesicles. The nanocarriers showed extremely high T_2-relaxivity when compared to commercially available SPION-based CAs. Moreover, in vivo studies in nude mice bearing HeLa tumor showed higher accumulation of folate-functionalized polymersomes in tumor site and significantly higher tumor size reduction when compared with non-targeted polymeric vesicles.

Other actively targeted systems include stem cell antigen antibodies ($scAb_{PSCA}$)-functionalized PEGylated PLGA polymersomes co-encapsulating docetaxel (DTXL) and SPIONs [68] or co-loaded with DTXL or silibinin and SPIONs [69] targeting breast cancer cells through LHRH (i.e., a decapeptide hormone secreted by hypothalamus upregulated on cancer cells). Lung cancer cells were also successfully targeted via lung cancer-targeting peptide (LCP)-presenting DOX/SPION co-loaded PEGylated PLA nanomicelles [70]. All these systems showed a promising theranostic capacity in vitro. In 2017, C26 colon carcinoma cells targeting aptamers were used to functionalize PLGA-based SPIONs/DOX co-loaded micelles demonstrating in vivo enhancement of MRI images in tumor site and higher tumor inhibition and animal survival in mice-bearing C26 colon carcinoma xenografts [71]. Finally, Situ et al. [72] coated SPION/DOX co-loaded PLGA micelles with a highly hydrophilic polymer, dextran, functionalized with A54 peptide. In this case, the in vivo MR imaging experiments validated the targeting capacity of the system and the antitumor activity was demonstrated in vitro and in vivo.

Smart polymers that suffer physicochemical changes with the intrinsic features of the tumor microenvironment have raised much attention in the last decades as promising vehicles for MRI CAs. For example, Yang et al. [54] developed folate-functionalized SPIO-loaded polymeric vesicles incorporating DOX through an acid-cleavable hydrazone bond in such a way that, at the acidic pH encountered in tumor microenvironments, the drug is released in a controlled manner. Li et al. [53] described DOX/SPIONs co-loaded micelles based on FA-PEG-b-poly[*N*-(*N′*,*N′*-diisopropylaminoethyl)glutamine] (FA-PEG-P[GA-DIP]) amphiphilic block copolymer that presents an acid-triggered core crosslinked. At acidic pH, the core becomes uncrosslinked and both the drug and the CA are delivered. Besides, ROS-responsive systems have also been described. In 2017, Yang et al. [73] designed nanovehicles based on poly(3-caprolactone)-*b*-glycopolypeptides (PCL–SS–GPPs) bearing disulfide bonds that encapsulate DOX and SPIONs. The drug and CA delivery were demonstrated to be enhanced in the presence of glutathione, a thiol-containing tripeptide highly abundant in the cytoplasm of cancer cells and capable of cleaving disulfide bonds by a redox reaction.

External stimuli including light or magnetic field-generated heat are also exploited for theranostic purposes. For example, a very recent paper describes the preparation of SPION/DOX co-loaded nanoparticles based on a photo-sensitive amphiphilic diblock copolymer poly(ethylene oxide)-*b*-poly(2-((((2-nitrobenzyl)oxy)carbonyl)amino)ethyl methacrylate) (PEO-*b*-PNBOC) [74]. Under UV radiation, the system underwent traceless crosslinking and triggered hydrophobic-to-hydrophilic transition leading to improved MRI contrast by avoiding SPION leakage and increasing water accessibility. Moreover, authors demonstrated DOX photoradiation-triggered release and therapeutic effect in vitro and in vivo. In another work, the development of magneto-thermo-responsive nanohydrogels based on poly(*N*-isopropylacrylamide) (PNIPA) encapsulating SPIONs and DOX was reported [75]. In this case, the system exhibited enhanced MRI contrast and drug release with radio frequency (RF) exposure as the hydrogel structure is compromised by the magnetic vibrations generated by the applied RF field.

Furthermore, some systems combine cancer thermal therapy and chemotherapy by temperature-triggered controlled release based on the hyperthermia capability that SPIONs show when high-frequency magnetic fields (HFMF) are applied [76, 77]. A good example is the system described by Chiang et al. [77], a hollow nanogel with an inner aqueous chamber containing SPIONs and a PEGylated PNIPA/PAA-based corona embedding DOX. The system delivered the cargo (i.e., SPIONs and DOX) in response to the acidic pH of tumor tissue and to the high temperature generated by SPIONs when subjected to HFMF. The authors demonstrated an enhanced in vitro T_2-type MRI contrast and in vitro high HeLa cells uptake, controlled drug release, and increased cytotoxicity with respect to the free drug.

Several T_1-weighed MRI CAs are described in the literature being gadolinium (Gd^{3+}) the most commonly used. Gd^{3+} ions are typically found forming complexes with diethylenetriamine pentaacetic acid (DTPA) or 1,4,7,10-tetraazacyclododecane-1,4,7,10-tetraacetic acid (DOTA).

As an example of successful passively targeted Gd-chelating theranostic systems, Vinh et al. developed in 2015 [11] polymeric micelles based on PEG-*b*-poly(glutamic acid) (PGA) copolymer that co-encapsulate DTPA-Gd complexes and the anticancer drug oxaliplatin with MRI imaging and chemotherapeutic purposes. The micelles exhibited strong and tumor-specific MRI contrast enhancement and significant suppression of tumor growth in an in vivo orthotropic rat model of hepatocellular carcinoma.

Actively targeted theranostic systems chelating Gd^{3+} have also been described. As in the case of T_2-weighed CAs, FA is the most common targeting ligand [7, 78, 79]. For example, Liu et al. [78] designed asymmetrical polymer vesicles based on a biodegradable copolymer (i.e., R-PGA-*b*-PCL (R=FA or DTPA)) that presented a cancer-targeting outer corona relying on FA and a Gd^{3+}-chelating and DOX-loading inner corona. The folate-decorated vesicles demonstrated to have in vitro and in vivo tumor-targeting capability, high T_1-relaxivity, and twofold better tumor activity than the free drug.

Nanoparticles based on tumor microenvironment-responsive polymers like enzyme-responsive [80], ROS-responsive [10], and mainly pH-responsive [6–8, 81, 82] are also widely described in the bibliography. For example, the (Gd-DTPA)-chelated cationic poly(urethane amide) (CPUAs) having redox-responsive disulfide bonds [10]. The system was demonstrated to possess transfection ability in vitro and to be efficient for a combined MRI imaging and gene therapy of cancer. Liu et al. [8] described a pH-sensitive nanospherical system based on PEG-*b*-poly(L-histidine) cationic copolymer that combined dual drug (sorafenib)/imaging agent (Gd-DTPA) loading, pH-sensitive release, and active targeting functions through vascular endothelial growth factor (VEGF). The MRI contrast agent and antitumor capacity were evaluated in vivo in an H22 hepatocarcinoma tumor-bearing mouse model. Finally, an example of enzyme-responsive systems, which are the less common in the literature, is the hydroxyl propylmethacylate (HPMA)-based nanoparticulated micelles having DOX covalently linked to the polymeric backbone through a lysozyme-cleavable peptide (GFLG) and Gd-DOTA also covalently attached to the polymer. In this case, the system exhibited promising T_1-relaxivity, stability, and antitumor capacity in vitro [80].

External stimuli like magnetic field, temperature, or light have also been used to achieve controlled drug delivery and MRI-enhanced contrast. For example, Li et al. [83] synthesized POEGMA-*b*-poly(NIPA-*co*-NBA-*co*-Gd) amphiphilic diblock copolymer-based nanoparticles encapsulating DOX (see Fig. 3). They confirmed that after UV radiation the hydrophobic–hydrophilic transition of NBA led to a quick drug release and enhanced MRI contrast in vitro. Furthermore, biocompatible polymeric micelles based on 1,4,7,10-tetraazacyclododecane-1,4,7,10-tetraacetic acid-poly(ethyleneglycol)-*block*-poly(acrylamide-*co*-acrylonitrile) (DOTA-PEG-*b*-poly(AAm-*co*-AN)) were prepared presenting chelated Gd on the hydrophilic corona and DOX encapsulated in the core. These theranostic micelles presented noticeably accelerated drug release, antitumor activity, and MRI contrast enhancement at high temperatures in vitro [84].

More recent works have exploited the Gd-chelating capacity of bovine serum albumin (BSA) to prepare nanoparticulated theranostic system that combines MRI imaging and cancer therapy. In 2014, PEG-*b*-PLGA-based nanoparticles were prepared and co-loaded with DOX and BSA-Gd complexes. They showed high T_1-relaxivity in vitro and in vivo and antitumor capacity in vitro. In addition, in 2016, BSA-based NPs with Gd chelated in the corona and encapsulating Ce6, a photosensitizer, showed very promising results both in vivo and in vitro [85]. MR images of the system in vivo showed enhanced passive targeting to tumor site and complete cure of tumor-bearing mice after intravenous injection.

However, Gd^{3+} has been associated with potential renal side effects and several safety concerns have arisen about its use in theranostic nanoparticulated systems [86]. Manganese (Mn^{2+}) has emerged as a promising alternative because of its favorable electronic configuration and biochemical features [87]. In particular, Mn chelation and MnO_2 encapsulation within theranostic nanoparticles have been the most studied strategies to achieve high T_2-type MRI-enhanced contrast.

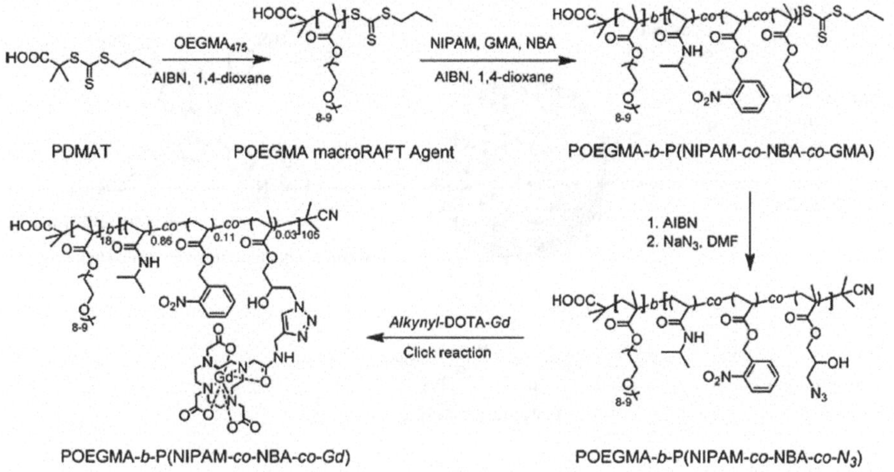

Fig. 3 Schematic representation of the preparation of POEGMA-*b*-poly(NIPA-*co*-NBA-*co*-Gd) amphiphilic diblock copolymer. Reprinted with permission from [83]. Copyright (2012) American Chemical Society

In 2013, Jing et al. [88] fabricated theranostic DOX-loaded PLA nanoparticles conjugating Mn porphyrin onto their surface (see Fig. 4). The authors demonstrated the MRI contrast enhancement with respect to commercially available Mn-based MRI CAs in vitro and in vivo, as well as the acidic pH-dependent DOX release and antitumor activity against HeLa cells and human colon adenocarcinoma HT-29 cells in vitro. More modern studies take advantage of the recently reported high photothermal conversion efficiency of polydopamine (PDA) and its ability of directly chelating metal ions, thanks to the rich functional groups it presents (i.e., catechol, carboxyl, and amino) [89]. For example, PEGylated Mn-chelated PDA nanoparticles have been successfully tested in vitro for photothermal ablation of HeLa cells at the time that they have exhibited MRI signal enhancement in vitro and in vivo [90]. More sophisticated systems like the PLGA-based nanovehicles described by Xi et al. [89] exhibited a synergistic effect by combining photothermal therapy and ion-chelating capacity of PDA and chemotherapy by encapsulating DOX. The system showed MRI contrast enhancement and therapeutic efficacy in vitro and in vivo. Zhao et al. [91] employed the same strategy incorporating DOX to PEGylated PDA nanoparticles but conferred the system redox-responsive MRI contrast enhancement by linking Mn to the particles through a disulfide-containing organic ligand. The linkage was cleaved in the presence of glutathione, an overexpressed enzyme in the tumor microenvironment, achieving targeted MRI contrast. In addition, the nanovehicles exhibited improved in vivo therapeutic outcome when compared to the free drug. Folate-functionalized PEGylated PDA nanoparticles encapsulating DOX have also shown very promising theranostic behavior in vitro and in vivo [92].

Another interesting approach is the one exploiting the synergistic effect of MnO_2 nanoparticles and photodynamic therapy. Hao et al. [93] described this year

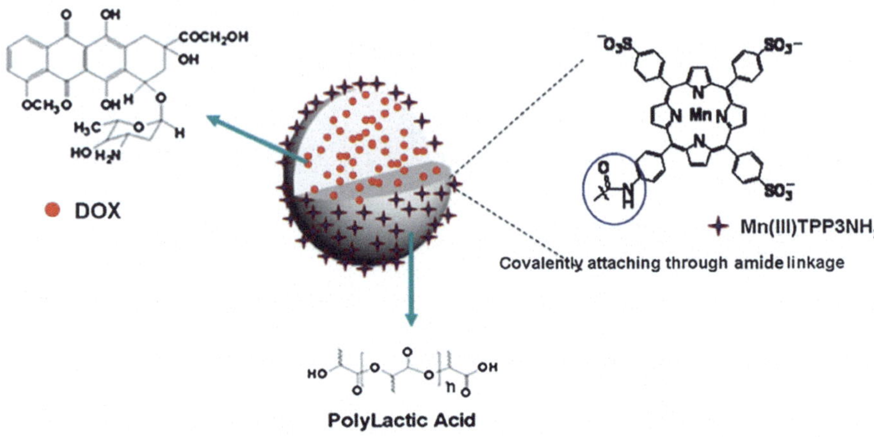

Fig. 4 Schematic representation of the preparation of PEGylated Mn^{2+}–PDA nanoparticles [88]. Copyright (2013), with permission from Elsevier

MnO_2-coated PLGA nanoparticles loaded with hematoporphyrin monomethyl ether (HMME), a new effective porphyrin-related photosensitizer. The idea is for MnO_2 (a) to perform a catalytic function promoting the production of O_2 from tumor-overexpressed H_2O_2 amplifying photodynamic therapeutic effect after irradiation; and (b) to be degraded in the presence of glutathione to Mn^{2+} ions that may serve as MRI CAs. The system was tested in tumor model mice exhibiting both MR image enhancement and anticancer activity after irradiation.

Moreover, Cu^{2+} has also been investigated as a good T_1-weighed CA in cancer theranostic nanoparticles. Very recently Ge et al. [28] have described Cu^{2+}-loaded PDA nanoparticles for thermo-chemotherapy employing Cu^{2+} as the CAs and the chemotherapeutic drug at the same time relying on its intrinsic cytotoxicity. The system exhibited a remarkable synergistic effect in eliminating tumors and MRI-enhanced contrast in vivo.

And to conclude with T_1-weighed MRI CAs, it is interesting to mention a new and not much studied alternative approach involving ^{19}F-MRI instead of the most common proton MRI. ^{19}F-MRI presents an important advantage which is the reduction of the signal-to-noise ratio as there is no endogenous ^{19}F-MRI signal from the body. However, for the image quality to be comparable to the one obtained with ^{1}H-MRI, the contrast agent should be at a high concentration in the tumor site [94]. Porsch et al. [95] described, in 2013, a functional breast cancer theranostic nanoparticulated system based on fluorinated polymers encapsulating DOX as model drug. In vitro testing of the vehicles showed high ^{19}F-MRI contrast and cytotoxicity effect when the NPs were loaded with DOX.

The bibliography even describes nanocomplexes that serve as dual-mode CAs (i.e., combine both types of contrasts T_1 and T_2). Chen et al. [57] developed a biodegradable coordination polymer that co-encapsulates DOX and PDA nanospheres conjugated to iron ions for T_1/T_2 dual-mode MRI-guided chemo-photothermal synergistic therapy. The system exhibited high photothermal conversion efficiency controlled drug release in response to the laser, which results in prominent chemo-photothermal synergistic cancer cell therapy both in vitro and in vivo.

4 X-ray Computed Tomography (CT)

X-ray computed tomography (CT) is one of the most commonly used noninvasive clinical imaging modalities in modern medicine, as a result of the high penetration ability of X-rays. Currently, more than half of all patients with cancer have been treated with this technique, also even in combination with other therapies. This methodology uses ionizing X-rays to generate images by rotating an X-ray tube and a detector located on opposite sides of a patient. One of the most interesting advantages of this technique is the ability to precisely depict anatomical features [96]. However, it is necessary to administer high intensities of radiation beams and elevated doses of CAs. Additionally, the unique features of different tumors are relevant factors that determine the efficacy and imaging quality of CT [97–99].

The most widely used contrast agents for CT are iodinated small molecules, basically derivatives of 1,3,5-triiodobenzene [96]. However, these molecules have a short circulation in blood, undergoing rapid clearance by the mononuclear phagocyte system. For that reason, these radioisotopes are unable to achieve therapeutic effect. In this sense, polymeric nanoparticles are an excellent alternative to avoid these drawbacks, increasing the pharmacokinetic properties of iodine molecules and decreasing their renal clearance [99, 100]. For this purpose, the amine or hydroxyl groups of different polymers have been used to conjugate iodinated benzoic acids. The covalent anchorage of these iodinated molecules prevented the loss of sensibility for imaging and diagnosis. Additionally, the presence of long chains of polymers such as PEG favors the long circulation time of this iodinated polymeric CT agent in the body, avoiding their elimination in the liver or spleen [98].

In addition to iodine molecules, heavy metal NPs based on Au, Bi, Ta, or Gd have been extensively explored as contrast agents for CT. These radioactive NPs can be used also for therapy because the interaction of X-rays with heavy metals produces Auger electrons and photoelectrons that can kill tumor cells [101]. Particularly, gold nanoparticles (AuNPs) represent one of the most representative examples of these CT agents, owing to the high atomic number ($Z=79$) and k-edge value (80.7 keV) of gold. Moreover, this heavy metal provides a higher contribution of photoelectron effect to X-ray attenuation in comparison to iodine molecules. Other important advantages of gold-based NPs are their high inertness, biocompatibility, and their ability to be easily functionalized. Moreover, AuNPs can be used as theranostics, increasing the effect of ionizing radiation against cancer cells [96–98].

Currently, research is focused on the development of polymeric nanovehicles of AuNPs in order to improve their tumor accumulation, their CT imaging contrast, also acting as a radiosensitizer. For example, Al Zaki et al. encapsulated AuNPs in the inner core of micelles based on amphiphilic diblock copolymer PEG-*b*-PCL. These NPs were exhaustively characterized, obtaining hydrodynamic diameters ranging from 25 to 150 nm. The encapsulation of AuNPs allowed to improve imaging quality by CT with precise delineation of tumor areas. Additionally, in vivo experiments demonstrated that the combination of the CT-guided radiation therapy and gold-mediated radiosensitization using these NPs increased the survival time of tumor-bearing mice in comparison to mice receiving radiation alone [102].

Currently, the real application of Au-based nanoparticulate CT contrast agents is limited due to the elevated cost of gold. For that reason, other nanoparticulate X-ray CT contrast agents have been explored in the last years. In particular, bismuth-based NPs have been explored as a CT contrast as a result of its high X-ray attenuation coefficient and its competitive price. Additionally, bismuth is well known for its low toxicity and is usually incorporated into pharmaceuticals and cosmetics [96, 98, 101]. Yao et al. successfully embedded hydrophobic Bi_2S_3 NPs into the inner core of PLGA nanocapsule by water/oil/water (W/O/W) emulsion. These nanocapsules with a hydrodynamic diameter of 755 nm in aqueous solution demonstrated a remarkable inhibition of prostate tumor growth under external X-ray irradiation [103].

The stability and toxicity of heavy metal NPs still need to be addressed. For that reason, different authors have incorporated chelating agents such as DTPA and DOTA that allow to reduce the toxicity of these metals [98, 104]. Particularly, Heckert et al. encapsulated Bi-DOTA complexes in a new hyperbranched polyester polymer with sulfur pendants groups (HBPE-S) in each branching point. Authors demonstrated that the presence of sulfur pendants in the polymeric formulation facilitated the incorporation of a high concentration of Bi, increasing the quality of X-ray images. Moreover, the hydrophobic core of the NPs allowed the encapsulation of taxol, and the carboxylic acid into their surface was used to conjugate folate receptor targeting folic acid using click chemistry. This actively targeted theranostic system produced more than 80% cell death after 24 h with little or no harm to nonmalignant cells (see Fig. 5) [105].

5 Radionuclide Imaging: PET and SPECT

Radionuclide imaging mainly includes the positron emission computed tomography imaging (PET) and single-photon emission computed tomography imaging (SPECT). PET is one of the most common molecular imaging techniques, owing to its high sensitivity and resolution. In fact, the high energy of γ rays resulting from positron annihilation increases the penetration capability in comparison with CT [106]. Currently, ^{18}F is the most characteristic radionuclide applied for PET. However, this radionuclide exhibits a relatively short half-life. To avoid this drawback, different researches are focused on the development of different polymeric nanovehicles of this molecule [107].

The direct labeling of micelles with ^{18}F requires high temperature, absence of water, and long reaction times. In this sense, Di Mauro et al. developed a new synthetic

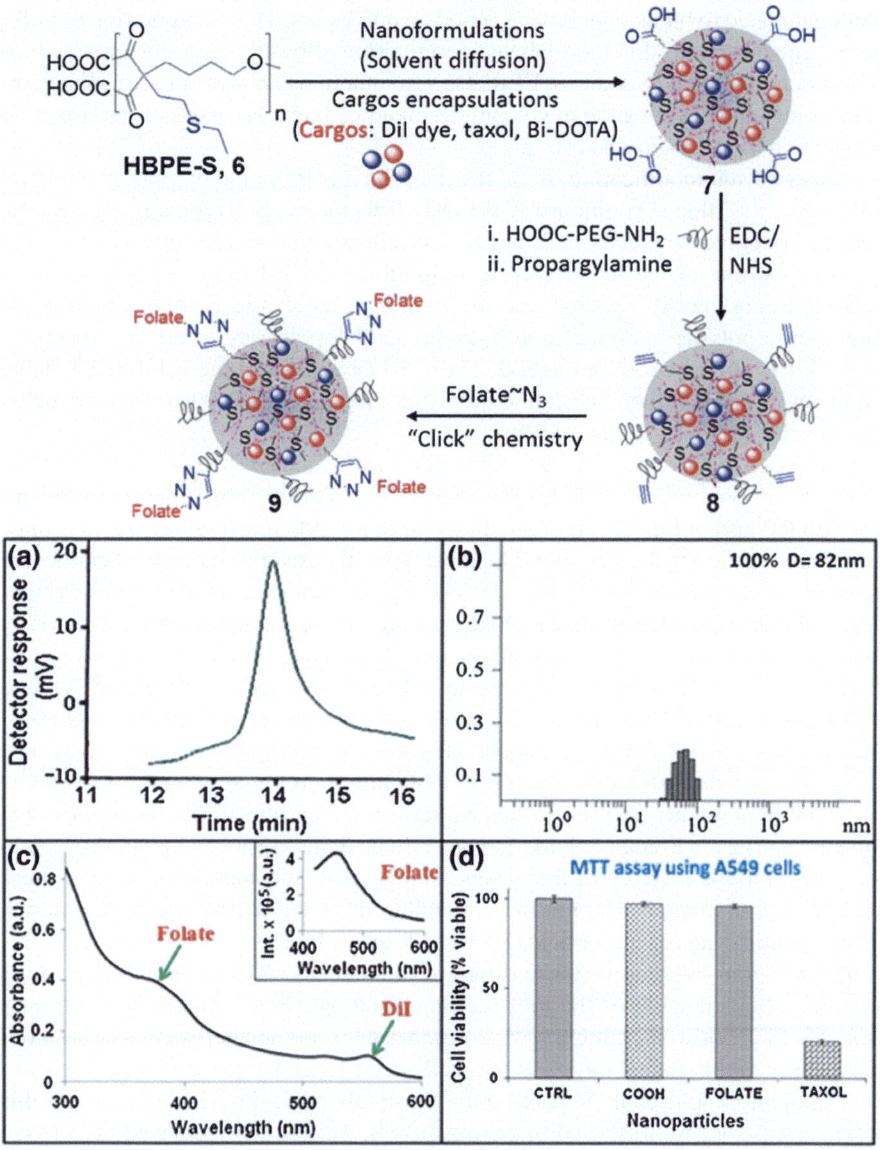

Fig. 5 Schematic representation of a new theranostic nanoformulation based on hyperbranched polyester where Bi-DOTA and taxol were physically entrapped in the inner core and the folate receptor targeting folic acid conjugated into their surface. Reprinted from [105]. Copyright (2017) American Chemical Society

strategy to label polymeric NPs. Specifically, these authors synthesized the radiolabeled active agent 4-[^{18}F] fluorobenzyl-2-bromoacetamide ([^{18}F]FBBA) which was later conjugated to block copolymers based on polyesters. The incorporation of polymeric sequences into this radiolabeled formulation allowed to significantly reduce their accumulation in liver due to EPR effect, maintaining an appropriate radiochemical stability and enabling the in vivo guide-imaging diagnosis and the quantification of biodistributed NPs in rats [107].

Other relevant modification of ^{18}F involves the incorporation of glucose to obtain 2-Deoxy-2-[^{18}F]fluoro-D-glucose (^{18}F-FDG). This molecule is typically used for the evaluation of glucose metabolism and also as radionuclide in oncology. For imaging applications, one of the most relevant limitations of ^{18}F-FDG is its low accumulation in tumors [108]. For that reason, Yamamoto et al. successfully modified an amphiphilic polydepsipeptide, poly(L-lactic acid)-*b*-poly(sarcosine) "Lactosome", with ^{18}F by a film hydration method. The ^{18}F-labeled lactosome exhibited a stable radioactivity and a higher circulation time after injection in comparison to unlabeled micelles due to the EPR effect [108].

In addition to the use of ^{18}F, different radiometals such as ^{68}Ga or ^{64}Cu have been attached to polymeric NPs. For this purpose, the use of bifunctional chelators is essential with the aim to bind the radionuclide and to provide reactive functional groups for the conjugation to the polymer. The most typically used bifunctional chelators are based on derivatives of DOTA and DTPA [107]. For example, Stigliano et al. synthesized polymeric micelles formed by hydrophobic PLGA polymeric core stabilized by an external monolayer of phospholipids and PEG chains. The phospholipid monolayer was successfully used to conjugate radioactive molecules ^{64}Cu(DOTA), and DTXL and curcumin were physically entrapped in the inner core of these spherical micelles (see Fig. 6). These theranostic NPs were tested in vitro and in vivo against glioblastoma cells with promising results. The tumor progression was monitored by PET with high quality and sensitivity. Moreover, the combinatorial therapy of both anticancer drugs in a same polymeric nanovehicle demonstrated a higher efficacy in comparison to the delivery of free drugs. In this sense, the appropriate size of these micelles, approximately 100 nm, was crucial to favor the release and accumulation of therapeutic drugs in tumor area due to EPR effect [109].

The real translocation of these different radiolabeled NPs is limited due to the relatively short half-life of the most common radionuclides. Among the currently available PET positron emitters, ^{124}I represents an excellent alternative as a result of the half-life higher than 4 days and the well-investigated behavior of iodine in vivo. For example, Simone et al. prepared poly(4-vinylphenol) NPs. The backbone of this polymer was successfully used to conjugate ^{124}I. This strategy allowed the in vivo and real-time tracking of NPs with high resolution and the complete study of pharmacokinetic parameters in animals [110].

Despite the progress in the development of radionuclide-based NPs for PET, ***Single-photon emission computed tomography (SPECT)*** is still the most extensively used imaging technique in clinic. In fact, more than 80% of all radio-diagnostic scans are performed by SPECT, owing to its sensitivity at nanomolar and even picomolar level [106]. The most applied gamma-emitting radionuclides are ^{123}I, ^{111}In, and ^{99m}Tc due to their optimal nuclear decay properties, availability, and rational price [106]. These different radionuclides have been incorporated into polymeric

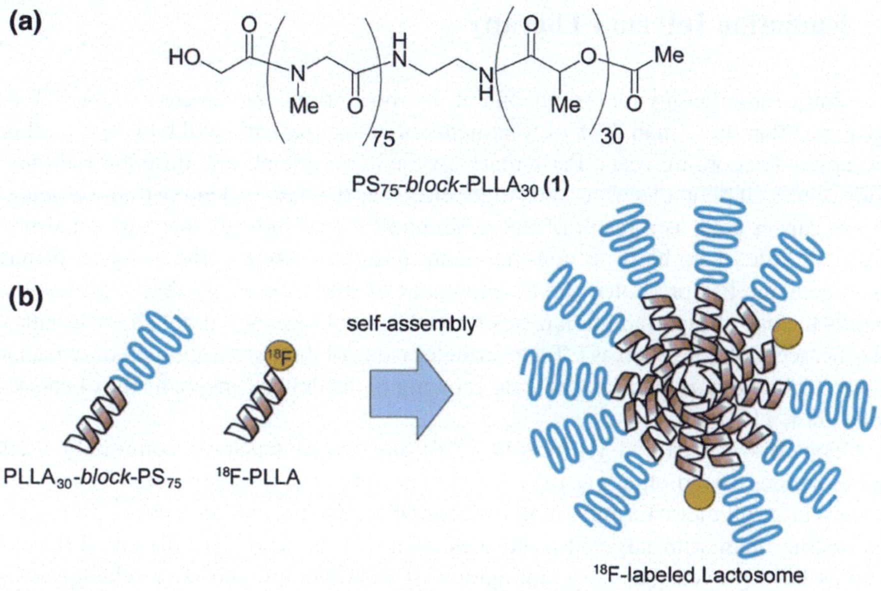

Fig. 6 **a** Molecular structure of amphiphilic block polymer. **b** Schematic representation of the preparation of ^{18}F-labeled Lactosome [108]. Copyright (2013), with permission from Elsevier

nanoformulations with the purpose to improve in vivo biodistribution due to the EPR effect. Miura et al. successfully synthesized poly(ethylene glycol)-*b*-poly(γ-benzyl L-glutamate) block copolymer where DOTA chelator was linked to aldehyde terminal functionalization of PEG. This block copolymer was able to self-assembly, and their inner core was used to encapsulate ^{111}In and/or Gd. The radionuclide NPs with a mean diameter of 30 nm were able to penetrate into tumor cells with a prolonged circulation time in blood [111]. In order to avoid the use of chelates that increase the synthetic procedures and can modify the size of NPs, Laan et al. entrapped ^{111}In in the inner core of polystyrene-*b*-poly(ethylene oxide) block copolymer micelles. For this purpose, tropolone as lipophilic ligand was successfully used as a result of its complexion with the radionuclide, without affecting the external shell of the NPs. These NPs were tested in vitro and in vivo, demonstrating an excellent radiolabeling efficiency and biodistribution with a prolonged circulation in blood post-injection [112]. More recently, rhenium-188 (^{188}Re) has emerged as a promising alternative to the most common radionuclide used for SPECT [113]. For example, Peng et al. prepared mPEG-*b*-PCL that were radiolabeled with ^{188}Re. Additionally, the core of these micelles was used to encapsulate an optical imaging agent, IR-780, obtaining an appropriate nanoformulation for targeting, imaging, and drug delivery [113].

6 Radiation Internal Therapy

Currently, radiotherapy represents one of the most promising cancer therapies. It is estimated that more than 70% of treatments of cancer patients will be based on this therapy in the coming years. Particularly, some heavy metals and different radionuclides can emit photoelectrons and Auger electrons that have a demonstrated efficacy to kill cancer cells as a result of the generation of free radicals that cause apoptosis, with little or no harm to non-malignant tissues. Currently, the progress of the nanomedicine has promoted the development of this technology that significantly avoids the use of high radiation doses for penetrating tissues that are characteristics of other technologies such as CT. Nevertheless, one of the most important drawbacks of this technology is the radioresistance, owing to the defense mechanisms of cancer cells [114, 115].

Copper-64 (^{64}Cu) and yttrium-90 (^{90}Y) are two of the most commonly used radionuclides for radiotherapy. In fact, different authors have incorporated these compounds into polymeric formulations, encapsulating also chemotherapeutics drugs and decorating them with targeted molecules such as folic acid. Specifically, different studies have proven that the combination of radiotherapy and chemotherapy can effectively decrease the radioresistance [97, 116]. For example, Rossin et al. synthesized micelles based on poly(acrylic acid-*b*-methyl acrylate) (PAA-*b*-PMA) block copolymer. Additionally, these micelles were successfully functionalized with folate and labeled with ^{64}Cu. In vivo experiments demonstrated the improved accumulation of these micelles in tumor as a result of the EPR effect and their interaction with folate receptor overexpressed in human nasopharyngeal epidermal carcinoma cell [117]. Moreover, Werner et al. prepared core–shell nanoparticles formed by PLGA and PEG that were effectively labeled with ^{90}Y and targeted with folate for the effective treatment of ovarian cancer. The inner core of the NPs was also used to entrap paclitaxel. Authors confirmed that folate-targeted NPs with ^{90}Y in their structure were the most effective for the treatment of ovarian cancer peritoneal metastasis [118]. In the same manner, Wang et al. encapsulated DTXL in the core of analogous NPs, chelating with ^{111}In and ^{90}Y as therapeutic radioisotopes. These biodegradable and biocompatible NPs exhibited a high therapeutic efficiency as a result of the simultaneous delivery of chemotherapy and radiotherapy, demonstrating the potential of these nanoformulations to be translocated to clinical [119].

Iodine radionuclides have been extensively used for radiotherapy as Auger electron-emitting nuclides. In particular, these molecules are quite interesting for this purpose because Auger electrons have short range, only some nanometers, in biological tissues, favoring their transportation almost solely to tumors, avoiding radiation exposure to the nonmalignant tissues [106]. Huang et al. successfully encapsulated DOX in the core of poly(ε-caprolactone-*co*-1,4,8-trioxa[4.6]spiro-9-undecanone)-poly(ethyleneglycol)-poly(ε-caprolactone-*co*-1,4,8 trioxa[4.6]spiro-9-undecanone) (PECT) triblock copolymer micelles. Additionally, these nanoassemblies were mixed with ^{131}I-labeled hyaluronic acid (^{131}I-HA), leading to an aqueous mixture that exhibited sol-to-gel transition around body temperature. This formu-

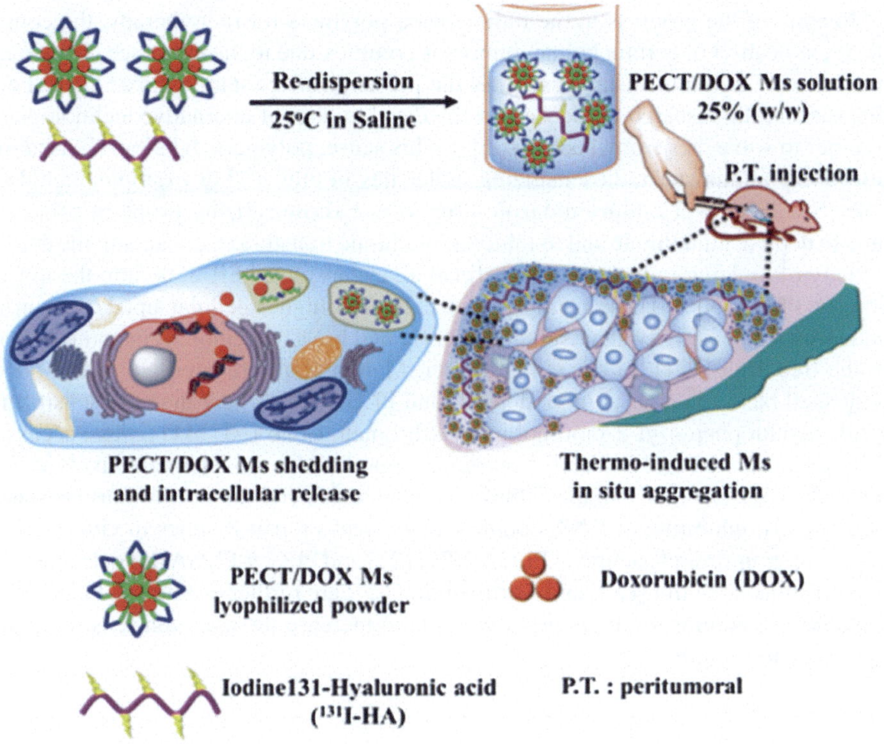

Fig. 7 Schematic illustration of theranostic NPs loaded with DOX and mixed with ^{131}I-labeled hyaluronic acid (^{131}I-HA) for PECT imaging and chemo-radiotherapy [116]. Copyright (2015), with permission from Elsevier

lation was internalized by tumor cells where DOX was completely released after 5 weeks at pH 7.4. In vitro and in vivo experiments evidenced the excellent antitumor activity of DOX-loaded NPs with a noticeable radiosensitization. In this sense, the radio-response of these NPs was significantly improved with the encapsulation of DOX that synergistically inhibited the biosynthesis of DNA, decreasing the repair of radiation-induced DNA damage by ^{131}I-HA (see Fig. 7) [116].

More recently, Schaal et al. reported the preparation of radioactive elastin-like polypeptide (ELP) micelles. In particular, both ^{125}I and ^{131}I were labeled to ELP using the iodogen oxidative reaction. These authors previously optimized the ELP that was based on pentapeptide sequence—Val-Pro-Gly-X_{aa}-Gly, derived from tropoelastin. Additionally, ELP exhibited a thermosensitivity with a lower critical solution temperature, LCST, of 21 °C. These micelles were structural stabilized into a hydrogel as a result of ^{131}I emission that induced covalent crosslinking within 24 h. Biological experiments were performed using prostate and pancreatic tumor models. In this context, ^{131}I-ELP micelles exhibited appropriated radioactivity, inhibiting tumor progression in both types of animal models [120].

In spite of the progress in the radiolabeled polymers for radiotherapy, the complete eradication of certain human tumors is complex due to the existence of several mechanisms of radioresistance such as the presence of resistant hypoxic cells. For that reason, the research is focused on the development of alternative technologies in order to solve these drawbacks [106]. In this sense, polymeric NPs can be used to effectively administer radiosensitizing molecules in tumors. For example, etanidazole (SR-2508) is a 2-nitroimidazole drug with radiosensitizing properties that is able to deplete glutathione and to inhibit glutathione transferase, enhancing the cytotoxicity of ionizing radiation. In fact, Jin et al. encapsulated SR-2508 into the inner core of PLGA NPs. These nanoparticles exhibited higher cellular uptake in vitro and radiation sensitivity against human cervix and breast carcinomas in comparison to the free drug [121]. In the same manner, Menon et al. prepared polymeric NPs with shell based on a poly(vinyl alcohol) and PLGA core where the radiosensitizer 8-dibenzothiophen-4-yl-2-morpholin-4-yl-chromen-4-one (NU7441) was successfully encapsulated and the cell-penetrating peptide R11 was also effectively conjugated. After external radiation, these micelles enhanced the uptake into prostate cancer cells, inhibiting of DNA double-strand breaks repair kinetics in vitro [122]. Finally, curcumin and paclitaxel PLGA NPs [123] and PEG-*b*-PLGA diblock copolymer NPs loaded with DTXL and wortmannin [124] are further examples of nanoformulations to administer drugs that were able to increase the radiation sensitivity of hypoxic cancer cells.

7 Ultrasound (US) Imaging

Ultrasound imaging is a noninvasive, safe, low–cost, and real-time imaging technique where sound waves at frequencies of 2 MHz or higher are sent to the patient's body and they echo off the tissue. These sound waves are reflected by the different tissues and received by a transducer that transforms this information into black and white images. Contrast is achieved because the waves are echoed in a dissimilar manner depending on the density of the tissue and the speed of the sound waves through this tissue [51] (i.e., acoustic impedance [2]).

The main drawbacks of US imaging modality are the low resolution (50 μm) and its inability to differentiate the boundary between soft tissues [4]. This is an important challenge in cancer diagnosis via US imaging because the acoustic impedance of cancer tissue and soft tissue is extremely similar as they both are mainly composed of water [2]. Therefore, for an accurate cancer diagnosis using this imaging modality, CAs are needed. Microbubbles and nanobubbles (NBs) are the most commonly used ones because they present high acoustic impedances when compared to most tissues [1]. Gases different from air (e.g., perfluorocarbons) are chosen for this purpose because they should present low solubility in biological media to improve contrast [2].

The most commonly described strategy for cancer theranostics combining cancer therapy and US imaging implies the use of passively targeted biodegradable polymers like PCL, PLGA, or PLLA. For example, in 2008, Gao and collaborators [125] prepared DOX-loaded nanoparticles based on biodegradable diblock copolymers PEG-*b*-PCL and PEG-*b*-PLLA. Perfluoropentane (PFP) was added to obtain a mixture of DOX-loaded micelles and DOX-loaded, PFP-encapsulating NBs. The mixture was demonstrated to passively accumulate in cancer sites in an in vivo model of breast cancer and, when subjected to US waves, microbubbles were formed leading to a controlled drug delivery and US contrast enhancement. Some years later, Ji et al. [126] described a PEG-*b*-PCL-based nanoparticulated system also encapsulating PFP, but loaded with curcumin as an anticancer drug. Authors showed in in vivo tumor-bearing mice that upon heating NBs became microbubbles due to PFP vaporization within the droplet walls improving US imaging contrast. And moreover, they demonstrated a significant anticancer activity under US because the nano/microbubbles cavitated and collapsed leading to curcumin-controlled release. DOX-loaded PLGA NBs formed by entrapping air during a conventional double emulsion preparation method were also described by Yang et al. [127]. In this case, the surface of the nanoparticles was modified with PEI to favor the incorporation of P-gp siRNA into the system by electrostatic interactions in order to combine gene therapy and US imaging. The nanotheranostic system showed interesting anticancer capacity and US image contrast enhancement in vitro. In 2016, Meng et al. [128] encapsulated DOX in PLGA NBs but wrapping carbon tetrafluoride to combine chemotherapy and US imaging. Authors confirmed the enhanced ultrasonic function and antitumor properties of the system in vivo.

High intensity-focused ultrasound (HIFU) is a noninvasive technique that is being used for the treatment of cancer as can induce coagulative necrosis at a precise focal point by thermal effect. However, this technique is not able to avoid cancer recurrence and metastasis. Zhang et al. [34] described multifunctional targeted PLGA NBs that combine the properties of an efficient ultrasound CA and a targeted anticancer drug nanocarrier. Methotrexate anticancer drug was loaded into the NBs and the active targeting monoclonal antibody human leukocyte antigen-G (anti-HLA-G) was conjugated on the surface. The chemotherapy was combined with HIFU for tumor ablation showing a synergistic anticancer effect that can kill the residual cancer cells and inhibit the reoccurrence of tumors during and after HIFU ablation.

Other authors have described passively targeted systems based on natural polymers like alginate [129] or human serum albumin (HSA) [130]. For example, Baghbani and collaborators described DOX-loaded alginate stabilized perfluorohexane (PFH) nanodroplets [129]. The idea was the same, microbubbles formation by acoustic vaporization of the nanodroplets, and particle cavitation leading to controlled drug release when subjected to US waves. Excellent anticancer effects and improved US contrast were demonstrated in vivo.

Another interesting strategy described in the literature implies the chemical generation of gases within the body in response to hallmarks of the tumor microenvironment. Min et al. [9] described DOX-loaded pH-responsive gas-generating mineralized NPs that are able to produce CO_2 gas through acidic pH-catalyzed reactions from $CaCO_3$ (see Fig. 8). The high theranostic capacity of the system was shown in vitro and in vivo. A year later, the same group designed another anticancer drug-

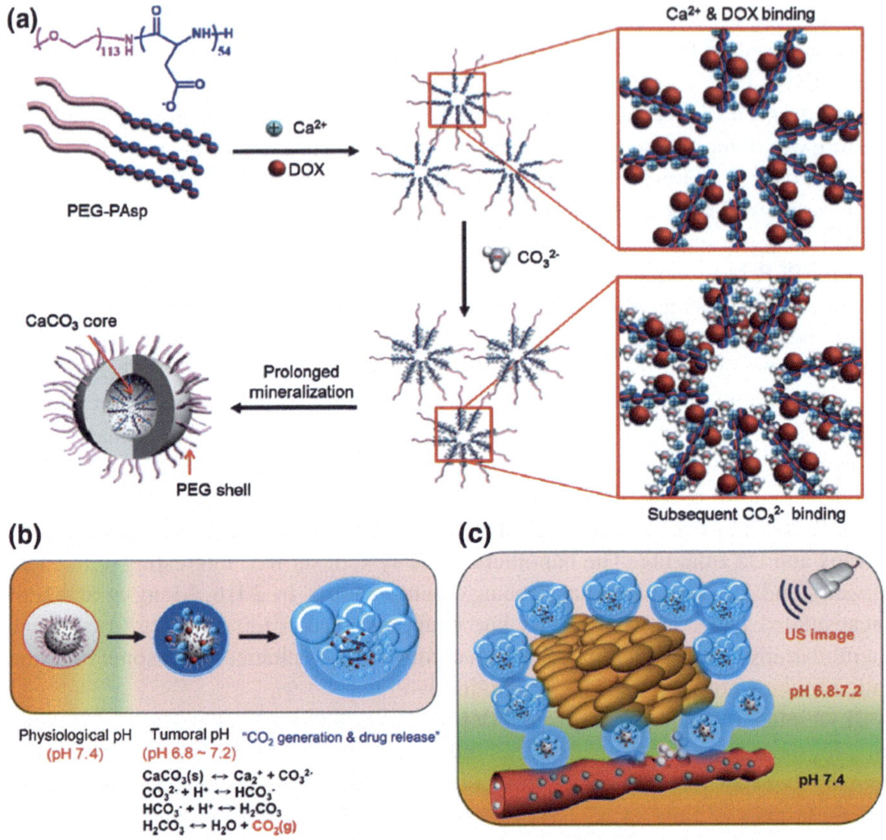

Fig. 8 Schematic illustration of the strategy described by Min et al. **a** fabrication process of DOX-$CaCO_3$-MNPs, **b** mechanism of CO_2 generation and drug release, and **c** bubble generation and drug release after accumulation of DOX-$CaCO_3$-MNPs at tumor tissues. Reprinted with permission from [9]. Copyright (2015) American Chemical Society

loaded CO_2-generating nanoparticles based on polycarbonate copolymer [131]. The nanoparticles showed great innovation and therapeutic applicability in tumor-targeted US imaging and US-triggered drug delivery.

Finally, there are also smart polymer-based systems described in the bibliography like the one designed by Yang et al. [132] entrapping PFH that allowed DOX release in response to the US by collapse and cavitation; ROS by incorporating a disulfide bond in their structure and; acidic pH by diminishing the electrostatic interaction between carboxylic groups of PMMA and DOX.

8 Photoacoustic Imaging (PAI)

Photoacoustic imaging (PAI) is the most recently discovered and developed imaging technique [133]. It is based on the PA effect which consists on the production of localized heat and thermal expansion which generates thermoelastic stress waves (i.e., US waves) when tissues absorb an optical pulse of a few nanoseconds [2]. It is a noninvasive technique that offers higher contrast than US imaging and deeper tissue penetration and higher spatial resolution than optical imaging for theranostics [133].

In this case, image contrast comes from the difference in optical absorption of the incident light [133] and CAs can be either endogenous (e.g., melanin or hemoglobin) or exogenous like inorganic or organic nanomaterials (e.g., gold or semiconducting polymers, respectively). Furthermore, the generation of local heat produced by light irradiation required for image generation makes it easy to combine PA imaging with cancer photothermal therapy (PAI/PTT) [134].

To date, researchers mainly described inorganic nanomaterials as PAI/PTT agents [135–137]. However, there are increasing concerns about their lack of degradability and photostability, and possible bio-toxicity. To solve biocompatibility and biodegradability problems, polymer-based systems seem to be the best alternative and the recent literature focuses on them [134, 138–141].

Some examples of recent works taking advantage of endogenous CAs and showing good PTT efficacy in vivo include passively targeted HSA-based NPs encapsulating melanin [138, 139] and heteropoly blue (HPB)-doped poly(4-vinyl pyridine-*b*-ethylene oxide) NPs (HPB/P4VP-*b*-PEO NPs) [139]. HPB is a small molecule dye with significant heat stability, acid–base stability, and strong absorption with high potential as CA for PA imaging but presenting short circulation times [139].

Regarding exogenous CAs, Wang et al. [142] described the simplest solution, a theranostic system combining a polymeric corona of PLGA and encapsulating PFH and Au NPs in the core. The polymeric coating showed improved biocompatibility in vitro as well as enhanced anticancer therapy.

In addition, the newest strategy described in bibliography involves exogenous CAs and, in particular, semiconducting polymer NPs prepared from hydrophobic semiconducting polymer building blocks [134]. Semiconducting polymers (SPs) show superior biocompatibility, biodegradability, and photostability being the perfect candidates for the preparation of PAI/PTT systems. Some examples include the biocompatible PEGylated SP-NPs designed by Zhang et al. [134] that presented a light-harvesting unit to enhance PA brightness and a conjugated electron donor–acceptor for heat generation in response to light enabling PTT. In vitro and in vivo results demonstrated the biocompatibility and remarkable anticancer capacity of the NPs with 100% tumor elimination achieved.

Lyu et al. [141] have also recently described a theranostic system with similar characteristics. In this case, an amphiphilic triblock polymer (PEG-*b*-PPG-*b*-PEG) is used as coating and poly[2,6-(4,4-bis(2-ethylhexyl)-4*H*-cyclopenta-[2,1-b;3,4-b′]dithiophene)-*alt*-4,7-(2,1,3 benzothiadiazole)] (PCPDTBT) and (6,6)-phenyl-

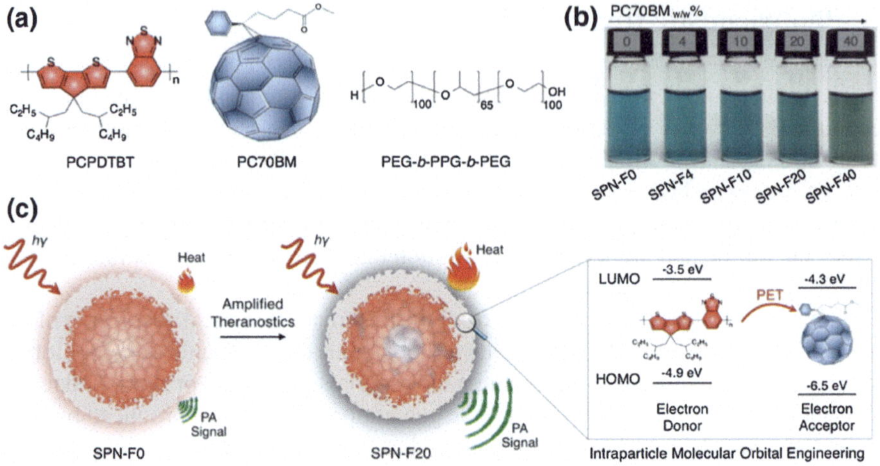

Fig. 9 Schematic illustration of intraparticle molecular orbital engineering design approach for amplified photoacoustic (PA) imaging and photothermal therapy (PTT). Reprinted with permission from [141]. Copyright (2016) American Chemical Society

C71-butyric acid methyl ester (PC70BM) were chosen as the primary component and the secondary dopant, respectively. The system showed amplified PA brightness and PTT efficacy against HeLa cells in vitro and in 4T1-tumor-bearing mice (see Fig. 9). Moreover, polypyrrole-based nanoparticles stabilized with astaxanthin-conjugated BSA have been recently described by Bharathuraja and collaborators [140]. This system showed in vitro a combination of improved PA imaging contrast and PTT capabilities, thanks to polypyrrole, an organic conductive polymer, and efficient PDT mediated by astaxanthin, a photosensitizer.

9 Multimodal Imaging

Multimodal imaging theranostics can be obtained by the combination of two or more imaging moieties in a single system [2, 4]. The numerous tools and the great versatility of organic chemistry make easier the incorporation of multiple imaging moieties into polymeric nanoparticles if compared to those based on inorganic materials [4].

Bimodal combination of PAI and optical imaging is commonly found in the literature. In some cases, the same molecule is used as CA for both imaging modalities like cyanine [143] and polypyrrole [144]. That is the case of the heptamethine-loaded PEG_{2k}-*b*-PLA_{2k} nanoparticles described by Lin et al. [143]. In this work, poly(heptamethine) was synthesized via multicomponent Passerini reaction, and then assembled into nanoparticles in the presence of PEG_{2k}-*b*-PLA_{2k} reaching high heptamethine loading (>50%). The light absorption of the cyanine derivative enhanced

PA and fluorescent image contrast and the heat generated after laser irradiation inhibited cervical carcinoma growth in mice.

Another bimodal PA and optical imaging CA is Ce6 that also works as a photosensitizer. Li et al. [145] designed theranostic nanomicelles based on hyaluronic acid (HA), a natural polymer, and active ligand of CD44 (i.e., a commonly overexpressed cancer cells surface receptor). The Ce6-conjugated HA-based system presented enzyme-responsive disassembling in the presence of hyaluronidase (i.e., an enzyme commonly overexpressed in cancer sites) and exhibited more effective tumor suppression via photodynamic therapy than the free Ce6. Additionally, in 2014, Gong et al. [38] described the use of Ce6 as a chelating agent for Gd^{3+} for the first time. They described PEGylated poly(maleic anhydride-*alt*-1-octadecene) (C18PMH) nanomicelles for triple modal photoacoustic imaging, magnetic resonance (MR), and fluorescence of tumors. They used Ce6 as PA and optical imaging CA and as chelator for Gd^{3+} to also achieve MRI T_1-weighed contrast, together with the NIR fluorescent dye IR825. The system presented photothermal and photodynamic properties achieving a synergistic antitumor effect in vitro and in vivo. Two years later, Wang et al. [146] designed an intracellular acid-switchable nanotheranostic system based on a pH-responsive diblock copolymer poly(ethylene glycol)-*b*-poly(diisopropanol amino ethyl methacrylate cohydroxyl methacrylate) (PEG-*b*-PDP), a DOX-prodrug, and Ce6 as photosensitizer. Gd^{3+} was chelated to Ce6 so that the nanoparticles combined pH-controlled multimodal imaging characteristics, pH-controlled delivery of DOX, and photodynamic and photothermal properties upon NIR irradiation.

MRI and X-ray present high resolution but low sensitivity, while optical imaging presents excellent sensitivity and low resolution; therefore, the combination of MRI or X-ray with optical imaging can result in highly enhanced images with the best features of each of the techniques.

A relevant example of X-ray/optical bimodal imaging theranostic system was obtained by the conjugation of functional molecules including polyethylene glycol (PEG), NIR fluorescent cyanine dye Cy7, DOX, and HA onto the surface of the TaOx NPs (TaOx@Cy7-DOX-PEG-HA NPs) for actively targeting delivery, pH-responsive drug release, and NIR fluorescence/X-ray CT bimodal imaging [147]. The system showed simultaneous enhancement of X-ray imaging, thanks to the tantalum oxide (TaOx) and optical imaging due to the presence of Cy7. Besides, the system actively targeted CD44 enhancing the accumulation of cancerous tissue and exhibiting an excellent tumor growth inhibition in vivo.

Other examples using the most common MRI CAs (i.e., Mn^{2+} [148], Gd^{3+} [149], fluoropolymers [150], and SPIONs [151, 152]) can be found in the bibliography. Regarding T_2-weighed MRI CAs, Park et al. [151] described a T-responsive nanogel based on poly(*N*-isopropylacrylamide-*co*-acrylic acid (poly(NiPA-*co*-AAc)) carrying amine-functionalized iron oxide-coated magnetic nanoparticles as MR CA, and D282 a lipophilic indocarbocyanine fluorescent dye for optical imaging. The system was complexed with different genes for gene therapy and showed high transfection efficiency as well as improved MR and optical image contrast.

Nevertheless, smart monomers responding to pH (i.e., methacrylic Liu et al. [149]), copolymerized methacrylic acid, *N*-isopropylacrylamide, and styrene, and the resulted polymer was reacted with an europium(III) derived monomer with optimum optical features. The pH and temperature-responsive amphiphilic polymer were used to encapsulate DOX as anticancer drug and gadolinium oxide as T_1-type MR contrast. Authors demonstrated that the paramagnetic, luminescent, and temperature/pH-responsive polymeric particles (PLTPPs) presented a better T_1-weighted relaxivity than clinical Magnevist® and MR and optical imaging demonstrated that are effective as dual probes in vivo.

Abbasi and collaborators [148] co-loaded MnO nanoparticles and DTXL into the matrix of fluorescent dye labeled amphiphilic block copolymer presenting a hydrophilic backbone (i.e., poly(isobutylene-*alt*-maleic anhydride) and hydrophobic side chains (i.e., dodecylamine). The system exhibited higher cytotoxicity than free drug, enhanced MRI contrast, and stained tumor tissue by fluorescence in an orthotropic MDA-MB-231 breast tumor model.

In a recent work, Vu-Quang et al. [150] presented a theranostic nanoparticulated system combining ^{19}F-MRI and optical imaging modalities. The actively targeted nanoparticles were based on PLGA-PEG-folate and co-encapsulated DOX as chemotherapeutic drug, perfluorooctyl bromide (PFOB) as ^{19}F-MRI CA, and indocyanine green (ICG) as NIR fluorescent dye. The active targeting system showed improved in vivo MR and optical image contrast and the in vitro experiments revealed higher cytotoxicity than non-folate-conjugated particles.

Besides, less common combinations of imaging modalities have been also reported like X-ray and PA imaging [153], MRI and US [154], or MRI/US/optical imaging [155].

10 Conclusions and Future Trends

As described in this chapter, polymeric nanoassemblies have demonstrated a great potential and several advantages to develop nanoplatforms for cancer molecular imaging and theranostics. Several approaches are currently available to create targeted polymeric NPs that are able to chemically by physically incorporate a wide range of theranostic molecules [156].

Actually, the number of FDA approvals and clinical trials using NPs has significantly increased in last years. However, clinical trials and most of the approved nanomedicines involve relatively simple nanovehicles based on well-described polymeric formulations and approved drugs. Additionally, the majority of approved nanoplatforms are based on PEGyated liposomes. Exclusively, PEG-*b*-poly(D,L-lactic acid) block copolymers micelles (Genexol-PM) that entrap paclitaxel have been approved for clinical cancer treatment in Korea in 2007 [157]. For theranostic purposes, AGuIX nanoassemblies are being evaluated in Phase I trials for tumor radiotherapy. This formulation is composed of polysiloxane network surrounded by Gd-DOTA. In particular, AGuIX nanoassemblies have been effectively applied for MR imaging-guided radiotherapy of different tumor models such as glioma, brain, or melanoma metastasis [158].

Research in this field still requires to improve the current developments of theranostic and polymeric NPs and overcome their drawbacks [51, 159]. In this sense, the implantation of controlled synthetic routes of polymers, such as RAFT polymerization and ATRP, and the continuous development of smart platforms using physicochemical or even biological stimuli are promising strategies to reach clinical transfer of theranostic NPs. Additionally, clinical studies with these NPs imply the enhancement of their biodistribution, especially regarding pharmacokinetic features of polymeric nanomaterials, and challenges with the aim to improve their biocompatibility, clearance, and selectivity toward cancer. All these goals involve an extensive work with multidisciplinary experts that work together as a single brain to understand the imaging techniques, the physicochemical behavior of the nanosystems and the biological responses of the organism.

Finally, it should also be pointed out that the complexity of the systems makes their regulatory approval and their manufacture extremely difficult. The regulatory agencies are making great efforts in this sense; however, the complexity of the nanosystems makes each one unique and the already established rules cannot be applied, since at the nanoscale unexpected properties appear that are not observed at the macroscale. Moreover, the described polymeric NPs for cancer therapy and bioimaging must be manufactured under the conditions of good manufacturing practice (GMP), and there are a limited number of certified companies specialized in nanosystems [106].

References

1. Luk BT, Zhang L (2014) Current advances in polymer-based nanotheranostics for cancer treatment and diagnosis. ACS Appl Mater Interfaces 6(24):21859–21873
2. Methachan B, Thanapprapasr K (2017) Polymer-based materials in cancer treatment: from therapeutic carrier and ultrasound contrast agent to theranostic applications. Ultrasound Med Biol 43(1):69–82
3. Fang J, Nakamura H, Maeda H (2011) The EPR effect: unique features of tumor blood vessels for drug delivery, factors involved, and limitations and augmentation of the effect. Adv Drug Deliv Rev 63(3):136–151
4. Mi P et al (2017) Molecular cancer imaging with polymeric nanoassemblies: from tumor detection to theranostics. Macromol Biosci 17(1):1600305
5. Aguilar MR, San Román J (2014) 1—Introduction to smart polymers and their applications, in smart polymers and their applications. Woodhead Publishing, pp 1–11
6. Gao M et al (2017) Tumor acidity-activatable TAT targeted nanomedicine for enlarged fluorescence/magnetic resonance imaging-guided photodynamic therapy. Biomaterials 133:165–175
7. Liu T et al (2012) Multifunctional pH-disintegrable micellar nanoparticles of asymmetrically functionalized beta-cyclodextrin-based star copolymer covalently conjugated with doxorubicin and DOTA-Gd moieties. Biomaterials 33(8):2521–2531
8. Liu Y et al (2014) Multifunctional pH-sensitive polymeric nanoparticles for theranostics evaluated experimentally in cancer. Nanoscale 6(6):3231–3242
9. Min KH et al (2015) pH-controlled gas-generating mineralized nanoparticles: a theranostic agent for ultrasound imaging and therapy of cancers. ACS Nano 9(1):134–145
10. Gao X et al (2016) Biodegradable gadolinium-chelated cationic poly(urethane amide) copolymers for gene transfection and magnetic resonance imaging. Mater Sci Eng C Mater Biol Appl 65:181–187

11. Vinh NQ et al (2015) MRI-detectable polymeric micelles incorporating platinum anticancer drugs enhance survival in an advanced hepatocellular carcinoma model. Int J Nanomed 10:4137–4147
12. Liu, T.-M., et al (2017) Smart NIR linear and nonlinear optical nanomaterials for cancer theranostics: prospects in photomedicine. Progr Mat Sci
13. Song X, Chen Q, Liu Z (2015) Recent advances in the development of organic photothermal nano-agents. Nano Res 8(2):340–354
14. Luo S et al (2011) A review of NIR dyes in cancer targeting and imaging. Biomaterials 32(29):7127–7138
15. Yuan A et al (2013) Application of near-infrared dyes for tumor imaging, photothermal, and photodynamic therapies. J Pharm Sci 102(1):6–28
16. Shi Y et al (2017) Recent progress and development on polymeric nanomaterials for photothermal therapy: a brief overview. J Mat Chem B 5(2):194–206
17. Sortino S (2016) Light-responsive nanostructured systems for applications in nanomedicine. Springer, Berlin
18. Zheng M et al (2014) Robust ICG theranostic nanoparticles for folate targeted cancer imaging and highly effective photothermal therapy. ACS Appl Mater Interfaces 6(9):6709–6716
19. Wan Z et al (2014) Highly efficient hierarchical micelles integrating photothermal therapy and singlet oxygen-synergized chemotherapy for cancer eradication. Theranostics 4(4):399
20. Zheng M et al (2013) Single-step assembly of DOX/ICG loaded lipid–polymer nanoparticles for highly effective chemo-photothermal combination therapy. ACS Nano 7(3):2056–2067
21. Zhu A et al (2015) Dually pH/reduction-responsive vesicles for ultrahigh-contrast fluorescence imaging and thermo-chemotherapy-synergized tumor ablation. ACS Nano 9(8):7874–7885
22. Zhuoxuan Lu, H FY, Cao Rong, Lin Ying-Ying, Zhou Songlin, Zhao Huange, Yonghao Huang AG-HT, Zhang Liming (2017) Synthesis of mPEG-dBSA-Cy5.5 nanoparticles for tumor imaging and drug delivery. Nanosci Nanotechnol Lett 9:184–189
23. Duan Z et al (2017) Stimuli-sensitive biodegradable and amphiphilic block copolymer-gemcitabine conjugates self-assemble into a nanoscale vehicle for cancer therapy. ACS Appl Mater Interfaces 9(4):3474–3486
24. Hill TK et al (2016) Near infrared fluorescent nanoparticles derived from hyaluronic acid improve tumor contrast for image-guided surgery. Theranostics 6(13):2314–2328
25. Lee Y-H, Lai Y-H (2016) Synthesis, characterization, and biological evaluation of anti-HER2 indocyanine green-encapsulated PEG-coated PLGA nanoparticles for targeted phototherapy of breast cancer cells. PLoS ONE 11(12):e0168192
26. Hung C-C et al (2017) Active tumor permeation and uptake of surface charge-switchable theranostic nanoparticles for imaging-guided photothermal/chemo combinatorial therapy: erratum. Theranostics 7(3):559
27. Fu L et al (2016) Folic acid targeted pH-responsive amphiphilic polymer nanoparticles conjugated with near infrared fluorescence probe for imaging-guided drug delivery. RSC Adv 6(46):40312–40322
28. Ge R et al (2017) Cu2+ loaded polydopamine nanoparticles for magnetic resonance imaging-guided pH- and near-infrared-light-stimulated thermochemotherapy. ACS Appl Mater Interfaces 9(23):19706–19716
29. Fernandez-Fernandez A et al (2014) Covalent IR820-PEG-diamine nanoconjugates for theranostic applications in cancer. Int J Nanomed 9:4631
30. Duong T et al (2017) Phototheranostic nanoplatform based on a single cyanine dye for image-guided combinatorial phototherapy. Nanomed Nanotechnol Biol Med 13(3):955–963
31. Conceição DS, Ferreira DP, Ferreira LFV (2013) Photochemistry and cytotoxicity evaluation of heptamethinecyanine near infrared (NIR) dyes. Int J Mol Sci 14(9):18557–18571
32. Tan X et al (2012) A NIR heptamethine dye with intrinsic cancer targeting, imaging and photosensitizing properties. Biomaterials 33(7):2230–2239
33. Wang Y et al (2014) Preferential accumulation of the near infrared heptamethine dye IR-780 in the mitochondria of drug-resistant lung cancer cells. Biomaterials 35(13):4116–4124

34. Zhang X et al (2014) Methotrexate-loaded PLGA nanobubbles for ultrasound imaging and synergistic targeted therapy of residual tumor during HIFU ablation. Biomaterials 35(19):5148–5161
35. Yuan A et al (2015) Self-assembled PEG-IR-780-C13 micelle as a targeting, safe and highly-effective photothermal agent for in vivo imaging and cancer therapy. Biomaterials 51:184–193
36. Yue C et al (2013) IR-780 dye loaded tumor targeting theranostic nanoparticles for NIR imaging and photothermal therapy. Biomaterials 34(28):6853–6861
37. Palao-Suay R, et al (2017) Photothermal and photodynamic activity of polymeric nanoparticles based on α-tocopheryl succinate-RAFT block copolymers conjugated to IR-780. Acta Biomat
38. Gong H et al (2014) Engineering of multifunctional nano-micelles for combined photothermal and photodynamic therapy under the guidance of multimodal imaging. Adv Func Mater 24(41):6492–6502
39. Guo F et al (2015) Smart IR780 theranostic nanocarrier for tumor-specific therapy: hyperthermia-mediated bubble-generating and folate-targeted liposomes. ACS Appl Mater Interfaces 7(37):20556–20567
40. Hamon CL et al (2016) Near-infrared fluorescent aza-BODIPY dye-loaded biodegradable polymeric nanoparticles for optical cancer imaging. J Nanopart Res 18(7):207
41. Ruan Z et al (2017) NIR imaging-guided combined photodynamic therapy and chemotherapy by a pH-responsive amphiphilic polypeptide prodrug. Biomat Sci 5(2):313–321
42. Pellosi DS et al (2017) Multifunctional theranostic pluronic mixed micelles improve targeted photoactivity of verteporfin in cancer cells. Mater Sci Eng, C 71:1–9
43. Chen R et al (2013) Near-IR-triggered photothermal/photodynamic dual-modality therapy system via chitosan hybrid nanospheres. Biomaterials 34(33):8314–8322
44. Oh I-H et al (2013) Cancer cell-specific photoactivity of pheophorbide a-glycol chitosan nanoparticles for photodynamic therapy in tumor-bearing mice. Biomaterials 34(27):6454–6463
45. Cohen EM et al (2010) Polymeric micelle nanoparticles for photodynamic treatment of head and neck cancer cells. Otolaryngology-Head and Neck Surgery 143(1):109–115
46. Conte C et al (2013) Biodegradable core-shell nanoassemblies for the delivery of docetaxel and Zn (II)-phthalocyanine inspired by combination therapy for cancer. J Controlled Release 167(1):40–52
47. Lee DJ et al (2012) Multifunctional poly (lactide-co-glycolide) nanoparticles for luminescence/magnetic resonance imaging and photodynamic therapy. Int J Pharm 434(1):257–263
48. Gupta A et al (2012) Multifunctional nanoplatforms for fluorescence imaging and photodynamic therapy developed by post-loading photosensitizer and fluorophore to polyacrylamide nanoparticles. Nanomed Nanotechnol Biol Med 8(6):941–950
49. Lamch Ł et al (2014) Polymeric micelles for enhanced Photofrin II® delivery, cytotoxicity and pro-apoptotic activity in human breast and ovarian cancer cells. Photodiagn Photodyn Ther 11(4):570–585
50. Sahu A et al (2013) Graphene oxide mediated delivery of methylene blue for combined photodynamic and photothermal therapy. Biomaterials 34(26):6239–6248
51. Vijayan VM, Muthu J (2017) Polymeric nanocarriers for cancer theranostics. Polym Adv Technol
52. Liu Q et al (2017) A superparamagnetic polymersome with extremely high T2 relaxivity for MRI and cancer-targeted drug delivery. Biomaterials 114:23–33
53. Li X et al (2013) Acid-triggered core cross-linked nanomicelles for targeted drug delivery and magnetic resonance imaging in liver cancer cells. Int J Nanomed 8:3019
54. Yang X et al (2010) Multifunctional stable and pH-responsive polymer vesicles formed by heterofunctional triblock copolymer for targeted anticancer drug delivery and ultrasensitive MR imaging. ACS Nano 4(11):6805–6817
55. Sun W et al (2017) Gadolinium-loaded poly(*N*-vinylcaprolactam) nanogels: synthesis, characterization, and application for enhanced tumor MR imaging. ACS Appl Mater Interfaces 9(4):3411–3418

56. Caro C, García-Martín ML, Pernia Leal M (2017) Manganese-based nanogels as pH switches for magnetic resonance imaging. Biomacromol 18(5):1617–1623
57. Chen Y et al (2016) Polydopamine-based coordination nanocomplex for T1/T2 dual mode magnetic resonance imaging-guided chemo-photothermal synergistic therapy. Biomaterials 77:198–206
58. Cao Y et al (2017) Poly(glycerol) used for constructing mixed polymeric micelles as T1 MRI contrast agent for tumor-targeted imaging. Biomacromol 18(1):150–158
59. Dong X et al (2017) Preparation of MRI-visible gadolinium methacrylate nanoparticles with low cytotoxicity and high magnetic relaxivity. J Mat Sci 52(13):7625–7636
60. Esser L et al (2016) Gadolinium-functionalized nanoparticles for application as magnetic resonance imaging contrast agents via polymerization-induced self-assembly. Polym Chem 7(47):7325–7337
61. Sanson C et al (2011) Doxorubicin loaded magnetic polymersomes: theranostic nanocarriers for MR imaging and magneto-chemotherapy. ACS Nano 5(2):1122–1140
62. Wang C et al (2012) Dual-purpose magnetic micelles for MRI and gene delivery. J Control Release 163(1):82–92
63. Chiang WH et al (2013) Functionalized polymersomes with outlayered polyelectrolyte gels for potential tumor-targeted delivery of multimodal therapies and MR imaging. J Control Release 168(3):280–288
64. Qin J et al (2015) Rationally separating the corona and membrane functions of polymer vesicles for enhanced T(2) MRI and drug delivery. ACS Appl Mater Interfaces 7(25):14043–14052
65. Schleich N et al (2013) Dual anticancer drug/superparamagnetic iron oxide-loaded PLGA-based nanoparticles for cancer therapy and magnetic resonance imaging. Int J Pharm 447(1–2):94–101
66. Yoon HY et al (2012) Hydrotropic magnetic micelles for combined magnetic resonance imaging and cancer therapy. J Control Release 160(3):692–698
67. Ao L et al (2014) A folate-integrated magnetic polymer micelle for MRI and dual targeted drug delivery. Nanoscale 6(18):10710–10716
68. Ling Y et al (2011) Dual docetaxel/superparamagnetic iron oxide loaded nanoparticles for both targeting magnetic resonance imaging and cancer therapy. Biomaterials 32(29):7139–7150
69. Dadras P et al (2017) Formulation and evaluation of targeted nanoparticles for breast cancer theranostic system. Eur J Pharm Sci 97:47–54
70. Guthi JS et al (2010) MRI-visible micellar nanomedicine for targeted drug delivery to lung cancer cells. Mol Pharm 7(1):32–40
71. Mosafer J et al (2017) In vitro and in vivo evaluation of anti-nucleolin-targeted magnetic PLGA nanoparticles loaded with doxorubicin as a theranostic agent for enhanced targeted cancer imaging and therapy. Eur J Pharm Biopharm 113:60–74
72. Situ J-Q et al (2016) Multifunctional SPIO/DOX-loaded A54 homing peptide functionalized dextran-g-PLGA micelles for tumor therapy and MR imaging. Sci Rep 6:35910
73. Yang H-K et al (2017) Bioreducible amphiphilic block copolymers based on PCL and glycopolypeptide as multifunctional theranostic nanocarriers for drug delivery and MR imaging. RSC Adv 7(34):21093–21106
74. Zhu K et al (2017) Photoregulated cross-linking of superparamagnetic iron oxide nanoparticle (SPION) loaded hybrid nanovectors with synergistic drug release and magnetic resonance (MR) imaging enhancement. Macromolecules 50(3):1113–1125
75. Jaiswal MK et al (2014) Thermoresponsive magnetic hydrogels as theranostic nanoconstructs. ACS Appl Mater Interfaces 6(9):6237–6247
76. Chiang W-H et al (2013) Functionalized polymersomes with outlayered polyelectrolyte gels for potential tumor-targeted delivery of multimodal therapies and MR imaging. J Controlled Release 168(3):280–288
77. Chiang W-H et al (2013) Superparamagnetic hollow hybrid nanogels as a potential guidable vehicle system of stimuli-mediated MR imaging and multiple cancer therapeutics. Langmuir 29(21):6434–6443

78. Liu Q et al (2015) An asymmetrical polymer vesicle strategy for significantly improving T1 MRI sensitivity and cancer-targeted drug delivery. Macromolecules 48(3):739–749
79. Li X et al (2011) Amphiphilic multiarm star block copolymer-based multifunctional unimolecular micelles for cancer targeted drug delivery and MR imaging. Biomaterials 32(27):6595–6605
80. Zarabi B et al (2008) HPMA copolymer–doxorubicin–gadolinium conjugates: synthesis, characterization, and in vitro evaluation. Macromol Biosci 8(8):741–748
81. Liu R et al (2015) Paramagnetic, pH and temperature-sensitive polymeric particles for anticancer drug delivery and brain tumor magnetic resonance imaging. RSC Adv 5(106):87512–87520
82. Li Y et al (2015) Nanoparticles based on star polymers as theranostic vectors: endosomal-triggered drug release combined with mri sensitivity. Adv Healthcare Mat 4(1):148–156
83. Li Y et al (2012) Light-triggered concomitant enhancement of magnetic resonance imaging contrast performance and drug release rate of functionalized amphiphilic diblock copolymer micelles. Biomacromol 13(11):3877–3886
84. Tong G et al (2016) Gadolinium/DOTA functionalized poly(ethylene glycol)-block-poly(acrylamide-co-acrylonitrile) micelles with synergistically enhanced cellular uptake for cancer theranostics. RSC Adv 6(56):50534–50542
85. Wen Y et al (2016) Nano-assembly of bovine serum albumin driven by rare-earth-ion (Gd) biomineralization for highly efficient photodynamic therapy and tumor imaging. J Mat Chem B 4(4):743–751
86. Chen AY, Zirwas MJ, Heffernan MP (2010) Nephrogenic systemic fibrosis: a review. J Drugs Dermatol 9(7):829–834
87. Pan D et al (2011) Manganese-based MRI contrast agents: past, present and future. Tetrahedron 67(44):8431–8444
88. Jing L et al (2013) Covalent attachment of Mn-porphyrin onto doxorubicin-loaded poly(lactic acid) nanoparticles for potential magnetic resonance imaging and pH-sensitive drug delivery. Acta Biomater 9(12):9434–9441
89. Xi J et al (2017) Mn(2+)-coordinated PDA@DOX/PLGA nanoparticles as a smart theranostic agent for synergistic chemo-photothermal tumor therapy. Int J Nanomed 12:3331–3345
90. Miao Z-H et al (2015) Intrinsically Mn2+-chelated polydopamine nanoparticles for simultaneous magnetic resonance imaging and photothermal ablation of cancer cells. ACS Appl Mater Interfaces 7(31):16946–16952
91. Dong Z et al (2016) Polydopamine nanoparticles as a versatile molecular loading platform to enable imaging-guided cancer combination therapy. Theranostics 6(7):1031–1042
92. Ding X et al (2016) Polydopamine coated manganese oxide nanoparticles with ultrahigh relaxivity as nanotheranostic agents for magnetic resonance imaging guided synergetic chemo-/photothermal therapy. Chem Sci 7(11):6695–6700
93. Hao Y et al (2017) Multifunctional nanoplatform for enhanced photodynamic cancer therapy and magnetic resonance imaging. Colloids Surf, B 151:384–393
94. Ruiz-Cabello J et al (2011) Fluorine ((19)F) MRS and MRI in biomedicine. NMR Biomed 24(2):114–129
95. Porsch C et al (2013) In vitro evaluation of non-protein adsorbing breast cancer theranostics based on 19f-polymer containing nanoparticles. Part Part Syst Charact 30(4):381–390
96. Liu Y, Ai K, Lu L (2012) Nanoparticulate X-ray computed tomography contrast agents: from design validation to in vivo applications. Acc Chem Res 45(10):1817–1827
97. Song G, et al (2017) Emerging nanotechnology and advanced materials for cancer radiation therapy. Adv Mat
98. Lee N, Choi SH, Hyeon T (2013) Nano-sized CT contrast agents. Adv Mater 25(19):2641–2660
99. Anton N, Vandamme TF (2014) Nanotechnology for computed tomography: a real potential recently disclosed. Pharm Res 31(1):20–34
100. Li X et al (2013) Iodinated α-tocopherol nano-emulsions as non-toxic contrast agents for preclinical X-ray imaging. Biomaterials 34(2):481–491

101. Du F et al (2017) Hyaluronic acid-functionalized bismuth oxide nanoparticles for computed tomography imaging-guided radiotherapy of tumor. Int J Nanomed 12:5973
102. Al Zaki A et al (2014) Gold-loaded polymeric micelles for computed tomography-guided radiation therapy treatment and radiosensitization. ACS Nano 8(1):104–112
103. Yao M-H et al (2014) Multifunctional Bi 2 S 3/PLGA nanocapsule for combined HIFU/radiation therapy. Biomaterials 35(28):8197–8205
104. Zhao J, Zhou M, Li C (2016) Synthetic nanoparticles for delivery of radioisotopes and radiosensitizers in cancer therapy. Cancer Nanotechnol 7(1):9
105. Heckert B et al (2017) Design and synthesis of new sulfur-containing hyperbranched polymer and theranostic nanomaterials for bimodal imaging and treatment of cancer. ACS Macro Lett 6(3):235–240
106. Pant K, et al (2017) Radiolabelled polymeric materials for imaging and treatment of cancer: Quo Vadis? Adv Healthcare Mat
107. Di Mauro PP et al (2015) Novel 18F labeling strategy for polyester-based NPs for in vivo PET-CT imaging. Bioconjug Chem 26(3):582–592
108. Yamamoto F et al (2013) Radiosynthesis and initial evaluation of 18 F labeled nanocarrier composed of poly (l-lactic acid)-block-poly (sarcosine) amphiphilic polydepsipeptide. Nucl Med Biol 40(3):387–394
109. Stigliano C et al (2015) Radiolabeled polymeric nanoconstructs loaded with docetaxel and curcumin for cancer combinatorial therapy and nuclear imaging. Adv Func Mater 25(22):3371–3379
110. Simone EA et al (2012) Endothelial targeting of polymeric nanoparticles stably labeled with the PET imaging radioisotope iodine-124. Biomaterials 33(21):5406–5413
111. Miura Y et al (2015) Polymeric micelle platform for multimodal tomographic imaging to detect scirrhous gastric cancer. ACS Biomat Sci Eng 1(11):1067–1076
112. Laan AC et al (2016) Radiolabeling polymeric micelles for in vivo evaluation: a novel, fast, and facile method. EJNMMI Res 6(1):12
113. Peng C-L et al (2011) Multimodal image-guided photothermal therapy mediated by 188Re-labeled micelles containing a cyanine-type photosensitizer. ACS Nano 5(7):5594–5607
114. Zhang L et al (2010) Delivery of therapeutic radioisotopes using nanoparticle platforms: potential benefit in systemic radiation therapy. Nanotechnol, Sci Appl 3:159
115. Mi Y et al (2016) Application of nanotechnology to cancer radiotherapy. Cancer Nanotechnol 7(1):11
116. Huang P et al (2015) Co-delivery of doxorubicin and 131 I by thermosensitive micellar-hydrogel for enhanced in situ synergetic chemoradiotherapy. J Controlled Release 220:456–464
117. Rossin R et al (2005) 64Cu-labeled folate-conjugated shell cross-linked nanoparticles for tumor imaging and radiotherapy: synthesis, radiolabeling, and biologic evaluation. J Nucl Med 46(7):1210–1218
118. Werner ME et al (2011) Folate-targeted nanoparticle delivery of chemo-and radiotherapeutics for the treatment of ovarian cancer peritoneal metastasis. Biomaterials 32(33):8548–8554
119. Wang AZ et al (2010) ChemoRad nanoparticles: a novel multifunctional nanoparticle platform for targeted delivery of concurrent chemoradiation. Nanomedicine 5(3):361–368
120. Schaal JL et al (2016) Injectable polypeptide micelles that form radiation crosslinked hydrogels in situ for intratumoral radiotherapy. J Controlled Release 228:58–66
121. Jin C et al (2008) Cellular uptake and radiosensitization of SR-2508 loaded PLGA nanoparticles. J Nanopart Res 10(6):1045–1052
122. Menon JU et al (2015) Polymeric nanoparticles for targeted radiosensitization of prostate cancer cells. J Biomed Mater Res, Part A 103(5):1632–1639
123. Yallapu MM et al (2010) Curcumin induces chemo/radio-sensitization in ovarian cancer cells and curcumin nanoparticles inhibit ovarian cancer cell growth. J Ovarian Res 3(1):11
124. Au KM et al (2015) Improving cancer chemoradiotherapy treatment by dual controlled release of wortmannin and docetaxel in polymeric nanoparticles. ACS Nano 9(9):8976–8996

125. Gao Z et al (2008) Drug-loaded nano/microbubbles for combining ultrasonography and targeted chemotherapy. Ultrasonics 48(4):260–270
126. Ji G, Yang J, Chen J (2014) Preparation of novel curcumin-loaded multifunctional nanodroplets for combining ultrasonic development and targeted chemotherapy. Int J Pharm 466(1–2):314–320
127. Yang H et al (2015) Multifunctional PLGA nanobubbles as theranostic agents: combining doxorubicin and P-gp siRNA Co-Delivery into human breast cancer cells and ultrasound cellular imaging. J Biomed Nanotechnol 11(12):2124–2136
128. Meng M et al (2016) Doxorubicin nanobubble for combining ultrasonography and targeted chemotherapy of rabbit with VX2 liver tumor. Tumour Biol 37(7):8673–8680
129. Baghbani F, et al (2016) Novel alginate-stabilized doxorubicin-loaded nanodroplets for ultrasounic theranosis of breast cancer. Int J Biol Macromol 93(Pt A):512–519
130. Lee JY et al (2015) Nanoparticle-loaded protein-polymer nanodroplets for improved stability and conversion efficiency in ultrasound imaging and drug delivery. Adv Mater 27(37):5484–5492
131. Min HS et al (2016) Chemical gas-generating nanoparticles for tumor-targeted ultrasound imaging and ultrasound-triggered drug delivery. Biomaterials 108:57–70
132. Yang P et al (2014) Stimuli-responsive biodegradable poly(methacrylic acid) based nanocapsules for ultrasound traced and triggered drug delivery system. Biomaterials 35(6):2079–2088
133. Valluru KS, Willmann JK (2016) Clinical photoacoustic imaging of cancer. Ultrasonography 35(4):267–280
134. Zhang J, et al (2017) Biocompatible D–A semiconducting polymer nanoparticle with light-harvesting unit for highly effective photoacoustic imaging guided photothermal therapy. Adv Func Mat 27(13):1605094-n/a
135. Zhou W et al (2016) Aromatic cyanoalkylation through double C-H activation mediated by Ni(III). J Am Chem Soc 138(18):5777–5780
136. Liu Y et al (2016) Human induced pluripotent stem cells for tumor targeted delivery of gold nanorods and enhanced photothermal therapy. ACS Nano 10(2):2375–2385
137. Cheng L et al (2014) PEGylated WS2 nanosheets as a multifunctional theranostic agent for in vivo dual-modal CT/photoacoustic imaging guided photothermal therapy. Adv Mater 26(12):1886–1893
138. Sim C et al (2015) Photoacoustic-based nanomedicine for cancer diagnosis and therapy. J Control Release 203:118–125
139. Yan Y et al (2017) Heteropoly blue doped polymer nanoparticles: an efficient theranostic agent for targeted photoacoustic imaging and near-infrared photothermal therapy in vivo. J Mat Chem B 5(2):382–387
140. Bharathiraja S et al (2017) Astaxanthin conjugated polypyrrole nanoparticles as a multimodal agent for photo-based therapy and imaging. Int J Pharm 517(1):216–225
141. Lyu Y et al (2016) Intraparticle molecular orbital engineering of semiconducting polymer nanoparticles as amplified theranostics for in vivo photoacoustic imaging and photothermal therapy. ACS Nano 10(4):4472–4481
142. Wang YJ, Strohm EM, Kolios MC (2016) Biodegradable polymer based theranostic agents for photoacoustic imaging and cancer therapy. In: SPIE BiOS. SPIE
143. Lin W et al (2016) Near-infrared polymeric nanoparticles with high content of cyanine for bimodal imaging and photothermal therapy. ACS Appl Mater Interfaces 8(37):24426–24432
144. Song X et al (2014) Ultra-small iron oxide doped polypyrrole nanoparticles for in vivo multimodal imaging guided photothermal therapy. Adv Func Mater 24(9):1194–1201
145. Li W et al (2016) Smart hyaluronidase-actived theranostic micelles for dual-modal imaging guided photodynamic therapy. Biomaterials 101:10–19
146. Wang T et al (2016) Intracellularly acid-switchable multifunctional micelles for combinational photo/chemotherapy of the drug-resistant tumor. ACS Nano 10(3):3496–3508
147. Jin Y et al (2015) Hyaluronic acid modified tantalum oxide nanoparticles conjugating doxorubicin for targeted cancer theranostics. Bioconjug Chem 26(12):2530–2541

148. Abbasi AZ et al (2015) Manganese oxide and docetaxel co-loaded fluorescent polymer nanoparticles for dual modal imaging and chemotherapy of breast cancer. J Control Release 209:186–196
149. Liu R et al (2016) Smart polymeric particle encapsulated gadolinium oxide and europium: theranostic probes for magnetic resonance/optical imaging and antitumor drug delivery. J Mat Chem B 4(6):1100–1107
150. Vu-Quang H et al (2016) Theranostic tumor targeted nanoparticles combining drug delivery with dual near infrared and 19F magnetic resonance imaging modalities. Nanomedicine 12(7):1873–1884
151. Park JS et al (2013) Poly(N-isopropylacrylamide-co-acrylic acid) nanogels for tracing and delivering genes to human mesenchymal stem cells. Biomaterials 34(34):8819–8834
152. Asem H et al (2016) Biodistribution of biodegradable polymeric nano-carriers loaded with busulphan and designed for multimodal imaging. J Nanobiotechnol 14:82
153. Jin Y et al (2014) Encapsulating tantalum oxide into polypyrrole nanoparticles for X-ray CT/photoacoustic bimodal imaging-guided photothermal ablation of cancer. Biomaterials 35(22):5795–5804
154. Niu C et al (2013) Doxorubicin loaded superparamagnetic PLGA-iron oxide multifunctional microbubbles for dual-mode US/MR imaging and therapy of metastasis in lymph nodes. Biomaterials 34(9):2307–2317
155. Yang PS et al (2014) A novel bubble-forming material for preparing hydrophobic-agent-loaded bubbles with theranostic functionality. Acta Biomater 10(8):3762–3774
156. Coelho SC et al (2015) Supramolecular nanoscale assemblies for cancer diagnosis and therapy. J Controlled Release 213:152–167
157. Bobo D et al (2016) Nanoparticle-based medicines: a review of FDA-approved materials and clinical trials to date. Pharm Res 33(10):2373–2387
158. Mi P, et al (2017) Molecular cancer imaging with polymeric nanoassemblies: from tumor detection to theranostics. Macromol Biosci 17(1)
159. Thakor AS, Gambhir SS (2013) Nanooncology: the future of cancer diagnosis and therapy. CA Cancer J Clin 63(6):395–418

Imaging and Treating Cancer with Carbon Nanotube Technology

Nicholas G. Zaibaq, Sakineh E. Moghaddam and Lon J. Wilson

Abstract Carbon nanotubes (CNTs) are one-dimensional cylindrical tubes of carbon atoms bonded together, similar to graphene and have drawn great interest for their use as medical agents due to their potential for a variety of clinical applications. Researchers have used CNTs as therapeutic agents, both by taking advantage of their powerful light-driven effects and using them as carriers for peptide and small molecule drugs. As progress toward these applications continues, monitoring CNTs in vivo will also be important to evaluate their potential toxicity, therapeutic activity, and any physiological side effects that the material may cause in various organs. Furthermore, CNTs provide a versatile platform for combining therapeutic and diagnostic capabilities for so-called theranostic agents, a field which has seen a surge of activity in recent years. Here, we focus on CNT-based materials functionalized for therapeutic, diagnostic, and theranostic applications in cancer, while also discussing the future directions and challenges of CNT medical agents in general. As evidenced by the body of work presented in this Chapter, CNTs may have a future as multimodal anti-cancer agents, thereby increasing the efficacy of current therapies and imaging techniques while also decreasing harmful side effects that often impede effective patient care.

Keywords Carbon nanotubes · Medical imaging · Cancer therapy · Theranostic agents · Drug delivery

1 Introduction

In order to diagnose and track the progression of diseases, physicians rely on non-invasive imaging to guide their therapeutic strategies. Medical images most often elucidate anatomical information using such techniques as magnetic resonance imag-

N. G. Zaibaq · S. E. Moghaddam · L. J. Wilson (✉)
Department of Chemistry and the Smalley-Curl Institute, Rice University, Houston, TX, USA
e-mail: durango@rice.edu

G. Gonçalves and G. Tobias (eds.), *Nanooncology*, Nanomedicine and Nanotoxicology, https://doi.org/10.1007/978-3-319-89878-0_5

ing (MRI), computed tomography/X-ray (X-ray CT), and ultrasound. Other nuclear imaging modalities, such as positron emission tomography (PET) and single-photon emission computed tomography (SPECT), are used to reveal functional information. Fluorescence imaging is an optical technique that can be used to provide real-time, wide-field images with high spatial resolution. While MRI and fluorescent imaging methods offer superior resolution without using ionizing radiation, they suffer from low sensitivity and poor penetration depth, respectively. And whereas nuclear medicine techniques such as PET and SPECT offer great sensitivity, they suffer from low resolution. By using some combination of these imaging techniques, physicians can make accurate diagnoses prior to developing a treatment plan.

For cancer therapy, the main methods used today include surgery, radiation therapy, chemotherapy, or some combination of the three, depending on the type, location, and nature of the disease. The goal of any treatment is to destroy malignant cells while leaving normal cells unaffected. Unfortunately, however, this is not the usual case for most of the current therapies. Surgery is invasive and risks leaving positive malignant margins and is not suitable for all patients. Radiation and chemotherapy, although noninvasive, are rarely sufficiently targeted to solely destroy cancerous tissue, and often come with severe side effects that become a limiting factor in patient care. Since chemotherapy has the potential to be effective without harming healthy tissue, there has been intense effort to create formulations in which small molecule drugs are protected and/or targeted such that side effects are minimized.

Carbon nanotubes (CNTs) can be visualized as single sheets of graphene, rolled up in such a way that the roll-up vector determines the diameter and chirality of the CNTs, thereby determining the material's properties, as shown in Fig. 1 [1]. The diameters of CNTs can range from less than one nanometer to a few nanometers, while lengths are often in the order of a few microns [2, 3]. CNTs can be either single-walled (SWCNTs) or multi-walled (MWCNTs), with MWCNTs containing between 2 and 30 concentric SWCNTs with outer diameters between 10 and 30 nm. Of the many nanoparticles (NPs) that can be used for different medical applications, CNTs are extremely versatile and have been studied extensively in cancer imaging and therapy with the goal of overcoming the challenges mentioned above. The present chapter is comprised of the following four interrelated sections: (A) CNT agents for cancer imaging, (B) CNT agents for cancer therapy, (C) CNT agents for theranostic medicine, and (D) Future challenges for clinical applications.

2 CNT Agents for Cancer Imaging

Fluorescence Imaging. The biological optical imaging process begins by excitation photons from a given source traveling through layers of tissue to excite a probe that emits fluorescence photons. These photons then travel back through the various tissues and are received by a detector, which then generates an image [4]. Throughout this process, excitation and fluorescence photons can be scattered due to changes in indices of refraction between air, water, and the various tissues through which

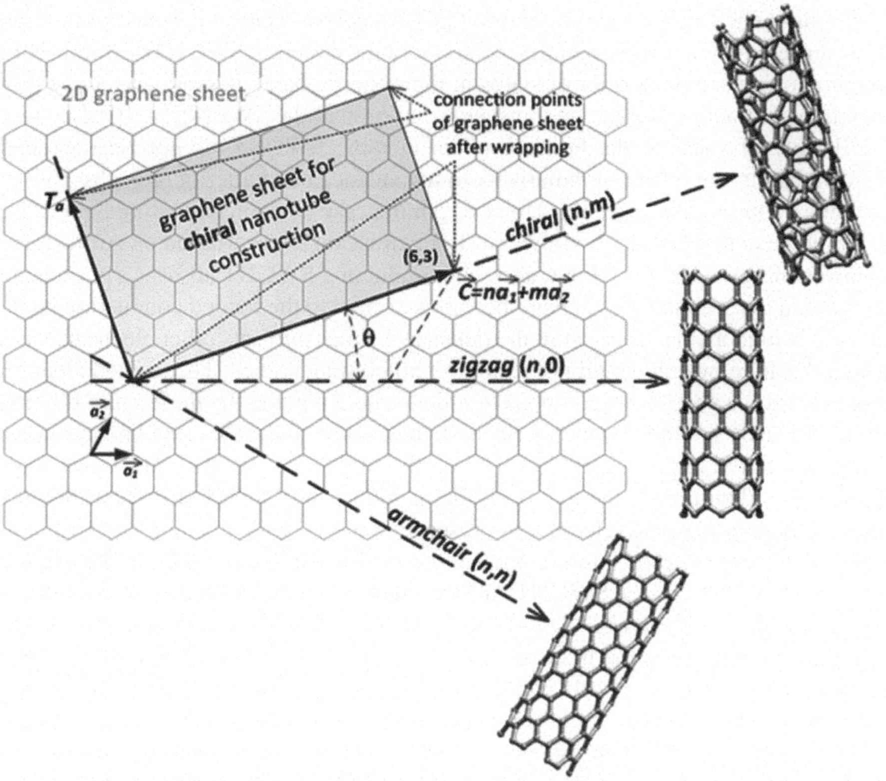

Fig. 1 Schematic diagrams showing how the angle at which graphene is rolled determines CNT chirality [1]

they travel. Photons can also be absorbed and dispersed as heat by naturally occurring chromophores in the organism, such as hemoglobins, flavins, aromatic amino acids, water, and many others [5]. Water has vibrational bands with maxima at 970, 1200, 1450, and >1800 nm, setting a boundary where fluorescence is possible for deep tissue imaging [6]. Both absorption and scattering processes reduce the signal that is detected and increase the background noise. Another factor that contributes to a low signal-to-noise ratio is the inherent non-specific tissue autofluorescence background. Additionally, fluorescence imaging in the visible region of the spectrum (400–700 nm) is limited, since mammalian tissue is opaque to wavelengths of light in that region [7]. Therefore, researchers have developed probes that are active in the near-infrared regions of the spectrum (NIR I: 700–900 nm and NIR II: 1000–1700 nm). Apart from advances in instrumentation to improve the efficacy of deep tissue optical imaging [8], there has been a drive to create fluorescent probes with enhanced quantum yield, photostability, and biocompatibility for the purpose of clearly discriminating between malignant and healthy tissue.

Single-walled carbon nanotubes (SWCNTs) have two main chiralities, which lead to either semiconducting or metallic behavior of the material. SWCNTs exhibit sharp maxima of electronic density of states in their energy band diagrams, as a result of quantum confinement along the transverse direction of the SWCNT [9, 10]. Because the density of states at the Fermi level in metallic SWCNTs is not zero, excited state electrons can relax nonradiatively to the valence band after photon absorption, yielding no photoluminescence. However, in the case of semiconducting SWCNTs, a gap in the density of states allows for light emission as the excited electrons relax to lower energy states [11]. First, as a semiconducting SWCNT absorbs a photon, an electron in the second valence band becomes excited to the second conduction band, leaving behind a hole. The gap in the density of states prevents other electrons from filling this hole, which would not result in photoluminescence. The excited electron and hole form a pair known as an exciton that is held together by the exciton binding energy (usually around 400 meV). Second, the exciton undergoes rapid nonradiative relaxation to the lowest energy levels, i.e., the first valence and conduction bands. The exciton then travels the length of the nanotube until the electron and the hole radiatively recombine to release a fluorescence photon, with an energy corresponding to the energy gap, E_{11} minus the energy of the exciton [12–14]. If the exciton encounters a defect on the SWCNT surface, such as a chemically derivatized site, or simply the end of the SWCNT, the energy will be released in a nonradiative fashion, thus extinguishing the fluorescence. Therefore, little to no photoluminescence is seen for SWCNTs shorter than 100 nm [15]. Many groups have utilized the ability of semiconducting SWCNTs to absorb strongly in the near-infrared range (NIR I, 750–900 nm) and emit fluorescence in the NIR II region (1000–1400 nm), where there is little to no autofluorescence from biological tissue in order to image cancer in vivo [16, 17]. The quantum yield of SWCNTs also depends on their diameters, which is further dependent on the synthetic method used to produce the SWCNTs. Of the main methods used to produce SWCNTs, HiPco SWCNTs have the highest quantum yield, and therefore have been used the most for biological imaging in the NIR II window [16, 18–20].

Furthermore, pristine SWCNTs are essential for any biological imaging to ensure good quantum yields. It is because these effects are not seen with covalently functionalized or shortened SWCNTs due to the greater number of defect sites inhibiting exciton movement that surfactants are used to non-covalently suspend SWCNTs to ensure their fluorescent properties are maintained. For example, Dai and coworkers have performed extensive work on producing surfactant-wrapped, debundled SWCNTs for in vivo fluorescent imaging [17]. One of the methods they reported for functionalizing SWCNTs used is a surfactant exchange method that minimized the number of defect sites on the SWCNTs, hence increasing the quantum yield [16]. The authors first suspended the SWCNTs with sodium cholate by sonication for 1–6 h. Then, a phospholipid–polyethylene glycol (PL-PEG) derivative was introduced to the suspended SWCNTs and replaced the sodium cholate. The resulting material had a quantum yield in order of magnitude greater than SWCNTs directly suspended with the PL-PEG, while still maintaining biocompatibility. The SWCNTs that were suspended using solvent exchange (exchange SWCNTs) were also longer (averag-

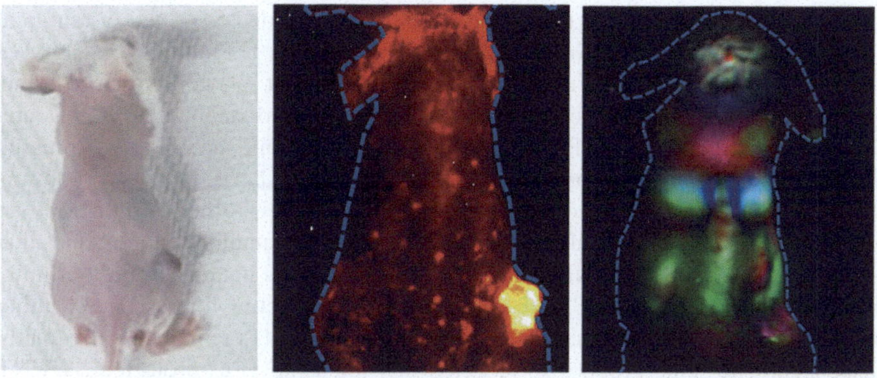

Fig. 2 Photograph of tumor-bearing mouse (left). NIR II imaging of a tumor (yellow) using SWCNTs (middle). NIR II image using PCA analysis (right) [17]

ing ~370 ± 310 nm) than SWCNTs that were suspended directly with PL-PEG (direct SWCNTs; averaging ~160 ± 85 nm). NIR images of the material were taken in vitro using malignant glioma (U87MG) cells and in vivo, with LS174T tumor-bearing mice using intravital microscopy (IVM) to image tumor angiogenesis. Using IVM, the authors could not only resolve small vessels a few micrometers beneath the skin, but could also resolve other structures in deeper organs in nontumor-bearing animals.

Another study used intravenously (i.v.)-injected SWCNTs to conduct video-rate imaging in mice that could clearly resolve anatomical features that are usually very difficult to image, such as the pancreas [21]. The study reported biodistribution data over time and could image SWCNTs in the vasculature and in organs of the reticuloendothelial system (RES), particularly the liver and spleen. Furthermore, the group used dynamic imaging through principal component analysis (PCA) in order to visualize organs that are not seen through real-time raw images [22]. This technique organizes pixels that vary similarly over time, which correspond to the same organ, as each organ would vary slightly differently in terms of blood flow and movement. This approach has the potential to image tumors buried underneath larger organs that would not be visualized using conventional optical methods. For example, PCA was used to image subcutaneous 4T1 breast cancer tumors in mice which found long blood circulation ($t_{1/2}$ of ~30 h) and high tumor uptake of polymer-wrapped SWCNTs, as shown in Fig. 2 [17]. Due to the difference in blood flow for normal and cancerous vasculature [23], PCA was able to discriminate between malignant and nonmalignant tissue. SWCNTs were seen in the tumor blood vessels as early as 20 s postinjection using dynamic contrast imaging through PCA. After 60 s postinjection, the tumor was clearly outlined separating it from surrounding healthy tissue. The authors also reported ex vivo imaging in order to do 3D reconstruction of the tumor to determine the depth of tumor penetration. The study found that the SWCNTs remained mainly in the highly vascularized regions around the edges of the tumor.

Another report described using bacteriophage-suspended SWCNTs as surgical NIR II dyes to detect submillimeter tumors during surgery. The authors used an M13 virus that can be genetically modified to express different relevant proteins. In this case, the M13 virus expressed some peptides to coat and stabilize SWCNTs and other peptides that bind to SPARC, a matricellular protein that is highly expressed in certain types of breast, prostate, and ovarian cancer [24]. The targeted probe was shown, using an orthotopic ovarian cancer mouse model, to be a useful tool for both pre-surgical imaging and image-guided resection of smaller tumor nodules that small molecule dyes in the NIR I and visible ranges could not detect. Furthermore, the study found that the SWCNT-based agent had long-lasting fluorescence without photobleaching, indicating that the virus-protected SWCNTs remained individually suspended. The same group also reported that the same virus-protected SWCNT probe can be targeted to prostate cancer [25]. Additionally, Heller and coworkers used NIR imaging on semiconducting SWCNTs to determine penetration of the SWCNTs in two different in vitro 3D spheroid tumor models, to predict the SWCNT's propensity to serve as a drug vector [26]. The study demonstrated that SWCNTs could penetrate into MCF-7 breast cancer spheroids better than it could SK-136 murine liver cancer spheroids, due to differences in cellular density and extracellular matrix (ECM) composition.

CNT/QD Conjugates. Quantum dots (QDs) are fluorescent inorganic nanoparticles that are usually comprised of group II–IV or III–V elements, such as CdTe, CdS, ZnSe, InP, or InAs [27]. QDs are of interest in biomedical research due to their tunable properties and their similar size to biological macromolecules (oligonucleotides, polypeptides, etc.). Furthermore, they have several advantages over organic fluorophores, such as increased quantum yield and photostability, broad absorption, and narrow emission spectra, as well as resistance to photobleaching and to changes in pH. These properties are all dependent on the QD size, shape, elemental composition, and structure, which can be studied under UV or NIR excitation [28]. QDs are usually made up of an inorganic core, whose diameter controls the wavelength of light that is emitted. Inorganic cores constructed from group III–V elements are preferred for medical applications, since these elements are generally less toxic than group II–IV elements. Covering the core is an inorganic shell, which contributes to the QD's photostability and increases the photoluminescence of the particle. The whole QD is then coated with a water-soluble polymer or surfactant, which allows the QD to be suspended in biological media and provides for conjugation to biomolecules and other nanoparticles for various biomedical applications. There is considerable interest in conjugating QDs to the large surface area provided by CNTs, due to the increase in possible functionalities that can be achieved [29, 30].

QDs may be attached to CNTs either covalently or non-covalently, through the use of an intermediary linker or directly onto the CNT surface. For covalent attachment, CNTs are usually first oxidized, producing hydroxyl and carboxylate groups on the surface, which serve as points for attachment. However, this may lead to an uneven distribution of functionalization, since CNTs usually preferentially oxidize at their tips [31]. Another issue with covalent conjugation arises from quenching of the luminescence of the QD from close contact with the CNT without the use of a molecular spacer [32]. In addition to using molecular linkers between the CNT surface and the QD, another conjugation method is to use polymeric surfactants to first coat the CNT, and then attach the QD to the polymer, either covalently or non-covalently. For example, one report used semiconductor QDs covalently attached to an oligonucleotide that was non-covalently wrapped around SWCNTs [33]. The authors reported that because they were able to visualize the CNT peaks in the UV/Vis/NIR spectrum, the QD/CNT hybrids remained well-suspended even after conjugation. This is the case because quenching occurs when the QD/CNT conjugates bundle from loss of the surfactant. Furthermore, the researchers reported redshifts in the CNT peaks, indicating charge transfer from the QDs to the SWCNTs, and therefore, successful conjugation. QDs have also been used to track intracellular delivery using an MWCNT-functionalized atomic force microscope (AFM) tip [34]. The MWCNTs acted as a nanoinjector that deposited the QD cargo into the cytosol and fluorescence imaging was then used to quantify the amount injected and to track the QD cargo inside HeLa cells. These examples show the practicality of using QD/CNT conjugates, since MWCNTs and covalently functionalized SWCNTs exhibit little or no photoluminescence. However, due to the reduced photoluminescence for QD/CNT conjugates [35] and the high toxicity of elements ordinarily used in QDs (such as Cd), biocompatibility of the material is of great concern which requires continued investigation. QD/CNT conjugates for biomedical applications have been reviewed in more detail elsewhere [30, 36].

Raman Imaging. Using Raman spectroscopy for biomedical imaging offers a good alternative to fluorescence imaging, since Raman is a scattering technique, as opposed to an absorption/emission technique, and there is no spectral overlap as there can be in fluorescence imaging. Also, Raman spectroscopy is extremely sensitive to small chemical and structural changes, allowing for narrow spectral lines that can be imaged with great multiplicity [37]. Raman effects are not subject to photobleaching or quenching, making imaging probes able to be visualized over longer timescales, such as months [38]. However, the Raman signal without any enhancement is fairly weak, which has led researchers to seek ways to incorporate surface-enhanced Raman spectroscopy (SERS) into biomedical imaging probes [39]. CNTs have many distinct Raman peaks, such as the radial breathing mode (100–300 cm^{-1}) and the tangential G band (1580 cm^{-1}), and their peaks are usually very sharp and intense, with a full width at half maximum usually less than 2 nm, which easily distinguishes them from background noise. Another means of enhancement is to use a laser energy equal to the energy required for an electron to transition from the valence band to the conduction band, known as resonance Raman scattering, and the resonance Raman peaks for CNTs are highly dependent on their diameter and chirality [40, 41].

The first report of using SWCNTs for biological Raman imaging used oligonucleotide DNA to wrap SWCNTs and image live 3T3 fibroblast and myoblast cells using fluorescence and Raman scattering [42]. The authors found that the Raman scattering was still visible through the H&E stain of the cells, whereas organic fluorophores and semiconducting nanoparticles are not, due to the high concentration of quenching agents in the stain. Furthermore, the scattering from Raman was consistently visible for much longer times than fluorescence imaging. Later, Zavaleta et al.

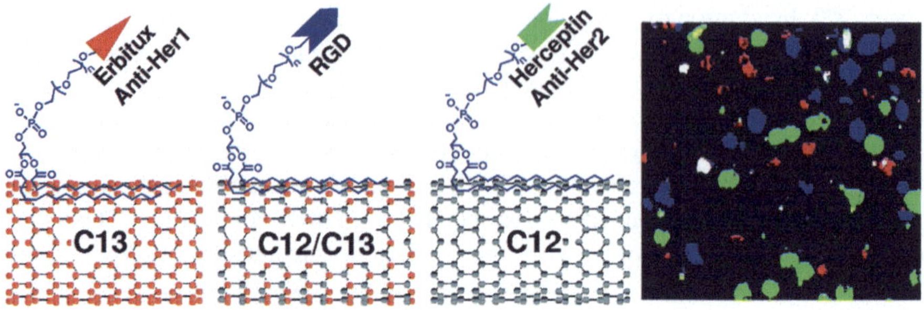

Fig. 3 Isotope-dependent Raman imaging [44]

showed that arginine-glycine-aspartic acid (RGD)-targeted SWCNTs were able to localize to U87MG tumors in a mouse model and were tracked using noninvasive Raman imaging [43]. Using a mock tissue sample, the authors determined that the penetration depth of their Raman microscope was 2 mm. The tumors were imaged up to 72 h postinjection, and pharmacokinetic data was acquired using dynamic imaging. Furthermore, it was found that there was significantly more SWCNT uptake in tumors with an RGD targeting moiety compared to non-targeted SWCNTs.

Due to the spectral overlap observed in fluorescence, it is impossible to image multiple objects simultaneously, a technique known as multiplexing. For example, isotopically different SWCNTs completely or partly populated with ^{13}C instead of ^{12}C have shown different Raman peaks, and therefore can be used to produce different "colors." The study imaged these nanotubes, which were targeted to one of three different types of cancers: Her1+ and Her2+ breast cancer and U87MG glioblastoma cells, as shown in Fig. 3 [44]. The group then incubated a mixture of cells with the three types of SWCNTs and recorded the Raman spectra at each pixel in the image before deconvoluting the intensities of the colors representing each different Raman spectra. They found that each type of SWCNT bound specifically to the type of cancer it was targeted to, while also giving a different Raman signal from a single excitation source. Another example of a CNT-based Raman imaging probe was functionalized non-covalently with a single-strand DNA and gold seeds that grew into nanoshells along the surface of the SWCNT, providing a SERS effect which decreased the image time by an order of magnitude [45]. The authors targeted cancer cells with folic acid (FA) and reported improved anti-cancer photothermal properties of the gold-covered SWCNTs.

Photoacoustic Imaging. Photoacoustic imaging (PAI; also known as optoacoustic imaging) was recently developed and has been used to provide structural and functional image data about biological systems. As the name suggests, PAI measures sound instead of light and converts sound into a 2D or 3D image. Briefly, nanosecond-pulsed laser energy is absorbed by either native molecules in the system or exogenous contrast agents (CAs) to generate a rapid thermoelastic expansion of the tissue, which generates a wideband ultrasound wave. This wave is then detected and a transducer converts the mechanical waves into electrical signals. The photoacoustic signal relays information about the position (time-of-flight) and the strength (signal intensity) of the absorbing species. The NIR region 700–1000 nm is a suitable window for PAI, since the penetration depth is in the order of tens of millimeters and absorption by tissues is modest, yet contrast between tissues is fairly significant. Hemoglobin absorbs strongly below 600 nm and water absorbs above 1000 nm, making those regions impractical for excitation for photoacoustic tissue imaging [46]. Contrast in PAI is determined by the differences in absorption of the laser energy by various tissues, and the intensity of the thermoelastic response is directly proportional to optical absorption [39, 47]. For example, since tumors have higher blood content and lower oxygen content than normal tissue [48], differences in the absorption of oxy- and deoxyhemoglobin can be seen using PAI, and this is often used for endogenous photoacoustic contrast. In fact, PAI can image tumor vasculature at comparable or better resolution than CT perfusion, PET, or dynamic-enhanced MRI. PAI can be combined with other imaging techniques, such as fluorescence [49], ultrasound [46], and magneto-acoustic imaging [50], in addition to applications with photodynamic therapy, in particular [51].

Because of their strong NIR absorbance, SWCNTs have been used as exogenous CAs for PAI. One study reported that SWCNTs wrapped with 1 wt% Pluronic® F127 surfactant were used to noninvasively image the sentinel lymph node (SLN) in vivo in a rat model [52]. Imaging the SLN using ultrasound is usually difficult when searching for metastasized breast cancer. Using chicken tissue to mimic human breast tissue, the authors found that the SWCNTs could be imaged at depths greater than 20 mm. The study also reported that they could image the SLN with a high contrast-to-noise ratio (CNR: 89) and with a resolution of about 500 μm. Another report enhanced the sensitivity of SWCNTs in PAI by conjugating indocyanine green dye through π–π stacking and targeted the material with RGD peptides toward $\alpha_v\beta_3$ integrins, which are associated with tumor angiogenesis, as shown in Fig. 4 [53]. When used in a tumor-bearing mouse model, the material produced 300 times greater PA contrast in living tissue than SWCNTs without the indocyanine green dye, and the material also showed greater tumor uptake than untargeted contrast agents. The authors also reported that the material was not taken up by the RES system, which also contributed to the greater PA signal.

PAI has also been combined with photothermal (PT) tomography for early breast cancer detection [54]. The hybrid system allows for good spatial resolution as well as enhanced soft tissue contrast, and can provide images describing the water/ion concentration, blood, and oxygen levels in tumors, leading to possible earlier diagnosis. The same group developed surfactant-wrapped SWCNTs as a multimodal contrast agent for PA/PT imaging, due to their high radiofrequency (RF) absorption, in addition to their high NIR absorption [55]. The authors conducted in vitro experiments and reported a greater than twofold enhancement in PT signal and a greater than sixfold improvement in the PA signal. Another example of an SWCNT-based multimodal PA/PT contrast agent is a gold-plated carbon nanotube developed by Kim and coworkers [56]. The material is composed of SWCNTs plated with a 4–8-nm-thick layer of gold which is about 100 nm in length, 11 nm in diameter, and had minimal gaps along the SWCNT surface to indicate complete gold coverage. These "golden

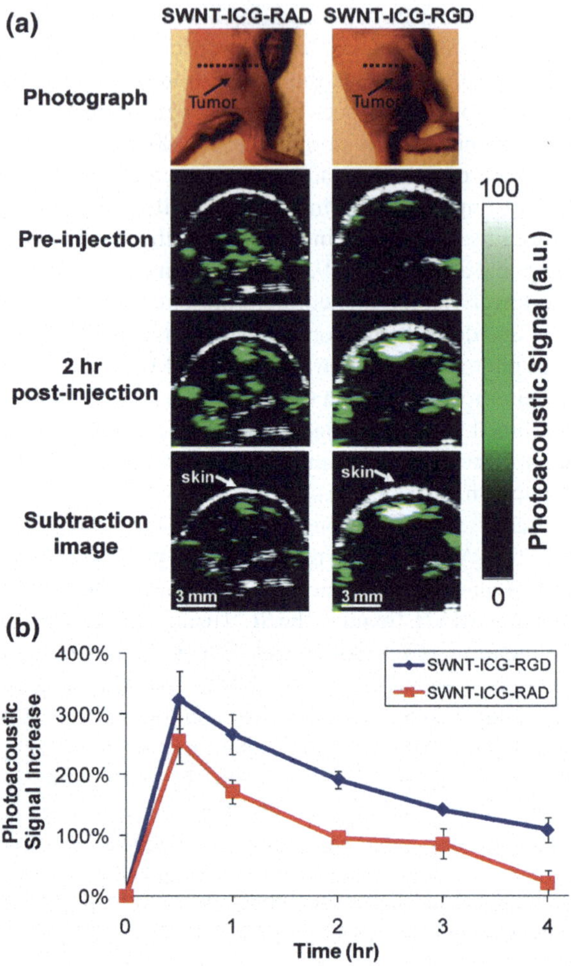

Fig. 4 **a** Ultrasound (gray) and photoacoustic (green) images of one vertical slice through the tumor (dotted black line) using both non-targeted control material (SWCNT–ICG–RAD) and targeted material (SWCNT–ICG–RGD). **b** Mice injected with SWNT–ICG–RGD showed higher PA signal than mice injected with the untargeted control ($p < 0.001$) [53]

SWCNTs" were targeted using an antibody specific for the lymphatic endothelial hyaluronan receptor-1 in order to map lymphatic endothelial cells in mice, which can be useful in studying tumor progression and metastasis. The material could be detected at pico- to possibly femtomolar concentrations, while still providing PA/PT signal improvement. It was also found to be nontoxic from preliminary cell viability and toxicity studies.

X-ray Computed Tomography. X-ray computed tomography (CT) is one of the most used clinical imaging techniques today, because it is inexpensive and useful in diagnosing many different diseases. Unlike the other imaging modalities discussed above, X-ray CT, along with PET, SPECT, and MRI are full-body-penetrating techniques that do not suffer from penetration depth issues. The CT technique uses X-rays that are directed through the body at multiple angles, and these projections are reconstructed using the differences in attenuation in various tissues to produce image contrast. For instance, because attenuation is a result of the photoelectron effect in which elements with higher atomic numbers attenuate more X-rays, hard tissue such as bone and cartilage appears lighter on CT images compared to soft tissue. Currently, the only contrast agents approved for human use are barium sulfate suspensions for gastrointestinal imaging and highly iodinated molecules for vascular imaging [57]. Although these agents are not optimal X-ray CAs since there are elements with higher atomic numbers that can provide greater attenuation, these compounds have good safety profiles and are cost effective. Other materials that have been shown to have good X-ray attenuation are gold and bismuth sulfide NPs [58–61].

Because CNTs do not inherently attenuate X-rays, they must be loaded with metals or other species that do absorb X-rays, and this has limited the amount of research in this area, since a large amount of contrast is needed to produce a good image and only a relatively low concentration of CNTs can be suspended in aqueous media [62]. Despite these limitations, researchers have loaded I_2 molecules into ultra-short (US) SWCNTs (US tubes; 20–80 nm in length) and demonstrated the attenuation ability of the material, although no biological experiments using the material were conducted [63]. Because of their greater X-ray attenuation and low toxicity profile [64], bismuth ions were later loaded into surfactant-wrapped US tubes by the same group who showed X-ray contrast with 2–3% Bi by weight in mesenchymal stem cells without inducing toxicity [65]. Later, a second-generation, high-performance Bi-based US-tube X-ray CA was produced with 20% Bi by weight using a Bi(III) oxosalicylate cluster containing four Bi^{3+} ions [66]. Although these reports did not specifically target cancer, these Bi-based CNT materials could also be utilized for oncological imaging using X-ray CT.

Magnetic Resonance Imaging. Magnetic resonance imaging (MRI) is a powerful, often-used clinical imaging modality that offers higher spatial resolution than X-ray CT without the use of ionizing radiation. The principle behind MRI originates from nuclear magnetic resonance (NMR) spectroscopy, which uses a high-intensity magnetic field that aligns the nuclear spins of water protons in the body in the case of MRI [67, 68]. An RF field is then applied perpendicular to the magnetic field which changes direction and spin of the aligned protons in the transverse plane. When the RF field is removed, the protons realign (or relax) with the magnetic field in an amount of time known as the relaxation time. This process along the longitudinal plane is referred to as spin–lattice relaxation, or T_1, and the recovery along the transverse plane is referred to as spin–spin relaxation, or T_2. Fourier transform is then used to convert the mathematical data from each point in the body into a 2D or 3D image which relies on differences in water content to distinguish between various tissues, although it is fairly common to inject external paramagnetic species to enhance the contrast of the image [69]. These CAs augment the MR image by influencing surrounding water molecules and shorten T_1 or T_2, using paramagnetic or superparamagnetic species, respectively. The efficiency of a CA is described by the agent's relaxivity, which is a measure of the shortening of the relaxation time.

T_1 CAs typically make images brighter (positive contrast), whereas T_2 CAs make images darker (negative contrast).

CNTs must be functionalized with MR-active species since they cannot produce any MR contrast on their own. Most clinically used agents are small molecule Gd^{3+}-ion chelates. The first MRI active CNT-based agent was produced by Wilson and coworkers [70], using surfactant-wrapped US tubes that were loaded with Gd^{3+}-ion clusters into the sidewall defects of the US tubes, and the authors coined the term "gadonanotubes" (GNTs) for the material. The material consisted of about 3–4 wt% Gd and exhibited 40–90 times greater relaxivity values over clinically used MRI CAs, such as Magnevist®. The stability of the ion clusters was found to be pH dependent, such that no Gd^{3+} ions leaked from the material at physiological pH values, which is important given the toxicity of free Gd^{3+} ions [71–73]. The GNTs have been used for many in vitro studies to label HeLa cervical cancer cells [74], as well as macrophages and mesenchymal stem cells to deliver about 10^9 Gd^{3+} ions per cell with no observed cytotoxicity [75–77]. GNTs have also been covalently functionalized using amino acid substituents to suspend the GNTs in aqueous media up to 2 mg/mL and to label MCF-7 breast cancer cells [78].

The same group also studied the effects of using chelated Gd^{3+} ions (instead of free Gd^{3+} ions) on material production and its CA performance. One report described the production of three US-tube-based CAs using three different Gd chelates [79]. The authors found that although there were more Gd^{3+} ions by weight in the US tubes using the chelated metal ions, the MR performance of the original GNTs was greater. This was determined by examining the number of coordinated water molecules on the Gd^{3+}-ion center (q), which is one of the factors governing inner sphere T_1 relaxivity of a CA. The GNTs had a q of about 3, while the chelated Gd^{3+}-ion materials had a q of 0 or 2, depending on the chelate. To further understand the structural foundation of the uniquely high-performing GNTs, a thorough extended X-ray absorption fine structure (EXAFS) investigation was conducted, which revealed detailed information about the structure of the Gd^{3+}-ion clusters in GNTs [80]. In efforts to afford the US tubes greater water suspendability, the surface of the US tubes was covalently functionalized with *p*-carboxyphenyldiazonium (PCP) tetrafluoroborate salt to attach benzoic acid groups to the US-tube surface [81]. Finally, these covalently functionalized US tubes were loaded with Gd^{3+}-ion chelates that the authors previously reported [79], which resulted in high water suspendability (35 mg/mL) and excellent MR contrast [82].

CNT-based MRI CAs have also been made by introducing Gd into the material at the stage of SWCNT growth. For example, Gd_2O_3 nanoparticles were employed as catalysts to produce SWCNTs through chemical vapor deposition for T_1-weighted MRI [83–86]. The material contained about 6.2 wt% of Gd and was non-covalently suspended in aqueous media using a PEGylated phospholipid polymer, PEG5400-DSPE (DSPE: 1,2-distearoyl-sn-glycero-3-phosphoethanolamine). The biodistribution of the material was determined ex vivo using bright-field optical microscopy, and it was found that uptake occurred in the lungs, liver, kidneys, brain, and spleen of rats after i.v. injection. After 24 h, the material left the vasculature and accumulated in the various tissues, and was totally cleared from the body within 10 days

[86]. MWCNTs were also used as MRI CAs, by conjugating metal chelates to the surface of the MWCNTs. Marangon et al. produced oxidized MWCNTs (20–30 nm in diameter and about 400 nm in length) and covalently attached diethylenetriamine-pentaacetic acid dianhydride (DTPA) to the carboxylic acid groups on the oxidized material. DTPA was then used to chelate Gd^{3+} ions, and the resulting material was tested for T_1- and T_2-weighted MRI contrast [87]. This CA material was imaged in vivo using nontumor-bearing mice which found significant uptake in the bladder, liver, and spleen. Using MR imaging in conjunction with histological data and transmission electron microscopy (TEM), it was determined that the material was cleared by the kidneys.

In addition to gadolinium, iron species such as superparamagnetic iron oxide nanoparticles (SPIONs) have been used to make CNT-based CAs. For example, Fe_3O_4 NPs were loaded in situ onto full-length MWCNTs using a solvothermal method. The resulting material had a T_2-weighted relaxivity of 175.5 $mM^{-1}s^{-1}$ with about 41.3% of the hybrid material consisting of Fe_3O_4 particles [88]. This material was loaded into MCF-7 human breast adenocarcinoma cells, and it was found that greater than 80% of the cells were viable after 24 h. Another report of a SPION-MWCNT conjugate was reported by Liu et al. using oxidized MWCNTs and poly(diallydimethylammonium chloride) (PDDA) to link the SPIONs to the MWCNTs using electrostatic interactions [89]. The material was also targeted to liver cancer and tracked using T_2-weighted MRI in an in vivo mouse orthotopic tumor model. A second-generation material with the SPIONs sequestered inside the cavities of the MWCNTs was also made by reducing an iron compound with hydrazine in a microwave reactor [90]. This polymer-wrapped material was used to image liver cancer in vivo, as shown in Fig. 5. Other CNT-based MRI CAs have also been reported for different imaging applications; however, the studies are beyond the scope of this chapter [91–94].

Nuclear Imaging. Nuclear imaging techniques like PET and SPECT are useful clinical modalities that provide functional information with great sensitivity [95]. SPECT generates an image by first detecting gamma (γ) rays from a decaying radionuclide within the patient. The energy from the detected rays is then converted into electrical signals from each point in the body and processed into an image. The one or two detectors usually used in SPECT can also rotate around the patient to generate projections from multiple angles that can be reconstructed into a 3D image. PET imaging also begins with a decaying radionuclide, but differs from that used in SPECT in the type of particle and its decay pathway. In PET, positrons (β^+) are emitted from the radionuclide, which interact with electrons in the vicinity and undergo an annihilation event, producing two γ rays emitted 180° with respect to one another of equal energies (511 keV). A ring of detectors surrounding the patient identifies coincident rays along a line of response from many points in the body and uses those points to create an image. CT is also used in conjunction with PET to provide an attenuation correction in addition to a structural image to complement the functional image. Because of the many angles from which PET can detect γ rays, 3D image reconstruction is also possible. Whereas MRI and CT offer high resolution with poor sensitivity, nuclear techniques offer high sensitivity, while suffering from poor res-

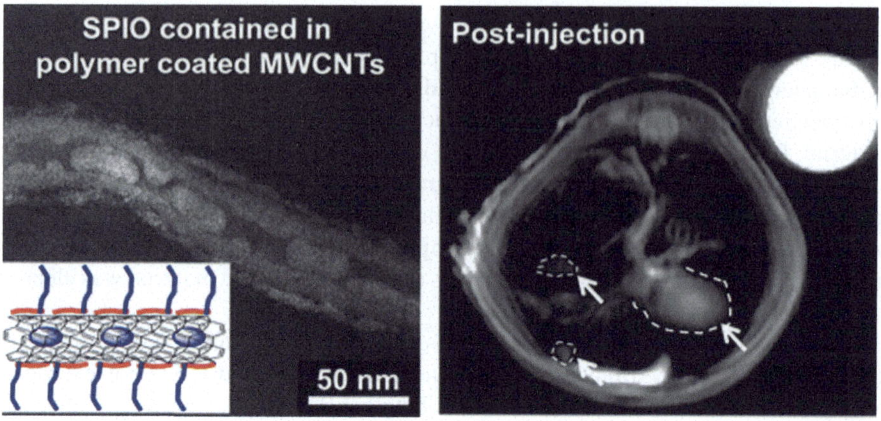

Fig. 5 TEM image of superparamagnetic iron oxide (SPIO) NPs loaded inside MWCNTs (left) and MR image of the material in mice, highlighted by the white arrows (right) with water shown as a control (top right corner) [90]

olution. Because of the greater number of detectors, PET is superior to SPECT in resolution (2–3 mm in PET compared to 6–8 mm in SPECT) and sensitivity, only requiring material concentrations in the nano- to picomolar range, while SPECT requires micromolar concentrations [96, 97].

SWCNTs were covalently functionalized with chelated metal ions for both radio-imaging and radiotherapy by McDevitt and coworkers [98] using $^{89}Zr^{4+}$ ($t_{1/2} = 78$ h) for PET imaging in one construct and $^{225}Ac^{3+}$ ($t_{1/2} = 10$ d) for therapy in another construct. Both the imaging and therapeutic materials were also targeted with tumor neovasculature-targeting antibody E4G10 and tested in a mouse xenograft model of human colon adenocarcinoma (LS174T). The study found that the therapeutic construct lowered tumor volumes and improved survival compared to controls. The study also showed that the imaging construct exhibited rapid blood clearance (less than 1 h) and high tumor uptake relative to the antibody alone. By using SWCNTs, the authors were able to increase the signal-to-noise ratio without compromising the reactivity of the targeting antibody. Bianco, Tobias, and coworkers also reported a method to load and seal metals within SWCNTs and covalently functionalizing the material with antibodies for either targeted radio-imaging, radiotherapy, or both [99]. Another example was reported by Liu et al. [100] that used non-covalently functionalized SWCNTs suspended in a PEGylated phospholipid that was attached to a chelated $^{64}Cu^{2+}$ ion ($t_{1/2} = 12.7$ h) for in vivo biodistribution with PET. The study was conducted using U87MG tumor-bearing animals and targeted the material using an RGD peptide. The group found that the material had a long circulation time, no uptake in RES organs, and high tumor uptake, which was confirmed ex vivo using the Raman signature of the SWCNTs.

CNTs have also been used for multimodal imaging, combining the strengths of multiple modalities to achieve the best image possible. For example, SWCNTs were

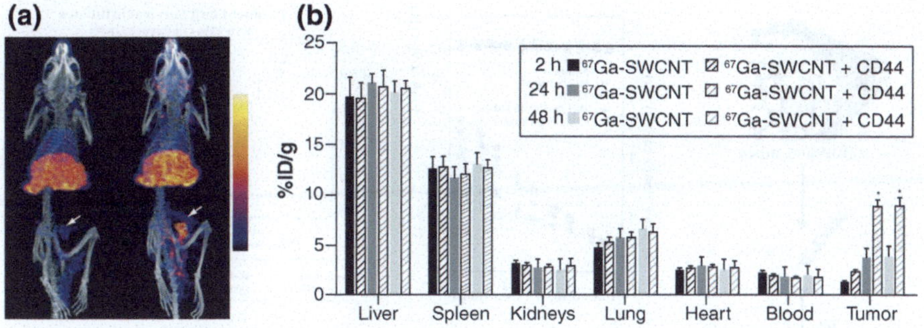

Fig. 6 A. Representative SPECT/CT images of tumor-bearing mice at 24 h postinjection of ^{67}Ga-SWCNT (left) and ^{67}Ga-SWCNT+CD44 (right), with white arrows showing the tumor site in the mammary fat pad. B. Quantitative radioactivity measurement of ^{67}Ga-SWCNTs and ^{67}Ga-SWCNT+CD44 in various tissues performed at 2, 24 and 48 h postinjection [101]

conjugated to SPIONs and ^{67}Ga ($t_{1/2}$ =3.26 d) for combination MRI and SPECT imaging [101]. The SWCNTs were first oxidized and shortened (~200 nm in length), PEGylated, and conjugated to SPIONs, a metal ion chelating agent, and CD44 monoclonal antibodies for breast cancer stem cell targeting. SPECT images of tumor-bearing mice showed uptake in RES organs, specifically the liver and spleen, which are shown in Fig. 6. Tumor uptake was enhanced with the use of the antibody compared to controls, and the SPECT data was confirmed using MRI and postmortem quantification of radioactivity using a gamma counter. The authors concluded that since the biodistributions at 2, 24, and 48 h postinjection were not significantly different, the material must be cleared very slowly from the body. An additional report of multimodal SWCNT-based CAs was performed using the Gd^{3+}-ion-filled US tubes (GNTs) discussed above [102]. The group co-loaded $^{64}Cu^{2+}$ and Gd^{3+} ions within US tubes (^{64}Cu@GNT) and acquired MR and PET phantom images. Biodistribution was also conducted in nontumor-bearing mice using PET and found that the material localized mainly in the lungs, liver, and somewhat in the kidneys with no transmetallation or loss of metal ions in vivo. However, no MRI images were obtained due to the high CA dose required for MRI (~1 mM) and the relatively low suspendability of the ^{64}Cu@GNT agent.

3 CNT Agents for Cancer Therapy

Over the past two decades, CNTs have attracted great attention as one of the most desirable nanocarriers for medical applications due to their high surface area, unique optical and electrical properties, enhanced cellular uptake and the potential to be easily paired with many drugs, biomolecules, and NPs for enhanced target specificity and diminished side effects [103]. In this section, we describe achievements related to the use of CNT agents for various cancer therapies.

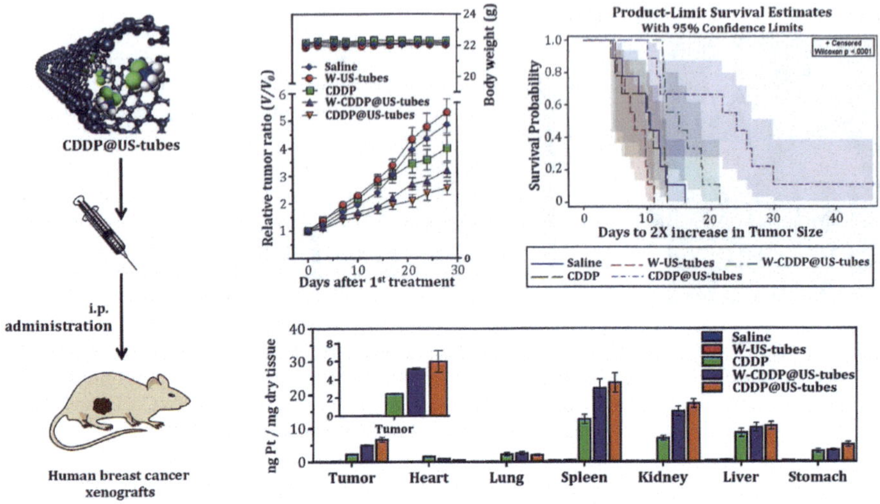

Fig. 7 The CDDP@US-tube platform for enhanced cancer chemotherapy [112]

Small Molecule Drug Vectors

Platinum agents. Pt-based anti-cancer drugs are commonly used to treat various types of cancers in the clinic today [104]. They form adducts with DNA which affects key cellular processes, such as transcription and replication, which triggers apoptosis [105, 106]. Pt(II) agents such as cisplatin (cis-dichlorodiammineplatinum (II), CDDP) conjugated to CNTs can reduce drug side effects by protecting CDDP from external reactive species [107–110]. Guven et al. developed a new US-tube-based drug delivery platform for the treatment of cancer, which was designed to be a cancer-specific enzyme-activated drug delivery platform that releases its Pt(II) drug cargo within cancer cells only [111, 112]. Their dialysis studies in phosphate-buffered saline (PBS) at 37 °C demonstrated that CDDP released from CDDP@US tubes can be controlled by wrapping the CDDP@US tubes with pluronic F108 surfactant. They evaluated the anti-cancer activity of pluronic-wrapped CDDP@US tubes against two different breast cancer cell lines, MCF-7 and MDA-MB-231, and observed that the material-enhanced cytotoxicity over free CDDP after 24 h. They also demonstrated that the CDDP@US tubes exhibited greater efficacy in suppressing tumor growth than free CDDP in both MCF-7 cell line xenograft and BCM-4272 patient-derived xenograft (PDX) models, as shown in Fig. 7. Platinum analysis of various tissues via inductively coupled plasma mass spectrometry (ICP-MS) revealed that the CDDP@US tubes also demonstrated a prolonged circulation time compared to free CDDP which resulted in significantly more CDDP accumulation in tumors.

An SWCNT–CDDP system called the "longboat" was developed by Dhar et al. in which CDDP was transported across the cell membrane by a non-covalently amine-functionalized SWCNT platform (SWCNTs–phospholipid–PEG–NH_2) [113]. To prepare the longboat, a platinum complex prodrug covalently linked FA and CDDP to the polymer-wrapped SWCNTs, which activated and released CDDP after intracellular reduction. The study found that there is an eightfold increase in activity using the SWCNTs as vectors compared to cisplatin alone.

Magnetic MWCNTs with magnetite (Fe_3O_4) NPs on the inner surface of CNTs was used as a carrier for anti-cancer drugs. Yang et al. developed this system and showed that chemotherapeutic agents, such as 5-fluorouracil and cisplatin, were incorporated into the pores of the MWCNTs [114]. The MWCNTs were non-covalently functionalized by phospholipids with a PEG moiety and an FA terminal group for enhancing drug delivery. These magnetic MWCNTs were then used for lymphatic tumor targeting. The authors performed an in vivo study involving the lymphatic distribution of the magnetic MWCNTs [115]. The MWCNTs functionalized with poly(acrylic acid) were subcutaneously administered to mice. The concentration of the administered magnetic MWCNTs and the degree of black staining of lymph nodes were found to be dose dependent. No local or systemic toxicity was observed in the study.

There are a variety of clinical Pt anti-cancer agents such as carboplatin (cis-diammine (1,1-cyclobutanedicarboxylato) platinum (II), CP) which has been conjugated to CNTs [103]. Hampel et al. encapsulated CP into MWCNTs using a wet chemical approach based on capillary forces after opening the CNTs [116]. Electron energy loss spectroscopy (EELS) and X-ray photoelectron spectroscopy (XPS) were used to confirm CP incorporation into the CNTs. In vitro analysis showed that the CP–MWCNTs stunted the growth of urinary bladder cancer cells, whereas unfilled MWCNTs had negligible effects on cancer cell growth.

Topoisomerase inhibitors. Another popular method of treating cancer targets topoisomerases, which are enzymes within the cell nucleus that participate in DNA replication. Here, topoisomerase inhibitors block the ligation step of the cell cycle, generating single- and double-stranded breaks in the DNA, which subsequently leads to apoptosis. There are two types of inhibitors, depending on the number of strands cut in one round of action [117]. In the clinic, topoisomerase I inhibitors include irinotecan, topotecan, and camptothecin (CPT) and topoisomerase II inhibitors include etoposide and teniposide. Some anti-cancer drugs such as the anthracyclines which include doxorubicin (DOX), epirubicin (EPI), and daunorubicin (DAU) can also inhibit topoisomerase II. Tian et al. synthesized a topoisomerase I inhibitor by loading CPT into MWCNTs and wrapping the material with a tri-block copolymer (Pluronic P123) through π–π stacking interactions [118]. Although in vitro cytotoxicity studies using Hela cells showed that this material had greater anti-tumor properties when compared to free CPT, no in vivo data were collected to confirm these results [119].

The chemotherapeutic drug, DOX, has received attention recently due to its relatively easy conjugation to CNTs. The loading and release of DOX from CNTs can be controlled by pH, CNT coatings, and temperature [120–128]. One study investigated a drug delivery system using PEG-functionalized SWCNTs attached to DOX [129]. PEGylated SWCNTs showed a particularly high loading capacity of DOX (~400%). It is notable that the CNT diameter-dependent binding and release of the drug from SWCNTs, as well as the conjugate's stability in physiological and acidic media, showed the potential for multiple avenues by which drugs can be delivered.

However, only cytotoxicity using U87MG and MCF-7 cell lines was studied. To enhance the therapeutic efficacy and the targeting properties of DOX, the drug can be conjugated with FA, estradiol, dexamethasone, or antibodies [129–134]. Zhang et al. investigated a material for targeted delivery using an SWCNT–DOX conjugate [135]. Polysaccharides were wrapped around the SWCNTs and remained bound at physiological pH. However, at lower pH, which is the situation often found in local tumor environments, the polysaccharides unraveled, which efficiently released DOX.

Because of the high surface area and hydrogen bonding interactions, carboxylated MWCNTs (c-MWCNTs) show higher adsorption efficiency and stability comparing to functionalized MWCNTs and SWCNTs. Chen and coworkers investigated the loading of anti-cancer agent, epirubicin hydrochloride (EPI) onto c-MWCNTs [136]. The adsorption capacity of EPI onto c-MWCNTs increased at higher pH. However, in vitro and in vivo studies of the EPI-loaded c-MWCNT materials were not conducted.

Anti-microtubules. Because they play an essential role in cell division, microtubules have become an attractive target for small molecule drugs to slow the growth and replication of cancer cells [137]. Among many anti-microtubule agents, paclitaxel (PTX) and docetaxel (DTX) have been studied extensively [138–144]. However, the poor solubility of PTX is a challenge which needs further attention. Although using Cremophor EL improves water solubility of PTX, it is not inert and increases toxicity of the formulation. One group developed a hyperbranched poly(citric acid) (PCA)-functionalized MWCNTs system with high hydrophilicity and conjugated it to PTX [145]. Their cytotoxicity studies, using A549 and SKOV3 cell lines, showed that PTX conjugated to the PCA-functionalized MWCNTs demonstrated greater cytotoxicity than the free drug over a shorter period of time. CNT materials are also good candidate for prolonging the circulation time of anti-cancer agents in the blood. For example, Liu et al. demonstrated that chemically functionalized SWCNT conjugated with PTX via branched PEG chains is a new approach for tumor accumulation of the drug with low systemic toxicity [146]. In in vivo experiments, PTX uptake into the tumor as a PTX–SWCNT conjugate was tenfold greater than PTX alone in a murine 4T1 breast cancer model, and the blood circulation time of PTX–SWCNT was longer than that of the free drug. Overall, this study showed that PTX–SWCNT constructs are highly efficient at suppressing tumor growth.

SWCNTs have also been used not only to deliver PTX but also used in conjunction to sensitize cells to chemotherapy, shown by a study that reported a synergistic effect of a CNT–PTX system on A549 and NCI-H460 lung cancer cell lines [147]. They discovered that SWCNTs and graphene oxide produce reactive oxygen species (ROS) in cells, which made the cells more susceptible to apoptosis by PTX. A similar study investigated the synergism between SWCNTs and PTX on ovarian cancer cells [148]. They concluded that PTX-conjugated SWCNTs sensitized human ovarian cancer OVCAR3 cells which resulted in greater cytotoxicity. These results indicate that co-treatment of SWCNTs and chemotherapeutic drugs might be a promising approach for improving cancer treatment.

DTX, which has less side effects than PTX, was also used for breast and lung cancer treatment with Asn–Gly–Arg (NGR)-linked SWCNTs. Wang et al. described how DTX–NGR–SWCNTs caused a greater suppression of tumor growth than free DTX in a murine S180 mouse cancer model [144]. The authors also irradiated the material using a NIR laser for combination drug delivery and photothermal therapy. The tumor size in this experiment decreased significantly under NIR irradiation compared to controls. This study demonstrated that the DTX–NGR–SWCNT drug delivery system is an effective anti-cancer therapy with more drug in the tumor, and hence less toxicity for off-target organs.

Gene Therapy

Gene therapy aims to introduce genetic material into cells to counteract abnormal genes or to generate beneficial proteins. The therapy has been used for a number of medical applications using a wide range of nucleic acid and peptide-based materials such as plasmid DNA (p-DNA), small interfering RNA (siRNA), antisense oligodeoxynucleotides (ODNs), aptamers, and proteins [149–161]. A number of gene therapy strategies are being evaluated in cancer patients using viral and non-viral vectors. In gene therapy, gene delivery to target cells across cellular barriers is challenging because of the hydrophilicity of DNA, the large molecular size of nucleic acids, and risk of destruction by nucleases before reaching the target. Despite being less efficient than viral delivery, the advantage of using nonviral vectors, including CNTs, is that they are safe and can deliver genes without size limitation [162, 163].

Plasmid DNA (p-DNA). A plasmid is a small piece of double-stranded DNA that can replicate independently of chromosomal DNA. Their application in genetic engineering as vector DNA make them good candidates as genes carriers [119]. Various functionalized CNTs have been designed and synthesized to deliver p-DNA using amine groups, polyethylenimine (PEI), polyamidoamine hybrids, cationic glycopolymers, and ethylenediamine. Pantarotto and coworkers developed ammonium-functionalized SWCNTs to deliver p-DNA that was internalized into mammalian cells with low cytotoxicity [164]. The gene expression level demonstrated by functionalized SWCNT containing p-DNA was tenfold greater than for DNA alone. Another study optimized amine-functionalized SWCNTs and MWCNTs, as well as lysine-functionalized SWCNTs as gene delivery vehicles for p-DNA [165]. The results showed that CNT surface area and charge density determined the electrostatic interaction between functionalized CNTs and DNA. Experiments using a mammalian cell line showed that all three CNT derivatives upregulated gene expression markers much better than DNA alone.

Similar amine-functionalized MWCNTs were used to also deliver small interfering RNA (siRNA) that showed good tumor suppression as well as good survival of tumor-bearing animals [166]. Also, polymers have been used to non-covalently bind SWCNTs and p-DNA. For example, it was reported that PEI-functionalized CNTs showed good stability and dispersion in aqueous media [167]. A p-DNA condensing test showed that CNTs were sufficiently potent as gene delivery agents with up to 19-fold better transfection than that of polymer alone as shown by in vivo studies. Another group employed ethylenediamine-functionalized SWCNTs, using the oncogene suppressor p53 for breast cancer treatment [168]. Their study showed that about 40% of MCF-7 cancer cells were dead after 72 h of treatment with the SWCNTs-p53 conjugate, which was fourfold better than cell death in the p53 and SWCNTs control groups. They also found significantly increased caspase-3 activity, indicating apoptosis, following treatment with the functionalized SWCNTs–p53 material. In addition to DNA, CNTs can also be used to deliver RNA in the same fashion. For instance, one group conjugated functionalized SWCNTs with telomerase reverse transcriptase siRNA to suppress tumor growth [169]. In vitro studies using murine tumor cells demonstrated reduced expression of telomerase reverse transcriptase and more importantly, tumor size reduction was observed after in vivo experiments in mice.

Antisense oligonucleotides. Antisense oligonucleotides (ODNs) are synthetic polymers in which the monomers are chemically modified nucleotides like those in DNA or RNA. Antisense therapy is known as an important technique for treating tumors or genetic disorders using antisense oligodeoxynucleotides (ASODNs), which inhibits gene expression at the protein level [170]. Pan et al. applied polyamidoamine dendrimer-modified MWCNTs as a platform for ASODNs to improve the delivery efficiency of genes [170]. These materials exhibited significant transfection efficiency and inhibition of MCF-7 and MDA-MB-435 breast cancer cells and HepG2 liver cancer cells.

Aptamers. Aptamers are RNA- or DNA-based oligonucleotides that can fold into a 3D conformation to bind to specific molecular targets. Aptamers can inhibit intracellular pathways by disrupting protein–protein interactions and are potential candidates for various therapies [160]. Van den Bossche et al. attached aptamers onto carboxylated CNTs as a vector to efficiently facilitate intracellular delivery into the cytosol of different cell types [160]. Another study examined the activity of RNA aptamers against the epithelial cell adhesion molecule (EpCAM) (EpDT3) and attached the aptamers to piperazine–PEI-conjugated SWCNTs [171]. The siRNA and DNA transfection efficiency of the SWCNTs–PEI–piperazine conjugate was investigated against the upregulation of BCL9 l, which is involved with colorectal and breast cancers. For EpCAM cells, the conjugate specifically induced apoptosis by more than 20% in EpCAM-positive cells. These results demonstrated the targeted silencing activity of the aptamer-containing SWCNT conjugates.

Protein carriers. CNTs have been studied as a protein delivery system for cancer therapy and as a tool for immunotherapy through antibody delivery and the production of artificial antigen-presenting cells (APCs). These therapeutic modalities induce, enhance, or suppress the immune response [172–179]. Although some proteins, such as streptavidin, show anti-cancer properties, their large molecular weight inhibits cellular uptake. One study reported streptavidin conjugated to an SWCNT–biotin transporter to internalize streptavidin into HL60 promyelocytic leukemia cells and human Jurkat T-cells by endocytosis [180]. CNTs also served as transporters for selective destruction of tumor cells, attacking them with the toxic protein recombined ricin A (RTA) [181]. The conjugate of MWCNTs and RTA was transfected into the cytoplasm of various cell lines, which led to cell death. Cell death rates for L-929, HeLa, HL7702, MCF-7, and COS-7 cells caused by the RTA–MWCNT conjugates were threefold higher than those achieved without the MWCNT vector. Moreover, selective destruction of certain breast cancer cells was obtained by coupling the

MWCNT–RTA construct to human epidermal growth factor receptor 2 (HER2) for targeting.

Immunotherapy is another potential approach for cancer therapy and CNT-based anti-tumor immunotherapy has been evaluated. What makes immunotherapy particularly difficult is the fact that cancer cells can secrete immunosuppressive proteins such as prostaglandins E and interleukin (IL)-10 in order to evade the host immune system [182–184]. To counter this, Meng and coworkers used oxidized MWCNTs (0.5-1 μm in length, 20–30 nm in diameter) to enhance immune system activity in an H22 hepatocellular carcinoma in vivo model [185]. MWCNTs were injected subcutaneously in the vicinity of the tumor and were phagocytized by macrophages. The authors report that since the MWCNTs act as a foreign pathogen to the mouse, the immune system was activated and attacked the tumor, even though no CNTs were seen in the cancer tissue after histological analysis. The same group also reported that tumor lysate proteins that were covalently conjugated to the oxidized MWCNTs enhanced the efficacy of immunotherapy of tumor cell vaccines [186]. Another example of CNT-based immunotherapy was reported by Fadel et al. where they produced a CNT–polymer composite loaded with magnetite particles and the T-cell growth factor IL-2 [187]. The material acted like an antigen-presenting cell and magnetic cell separation indicated an increase in the number of T-cells in circulation to be 1000-fold greater than for IL-2 alone as a result of the large surface area that the CNTs offered. The material was shown to stunt tumor growth in a murine melanoma model 2 weeks after tumor injection.

Remotely Stimulated Therapy

In addition to serving as carriers of molecular anti-cancer agents, the inherent properties of CNTs can also be leveraged for different therapeutic strategies. Photothermal therapy (PTT) and photodynamic therapy (PDT) are two methods that use the properties of CNTs to generate singlet oxygen through photochemistry or to generate significant amounts of heat upon excitation with NIR radiation, as shown schematically in Fig. 8. These two therapeutic techniques can be used against cancer with minimal side effects relative to current standards of care [188]. However, it has been difficult to identify materials that show both efficacy and safety, but some examples are discussed here.

Thermal and Photothermal Therapy. CNTs effectively absorb NIR radiation to generate heat, which can be transmitted to the body from an external light source [189], and because of this property, CNTs have been explored for the thermal ablation of cancer cells [190–192]. According to the antenna theory, surface defects can enhance this CNT–NIR interaction and various approaches have been used to achieve CNT sidewall defects for this purpose [193]. Gannon et al. discovered that SWCNTs also generate heat in an RF field [194]. They prepared water-suspendable SWCNTs using Kentera (a polyphenylene ethynylene-based polymer) and injected the material directly into hepatic Vx2 tumors in rabbits, followed by an immediate noninvasive RF treatment. After 48 h, tumors treated with the material were completely necrotic, whereas control groups injected with Kentera alone were completely viable after treatment with the RF field, as were SWCNT-treated tumors without RF

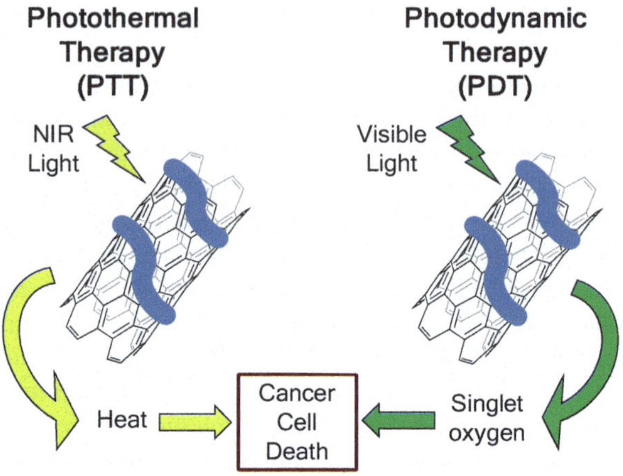

Fig. 8 The PTT and PDT processes which utilize functionalized CNTs

treatment. These results suggested that SWCNTs might permit noninvasive RF ablation of malignant cells. Later, Corr and coworkers studied separated metallic and semiconducting SWCNTs under RF radiation to show that both produced heat with the formation of assembled columns of aggregated SWCNTs, in agreement with the antenna theory for SWCNTs [195].

Another study used a photo-ablative approach using MWCNTs and NIR light for kidney cancer treatment [193]. The authors found that a successful NIR coupling effect occurs at MWCNT lengths exceeding half of the wavelength of the radiation, again demonstrating the antenna theory. This radiation heated the MWCNTs through induction and the material was able to continuously destroy cancer cells, even at low radiation doses. Their work showed that lengths between 700 and 1100 nm were the most effective for tumor ablation. Zhou and colleagues utilized SWCNTs of a uniform diameter (~0.81 nm) targeted with FA and a narrow absorption peak at 980 nm for selective PTT to ablate EMT6 breast cancer cells [196]. After conducting in vitro and in vivo experiments, it was shown that the material markedly improved the photothermal ablation of cancer cells while noticeably reducing the thermal effects to normal cells. The same group also reported a therapeutic model involving targeting mitochondria using PEGylated SWCNTs [197]. This photothermal method induced mitochondrial depolarization, cytochrome c release, and caspase-3 activation. The combination of radiation and SWCNTs showed notable efficacy for slowing breast cancer growth with tumor regression. Furthermore, the group developed an immunologically modified SWCNT system using an immunoadjuvant, glycated chitosan (GC), and SWCNTs [198]. Using GC–SWCNTs and the same laser system, they investigated the thermal and immunological effects on tumors both in vitro and in vivo. The combination of GC-SWCNTs and NIR irradiation exhibited significant growth suppression in animal models and led to complete tumor regression and long-term survival in many cases.

Wang et al. attached disialoganglioside (GD2) monoclonal antibody (anti-GD2) to oxidized MWCNTs to target neuroblastoma stNB-V1 cells [199]. To track the anti-GD2-bound MWCNTs, rhodamine B was attached to the carboxylated CNTs functionalized with and without anti-GD2. After incubation of antiGD2-conjugated MWCNTs with neuroblastoma cells, MWCNT-laden cells were irradiated with an 808-nm NIR laser, in a two-step irradiation process. The stNB-V1 cells became necrotic after the second laser treatment, while all non-GD2-expressing PC-12 cells remained viable. Based on these findings, MWCNTs bound to anti-GD2 showed promise as a PTT agent. Another study also used MWCNTs for NIR thermal ablation of kidney cancer, both in vitro and in vivo [200]. They measured the thermal effects using heat-shock protein-reactive immunohistochemistry and magnetic resonance temperature mapping. In these studies, the researchers found that MWCNTs ablated tumors at low laser power (3 W/cm^2) and a short treatment time (30 s), with minimal local damage to the surroundings. These treatment parameters resulted in cancer remission of 3.5 months in 80% of mice with 100 μg MWCNT dose per treatment. Another example was reported using NIR irradiation to destroy Ehrlich ascites carcinoma (EAC) cells and found that 95% of the EAC cells treated with MWCNTs were dead after 1.5 min of NIR treatment that generated an ablation temperature of ~50 °C [201].

Additional studies have shown that DNA encasement enhanced MWCNT heat emission by NIR irradiation and that DNA-encased MWCNTs can be used to destroy tumors in vivo [202]. At increasing temperatures, the performance of the material was threefold greater than that of non-wrapped MWCNTs. Treatment of PC-3 human prostate tumors using an intratumoral injection of MWCNTs followed by NIR treatment (1064 nm, 2.5 W/cm^2) completely ablated tumors in all mice. In addition, normal tissue did not sustain long-term damage following treatment. Moon et al. studied the SWCNT-induced PPT of solid tumors in vivo, as shown in Fig. 9 [190]. The tumors were completely destroyed in mice in the PTT treatment group, and no harmful side effects or tumor recurrence were observed over 6 months. However, tumors in only SWCNTs or only radiation control groups continued to grow until the animals eventually died. Another PPT study using SWCNTs in a squamous cell carcinoma in vivo model also used Raman to image the tissues and found that the SWCNT material remained in the tumor microenvironment for 3-month postinjection [191].

Photodynamic therapy. For PDT to occur, a light source, a photosensitizer, and molecular oxygen are required. First, the photosensitizer is taken up into a cell and activated by light at a specific wavelength. The excited photosensitizer then transmits energy to an oxygen molecule. The energy absorbed by the oxygen in the triplet ground state ($^3O_2$) excites it to the excited singlet state ($^1O_2$) which is cytotoxic at high concentrations [203]. CNTs show considerable potential as photosensitizers, as shown by a PDT agent that was developed using SWCNTs and spermine-conjugated zinc monocarboxyphenoxyphthalocyanine (ZnMCPPc) in MCF-7 breast cancer cells [204]. The ZnMCPPc–spermine–SWCNT conjugate increased the singlet oxygen quantum yields by 50% compared to ZnMCPPc alone. After incubation of ZnMCPPc, ZnMCPPc–spermine, and ZnMCPPc–spermine–SWCNTs with MCF-7 cancer cells for 24 h in the dark, all materials were found to be nontoxic. The PDT effect with ZnMCPPc showed only a 64% decrease in cell viability, but the PDT effect with

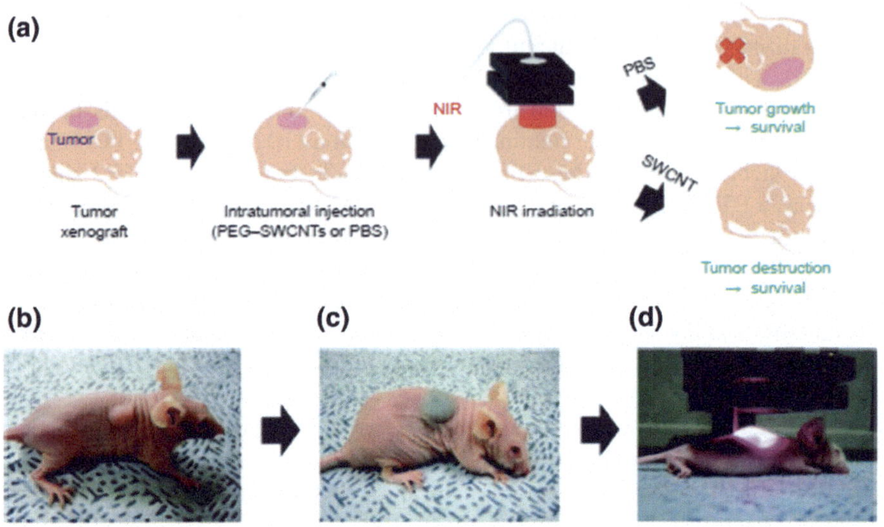

Fig. 9 **a** Schematic of the PTT procedure using PEG-SWCNTs in mice. **b** Photograph of the tumor-bearing mouse. **c**. Photograph of the tumor-bearing mouse after intratumoral injection of the material. **d** Photograph of NIR irradiation of the tumor region [190]

the ZnMCPPc–spermine and ZnMCPPc–spermine–SWCNT conjugates at the same concentrations resulted in a 97% decrease and a 95% decrease in cell viability, respectively.

Wang and associates produced two SWCNT derivatives, one covalently functionalized using polymerized PEI in the presence of amine-derivatized SWCNTs and a non-covalently functionalized SWCNT material wrapped in polyvinylpyrrolidone (PVPk30) for visible light-induced PDT both in vitro and in vivo [205]. The group found that both materials were taken up by mouse melanoma B16-F10 cells and that the materials could translocate to the nucleus. They also tested the DNA damage done by ROS generated by the conjugates, and it was demonstrated that the materials did not induce DNA damage without the use of visible light. In vivo studies demonstrated that PDT using the materials had a significant effect on shrinking tumor size, with the covalently functionalized SWCNT material showing the greatest effect.

Zhu et al. developed DNA aptamers attached to the heterocyclic molecule chlorin 6 (Ce6) and wrapped their aptamers in SWCNTs to control the generation of singlet oxygen [206]. In the absence of an aptamer target (e.g., thrombin), singlet oxygen generation (SOG) was stopped because the photosensitizing activity of Ce6 was quenched by the surrounding SWCNTs. When the aptamers were bound to a target, the SWCNTs no longer could quench the Ce6, and SOG occurred. Based on these experiments, addition of 2.0 μM thrombin generated a 13-fold increase of SOG. There have been other reports of using SWCNTs as vectors for photosensitizers that show promise for cancer PDT [207, 208]. There have also been reports of CNT-based

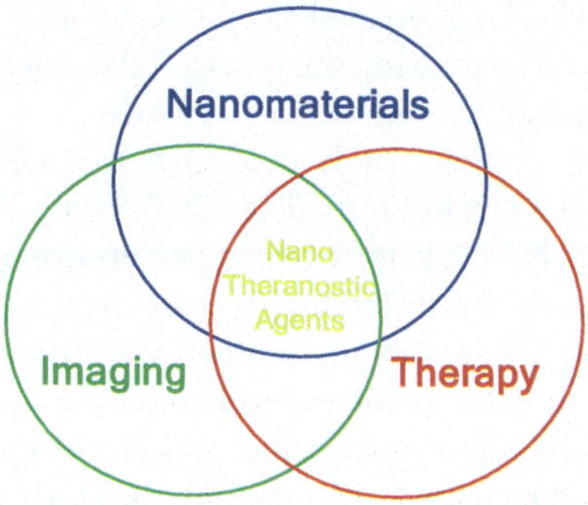

Fig. 10 The overlapping technologies needed for nanotheranostics agents

therapies using a combination of PTT and PDT, which have shown greater efficacy than for either of the two modalities alone [209–211].

4 CNT Agents for Theranostic Medicine

Theranostic nanomedicine is emerging as a promising therapeutic paradigm. It takes advantage of the high capacity of nanoplatforms to carry cargo for both diagnostic and therapeutic functions, shown in Fig. 10. The resulting nanosystems, capable of diagnosis, drug delivery, and monitoring of therapeutic response are expected to play an ever-growing role in the dawning era of personalized medicine, and considerable research effort has been devoted toward this goal [212]. There are many reasons for using CNTs for theranostic applications. They show interesting inherent optical properties, making them useful CAs for optical imaging and sensing. In addition, the excellent electrical properties of CNTs allow them to be extensively used in a wide range of biosensing platforms, and most importantly, they have ultra-high surface areas available for efficient drug loading and bioconjugation [213].

Magnetic MWCNTs were projected to act as both a structure for targeted cancer treatment and as a CA for MRI [214]. To functionalize the MWCNTs, iron NPs were loaded into the CNTs, and functionalized gadolinium was added to the sidewalls of the MWCNTs. The heating functionality of the Gd-doped magnetic CNTs was confirmed using a specific magnetic field. Their results indicated that functionalized CNTs have potential as both MRI CAs for cancer diagnosis and cancer therapy by magnetic hyperthermia. Another report described a CNT-based magnetic dual-targeted nanocarrier for drug delivery [215]. Under the guidance of a magnetic field and ligand–receptor interactions, dual-targeted delivery of DOX into cancer cells was successfully performed. DOX was loaded into MWCNTs and released into the cytoplasm with high efficiency to show enhanced cytotoxicity against U87 human glioblastoma cells. These results demonstrated the potential of using magnetic nanocarriers for targeted delivery of DOX against cancer. Wu et al. synthesized MWCNTs/cobalt ferrite ($CoFe_2O_4$) magnetic hybrids using a solvothermal method.

The DOX-loaded magnetic hybrids showed notable cytotoxicity for HeLa cancer cells following the intracellular release of DOX. The magnetic hybrids also showed a high T_2 relaxivity of 153 $mM^{-1}\ s^{-1}$ and significant negative contrast enhancement.

Another study used QDs to track intracellular delivery of antisense DNA using MWCNTs [216]. The CNTs were oxidized and then conjugated to various cationic polymers using carboxylate groups on the MWCNTs. When the MWCNTs wrapped with the different polymers were tested for cytotoxicity, it was shown that PEI exhibited the greatest cell viability compared to controls, and therefore it was used for subsequent experiments. The authors then attached CdTe QDs to an antisense oligodeoxynucleotide, which was then conjugated to the PEI–MWCNTs. Using the QDs to track the material, the study demonstrated that the nanoconstruct was effective at transporting the antisense oligonucleotide into the nucleus of HeLa cancer cells to produce significant anti-cancer activity over controls. Another example of fluorescence guidance used magnetic CNTs that were conjugated with silica-coated QDs and molecular targeting moieties to produce an image-guided drug delivery vector that was also tested using HeLa cells [217].

Al Faraj and colleagues produced a theranostic agent, using iron chloride salts and ammonium hydroxide to grow ferrous NPs on oxidized SWCNTs (200–300 nm in length) [218, 219]. With 40% by weight of the material consisting of the ferrous NPs, the material was used for MR contrast as well as magnetic targeting to breast cancer tumors in mice. The agent also consisted of mouse Endoglin/CD105 monoclonal antibody and DOX. An external magnet was placed over the tumor site, and tumor growth and progression were monitored using MRI. The authors found that DOX delivery was enhanced with the targeted SWCNT vector, as indicated by shrinking tumor size.

To increase water dispersibility and anti-tumor activity, 10-hydroxycamptothecin (HCPT), which is similar to CPT, was utilized to develop an MWCNT-based drug delivery platform [220]. The new conjugate showed superior anti-tumor activity compared to clinical HCPT formulations in both in vitro and in vivo experiments. In vivo SPECT and ex vivo gamma-scintillation counting analyses using ^{99m}Tc ($t_{1/2} = 6$ h) showed that the conjugates had a long circulation time (~3.6 h) in the blood and a high accumulation level in a tumor (~3.6% injected dose per gram of tissue). Finally, Xiao and coworkers prepared a HER2 IgY-SWCNT conjugate and examined the detection and selective destruction of cancer cells using the conjugate in an in vitro experiment [221]. Raman signals from the SWCNTs were detected using 785-nm laser excitation, and NIR irradiation was performed at 808 nm with 5 W/cm^2 for 2 min. Raman signals showed the selectivity of the conjugate-treated SK-BR-3 cells (HER2-positive), and NIR irradiation showed selective ablation of the conjugate-targeted breast cancer cells without damaging untargeted cells. Therefore, a combination of the intrinsic imaging and photothermal properties of SWCNTs and the targeting capabilities of the IgY antibody offered another method for simultaneous cancer detection and therapy.

From the examples discussed above, it is clear that CNT-based theranostic medicine will likely continue to grow in importance in the developing field of diagnostic and therapeutic nano-oncology. It should be an exciting future to behold.

5 Future Challenges for Clinical Applications

Assuming that the future direction of CNT-based agents for cancer detection and therapy continues to move toward multifunctional theranostic agents, there is still much work to be done to achieve their full potential. Some of the currently accepted criteria for such agents are that they (1) are safe (nontoxic), (2) have sufficiently high performance for the needed function, and (3) are stable long enough until biodegradation or eventual clearance from the body. In fact, recent studies have focused on peroxidase-induced biodegradation of CNTs, and other natural enzymatic processes [222–224]. Finally, and most importantly, the agents must be produced using good manufacturing practices as required by appropriate regulatory agencies. From this, it follows that SWCNTs are superior to MWCNTs for translational medicine simply because MWCNTs are a much less uniform material.

While several SWCNT-based materials have already satisfied most of the above criteria, two major problems remain: (1) more work is needed to create a more uniform NP distribution in both length and degree of functionalization, and (2) more preclinical data are needed to establish total excretion or safe degradation in the body of the agents over time. Although these challenges are formidable, they seem worthy of pursuit due to the promise already demonstrated by the CNT-based agents described in this chapter.

Acknowledgements This work was supported in part by the Welch Foundation (Grant C-0627; L.J.W.) and the U.S. National Science Foundation Graduate Research Fellowship Program (1450681; N.G.Z.).

References

1. Prasek J, Drbohlavova J, Chomoucka J et al (2011) Methods for carbon nanotubes synthesis—review. J Mater Chem 21:15872–15884. https://doi.org/10.1039/C1JM12254A
2. Odom TW, Huang J-L, Kim P, Lieber CM (1998) Atomic structure and electronic properties of single-walled carbon nanotubes. Nature 391:62–64. https://doi.org/10.1038/34145
3. Wilder JWG, Venema LC, Rinzler AG et al (1998) Electronic structure of atomically resolved carbon nanotubes. Nature 391:59–62. https://doi.org/10.1038/34139
4. Hong G, Antaris AL, Dai H (2017) Near-infrared fluorophores for biomedical imaging. Nat Biomed Eng 1:s41551–016–0010–016. https://doi.org/10.1038/s41551-016-0010
5. Lakowicz JR (2010) Principles of fluorescence spectroscopy, 3 ed, corr 4. print. Springer, New York, NY
6. Woźniak B, Dera J (2007) Light absorption in sea water. Springer, New York, NY
7. Johnsen S (2001) Hidden in plain sight: the ecology and physiology of organismal transparency. Biol Bull 201:301–318. https://doi.org/10.2307/1543609
8. Hoover EE, Squier JA (2013) Advances in multiphoton microscopy technology. Nat Photonics 7:93–101. https://doi.org/10.1038/nphoton.2012.361
9. Saito R, Fujita M, Dresselhaus G, Dresselhaus MS (1992) Electronic structure of chiral graphene tubules. Appl Phys Lett 60:2204–2206. https://doi.org/10.1063/1.107080
10. Hamada N, Sawada S, Oshiyama A (1992) New one-dimensional conductors: Graphitic microtubules. Phys Rev Lett 68:1579–1581. https://doi.org/10.1103/PhysRevLett.68.1579

11. Bachilo SM, Strano MS, Kittrell C et al (2002) Structure-assigned optical spectra of single-walled carbon nanotubes. Science 298:2361–2366. https://doi.org/10.1126/science.1078727
12. Hong G, Diao S, Antaris AL, Dai H (2015) Carbon nanomaterials for biological imaging and nanomedicinal therapy. Chem Rev 115:10816–10906. https://doi.org/10.1021/acs.chemrev.5b00008
13. Ma Y-Z, Valkunas L, Bachilo SM, Fleming GR (2005) Exciton binding energy in semiconducting single-walled carbon nanotubes. J Phys Chem B 109:15671–15674. https://doi.org/10.1021/jp053011t
14. Hertel T, Perebeinos V, Crochet J et al (2008) Intersubband decay of 1-D exciton resonances in carbon nanotubes. Nano Lett 8:87–91. https://doi.org/10.1021/nl0720915
15. Cognet L, Tsyboulski DA, Rocha J-DR et al (2007) Stepwise quenching of exciton fluorescence in carbon nanotubes by single-molecule reactions. Science 316:1465–1468. https://doi.org/10.1126/science.1141316
16. Welsher K, Liu Z, Sherlock SP et al (2009) A route to brightly fluorescent carbon nanotubes for near-infrared imaging in mice. Nat Nanotechnol 4:773–780. https://doi.org/10.1038/nnano.2009.294
17. Robinson JT, Hong G, Liang Y et al (2012) In-vivo fluorescence imaging in the NIR-II with long circulating carbon nanotubes capable of ultra-high tumor uptake. J Am Chem Soc 134:10664–10669. https://doi.org/10.1021/ja303737a
18. Hong G, Lee JC, Jha A et al (2014) Near-infrared II fluorescence for imaging hindlimb vessel regeneration with dynamic tissue perfusion measurement. Circ Cardiovasc Imaging 7:517–525. https://doi.org/10.1161/CIRCIMAGING.113.000305
19. Welsher K, Liu Z, Daranciang D, Dai H (2008) Selective probing and imaging of cells with single walled carbon nanotubes as near-infrared fluorescent molecules. Nano Lett 8:586–590. https://doi.org/10.1021/nl072949q
20. Hong G, Lee JC, Robinson JT et al (2012) Multifunctional in vivo vascular imaging using near-infrared II fluorescence. Nat Med 18:1841–1846. https://doi.org/10.1038/nm.2995
21. Welsher K, Sherlock SP, Dai H (2011) Deep-tissue anatomical imaging of mice using carbon nanotube fluorophores in the second near-infrared window. Proc Natl Acad Sci 108:8943–8948. https://doi.org/10.1073/pnas.1014501108
22. Hillman EMC, Moore A (2007) All-optical anatomical co-registration for molecular imaging of small animals using dynamic contrast. Nat Photonics 1:526–530. https://doi.org/10.1038/nphoton.2007.146
23. Shirinifard A, Gens JS, Zaitlen BL et al (2009) 3D multi-cell simulation of tumor growth and angiogenesis. PLoS ONE 4:e7190. https://doi.org/10.1371/journal.pone.0007190
24. Chen J, Wang M, Xi B, et al (2012) SPARC Is a key regulator of proliferation, apoptosis and invasion in human ovarian cancer. PLoS ONE 7. https://doi.org/10.1371/journal.pone.0042413
25. Yi H, Ghosh D, Ham M-H et al (2012) M13 phage-functionalized single-walled carbon nanotubes as nanoprobes for second near-infrared window fluorescence imaging of targeted tumors. Nano Lett 12:1176–1183. https://doi.org/10.1021/nl2031663
26. Jena PV, Shamay Y, Shah J et al (2016) Photoluminescent carbon nanotubes interrogate the permeability of multicellular tumor spheroids. Carbon 97:99–109. https://doi.org/10.1016/j.carbon.2015.08.024
27. Mansur HS (2010) Quantum dots and nanocomposites. Wiley Interdiscip Rev Nanomed Nanobiotechnol 2:113–129. https://doi.org/10.1002/wnan.78
28. Chan WCW, Maxwell DJ, Gao X et al (2002) Luminescent quantum dots for multiplexed biological detection and imaging. Curr Opin Biotechnol 13:40–46. https://doi.org/10.1016/S0958-1669(02)00282-3
29. Ghasemi Y, Peymani P, Afifi S (2009) Quantum dot: magic nanoparticle for imaging, detection and targeting. Acta Bio Medica Atenei Parm 80:156–165
30. Madani SY, Shabani F, Dwek MV, Seifalian AM (2013) Conjugation of quantum dots on carbon nanotubes for medical diagnosis and treatment. Int J Nanomedicine 8:941–950. https://doi.org/10.2147/IJN.S36416

31. Banerjee S, Wong SS (2003) In situ quantum dot growth on multiwalled carbon nanotubes. J Am Chem Soc 125:10342–10350. https://doi.org/10.1021/ja035980c
32. Si H-Y, Liu C-H, Xu H et al (2009) Shell-controlled photoluminescence in CdSe/CNT nanohybrids. Nanoscale Res Lett 4:1146–1152. https://doi.org/10.1007/s11671-009-9373-z
33. Zhou Z, Kang H, Clarke ML et al (2009) Water-soluble dna-wrapped single-walled carbon-nanotube/quantum-dot complexes. Small 5:2149–2155. https://doi.org/10.1002/smll.200801932
34. Chen X, Kis A, Zettl A, Bertozzi CR (2007) A cell nanoinjector based on carbon nanotubes. Proc Natl Acad Sci U S A 104:8218–8222. https://doi.org/10.1073/pnas.0700567104
35. Grzelczak M, Correa-Duarte MA, Salgueiriño-Maceira V et al (2006) Photoluminescence quenching control in quantum dot-carbon nanotube composite colloids using a silica-shell spacer. Adv Mater 18:415–420. https://doi.org/10.1002/adma.200501523
36. Tan A, Yildirimer L, Rajadas J et al (2011) Quantum dots and carbon nanotubes in oncology: a review on emerging theranostic applications in nanomedicine. Nanomed 6:1101–1114. https://doi.org/10.2217/nnm.11.64
37. Wagnieres GA, Star WM, Wilson BC (1998) In vivo fluorescence spectroscopy and imaging for oncological applications. Photochem Photobiol 68:603–632. https://doi.org/10.1111/j.1751-1097.1998.tb02521.x
38. Ryder AG (2005) Surface enhanced Raman scattering for narcotic detection and applications to chemical biology. Curr Opin Chem Biol 9:489–493. https://doi.org/10.1016/j.cbpa.2005.07.001
39. Gong H, Peng R, Liu Z (2013) Carbon nanotubes for biomedical imaging: The recent advances. Adv Drug Deliv Rev 65:1951–1963. https://doi.org/10.1016/j.addr.2013.10.002
40. Martin RM, Falicov LM (1983) Resonant Raman scattering. Light Scattering in Solids I. Springer, Berlin, Heidelberg, pp 79–145
41. Rao AM, Richter E, Bandow S et al (1997) Diameter-selective raman scattering from vibrational modes in carbon nanotubes. Science 275:187–191. https://doi.org/10.1126/science.275.5297.187
42. Heller DA, Baik S, Eurell TE, Strano MS (2005) Single-walled carbon nanotube spectroscopy in live cells: towards long-term labels and optical sensors. Adv Mater 17:2793–2799. https://doi.org/10.1002/adma.200500477
43. Zavaleta C, de la Zerda A, Liu Z et al (2008) Noninvasive Raman spectroscopy in living mice for evaluation of tumor targeting with carbon nanotubes. Nano Lett 8:2800–2805. https://doi.org/10.1021/nl801362a
44. Liu Z, Li X, Tabakman SM et al (2008) Multiplexed multicolor raman imaging of live cells with isotopically modified single walled carbon nanotubes. J Am Chem Soc 130:13540–13541. https://doi.org/10.1021/ja806242t
45. Wang X, Wang C, Cheng L et al (2012) Noble metal coated single-walled carbon nanotubes for applications in surface enhanced raman scattering imaging and photothermal therapy. J Am Chem Soc 134:7414–7422. https://doi.org/10.1021/ja300140c
46. Park S, Aglyamov SR, Scott WG, et al (2006) 1E-5 synergy and applications of combined ultrasound, elasticity, and photoacoustic imaging (Invited). In: 2006 IEEE Ultrasonics Symposium, pp 405–415
47. Mallidi S, Luke GP, Emelianov S (2011) Photoacoustic imaging in cancer detection, diagnosis, and treatment guidance. Trends Biotechnol 29:213–221. https://doi.org/10.1016/j.tibtech.2011.01.006
48. Folkman J (1990) What is the evidence that tumors are angiogenesis dependent? JNCI J Natl Cancer Inst 82:4–7. https://doi.org/10.1093/jnci/82.1.4
49. Li ML, Oh JT, Xie X et al (2008) Simultaneous molecular and hypoxia imaging of brain tumors in vivo using spectroscopic photoacoustic tomography. Proc IEEE 96:481–489. https://doi.org/10.1109/JPROC.2007.913515
50. Qu M, Mallidi S, Mehrmohammadi M, et al (2009) Combined photoacoustic and magneto-acoustic imaging. In: 2009 annual international conference of the IEEE engineering in medicine and biology society, pp 4763–4766

51. Xiang L, Xing D, Gu H et al (2007) Real-time optoacoustic monitoring of vascular damage during photodynamic therapy treatment of tumor. J Biomed Opt 12:014001. https://doi.org/10.1117/1.2437752
52. Pramanik M, Song KH, Swierczewska M et al (2009) In vivo carbon nanotube-enhanced non-invasive photoacoustic mapping of the sentinel lymph node. Phys Med Biol 54:3291. https://doi.org/10.1088/0031-9155/54/11/001
53. de la Zerda A, Liu Z, Bodapati S et al (2010) Ultrahigh sensitivity carbon nanotube agents for photoacoustic molecular imaging in living mice. Nano Lett 10:2168–2172. https://doi.org/10.1021/nl100890d
54. Pramanik M, Ku G, Li C, Wang LV (2008) Design and evaluation of a novel breast cancer detection system combining both thermoacoustic (TA) and photoacoustic (PA) tomography. Med Phys 35:2218–2223. https://doi.org/10.1118/1.2911157
55. Pramanik M, Swierczewska M, Green D, et al (2009) Single-walled carbon nanotubes as a multimodal-thermoacoustic and photoacoustic-contrast agent. J Biomed Opt 14:034018–034018–8. https://doi.org/10.1117/1.3147407
56. Kim J-W, Galanzha EI, Shashkov EV et al (2009) Golden carbon nanotubes as multimodal photoacoustic and photothermal high-contrast molecular agents. Nat Nanotechnol 4:688–694. https://doi.org/10.1038/nnano.2009.231
57. Yu SB, Watson AD (1999) Metal-based X-ray contrast media. Chem Rev 99:2353–2378
58. Popovtzer R, Agrawal A, Kotov NA et al (2008) Targeted gold nanoparticles enable molecular CT imaging of cancer. Nano Lett 8:4593–4596. https://doi.org/10.1021/nl8029114
59. Hainfeld JF, Slatkin DN, Focella TM, Smilowitz HM (2006) Gold nanoparticles: a new X-ray contrast agent. Br J Radiol 79:248–253. https://doi.org/10.1259/bjr/13169882
60. Rabin O, Manuel Perez J, Grimm J et al (2006) An X-ray computed tomography imaging agent based on long-circulating bismuth sulphide nanoparticles. Nat Mater 5:118–122. https://doi.org/10.1038/nmat1571
61. Kinsella JM, Jimenez RE, Karmali PP et al (2011) X-ray computed tomography imaging of breast cancer by using targeted peptide-labeled bismuth sulfide nanoparticles. Angew Chem Int Ed Engl 50:12308–12311. https://doi.org/10.1002/anie.201104507
62. Hernández-Rivera M, Zaibaq NG, Wilson LJ (2016) Toward carbon nanotube-based imaging agents for the clinic. Biomaterials 101:229–240. https://doi.org/10.1016/j.biomaterials.2016.05.045
63. Ashcroft JM, Hartman KB, Kissell KR et al (2007) Single-molecule I2@US-tube nanocapsules: a new X-ray contrast-agent design. Adv Mater 19:573–576. https://doi.org/10.1002/adma.200601424
64. Algethami M, Blencowe A, Feltis B, Geso M (2017) Bismuth sulfide nanoparticles as a complement to traditional iodinated contrast agents at various X-Ray computed tomography tube potentials. J Nanomater Mol Nanotechnol 2017. https://doi.org/10.4172/2324-8777.1000222
65. Rivera EJ, Tran LA, Hernández-Rivera M et al (2013) Bismuth@US-tubes as a potential contrast agent for X-ray imaging applications. J Mater Chem B 1:4792–4800. https://doi.org/10.1039/C3TB20742K
66. Hernández-Rivera M, Kumar I, Cho SY et al (2017) High-performance hybrid bismuth-carbon nanotube based contrast agent for X-ray CT imaging. ACS Appl Mater Interfaces 9:5709–5716. https://doi.org/10.1021/acsami.6b12768
67. Roch A, Muller RN, Gillis P (1999) Theory of proton relaxation induced by superparamagnetic particles. J Chem Phys 110:5403–5411. https://doi.org/10.1063/1.478435
68. Merbach A, Helm L, Tóth É (2013) The chemistry of contrast agents in medical magnetic resonance imaging: helm/the chemistry of contrast agents in medical magnetic resonance imaging. Wiley, Chichester, UK
69. Sethi R, Mackeyev Y, Wilson LJ (2012) The Gadonanotubes revisited: a new frontier in MRI contrast agent design. Inorganica Chim Acta 393:165–172. https://doi.org/10.1016/j.ica.2012.07.004
70. Sitharaman B, Kissell KR, Hartman KB, et al (2005) Superparamagnetic gadonanotubes are high-performance MRI contrast agents. Chem Commun 3915–3917. https://doi.org/10.1039/b504435a

71. Sitharaman B, Wilson LJ (2006) Gadonanotubes as new high-performance MRI contrast agents. Int J Nanomedicine 1:291–295
72. Hartman KB, Laus S, Bolskar RD et al (2008) Gadonanotubes as ultrasensitive pH-smart probes for magnetic resonance imaging. Nano Lett 8:415–419. https://doi.org/10.1021/nl0720408
73. Broome DR (2008) Nephrogenic systemic fibrosis associated with gadolinium based contrast agents: a summary of the medical literature reporting. Eur J Radiol 66:230–234. https://doi.org/10.1016/j.ejrad.2008.02.011
74. Holt BD, Law JJ, Boyer PD et al (2015) Subcellular partitioning and analysis of Gd3+ -loaded ultrashort single-walled carbon nanotubes. ACS Appl Mater Interfaces 7:14593–14602. https://doi.org/10.1021/acsami.5b04851
75. Tang AM, Ananta JS, Zhao H et al (2011) Cellular uptake and imaging studies of gadolinium-loaded single-walled carbon nanotubes as MRI contrast agents. Contrast Media Mol Imaging 6:93–99. https://doi.org/10.1002/cmmi.410
76. Tran LA, Krishnamurthy R, Muthupillai R et al (2010) Gadonanotubes as magnetic nanolabels for stem cell detection. Biomaterials 31:9482–9491. https://doi.org/10.1016/j.biomaterials.2010.08.034
77. Tran LA, Hernández-Rivera M, Berlin AN et al (2014) The use of gadolinium-carbon nanostructures to magnetically enhance stem cell retention for cellular cardiomyoplasty. Biomaterials 35:720–726. https://doi.org/10.1016/j.biomaterials.2013.10.013
78. Hassan AA, Chan BT-Y, Tran LA et al (2010) Serine-derivatized gadonanotubes as magnetic nanoprobes for intracellular labeling. Contrast Media Mol Imaging 5:34–38. https://doi.org/10.1002/cmmi.293
79. Law JJ, Guven A, Wilson LJ (2014) Relaxivity enhancement of aquated Tris(β-diketonate)gadolinium(III) chelates by confinement within ultrashort single-walled carbon nanotubes. Contrast Media Mol Imaging 9:409–412. https://doi.org/10.1002/cmmi.1603
80. Ma Q, Jebb M, Tweedle MF, Wilson LJ (2013) The gadonanotubes: structural origin of their high-performance MRI contrast agent behavior. J Mater Chem B 1:5791–5797. https://doi.org/10.1039/C3TB20870B
81. Gizzatov A, Dimiev A, Mackeyev Y et al (2012) Highly water soluble multi-layer graphene nanoribbons and related honey-comb carbon nanostructures. Chem Commun 48:5602–5604. https://doi.org/10.1039/C2CC31407J
82. Gizzatov A, Hernández-Rivera M, Keshishian V et al (2015) Surfactant-free Gd3+-ion-containing carbon nanotube MRI contrast agents for stem cell labeling. Nanoscale 7:12085–12091. https://doi.org/10.1039/C5NR02078F
83. Swierczewska M, Rusakova I, Sitharaman B (2009) Gadolinium and europium catalyzed growth of single-walled carbon nanotubes. Carbon 47:3139–3142. https://doi.org/10.1016/j.carbon.2009.07.021
84. Sitharaman B, Jacobson BD, Wadghiri YZ et al (2013) The magnetic, relaxometric, and optical properties of gadolinium-catalyzed single walled carbon nanotubes. J Appl Phys 113:134308. https://doi.org/10.1063/1.4796183
85. Avti PK, Caparelli ED, Sitharaman B (2013) Cytotoxicity, cytocompatibility, cell-labeling efficiency, and in vitro cellular magnetic resonance imaging of gadolinium-catalyzed single-walled carbon nanotubes. J Biomed Mater Res A 101:3580–3591. https://doi.org/10.1002/jbm.a.34643
86. Avti PK, Talukdar Y, Sirotkin MV et al (2013) Toward single-walled carbon nanotube–gadolinium complex as advanced MRI contrast agents: Pharmacodynamics and global genomic response in small animals. J Biomed Mater Res B Appl Biomater 101B:1039–1049. https://doi.org/10.1002/jbm.b.32914
87. Marangon I, Ménard-Moyon C, Kolosnjaj-Tabi J et al (2014) Covalent functionalization of multi-walled carbon nanotubes with a gadolinium chelate for efficient T1-weighted magnetic resonance imaging. Adv Funct Mater 24:7173–7186. https://doi.org/10.1002/adfm.201402234

88. Wu H, Liu G, Wang X et al (2011) Solvothermal synthesis of cobalt ferrite nanoparticles loaded on multiwalled carbon nanotubes for magnetic resonance imaging and drug delivery. Acta Biomater 7:3496–3504. https://doi.org/10.1016/j.actbio.2011.05.031
89. Liu Y, Hughes TC, Muir BW et al (2014) Water-dispersible magnetic carbon nanotubes as T2-weighted MRI contrast agents. Biomaterials 35:378–386. https://doi.org/10.1016/j.biomaterials.2013.09.079
90. Liu Y, Muir BW, Waddington LJ et al (2015) Colloidally stabilized magnetic carbon nanotubes providing mri contrast in mouse liver tumors. Biomacromol 16:790–797. https://doi.org/10.1021/bm501706x
91. Fain S, Schiebler ML, McCormack DG, Parraga G (2010) Imaging of lung function using hyperpolarized helium-3 magnetic resonance imaging: review of current and emerging translational methods and applications. J Magn Reson Imaging JMRI 32:1398–1408. https://doi.org/10.1002/jmri.22375
92. Al Faraj A, Fauvelle F, Luciani N et al (2011) In vivo biodistribution and biological impact of injected carbon nanotubes using magnetic resonance techniques. Int J Nanomedicine 6:351–361. https://doi.org/10.2147/IJN.S16653
93. Tucker-Schwartz JM, Hong T, Colvin DC et al (2012) Dual-modality photothermal optical coherence tomography and magnetic-resonance imaging of carbon nanotubes. Opt Lett 37:872–874
94. Vittorio O, Duce SL, Pietrabissa A, Cuschieri A (2011) Multiwall carbon nanotubes as MRI contrast agents for tracking stem cells. Nanotechnology 22:095706. https://doi.org/10.1088/0957-4484/22/9/095706
95. Lodge MA, Frey EC (2014) Nuclear medicine imaging devices. In: Bailey DL, American Association of Physicists in Medicine (eds) Nuclear medicine physics: a handbook for teachers and students. International Atomic Energy Agency, Vienna, pp 312–397
96. Zeglis BM, Lewis JS (2011) A practical guide to the construction of radiometallated bioconjugates for positron emission tomography. Dalton Trans 40:6168–6195. https://doi.org/10.1039/C0DT01595D
97. de Barros AB, Tsourkas A, Saboury B et al (2012) Emerging role of radiolabeled nanoparticles as an effective diagnostic technique. EJNMMI Res 2:39. https://doi.org/10.1186/2191-219X-2-39
98. Ruggiero A, Villa CH, Holland JP et al (2010) Imaging and treating tumor vasculature with targeted radiolabeled carbon nanotubes. Int J Nanomedicine 5:783–802. https://doi.org/10.2147/IJN.S13300
99. Spinato C, de Garibay APR, Kierkowicz M et al (2016) Design of antibody-functionalized carbon nanotubes filled with radioactivable metals towards a targeted anticancer therapy. Nanoscale 8:12626–12638. https://doi.org/10.1039/C5NR07923C
100. Liu Z, Cai W, He L et al (2007) In vivo biodistribution and highly efficient tumour targeting of carbon nanotubes in mice. Nat Nanotechnol 2:47–52. https://doi.org/10.1038/nnano.2006.170
101. Al Faraj A, Shaik AS, Al Sayed B et al (2015) Specific targeting and noninvasive imaging of breast cancer stem cells using single-walled carbon nanotubes as novel multimodality nanoprobes. Nanomed 11:31–46. https://doi.org/10.2217/nnm.15.182
102. Cisneros BT, Law JJ, Matson ML et al (2014) Stable confinement of positron emission tomography and magnetic resonance agents within carbon nanotubes for bimodal imaging. Nanomed 9:2499–2509. https://doi.org/10.2217/nnm.14.26
103. Wong BS, Yoong SL, Jagusiak A et al (2013) Carbon nanotubes for delivery of small molecule drugs. Adv Drug Deliv Rev 65:1964–2015. https://doi.org/10.1016/j.addr.2013.08.005
104. Wheate NJ, Walker S, Craig GE, Oun R (2010) The status of platinum anticancer drugs in the clinic and in clinical trials. Dalton Trans 39:8113–8127. https://doi.org/10.1039/C0DT00292E
105. Wang D, Lippard SJ (2005) Cellular processing of platinum anticancer drugs. Nat Rev Drug Discov 4:307–320. https://doi.org/10.1038/nrd1691
106. Lilley DMJ (1996) Cisplatin adducts in DNA: distortion and recognition. JBIC, J Biol Inorg Chem 1:189–191. https://doi.org/10.1007/s007750050042

107. Feazell RP, Nakayama-Ratchford N, Dai H, Lippard SJ (2007) Soluble single-walled carbon nanotubes as longboat delivery systems for platinum(IV) anticancer drug design. J Am Chem Soc 129:8438–8439. https://doi.org/10.1021/ja073231f
108. Bhirde AA, Patel V, Gavard J et al (2009) Targeted killing of cancer cells in vivo and in vitro with EGF-directed carbon nanotube-based drug delivery. ACS Nano 3:307–316. https://doi.org/10.1021/nn800551s
109. Bhirde AA, Sousa AA, Patel V et al (2009) Imaging the distribution of individual platinum-based anticancer drug molecules attached to single-wall carbon nanotubes. Nanomed 4:763–772. https://doi.org/10.2217/nnm.09.56
110. Tripisciano C, Kraemer K, Taylor A, Borowiak-Palen E (2009) Single-wall carbon nanotubes based anticancer drug delivery system. Chem Phys Lett 478:200–205. https://doi.org/10.1016/j.cplett.2009.07.071
111. Guven A, Rusakova IA, Lewis MT, Wilson LJ (2012) Cisplatin@US-tube Carbon Nanocapsules For Enhanced Chemotherapeutic Delivery. Biomaterials 33:1455–1461. https://doi.org/10.1016/j.biomaterials.2011.10.060
112. Guven A, Villares GJ, Hilsenbeck SG et al (2017) Carbon nanotube capsules enhance the in vivo efficacy of cisplatin. Acta Biomater 58:466–478. https://doi.org/10.1016/j.actbio.2017.04.035
113. Dhar S, Liu Z, Thomale J et al (2008) Targeted single wall carbon nanotube mediated Pt(IV) prodrug delivery using folate as a homing device. J Am Chem Soc 130:11467–11476. https://doi.org/10.1021/ja803036e
114. Yang F, Fu DL, Long J, Ni QX (2008) Magnetic lymphatic targeting drug delivery system using carbon nanotubes. Med Hypotheses 70:765–767. https://doi.org/10.1016/j.mehy.2007.07.045
115. Yang F, Hu J, Yang D et al (2009) Pilot study of targeting magnetic carbon nanotubes to lymph nodes. Nanomed 4:317–330. https://doi.org/10.2217/nnm.09.5
116. Hampel S, Kunze D, Haase D et al (2008) Carbon nanotubes filled with a chemotherapeutic agent: a nanocarrier mediates inhibition of tumor cell growth. Nanomed 3:175–182. https://doi.org/10.2217/17435889.3.2.175
117. Wang JC (1991) DNA topoisomerases: why so many? J Biol Chem 266:6659–6662
118. Tian Z, Yin M, Ma H et al (2011) Supramolecular assembly and antitumor activity of multiwalled carbon nanotube–camptothecin complexes. J Nanosci Nanotechnol 11:953–958
119. Son KH, Hong JH, Lee JW (2016) Carbon nanotubes as cancer therapeutic carriers and mediators. Int J Nanomedicine 11:5163–5185. https://doi.org/10.2147/IJN.S112660
120. Huang H, Yuan Q, Shah JS, Misra RDK (2011) A new family of folate-decorated and carbon nanotube-mediated drug delivery system: Synthesis and drug delivery response. Adv Drug Deliv Rev 63:1332–1339. https://doi.org/10.1016/j.addr.2011.04.001
121. Wen S, Liu H, Cai H et al (2013) Targeted and pH-responsive delivery of doxorubicin to cancer cells using multifunctional dendrimer-modified multi-walled carbon nanotubes. Adv Healthc Mater 2:1267–1276. https://doi.org/10.1002/adhm.201200389
122. Niu L, Meng L, Lu Q (2013) Folate-conjugated PEG on single walled carbon nanotubes for targeting delivery of doxorubicin to cancer cells. Macromol Biosci 13:735–744. https://doi.org/10.1002/mabi.201200475
123. Di Crescenzo A, Velluto D, Hubbell JA, Fontana A (2011) Biocompatible dispersions of carbon nanotubes: a potential tool for intracellular transport of anticancer drugs. Nanoscale 3:925–928. https://doi.org/10.1039/C0NR00444H
124. Zhao Q, Zhang J, Wang R, Cong W (2008) Use of a thermocouple for malignant tumor detection. IEEE Eng Med Biol Mag 27:64–66. https://doi.org/10.1109/MEMB.2007.913292
125. Mashat A, Deng L, Altawashi A et al (2012) Zippered release from polymer-gated carbon nanotubes. J Mater Chem 22:11503–11508. https://doi.org/10.1039/C2JM30454F
126. Ali-Boucetta H, Al-Jamal KT, McCarthy D, et al (2008) Multiwalled carbon nanotube-doxorubicin supramolecular complexes for cancer therapeutics. Chem Commun 459–461. https://doi.org/10.1039/b712350g

127. Li R, Wu R, Zhao L et al (2010) *P*-Glycoprotein antibody functionalized carbon nanotube overcomes the multidrug resistance of human leukemia cells. ACS Nano 4:1399–1408. https://doi.org/10.1021/nn9011225
128. Taghdisi SM, Lavaee P, Ramezani M, Abnous K (2011) Reversible targeting and controlled release delivery of daunorubicin to cancer cells by aptamer-wrapped carbon nanotubes. Eur J Pharm Biopharm 77:200–206. https://doi.org/10.1016/j.ejpb.2010.12.005
129. Liu Z, Sun X, Nakayama-Ratchford N, Dai H (2007) Supramolecular chemistry on water-soluble carbon nanotubes for drug loading and delivery. ACS Nano 1:50–56. https://doi.org/10.1021/nn700040t
130. Ji Z, Lin G, Lu Q et al (2012) Targeted therapy of SMMC-7721 liver cancer in vitro and in vivo with carbon nanotubes based drug delivery system. J Colloid Interface Sci 365:143–149. https://doi.org/10.1016/j.jcis.2011.09.013
131. Datir SR, Das M, Singh RP, Jain S (2012) Hyaluronate tethered, "smart" multiwalled carbon nanotubes for tumor-targeted delivery of doxorubicin. Bioconjug Chem 23:2201–2213. https://doi.org/10.1021/bc300248t
132. Das M, Singh RP, Datir SR, Jain S (2013) Surface chemistry dependent "switch" regulates the trafficking and therapeutic performance of drug-loaded carbon nanotubes. Bioconjug Chem 24:626–639. https://doi.org/10.1021/bc300598z
133. Lodhi N, Mehra NK, Jain NK (2013) Development and characterization of dexamethasone mesylate anchored on multi walled carbon nanotubes. J Drug Target 21:67–76. https://doi.org/10.3109/1061186X.2012.729213
134. Heister E, Neves V, Tîlmaciu C et al (2009) Triple functionalisation of single-walled carbon nanotubes with doxorubicin, a monoclonal antibody, and a fluorescent marker for targeted cancer therapy. Carbon 47:2152–2160. https://doi.org/10.1016/j.carbon.2009.03.057
135. Zhang X, Meng L, Lu Q et al (2009) Targeted delivery and controlled release of doxorubicin to cancer cells using modified single wall carbon nanotubes. Biomaterials 30:6041–6047. https://doi.org/10.1016/j.biomaterials.2009.07.025
136. Chen Z, Pierre D, He H et al (2011) Adsorption behavior of epirubicin hydrochloride on carboxylated carbon nanotubes. Int J Pharm 405:153–161. https://doi.org/10.1016/j.ijpharm.2010.11.034
137. Foley EA, Kapoor TM (2013) Microtubule attachment and spindle assembly checkpoint signaling at the kinetochore. Nat Rev Mol Cell Biol 14:25
138. Chen J, Chen S, Zhao X et al (2008) Functionalized single-walled carbon nanotubes as rationally designed vehicles for tumor-targeted drug delivery. J Am Chem Soc 130:16778–16785
139. Li C, Yu D, Inoue T et al (1996) Synthesis and evaluation of water-soluble polyethylene glycol-paclitaxel conjugate as a paclitaxel prodrug. Anticancer Drugs 7:642–648
140. Lay CL, Liu HQ, Tan HR, Liu Y (2010) Delivery of paclitaxel by physically loading onto poly (ethylene glycol)(PEG)-graftcarbon nanotubes for potent cancer therapeutics. Nanotechnology 21:065101
141. Lee Y, Geckeler KE (2012) Cellular interactions of a water-soluble supramolecular polymer complex of carbon nanotubes with human epithelial colorectal adenocarcinoma cells. Macromol Biosci 12:1060–1067
142. Moore TL, Pitzer JE, Podila R et al (2013) Multifunctional polymer-coated carbon nanotubes for safe drug delivery. Part Part Syst Charact 30:365–373
143. Wu C-H, Cao C, Kim JH et al (2012) Trojan-horse nanotube on-command intracellular drug delivery. Nano Lett 12:5475–5480
144. Wang L, Zhang M, Zhang N et al (2011) Synergistic enhancement of cancer therapy using a combination of docetaxel and photothermal ablation induced by single-walled carbon nanotubes. Int J Nanomedicine 6:2641
145. Sobhani Z, Dinarvand R, Atyabi F et al (2011) Increased paclitaxel cytotoxicity against cancer cell lines using a novel functionalized carbon nanotube. Int J Nanomedicine 6:705
146. Liu Z, Chen K, Davis C et al (2008) Drug delivery with carbon nanotubes for in vivo cancer treatment. Cancer Res 68:6652–6660

147. Arya N, Arora A, Vasu K et al (2013) Combination of single walled carbon nanotubes/graphene oxide with paclitaxel: a reactive oxygen species mediated synergism for treatment of lung cancer. Nanoscale 5:2818–2829
148. Zhang W, Zhang D, Tan J, Cong H (2012) Carbon nanotube exposure sensitize human ovarian cancer cells to paclitaxel. J Nanosci Nanotechnol 12:7211–7214
149. Inoue Y, Fujimoto H, Ogino T, Iwata H (2008) Site-specific gene transfer with high efficiency onto a carbon nanotube-loaded electrode. J R Soc Interface 5:909–918
150. Qin W, Yang K, Tang H et al (2011) Improved GFP gene transfection mediated by polyamidoamine dendrimer-functionalized multi-walled carbon nanotubes with high biocompatibility. Colloids Surf B Biointerfaces 84:206–213
151. Hao Y, Xu P, He C et al (2011) Impact of carbondiimide crosslinker used for magnetic carbon nanotube mediated GFP plasmid delivery. Nanotechnology 22:285103
152. Paul A, Shao W, Shum-Tim D, Prakash S (2012) The attenuation of restenosis following arterial gene transfer using carbon nanotube coated stent incorporating TAT/DNA Ang1+Vegf nanoparticles. Biomaterials 33:7655–7664
153. Wang X, Ren J, Qu X (2008) targeted rna interference of cyclin A2 mediated by functionalized single-walled carbon nanotubes induces proliferation arrest and apoptosis in chronic myelogenous leukemia K562 cells. ChemMedChem 3:940–945
154. Varkouhi AK, Foillard S, Lammers T et al (2011) SiRNA delivery with functionalized carbon nanotubes. Int J Pharm 416:419–425
155. Al-Jamal KT, Gherardini L, Bardi G et al (2011) Functional motor recovery from brain ischemic insult by carbon nanotube-mediated siRNA silencing. Proc Natl Acad Sci 108:10952–10957
156. Wang L, Shi J, Zhang H et al (2013) Synergistic anticancer effect of RNAi and photothermal therapy mediated by functionalized single-walled carbon nanotubes. Biomaterials 34:262–274
157. Liu Z, Winters M, Holodniy M, Dai H (2007) siRNA delivery into human T cells and primary cells with carbon-nanotube transporters. Angew Chem Int Ed 46:2023–2027
158. McCarroll J, Baigude H, Yang C-S, Rana TM (2009) Nanotubes functionalized with lipids and natural amino acid dendrimers: a new strategy to create nanomaterials for delivering systemic RNAi. Bioconjug Chem 21:56–63
159. Chen H, Ma X, Li Z et al (2012) Functionalization of single-walled carbon nanotubes enables efficient intracellular delivery of siRNA targeting MDM2 to inhibit breast cancer cells growth. Biomed Pharmacother 66:334–338
160. Van den Bossche J, Tian B, Nunes A et al (2010) Efficient receptor-independent intracellular translocation of aptamers mediated by conjugation to carbon nanotubes. Chem Commun 46:7379–7381
161. Bates K, Kostarelos K (2013) Carbon nanotubes as vectors for gene therapy: past achievements, present challenges and future goals. Adv Drug Deliv Rev 65:2023–2033
162. Fortunati E, Bout A, Zanta MA et al (1996) In vitro and in vivo gene transfer to pulmonary cells mediated by cationic liposomes. Biochim Biophys Acta BBA-Gene Struct Expr 1306:55–62
163. Ledley FD (1994) Non-viral gene therapy. Curr Opin Biotechnol 5:626–636
164. Pantarotto D, Singh R, McCarthy D et al (2004) Functionalized carbon nanotubes for plasmid DNA gene delivery. Angew Chem 116:5354–5358
165. Singh R, Pantarotto D, McCarthy D et al (2005) Binding and condensation of plasmid DNA onto functionalized carbon nanotubes: toward the construction of nanotube-based gene delivery vectors. J Am Chem Soc 127:4388–4396
166. Podesta JE, Al-Jamal KT, Herrero MA et al (2009) Antitumor activity and prolonged survival by carbon-nanotube-mediated therapeutic siRNA silencing in a human lung xenograft model. Small 5:1176–1185
167. Behnam B, Shier WT, Nia AH et al (2013) Non-covalent functionalization of single-walled carbon nanotubes with modified polyethyleneimines for efficient gene delivery. Int J Pharm 454:204–215
168. Karmakar A, Bratton SM, Dervishi E et al (2011) Ethylenediamine functionalized-single-walled nanotube (f-SWNT)-assisted in vitro delivery of the oncogene suppressor p53 gene to breast cancer MCF-7 cells. Int J Nanomedicine 6:1045

169. Zhang Z, Yang X, Zhang Y et al (2006) Delivery of telomerase reverse transcriptase small interfering RNA in complex with positively charged single-walled carbon nanotubes suppresses tumor growth. Clin Cancer Res 12:4933–4939
170. Pan B, Cui D, Xu P et al (2009) Synthesis and characterization of polyamidoamine dendrimer-coated multi-walled carbon nanotubes and their application in gene delivery systems. Nanotechnology 20:125101
171. Mohammadi M, Salmasi Z, Hashemi M et al (2015) Single-walled carbon nanotubes functionalized with aptamer and piperazine–polyethylenimine derivative for targeted siRNA delivery into breast cancer cells. Int J Pharm 485:50–60
172. Hussey SL, Peterson BR (2002) Efficient delivery of streptavidin to mammalian cells: clathrin-mediated endocytosis regulated by a synthetic ligand. J Am Chem Soc 124:6265–6273
173. Parra J, Abad-Somovilla A, Mercader JV et al (2013) Carbon nanotube-protein carriers enhance size-dependent self-adjuvant antibody response to haptens. J Controlled Release 170:242–251
174. McDevitt MR, Chattopadhyay D, Kappel BJ et al (2007) Tumor targeting with antibody-functionalized, radiolabeled carbon nanotubes. J Nucl Med 48:1180–1189
175. Fan H, Zhang I, Chen X et al (2012) Intracerebral CpG immunotherapy with carbon nanotubes abrogates growth of subcutaneous melanomas in mice. Clin Cancer Res 18:5628–5638. https://doi.org/10.1158/1078-0432.ccr-12-1911
176. Zhao D, Alizadeh D, Zhang L et al (2011) Carbon nanotubes enhance CpG uptake and potentiate antiglioma immunity. Clin Cancer Res 17:771–782
177. Ouyang M, White EE, Ren H et al (2016) Metronomic doses of temozolomide enhance the efficacy of carbon nanotube CpG immunotherapy in an invasive glioma model. PLoS ONE 11:e0148139
178. Villa CH, Dao T, Ahearn I et al (2011) Single-walled carbon nanotubes deliver peptide antigen into dendritic cells and enhance IgG responses to tumor-associated antigens. ACS Nano 5:5300–5311
179. Sacchetti C, Rapini N, Magrini A et al (2013) In vivo targeting of intratumor regulatory T cells using PEG-modified single-walled carbon nanotubes. Bioconjug Chem 24:852–858
180. Shi Kam NW, Jessop TC, Wender PA, Dai H (2004) Nanotube molecular transporters: internalization of carbon nanotube–protein conjugates into mammalian cells. J Am Chem Soc 126:6850–6851
181. Weng X, Wang M, Ge J et al (2009) Carbon nanotubes as a protein toxin transporter for selective HER2-positive breast cancer cell destruction. Mol BioSyst 5:1224–1231
182. Vinay DS, Ryan EP, Pawelec G et al (2015) Immune evasion in cancer: Mechanistic basis and therapeutic strategies. Semin Cancer Biol 35:S185–S198. https://doi.org/10.1016/j.semcancer.2015.03.004
183. Liu Y, Ng K, Lillehei KO (2003) Cell-mediated immunotherapy: a new approach to the treatment of malignant glioma. Cancer Control 10:138–147
184. Parney IF, Hao C, Petruk KC (2000) Glioma immunology and immunotherapy. Neurosurgery 46:778–792
185. Meng J, Yang M, Jia F et al (2010) Subcutaneous injection of water-soluble multi-walled carbon nanotubes in tumor-bearing mice boosts the host immune activity. Nanotechnology 21:145104
186. Meng J, Duan J, Kong H et al (2008) Carbon nanotubes conjugated to tumor lysate protein enhance the efficacy of an antitumor immunotherapy. Small 4:1364–1370
187. Fadel TR, Sharp FA, Vudattu N et al (2014) A carbon nanotube–polymer composite for T-cell therapy. Nat Nanotechnol 9:639–647
188. Li Q, Ruan H, Li H (2014) Nanocarbon materials for photodynamic therapy and photothermal therapy. Pharm Nanotechnol 2:58–64
189. Brennan ME, Coleman JN, Drury A et al (2003) Nonlinear photoluminescence from van Hove singularities in multiwalled carbon nanotubes. Opt Lett 28:266–268
190. Moon HK, Lee SH, Choi HC (2009) In vivo near-infrared mediated tumor destruction by photothermal effect of carbon nanotubes. ACS Nano 3:3707–3713

191. Huang N, Wang H, Zhao J et al (2010) Single-wall carbon nanotubes assisted photothermal cancer therapy: animal study with a murine model of squamous cell carcinoma. Lasers Surg Med 42:798–808
192. Liu X, Tao H, Yang K et al (2011) Optimization of surface chemistry on single-walled carbon nanotubes for in vivo photothermal ablation of tumors. Biomaterials 32:144–151
193. Torti SV, Byrne F, Whelan O et al (2007) Thermal ablation therapeutics based on CNx multi-walled nanotubes. Int J Nanomedicine 2:707
194. Gannon CJ, Cherukuri P, Yakobson BI et al (2007) Carbon nanotube-enhanced thermal destruction of cancer cells in a noninvasive radiofrequency field. Cancer 110:2654–2665
195. Corr SJ, Raoof M, Cisneros BT et al (2015) Radiofrequency electric-field heating behaviors of highly enriched semiconducting and metallic single-walled carbon nanotubes. Nano Res 8:2859–2870. https://doi.org/10.1007/s12274-015-0791-1
196. Zhou F, Xing D, Ou Z, et al (2009) Cancer photothermal therapy in the near-infrared region by using single-walled carbon nanotubes. J Biomed Opt 14:021009–021009–7
197. Zhou F, Wu S, Wu B et al (2011) Mitochondria-targeting single-walled carbon nanotubes for cancer photothermal therapy. Small 7:2727–2735
198. Zhou F, Wu S, Song S et al (2012) Antitumor immunologically modified carbon nanotubes for photothermal therapy. Biomaterials 33:3235–3242
199. Wang C-H, Huang Y-J, Chang C-W et al (2009) In vitro photothermal destruction of neuroblastoma cells using carbon nanotubes conjugated with GD2 monoclonal antibody. Nanotechnology 20:315101
200. Burke A, Ding X, Singh R et al (2009) Long-term survival following a single treatment of kidney tumors with multiwalled carbon nanotubes and near-infrared radiation. Proc Natl Acad Sci 106:12897–12902
201. Burlaka A, Lukin S, Prylutska S et al (2010) Hyperthermic effect of multi-walled carbon nanotubes stimulated with near infrared irradiation for anticancer therapy: in vitro studies. Exp Oncol 32:48–50
202. Ghosh S, Dutta S, Gomes E et al (2009) Increased heating efficiency and selective thermal ablation of malignant tissue with DNA-encased multiwalled carbon nanotubes. ACS Nano 3:2667–2673
203. Liou G-Y, Storz P (2010) Reactive oxygen species in cancer. Free Radic Res 44. https://doi.org/10.3109/10715761003667554
204. Ogbodu RO, Limson JL, Prinsloo E, Nyokong T (2015) Photophysical properties and photodynamic therapy effect of zinc phthalocyanine-spermine-single walled carbon nanotube conjugate on MCF-7 breast cancer cell line. Synth Met 204:122–132
205. Wang L, Shi J, Liu R et al (2014) Photodynamic effect of functionalized single-walled carbon nanotubes: a potential sensitizer for photodynamic therapy. Nanoscale 6:4642–4651
206. Zhu Z, Tang Z, Phillips JA et al (2008) Regulation of singlet oxygen generation using single-walled carbon nanotubes. J Am Chem Soc 130:10856–10857. https://doi.org/10.1021/ja802913f
207. Erbas S, Gorgulu A, Kocakusakogullari M, Akkaya EU (2009) Non-covalent functionalized SWNTs as delivery agents for novel Bodipy-based potential PDT sensitizers. Chem Commun 4956–4958
208. Huang P, Lin J, Yang D et al (2011) Photosensitizer-loaded dendrimer-modified multi-walled carbon nanotubes for photodynamic therapy. J Controlled Release 152:e33–e34
209. Shi J, Ma R, Wang L et al (2013) The application of hyaluronic acid-derivatized carbon nanotubes in hematoporphyrin monomethyl ether-based photodynamic therapy for in vivo and in vitro cancer treatment. Int J Nanomedicine 8:e73
210. Zhang P, Huang H, Huang J et al (2015) Noncovalent ruthenium (II) complexes–single-walled carbon nanotube composites for bimodal photothermal and photodynamic therapy with near-infrared irradiation. ACS Appl Mater Interfaces 7:23278–23290
211. Marangon I, Ménard-Moyon C, Silva AK et al (2016) Synergic mechanisms of photothermal and photodynamic therapies mediated by photosensitizer/carbon nanotube complexes. Carbon 97:110–123

212. Xie J, Lee S, Chen X (2010) Nanoparticle-based theranostic agents. Adv Drug Deliv Rev 62:1064–1079. https://doi.org/10.1016/j.addr.2010.07.009
213. Liu Z, Liang X-J (2012) Nano-carbons as theranostics. theranostics 2:235–237. https://doi.org/10.7150/thno.4156
214. Peci T, Dennis TJS, Baxendale M (2015) Iron-filled multiwalled carbon nanotubes surface-functionalized with paramagnetic Gd (III): A candidate dual-functioning MRI contrast agent and magnetic hyperthermia structure. Carbon 87:226–232. https://doi.org/10.1016/j.carbon.2015.01.052
215. Lu Y-J, Wei K-C, Ma C-CM et al (2012) Dual targeted delivery of doxorubicin to cancer cells using folate-conjugated magnetic multi-walled carbon nanotubes. Colloids Surf B Biointerfaces 89:1–9. https://doi.org/10.1016/j.colsurfb.2011.08.001
216. Jia N, Lian Q, Shen H et al (2007) Intracellular delivery of quantum dots tagged antisense oligodeoxynucleotides by functionalized multiwalled carbon nanotubes. Nano Lett 7:2976–2980. https://doi.org/10.1021/nl071114c
217. Chen M-L, He Y-J, Chen X-W, Wang J-H (2012) Quantum dots conjugated with Fe_3O_4-filled carbon nanotubes for cancer-targeted imaging and magnetically guided drug delivery. Langmuir 28:16469–16476. https://doi.org/10.1021/la303957y
218. Al Faraj A, Shaik AP, Shaik AS (2015) Magnetic single-walled carbon nanotubes as efficient drug delivery nanocarriers in breast cancer murine model: noninvasive monitoring using diffusion-weighted magnetic resonance imaging as sensitive imaging biomarker. Int J Nanomedicine 10:157–168. https://doi.org/10.2147/IJN.S75074
219. Al Faraj A, Shaik AS, Al Sayed B (2015) Preferential magnetic targeting of carbon nanotubes to cancer sites: noninvasive tracking using MRI in a murine breast cancer model. Nanomed 10:931–948. https://doi.org/10.2217/nnm.14.145
220. Wu W, Li R, Bian X et al (2009) Covalently combining carbon nanotubes with anticancer agent: preparation and antitumor activity. ACS Nano 3:2740–2750. https://doi.org/10.1021/nn9005686
221. Xiao Y, Gao X, Taratula O et al (2009) Anti-HER2 IgY antibody-functionalized single-walled carbon nanotubes for detection and selective destruction of breast cancer cells. BMC Cancer 9:351
222. Allen BL, Kichambare PD, Gou P et al (2008) Biodegradation of single-walled carbon nanotubes through enzymatic catalysis. Nano Lett 8:3899–3903. https://doi.org/10.1021/nl802315h
223. Kagan VE, Konduru NV, Feng W et al (2010) Carbon nanotubes degraded by neutrophil myeloperoxidase induce less pulmonary inflammation. Nat Nanotechnol 5:354–359. https://doi.org/10.1038/nnano.2010.44
224. Kotchey GP, Zhao Y, Kagan VE, Star A (2013) Peroxidase-mediated biodegradation of carbon nanotubes in vitro and in vivo. Adv Drug Deliv Rev 65:1921–1932. https://doi.org/10.1016/j.addr.2013.07.007

Micellar-Based Nanoparticles for Cancer Therapy and Bioimaging

Fernanda Andrade, Andreia Almeida, Diana Rafael, Simo Schwartz, Jr and Bruno Sarmento

Abstract Micelles are versatile nanosized systems composed by amphiphilic molecules. Their small size and capacity to encapsulate both hydrophilic and hydrophobic compounds, as well as their easier functionalization, are some of the characteristics responsible for their multifunctionality, and their potential use in different clinical settings. In fact, micelles have important applications in cancer therapy because of their capacity to deliver hydrophobic anticancer drugs to tumor sites. In recent years, applications beyond the delivery of hydrophobic drugs have been explored. In this chapter, we will discuss the main features of micelles that make them good candidates in the development of systems for cancer therapy and bioimaging. The state-of-the-art and recent advances in academic research and in clinical applications will be discussed

Keywords Polymeric micelles · Cancer · Drug delivery · Gene therapy · Bioimaging

---

F. Andrade · D. Rafael · S. Schwartz, Jr
Drug Delivery and Targeting Group, CIBBIM-Nanomedicine, Vall D'Hebron Institut de Recerca (VHIR), Vall D'Hebron Barcelona Hospital Campus, 08035 Barcelona, Spain

F. Andrade (✉) · A. Almeida · B. Sarmento (✉)
i3S—Instituto de Investigação e Inovação em Saúde, Universidade do Porto, Rua Alfredo Allen 208, 4200-180 Porto, Portugal
e-mail: fersilandrade@gmail.com

B. Sarmento
e-mail: bruno.sarmento@ineb.up.pt

F. Andrade · A. Almeida · B. Sarmento
INEB—Instituto Nacional de Engenharia Biomédica, Universidade do Porto, Rua Alfredo Allen 208, 4200-180 Porto, Portugal

A. Almeida
ICBAS—Instituto Ciências Biomédicas Abel Salazar, Universidade do Porto, 4150-180 Porto, Portugal

S. Schwartz, Jr
Networking Research Center on Bioengineering, Biomaterials and Nanomedicine (CIBER-BBN), 08035 Barcelona, Spain

G. Gonçalves and G. Tobias (eds.), *Nanooncology*, Nanomedicine and Nanotoxicology, https://doi.org/10.1007/978-3-319-89878-0_6

1 Amphiphilic Compounds and Micelles: Characteristics that Make Them Suitable for Clinical Applications

Amphiphilic compounds are heterogeneous composites of both hydrophilic and hydrophobic units, that could be of lipidic (phospholipids) or polymeric (amphiphilic polymers) nature. Changing characteristics like unit composition, chain length, and conformation, it is possible to modulate polymer properties and consequently, their suitability for a given pharmaceutical application [1, 2]. Amphiphilic compounds have been widely used as excipients such as emulsifiers, wetting, thickening and stabilizing agents of suspensions, colloidal dispersions, and gel-forming agents. In the last decades, however, they had risen an increased interest as new drug delivery systems (DDS) sustained by the progresses seen in pharmaceutical sciences and nanomedicine [2, 3]. Phospholipids are successfully used in the clinical practice as main components of liposomes, emulsions, and some lipidic/surfactant micelles [4], while amphiphilic polymers have been highlighted, with some products already on the market, due to their ability to form nanoscopic structures of different morphologies, e.g., polymersomes, nanocapsules, nanospheres, and nanogels, even though polymeric micelles (PM) are the most commonly used and investigated (Fig. 1) [3, 5, 6].

Micelles are spherical nanosized colloidal dispersions having a hydrophobic core coated by a hydrophilic shell arising from the self-assembly of amphiphilic molecules above the critical micelle concentration (CMC) and critical micellization temperature (CMT) (Fig. 2) [7, 8]. This process is driven by an increase in entropy of the solvent molecules in contact with the hydrophobic units of the phospholipid/polymer and a consequent decrease of free energy ($\Delta°Gm$) of the system as the hydrophobic components are withdrawn from the aqueous media originating the micelle core [9]. $\Delta°Gm$ is given by Eq. 1 where R is the gas constant and T is the temperature of the system.

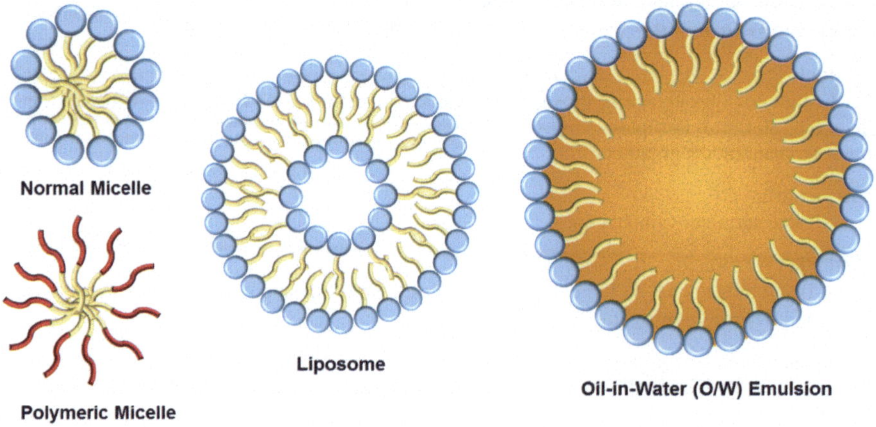

Fig. 1 Examples of nanosized systems obtained using amphiphilic compounds

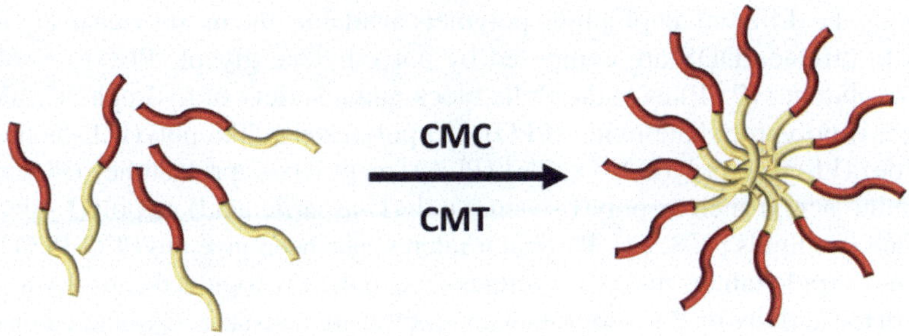

Fig. 2 Schematic representation of a micelle and its formation through self-assembling of polymer monomers above CMC and CMT

$$\Delta°\mathrm{Gm} = \mathrm{RT}\ln \mathrm{CMC} \quad (1)$$

Different parameters have an influence on the self-assembly, namely chemical composition, molecular weight and architecture (linear versus branched, double bounds, etc.), concentration, temperature, solvent–polymer interactions, or salt concentration. By increasing the temperature of the system, the solvency of hydrophilic unit as well as the CMC value will decrease, promoting the micelle formation. In the same way, this phenomenon is favored when the attractive hydrophobic interactions increase as result of a gain in molecular weight of the hydrophobic domain [10–12].

Micelles have been developed to modify several major intrinsic characteristics of drugs, including drug aqueous solubility, release pattern, pharmacokinetics, biodistribution, and in vivo stability [7, 13]. They allow the formulation and administration of highly hydrophobic drugs, otherwise often withdraw from drug development at an early stage [14], as well as the elimination of excipients such as Cremophor EL® which causes severe side effects [15]. Despite being mainly proposed to formulate hydrophobic drugs, micelles, especially PM, permit the encapsulation of drugs with different polarities. Hydrophobic drugs are incorporated into the micelle core being the solubilization capacity of drugs proportional to its hydrophobicity. Further, water-soluble drugs are adsorbed on the micelle shell and/or its surface. Drugs with intermediate polarity are distributed along the amphiphilic molecules [16, 17].

PM have higher core hydrophobicity and viscosity then surfactant micelles, conferring them higher thermodynamic and kinetic stability, thus presenting slower and delayed disintegration and drug release in circulation, even upon dilution below CMC value [2]. Due to their small size, generally lower than 200 nm, and hydrophilic surface, micelles are poorly recognized by the reticuloendothelial system (RES), presenting high blood circulation half-time and enhanced permeability and retention (EPR) effect in neovascularized tumors [18–20]. Additionally, they present high encapsulation efficiency which together with the possibility of being sterilized by filtration make these systems an interesting alternative for drug (with highlight to anticancer agents) [7, 16, 21], genetic material [22], and diagnostic agents [23] delivery.

Among the different amphiphilic polymers available, the most extensively investigated to produce DDS are composed by polyethylene glycol (PEG) or poly(*N*-vinylpyrrolidone) (PVP) as hydrophilic block and a variety of hydrophobic blocks, namely: (i) polypropylene oxide (PPO); (ii) poly(ester)s like poly(D,L-lactide-co-glycolide) (PLGA), poly(D,L-lactide) (PLA), or poly(ε-caprolactone) (PCL); (iii) poly(amino acid)s such as poly(L-lysine) poly(L-aspartic acid) or poly(L-glutamic acid); and (iv) lipids [7, 8, 24]. Particular interest has been given to PEG-PPO block copolymers (poloxamers and poloxamines) due to their recognized capacity to interfere with the activity of P-glycoprotein and multidrug resistance-associated proteins (MRP), which increases the therapeutic index of anticancer agents in multidrug-resistant cancer. Not to mention their capacity to increase the transfection efficiency of genetic material [25–30].

The main characteristics of the majority of amphiphilic compounds that make them suitable for clinical application are water solubility, biodegradability, biocompatibility, low immunogenicity, surface hydrophilicity, stimuli-responsiveness, among others [2, 3].

Apart from the encapsulation and protection of drugs, the hydrophobic core of micelles provides appropriate mechanical properties for the desired application whereas the hydrophilic shell conceals the particle from the biological environment enhancing its stability [31, 32]. The presence of a hydrophilic layer on the surface of particles provides them with stealth properties by reducing or delaying opsonization via steric repulsion forces and macrophage recognition. This increases the plasma circulation time of particles and the half-life of drugs [33, 34]. PEG and poloxamers are commonly used as hydrophilic polymers to decorate the surface of many nanoDDS creating the so-called stealth systems [35]. Moreover, the resultant functionalization potential by either chemical modification or bioconjugation could be used to bind specific targeting moieties to particles surface, allowing a better control of their biodistribution and increasing the therapeutic index of drugs [36]. A variety of ligands such as peptides, antibodies, folate, transferrin, or polysaccharides can be used accordingly to the required application. As an example, transferrin and arginine-glycine-aspartic acid (RGD) peptide and transferrin improved the cytotoxicity of paclitaxel-loaded micelles against transferrin and $\alpha_v\beta_3$ integrin overexpressed human cells, respectively [37, 38].

Stimuli-responsiveness grabs high attention among all the characteristics presented by some amphiphilic polymers for the development of advanced controlled DDS [39, 40]. Stimuli-responsive polymers are a class of "smart" polymers that undergo chemical or physical modifications in response to specific stimulus like temperature, pH, redox potential, enzymatic, magnetic field, and light [40–42]. Acrylate/methacrylate derivatives are commonly used in the production of pH-responsive polymers, while poloxamers and poly(*N*-alkylacrylamide) derivatives produce thermo-responsive systems [40, 43, 44]. Multi-stimuli-responsive polymers with possible applications in the biomedical field have been also developed [45]. For example, poly(2-(dimethylamino)ethyl methacrylate-co-6-O-methacryloyl-1,2,3,4-di-O-isopropylidene-D-galactopyranose)-b-poly(4-(4-methoxyphenylazo)phenoxy methacrylate) [P(DMAEMA-co-MAIpGP)-b-PMAZO] is an amphiphilic glycopolymer presenting triple-stimuli-responsive properties designed for delivery of biomolecules in the form of micelles [45]. The poly(2-(dimethylamino) ethyl methacrylate) segments garnish the polymer with thermo- and pH-responsiveness while the azobenzene moiety is the responsible for light-responsiveness. In the oncology field, temperature and pH are the main stimuli explored due to the physiologic differences observed between tumoral and healthy tissues.

2 Micelle-Based Anticancer Drug Delivery Systems

Cancer is becoming the principal cause of death in the Western world and, unfortunately, recent estimates point toward an increase in its worldwide incidence in the upcoming years [46]. Cell malignant transformation emerges from the conversion of healthy cells into tumor cells through several complex molecular alterations. Systemic chemotherapy combined with surgical resection or radiation are the most used approaches in cancer treatment. Despite that, most anticancer drugs used in the clinics are quite effective against tumor suppression, they often present drastic life-threatening side effects due to their lack of cell/tissue specificity and also to their low solubility in biological fluids, impairing its accumulation at desired targets [47].

The use of nanomedicines in the treatment of solid tumors, particularly, micelles to deliver anticancer drugs, has been widely investigated in recent years. The main advantages of these carriers over conventional therapies are the possibility to (i) deliver poorly water-soluble drugs; (ii) use ligands in their surface with high affinity and specificity for target cells; (iii) induce an efficient cellular endosomal escape and transport of the drugs to specific organelles; and (iv) circumvent drug resistance mechanisms [48, 49]. Overall, these systems may provide similar or better therapeutic profiles at lower drug concentrations when compared with classical drug formulations, while reducing side effects at the same time.

As referred above, micelles are amphiphilic molecules with the ability to incorporate both hydrophobic and hydrophilic drugs and can be administered by several routes, such as oral, intravenous, or rectal. The inner core is where hydrophobic drugs are encapsulated, improving the stability of the system, providing drug protection from degradation, and defining drug release kinetics. The outer shell, on the other hand, is responsible for the in vivo pharmacokinetic behavior of the system and can be specifically targeted through surface modifications to overexpressed molecules (receptors/ligands) of targeted cells. Some systems are also able to cross biological and mucosal barriers (Fig. 3) [49, 50]. Such moieties allow specific binding to tumor cells achieving active targeting. However, for a successful delivery, molecules targeted at the tumor cells surface must be homogeneously overexpressed [49].

Most anticancer drugs show poor water solubility and short blood circulation time. Certainly, the use of micellar solutions can contribute to overcome these drawbacks. PEG is the most used polymer to produce PM due to their hydrophilic character, which can improve water solubility, and reduce protein adsorption and RES recognition, which increases the blood circulation time of the system [51].

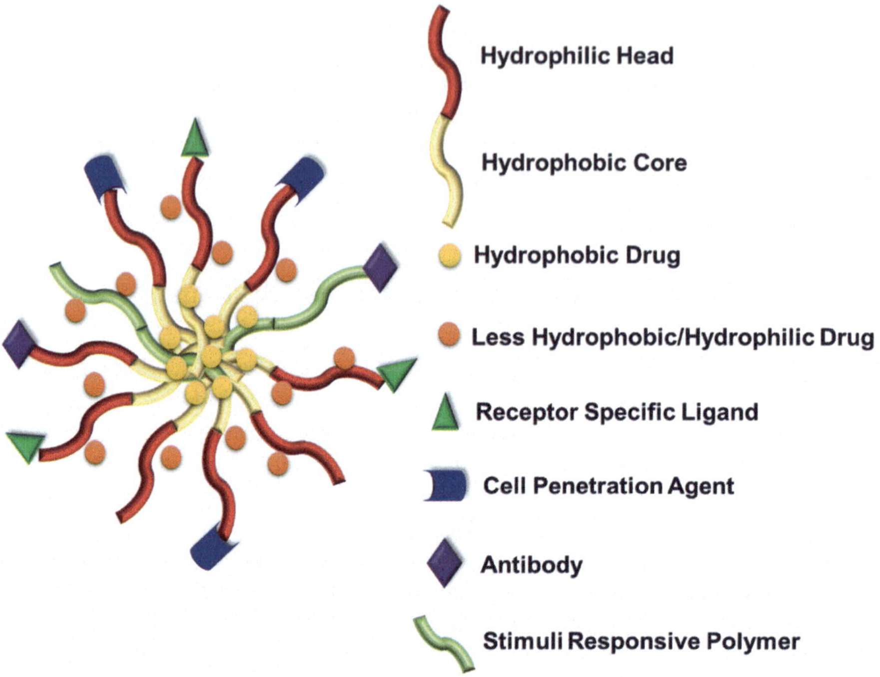

Fig. 3 Schematic representation of a multifunctional micelle for drug delivery

Lv et al. [52] designed micelles based on methoxy poly(ethylene glycol)-*b*-poly(L-glutamic acid)-*b*-poly(L-lysine) decorated with deoxycholate (mPEG-*b*-PLG-*b*-PLL/DOCA) to co-delivery doxorubicin (DOX) and paclitaxel (PTX), which are the most common chemotherapeutic drugs with different mechanisms of action as well as different water solubility. The micellar system with both anticancer drugs was produced by self-assembly in aqueous solution, where DOX and PTX were encapsulated through electrostatic and hydrophobic interactions, respectively [53]. In vivo antitumor efficacy was tested on A549 human lung tumor-bearing nude mice. The results showed that the combination of both drugs (free or encapsulated) was more effective in tumor growth suppression than the single drugs [52]. However, the stronger tumor growth suppression was seen when DOX and PTX were encapsulated in mPEG-*b*-PLG-*b*-PLL/DOCA micelles. In addition, co-delivery of DOX and PTX demonstrated reduced systemic toxicity and higher tumor accumulation, which may indicate that the use of micelles to co-delivery anticancer drugs is less toxic and more effective against cancer cells [52].

PM-based on polysaccharides have also attracted attention due to their low toxicity and biodegradable nature. It is known that i.v. administration of PTX presents serious drawbacks such as low drug stability and high toxicity caused by nonspecific delivery. Also, oral administration is the choice preferred by most patients. Thus, oral delivery has been investigated to replace the i.v. administration of drugs, however, the most used anticancer drugs do not have oral bioavailability [54]. Mo et al. [55] synthesized *N*-octyl-*O*-sulfate chitosan (NOSC) micelles to improve the oral absorption of PTX. In this work, the authors demonstrated significant enhancement of PTX oral bioavailability compared to Taxol® based on the pharmacokinetic studies performed in vivo [55]. Also, uptake studies in Caco-2 cells revealed a significantly higher uptake of NOSC micelles compared to Taxol®, being almost 10-fold higher than Taxol®. Moreover, the potential permeability of PTX across the intestinal barrier indicated an improvement on PTX transport by NOSC micelles compared to the commercially available product, in Caco-2 monolayers [51].

The design of multifunctional micelles combining different functional capacities in a single micelle such as active targeting and stimuli-responsiveness is highly regarded in order to improve selectivity and efficacy of delivered drugs (Fig. 3). Recently, Zhang et al. [53] have developed a micellar multifunctional system mixing Pluronic® P123 and Pluronic® F127 to deliver PTX (PF-PTX). Further, the same micelles were decorated with folic acid (FPF-PTX) to achieve active targeting to increase selectivity. In vitro cellular uptake demonstrated higher uptake from FPF-PTX than PF-PTX in KB and KBv cells. Additionally, in A-549 cells, no differences were observed, which may indicate that FPF-PTX micelles could enter cells via receptor-mediated endocytic pathways [53]. In vivo pharmacokinetic studies showed larger area under the curve (AUC) and longer half-life of the micellar system, exhibiting delayed blood clearance, when compared with Taxol®. To assess the antitumor activity, KBv cells were subcutaneously injected in the right flank of mice and tumor growth observed. Volume of tumors treated with FPF-PTX were 1.59-fold and 2.12-fold smaller than those treated with PF-PTX and Taxol, respectively, showing successful efficacy from the multifunctionality of the system [53].

Several stimuli-responsiveness polymeric micellar systems have been explored based on internal stimuli, mainly, temperature [56, 57] and pH [58, 59]. Recently, Guan et al. [60] developed a pH-responsiveness micellar system based on poly(lactic acid)–poly(L-lysine) (PLA–PLL) with its surface modified with folic acid (FA) to achieve active targeting. A citric acid (CA) group was also added to obtain a "negative-to-positive charge reversal system" that favors intracellular uptake and delivery of DOX (FA–PLL(CA)–PLA/DOX). As known, FA is a ligand with high binding affinity for the FA receptor which is found overexpressed in many cancer cells surface [61]. After the uptake of the system by cancer cells, the acidic environment found in endosomes and lysosomes disintegrates the system and releases the drug. The results of this study demonstrated that this system was highly sensitive to pH changes. Further, a negative-to-positive charge reversal was observed with pH decrease, and cumulative DOX release was observed from the micelles [60]. In vitro antitumor activity was higher for micelles with FA on the surface, which may be attributed to the internalization enhancement due to the presence of FA [60]. The in vivo distribution performed in HeLa tumor-bearing BALB/c nude mice showed higher accumulation of FA–PLL(CA)–PLA/DOX micelles at the tumor sites than micelles without FA or CA (Fig. 4). This was attributed to the active targeting effect and the electrostatic interaction between micelles and cells as a result of the positive-to-negative charge reversal in the acidic pH of tumor tissue [60]. The

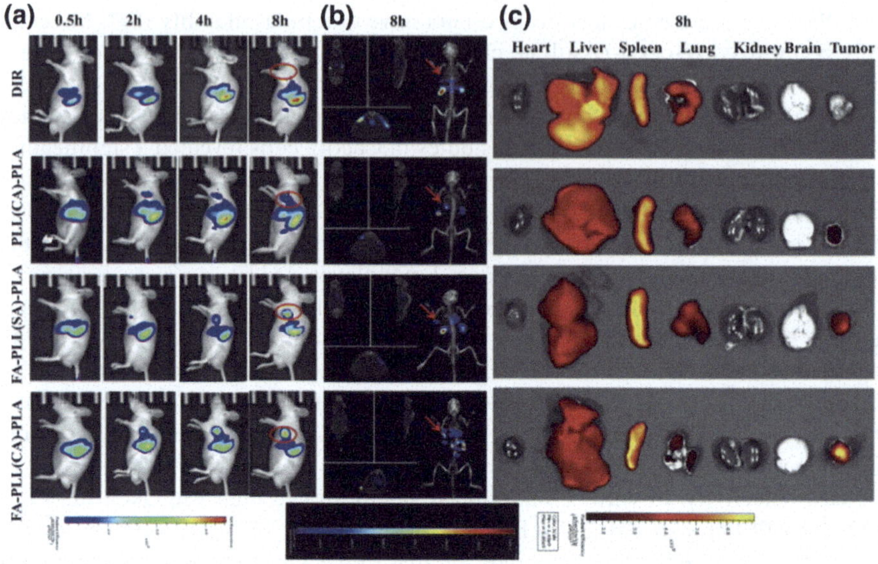

Fig. 4 In vivo biodistribution of different micelles (PLL(CA)–PLA, FA—PLL(SA)–PLA and FA—PLL(CA)–PLA) labeled by fluorescent probe DIR in HeLa tumor-bearing BALB/c nude mice. **a** Fluorescence biodistribution profile at different times postinjection. Tumor sites were enclosed by red circles. **b** 3D images of in vivo fluorescence biodistribution. Tumor sites were indicated by red arrows. **c** Ex vivo fluorescence imaging of the tissues and organs excised from sacrificial mice at 8 h post injection. Reprinted with permission of Elsevier Limited from Ref. [60]

antitumor activity was also assessed. Tumor volumes were significantly lower when treated with FA-targeted micelles compared to saline, DOX or micelles without FA ligands, which might also indicate an active targeting effect. Furthermore, micelles with both FA ligands and pH-responsiveness were the system showing better tumor growth suppression and higher mice survival ratio, proving the effectiveness of this multifunctional system [60].

Due to their promising therapeutic potentiality demonstrated both in vitro and in vivo, some PM-based formulations have advanced from preclinical assessment to the enrolled clinical trials (Table 1). In this context, NK911 (Nippon Kayaku, Co.) was the first micellar system to proceed into clinical evaluation in 2001 [62]. Preclinical data showed longer blood circulation of the micelles with a 29-fold higher AUC in plasma than free DOX, and 3.4-fold higher accumulation in tumors compared to free DOX in mice models of sarcoma, lung, breast, and colon cancer [63]. Based on the results obtained in Phase I, NK911 micelles had proceeded into Phase II clinical trials against metastatic pancreatic cancer.

Furthermore, Genexol-PM® (Samyang Co.), first approved in 2007 in Korea, is now being evaluated/marketed by the European Union and America under the name of Cynviloq® for the treatment of breast, ovary, and non-small cell lung cancers [64–66]. Genexol®-PM is a polymeric micellar system loaded with PTX based

Table 1 Examples of self-assembled particles under clinical trials evaluation for cancer treatment

Formulation	Copolymer	Drug	Indication	Clinical phase	Reference
SP1049C	Pluronic® L61 and F127	DOX	Advanced adenocarcinoma of the esophagus, gastroesophageal junction, and stomach	III	[95]
Genexol-PM®/Cynviloq®	PEG-PLA	PTX	Breast cancer, non-small cell lung cancer, advanced pancreatic cancer, and ovarian cancer	II/III/IV	[96–100]
NK012	PEG-poly(L-glutamic acid)-SN-38 conjugated	SN-38	Advanced breast cancer and small cell lung cancer	II	[101, 102]
NK105	PEG-poly(L-aspartic acid)	PTX	Advanced or recurrent breast cancer	III	[103]
NK-911	PEG–poly(aspartic acid)-doxorubicin conjugated	DOX	Solid tumors	I	[104]
NC-4016	PEG-DACH-platin	Oxaliplatin	Advanced solid tumors or lymphoma	I	[105]
NC-6300	PEG-b-poly(aspartate-hydrazone)	Epirubicin	Advanced solid tumors or soft tissue sarcoma	I/II	[106]
NC-6004 (Nanoplatin®)	PEG-poly(L-glutamic acid)-cisplatin conjugated	Cisplatin	Solid tumors, non-small cell lung, biliary, and bladder cancer, pancreatic cancer	II/III	[107, 108]
IT-101	Polymer-cyclodextrin-camptothecin conjugated	Camptothecin	Advanced solid tumors, ovarian cancer	I/II	[109, 110]
BIND-014	PEG-PLA	Docetaxel	Non-small cell lung cancer, prostate cancer, advanced solid tumors	I/II	[111–114]

DACH-platin diaminocyclohexane platinum, *DOX* doxorubicin, *PEG* polyethylene glycol, *PLA* poly(D,L-lactide), *PTX* paclitaxel on monomethoxy poly(ethylene glycol)-block-poly(D,L-lactide) (mPEG-PDLLA). This system presents advantages compared to Taxol® such as higher drug accumulation in the tumor, low toxicity levels, and effective P-gp inhibition effect [67]. Also, a Pluronic®-based micellar system to delivery DOX, SP1049C (Surpatek Pharma, Inc.) has been granted orphan drug designation by FDA for the treatment of gastric cancer. Even though this system presents similar biodistribution compared to DOX, SP1049C showed to be more efficient in tumor suppression [68].

NK105 is a PTX-loaded polymeric micellar system based on poly(ethylene glycol)-poly(aspartic acid) [PEG-P(Asp)] modified with 4-phenyl-1-butanol in order to increase its hydrophobicity [69]. Preclinical results of NK105 showed a plasma AUC of micellar-PTX 50-fold higher than free PTX. Further, tumoral AUC of NK105 in subcutaneous human adenocarcinoma tumor models (HT-29) was 25-fold higher than free PTX [69]. Currently, NK105 has advanced to Phase III clinical trials in order to compare NK105 with free PTX in patients with metastatic as well as recurrent breast cancer, and to determine the survival and therapeutic benefits of NK105.

In the future, micelle-based medicines may change the way the anticancer drugs are used in cancer therapy. By considering fundamental parameters such as the choice of the polymer, drug encapsulation, or micelle coating decoration, it is possible to optimize the solubility, biocompatibility, release profile, and pharmacokinetic behavior of each system. In this way, PM may provide efficient solutions for the treatment of different cancers.

3 Micelles as Vectors for Gene Therapy

Gene therapy has caught the attention as an opportunity to treat both acquired and innate diseases. Delivery and subsequent expression of an exogenous oligonucleotide encoding for a missing or defective gene or to achieve silencing of a particular gene of interest are different approaches that bring new treatment alternatives for several diseases, including hematological, immunological, ocular, and neurodegenerative diseases, and cancer [70].

The use of gene therapy in cancer might be an attractive alternative to overcome well-known drug-related adverse effects, chemoresistance, and toxicity problems associated with radiotherapy. However, the clinical use of in vivo delivery of oligonucleotides (OGN) as gene-based therapies is strongly precluded because of their vulnerability to enzymatic degradation, poor cellular uptake, and rapid body clearance. Therefore, successful clinical applications of gene therapy demand new tools to enable delivery of negatively charged OGN into target cells able to cross the hydrophobic cell membrane and reach cellular cytoplasm and nucleus (Fig. 5) [71–73]. Even though nowadays viral vectors are still considered the most efficient and commonly used delivery system for gene transfer in both pre-clinical and clinical research [70], only Glybera® and Kymriah® have been clinically approved so far. Glybera® (alipogene tiparvovec using an adeno-associated virus as vector) was granted market authorization by European Commission in late 2012 to treat lipopro-

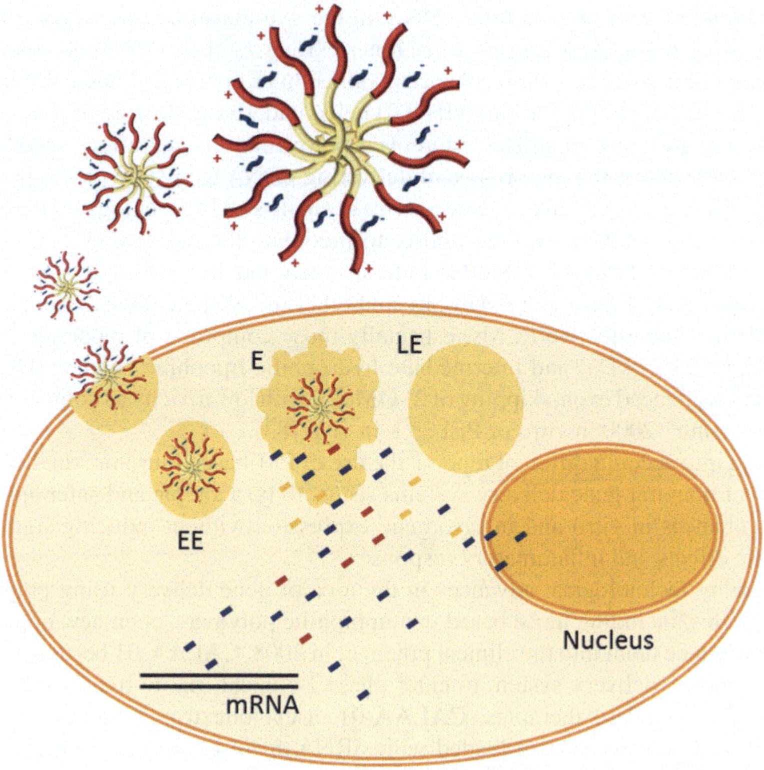

Fig. 5 Schematic representation of OGN-loaded PM endocytosis and endosome release of the OGN before reach the LE. EE: early endosome; E: endosome; LE: late endosome

tein lipase deficiency [74], and Kymriah® (tisagenlecleucel using a lentivirus as a vector) was approved by FDA in mid-2017 to treat B-cell acute lymphoblastic leukemia.

The well-known drawbacks related with viral-based vectors such as their immunogenicity, mutagenesis, carcinogenesis, limited cargo loading, and time-consuming/high-cost procedures have boosted the development of safer vehicles using a wide range of lipids and polymers [75, 76]. Indeed, an enormous outbreak of nonviral vectors has been seen in recent years [75, 77–80].

The application of amphiphilic polymers as nucleic acid delivery systems is particularly exciting, after they have demonstrated to improve biological activity while protecting nucleic acids from degradation and clearance by serum nucleases and the immune system [2, 3]. With special attention to Pluronic® amphiphilic block copolymers that have shown high stability and capacity to significantly improve the transfection efficiency and gene expression via different delivery routes [30, 81, 82]. The mechanism by which poloxamers enhance gene expression is not clearly understood but seems to be related to its ability to bind cellular membranes and further activation of cell signaling pathways, such as the Nf-κB pathway. Nf-κB is a transcription factor able to bind pDNA in the cytoplasm and transport it to the nucleus using the nuclear transport machinery. Yang Z, et al. (2008) demonstrated for the first time that Pluronic® enhances both cellular uptake and nuclear transport of polyplex-delivered pDNA into NIH 3T3 cells, underlying the role of poloxamers as biological response modifiers. Moreover, poloxamer did not affect naked DNA uptake, emphasizing the importance of delivering pDNA using polyplexes [81].

Also, Wang et al. [82] investigated a series of small-size PEI-conjugated Pluronic® polycarbamates (PCMs) for their ability to modulate the delivery of 2′-O-methyl phosphorothioate RNA (2′-OMePS) both in vitro and in dystrophic (mdx) mice for the treatment of muscular dystrophy and other myodegenerative diseases. They conclude that the effective PCMs, especially those composed of moderate molecular weight(2 k–5 kDa) and intermediate hydrophilic-lipophilic balance (HLB) of Pluronic®, enhanced exon-skipping of 2′-OMePS with low toxicity as compared with Lipofectamine® 2000 in vitro or PEI 25 k in vivo [83].

Taking into account different reports, the use of PM-based Pluronic® in the development of nonviral gene delivery systems seems to be a simple and safer approach which enhances in vitro and in vivo gene expression without inducing significant cytotoxic effects and inflammatory responses.

Emerging technological advances in the area of gene delivery using polymeric nanoparticles, including those based in amphiphilic polymers, open new opportunities to introduce them into the clinical practice. In 2008, CALAA-01 became the first targeted siRNA delivery system to enter phase I clinical trial to treat solid tumors refractory to standard therapies. CALAA-01, a cyclodextrin-containing polymer (CDP)-based delivery system loaded with siRNA against ribonucleotide reductase subunit 2 (RRM2), was designed to inhibit tumor growth. Its formulation also contains PEG as stabilizing agent and human transferrin (Tf) as targeting ligand to bind overexpressed transferrin receptors (TfR) in cancer cells [84]. Besides CALAA-01, Calando Pharmaceuticals already announced the development of CALAA-02, based on the same platform but encapsulating a siRNA against HIF-2alpha.

In the upcoming years, new RNAi and DNA therapeutic strategies based on polymeric nonviral vectors using amphiphilic polymers will reach clinical evaluation. This will contribute to expand the use of gene therapy in the clinical setting. Table 2 summarizes some of the most significant work involving polymeric nanoparticles for gene delivery and their stage of development.

Table 2 Examples of PM-based systems for gene delivery at different stages of development

Polymer vector	Genetic material	Utility	State	Reference
Dendritic PLL-block-poly(L-lactide)-block-dendritic PLL	pGFP	GFP expression	In vitro	[115]
PEI-PEG-TAT peptide	pGFP-N3	GFP expression	In vitro	[116]
PDMAEMA-PCL-PDMAEMA triblock copolymers	VEGF siRNA	Anticancer activity (in combination with the Paclitaxel)	In vitro	[117]
PEG/PCL	Bcl2 siRNA	Anticancer activity (in combination with the Paclitaxel)	In vitro	[118]
PLGA/F127	miR-542-3P	Anticancer activity (in combination with the Doxorrubicin)	In vitro	[119]
PLGA/PEI	Bcl-XL shRNA	Anticancer activity (in combination with the Doxorrubicin)	In vitro	[120]
PLGA-b.PEG	Antisense-miR-10b and antisense-miR-21	Anticancer activity	In vivo	[121]
PEG- PAsp(AED)-PDPA	Bcl2 siRNA	Anticancer activity (in combination with the Doxorubicin)	In vivo	[122, 123]
PDP-PDHA	Surviving shRNA	Anticancer activity (in combination with the Doxorubicin)	In vivo	[124]
POSS-PDMAEMA-PMPD	p53 plasmid	Anticancer activity (in combination with the Doxorubicin)	In vivo	[125]
PEG-PLL	sFlt-1 pDNA	Solid tumors	In vivo	[126]
PEG-poly[N'-[*N*-(2-aminoethyl)-2-aminoethyl]aspartamide]	pGL4.13 and sFlt-1 pDNA	Solid tumors	In vivo	[127, 128]
Pluronic® P85	pGFP	GFP expression	In vivo	[129]
CALAA-01	siRNA against ribonucleotide reductase M2	Melanoma	Phase I	[130, 131]
Poly(2-ethyl-2-oxazoline)–PLA-g–PEI	mcDNA	Anticancer activity (co-delivery with Doxorubicin)	In vivo	[132]
chitosan-graft-polyethyleneimine-candesartan	wild type p53 (wt-p53) gene	Tumor therapy by antiangiogenesis effect	In vivo	[133]

(continued)

Table 2 (continued)

Polymer vector	Genetic material	Utility	State	Reference
HA-ss-(OA-g-bPEI	AURKA specific siRNA	Anticancer activity (in combination with the Paclitaxel)	In vivo	[134]
poly(DMAEMA-co-PPGMA)	GFP Plasmid DNA	Anticancer activity (in combination with the Paclitaxel)	In vitro	[135]
MPEG-PCL-g-PEI	Msurvivin T34A gene	Anticancer activity (in combination with the Doxorubicin)	In vivo	[136]
PLA-b-PDMAEMA	miR-21 inhibitor	Anticancer activity (in combination with the Doxorubicin)	In vivo	[137]
PEI–SS–PCL–SS–PEI	p53 plasmid DNA	Anticancer activity (in combination with the Doxorubicin)	In vitro	[138]
PSMA micelles	PLK-1 siRNA	Anticancer activity (in combination with the Doxorubicin)	In vivo	[139]
MMP2-sensitive copolymer (PEG-pp-PEI-PE)	GFP siRNA	Anticancer activity (in combination with the Paclitaxel)	In vivo	[140]

CS chitosan, *CTAB* cetyltrimethylammonium bromide, *GFP* green fluorescence protein, *HA* Hyaluronic acid, *mcDNA* minicircle DNA, *MMP2* metalloproteinase 2, *PAsp(AED)* poly(N-(2,2′-dithiobis(ethylamine) aspartamide, *PCL* Polycaprolactone, *PDMAEMA* Poly(2-(dimethylamino)ethyl methacrylate), *PDP:PDHA* Poly[(1,4-butanediol)-diacrylate-β-5-polyethylenimine]-block-poly[(1,4-butanediol)-diacrylate-β-5-hydroxy amylamine], *PDPA* poly(2-(diisopropyl amino) ethyl methacrylate), *PEG* Poly(ethylene glycol), *PEI* Polyethylenimine, *PF127* Poloxamer F127, *PLGA* Poly(Lactide-co-Glycolide), *PLL* poly(L-lysine), *POSS* polyhedral oligomeric silsesquioxane, *PSMA* styrene-alt-maleic anhydride

4 Micelles as Platforms for Bioimaging Systems

An additional clinical application of micelles is the targeted delivery of diagnostic imaging agents for nuclear imaging, magnetic resonance imaging (MRI), X-ray computed tomography (CT), and ultrasonography, among other technologies, enabling both diagnosis and evaluation of the therapeutic response [24, 85].

For nuclear imaging, gamma-emitting isotopes, such as technetium-99 m (^{99m}Tc), iodine-125 (^{125}I) and indium-111 (^{111}In), or positron-emitting isotopes like fluorine-18 (^{18}F), zirconium-89 (^{89}Zr), and copper-64 (^{64}Cu) are used. In the case of MRI, the choice relies on paramagnetic elements, such as iron (Fe), manganese (Mn), holmium (Ho), gadolinium (Gd), or fluorine-19 (^{19}F), while for CT heavy elements such as iodine (I), barium (Ba), and bromine (Br) are the appropriate contrast agents [85]. PM-based systems have been explored to deliver such compounds. Through both, passive

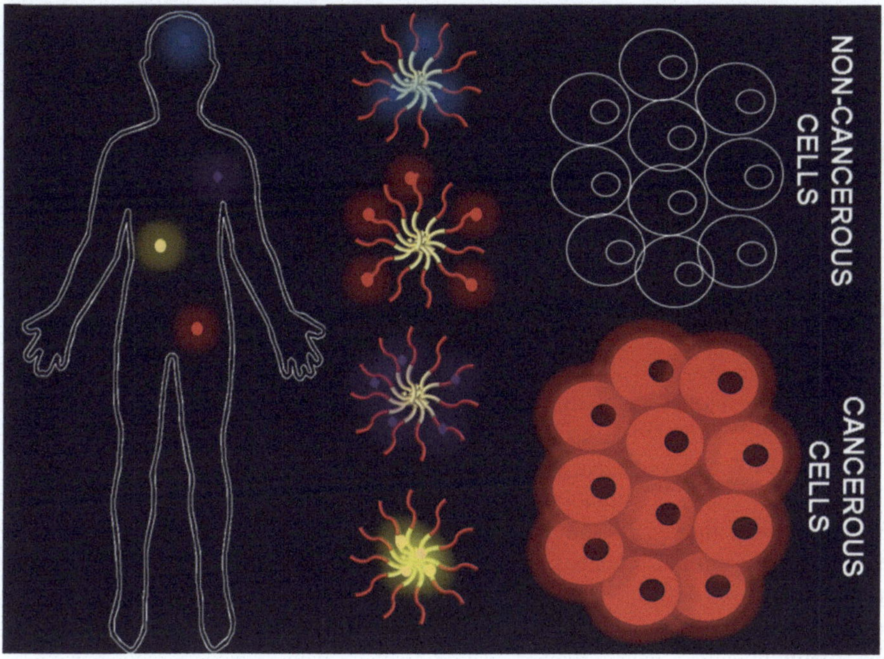

Fig. 6 Schematic representation of the use of PM as systems for bioimaging. PM accumulate preferentially in tumor tissues, allowing in vivo visualization of small tumors and improving diagnostic resolution

and active targeting, PM could increase the contrast specificity between healthy and tumor tissues, improving diagnostic resolution and supporting the choice of an appropriate therapeutic strategy (Fig. 6). Moreover, these systems could be used to follow the progress of the disease and assess the efficacy of the therapeutic scheme [24].

The incorporation of diethylenetriaminepentaacetic acid gadolinium (III) (Gd-DTPA) into 100 nm poly(ethylene glycol)-b-poly(aspartic acid) (PEG-b-Pasp)-based micelles allowed increased accumulation of Gd-DTPA in tumors by the EPR effect, in subcutaneous in vivo models of colon cancer [86]. Moreover, the relaxivity of Gd-DTPA increased nearly six times from 3.48 mM^{-1} s^{-1} of free Gd-DTPA to 22.19 mM^{-1} s^{-1} of encapsulated Gd-DTPA, contributing to a higher signal enhancement and contrast ability. In vivo studies, no signal enhancement was observed for free Gd-DTPA even when administrated with four times higher dose than that of Gd-DTPA-loaded micelles [86]. These results evidence the importance of the Gd-DTPA encapsulation in the development of an efficient MRI contrast agent for in vivo diagnosis.

pH-responsive micelles of 40 nm consisting of methoxy poly(ethylene glycol)-b-poly(l-histidine) [PEG-p(l-His)] and methoxy poly(ethylene glycol)-b-poly(l-lactic acid)-diethylenetriaminopenta acetic acid dianhydride-gadolinium chelate (PEG-p (l-LA)-DTPA-Gd) intended for MRI were developed by Kim et al. (2014). They exhibited a highly effective contrast enhancement in an acidic tumor environment enabling within minutes the in vivo detection of small tumors of ~3 mm^3 [87]. In another study, [88] developed targeted micelles with a size inferior to 140 nm composed by folate-conjugated *N*-palmitoyl chitosan encapsulating superparamagnetic iron oxide nanoparticles (SPIONs). The micelles accumulated predominantly in established folate receptor-positive HeLa-derived tumors in mice, proving to be a safe and efficient tool to diagnose folate overexpressing tumors [88]. Similar results were obtained with 44 nm folate-decorated SPIONs-loaded PEG-PCL micelles [89].

One of the main interesting features of micelles is their multifunctionality and applicability in theranostics, through the combination of imaging agents and drugs, and in some cases targeting agents (Table 3).

Kaida et al. [90] encapsulated Gd-DTPA and the anticancer agent (1,2-diaminocyclohexane)platinum(II) (DACHPt) into poly(ethylene glycol)-*b*-poly(glutamic acid) [PEG-b-P(Glu)] micelles [90]. The relaxivity of Gd-DTPA increased 24 times when encapsulated, improving the detection of solid tumors. Due to the higher accumulation of the micelles into the tumors through the EPR effect, a strong therapeutic efficacy was observed in subcutaneous colon and pancreatic cancer murine models [90]. The colocalization of Gd-DTPA and DACHPt at the tumor site confirms the potentiality of Gd-DTPA/DACHPt–loaded micelles to assess the distribution of the anticancer drug and to follow the clinical outcome of the treatment.

Lee et al. [91] proposed the development of a multifunctional system for simultaneous colon cancer therapy and imaging. PEG-PCL and PDMA-block-poly(ε-caprolactone) (PDMA-b-PCL) mixed micelles encapsulating SN-38, siRNA against VEGF and ultrasmall superparamagnetic iron oxide nanoparticles (USPIO) were able to accumulate in tumors by passive targeting and reduce the tumor size in xenografted LS174T tumor-bearing mice model. In addition to the double anticancer therapeutic effect provided by SN-38 and siRNA against VEGF, USPIO granted the system with effective T2-weighted MRI capability, which permit to use the micelles for diagnostic, treatment monitoring, and disease follow-up (Fig. 7) [91]. In another study, $\alpha_v\beta_3$-targeting micelles decorated with cyclo(Arg-Gly-Asp-D-Phe-Cys) peptide (cRGD) and composed of maleimide-terminated poly(ethylene glycol)-block-poly(D,L-lactide) (MAL-PEG-PLA) and methoxy-terminated poly-(ethylene glycol)-block-poly(D,L-lactide) (MPEG-PLA) were developed for theranostics through the encapsulation of SPIONs and doxorubicin [92]. The system showed improved therapeutic activity of doxorubicin and preferential uptake by $\alpha v\beta_3$-expressing cells [92]. Moreover, it showed higher accumulation in human lung cancer mice models. The clearly contrasted differences provided by SPIONs allowed better spatial resolution mapping and quantification of tumor angiogenic vasculature, thus garnishing the system with promising therapeutic and diagnostic properties [93].

Hoang et al. [94] developed ^{111}In-labeled PEG-PCL micelles targeting the nucleus of HER2-overexpressing breast cancer cells due to surface functionalization with Trastuzumab Fab fragments and 13-mer nuclear localization signal peptides [94]. In vivo studies showed higher accumulation of the system in a BT-474 tumor model compared to cells with low levels of HER2. Further, a 43% of the PM were found

Table 3 Examples of PM-based systems for bioimaging and theragnostic

Polymer	Imaging agent	Drug	Targeting moiety	Development state	Reference
H40-P(LG-Hyd-DOX)-b-PEG)	^{64}Cu	Doxorubicin	cRGD	In vivo	[141]
MAL-PEG-PLA and MPEG-PLA	SPIONs	Doxorubicin	cRGD	In vivo	[92]
MPEG-iodolysine	I			In vivo	[142]
N-palmitoyl chitosan	SPIONs		Folate	In vivo	[88]
PAMAM-PLA-b-PEG	^{64}Cu	Doxorubicin	anti-CD105 monoclonal antibody	In vivo	[143]
PDMA-b-PCL and PEG-PCL	USPIO	SN-38 and VEGF siRNA		In vivo	[91]
PEG-b-P(Glu)	Gd-DTPA	DACHPt		In vivo	[90]
PEG-b-Pasp	Gd-DTPA			In vivo	[86]
PEG-*b*-PPA	^{111}In	DNA		In vivo	[144]
PEG-p(l-His) and PEG-p (l-LA)-DTPA-Gd	Gd			In vivo	[87]
PEG-P[GA-DIP]	SPIONs	Doxorubicin	Folate	In vitro	[145]
PEG-PCL	SPIONs		Folate	In vivo	[89]
PEG-PCL	SPIONs	Bulsuphan		In vivo	[146]

(continued)

Table 3 (continued)

Polymer	Imaging agent	Drug	Targeting moiety	Development state	Reference
PEG-PCL	SPIONs	Doxorubicin	Folate	In vitro	[147]
PEG-PCL	^{111}In		Trastuzumab Fab fragments and 13-mer nuclear localization signal peptides	In vivo	[94]
PEG-PCL	SPIONs	Doxorubicin	Cetuximab	In vitro	[148]
PHEMA-PLLA-PEG	^{64}Cu	Doxorubicin	anti-CD105 monoclonal antibody	In vivo	[149]

cRGD cyclo(Arg-Gly-Asp-D-Phe-Cys) peptide, *DACHPt* (1,2-diaminocyclohexane)platinum(II), *Gd-DTPA* diethylenetriaminepentaacetic acid gadolinium (III), *H40-P(LG-Hyd-DOX)-b-PEG)* Boltorn® H40-poly(L-glutamate-hydrazone-doxorubicin)-*b*-poly(ethylene glycol), *MAL-PEG-PLA* maleimide-terminated poly(ethylene glycol)-block-poly(D,L-lactide), *MPEG-iodolysine* iodine-substituted poly-l-lysine, *MPEG-PLA* methoxy-terminated poly-(ethylene glycol)-block-poly(D,L-lactide), *PAMAM-PLA*-b-*PEG* poly(amidoamine)-poly(L-lactide)-b-poly(ethylene glycol), *PDMA*-b-*PCL* PDMA-block-poly(ε-caprolactone), *PEG*-b-*P(Glu)* poly(ethylene glycol)-b-poly(glutamic acid), *PEG-b-Pasp* poly(ethylene glycol)-b-poly(aspartic acid), *PEG*-b-*PPA* polyethylene glycol-b-polyphosphoramidate, *PEG-p (l-LA)-DTPA-Gd* methoxy poly(ethylene glycol)-b-poly(l-lactic acid)-diethylenetriaminopenta acetic acid dianhydride-gadolinium chelate, *PEG-p(l-His)* methoxy poly(ethylene glycol)-b-poly(l-histidine), *PEG-P[GA-DIP]* poly(ethylene glycol)-b-poly[*N*-(*N′*,*N′*-diisopropylaminoethyl) glutamine], *PEG-PCL* poly(ethylene glycol)- poly(ε-caprolactone), *PHEMA-PLLA-PEG* poly(2-hydroxyethyl methacrylate)-poly(l-lactide)-poly(ethylene glycol), *SPIONs* Superparamagnetic iron oxide nanoparticles, *USPIO* ultrasmall superparamagnetic iron oxide nanoparticles, *VEGF siRNA* small interfering RNA (siRNA) targeting human vascular endothelial growth factor (VEGF)

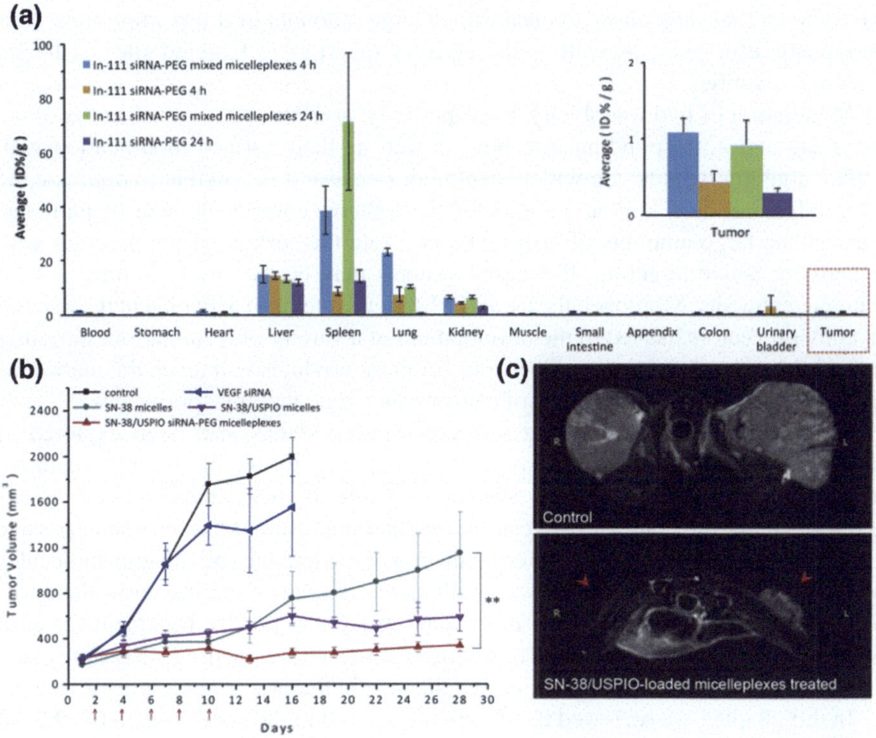

Fig. 7 The in vivo biodistribution and antitumor efficacy of mixed micelleplexes. **a** Biodistribution of free 111In-labeled siRNA-PEG conjugates and mixed micelleplexes at 4 and 24 h after i.v injection. The chart in the inset shows that the accumulation of mixed micelleplexes in the tumor was higher than that of free siRNA-PEG. Each column represents the mean±SD. **b** SN-38/USPIO-loaded siRNA-PEG mixed micelleplexes exhibited significant tumor suppression. The red arrows indicate the times at which the five administrations were given. (**: $P < 0.01$) **c** MRI images of representative mice from control and treated group on day 7 show that tumor growth was significantly inhibited by multidose treatment. Reprinted with permission of Elsevier Limited from Ref. [91]

within the nucleus, thus indicating the possible application of the system in diagnostics and in delivery of compounds with a nuclear or perinuclear therapeutic action, such as DNA or DNA-intercalating chemotherapeutic drugs [94].

5 Concluding Remarks and Future Perspectives

The scientific community is fully aware of the challenges rise by current cancer treatment failures and pitfalls. The progress of nanoscale formulations, particularly amphiphilic molecules with the possibility to form micelles by self-assembly in aqueous solution, has undergone great development in recent years. Micellar systems have shown that they allow to encapsulate large amounts of drugs improving their therapeutic efficiency, as well as the delivery of drugs to targeted sites, reducing systemic toxicities.

Modulation of hydrophobicity, hydrophilicity, and stimuli-responsiveness of the polymers and corresponding micelles, as well as their surface modification with active targeting ligands, are widely explored. Because it is possible to optimize the thermodynamic and kinetic stability of the system, control the size of particles, encapsulate large amounts of drugs, and decorate the surface of the particles with ligands for active targeting, PM-based systems may be the key to improved solid tumors treatments. Moreover, the use of PM-based systems to deliver genetic material as nonviral vectors has led to the development of a variety of formulations currently under evaluation. It is expected that some of these products will reach the market, as new treatments of different types of cancer using gene therapy approaches.

The incorporation of imaging molecules into PM has also been explored to improve diagnostic techniques as well as the monitoring of disease treatment responses and follow-ups. These systems are able to encapsulate imaging agents and to increase the contrast between injured and non-injured tissues due to passive or active targeting. In addition, the combination of therapeutic and imaging molecules within the same formulation allows simultaneous diagnosis and therapy—theranostics. In this way, the choice of a more appropriate and precise treatment for each patient can be achieved. However, translation into the clinical setting will be only successful if dose optimization, toxicity, and diagnostic/therapy efficacy are guaranteed.

In this chapter, we reviewed the progress observed in the development of PM and the great interest for these systems in the oncology field. Promising preclinical and clinical results, combined with the large translational potential of targeted delivery of anticancer agents, genetic material, and diagnostic agents, have culminated in the clinical approval of some formulations, with many others currently under clinical evaluation. Therefore, it is certainly expected that micellar-based nanoparticles will be able to change the paradigm of cancer treatment and will become a common reality in the near future, improving the quality of life of our patients.

Acknowledgements This chapter is a result of the project NORTE-01-0145-FEDER-000012, supported by Norte Portugal Regional Operational Programme (NORTE 2020), under the PORTUGAL 2020 Partnership Agreement, through the European Regional Development Fund (ERDF).
This work was partially funded by FEDER—Fundo Europeu de Desenvolvimento Regional funds through the COMPETE 2020—Operacional Programme for Competitiveness and Internationalisation (POCI), Portugal 2020, and by Portuguese funds through FCT—Fundação para a Ciência e a Tecnologia/Ministério da Ciência, Tecnologia e Ensino Superior in the framework of the project "Institute for Research and Innovation in Health Sciences" (POCI-01-0145-FEDER-007274). Andreia Almeida (grant SFRH/BD/118721/2016) and Fernanda Andrade (grant SFRH/BPD/120849/2016) would like to thank Fundação para a Ciência e a Tecnologia (FCT), Portugal for financial support. This research was also partially supported by CESPU/IINFACTS under the project MicelCampt-CESPU-2017.

References

1. Kataoka K, Harada A, Nagasaki Y (2001) Block copolymer micelles for drug delivery: design, characterization and biological significance. Adv Drug Deliv Rev 47(1):113–131
2. Adams ML, Lavasanifar A, Kwon GS (2003) Amphiphilic block copolymers for drug delivery. J Pharm Sci 92(7):1343–1355
3. Xiong XB, Binkhathlan Z, Molavi O, Lavasanifar A (2012) Amphiphilic block co-polymers: preparation and application in nanodrug and gene delivery. Acta Biomater 8(6):2017–2033
4. Li J, Wang X, Zhang T, Wang C, Huang Z, Luo X et al (2015) A review on phospholipids and their main applications in drug delivery systems—ScienceDirect. Asian J Pharm Sci 10(2):81–98
5. Urbani CN, Bell CA, Lonsdale D, Whittaker MR, Monteiro MJ (2008) Self-assembly of amphiphilic polymeric dendrimers synthesized with selective degradable linkages. Macromolecules 41:76–86
6. Letchford K, Burt H (2007) A review of the formation and classification of amphiphilic block copolymer nanoparticulate structures: micelles, nanospheres, nanocapsules and polymersomes. Eur J Pharm Biopharm 65(3):259–269
7. Torchilin V (2007) Micellar nanocarriers: pharmaceutical perspectives. Pharm Res 24(1):1–16
8. Andrade F, Videira M, Ferreira D, Sarmento B (2011) Micelle-based systems for pulmonary drug delivery and targeting. Drug Deliv Lett 1(2):171–185
9. Myers D (2006) Surfactant science and technology, 3rd edn. Wiley-Interscience, USA, p 380
10. Malmsten M (2002) Surfactants and polymers in drug delivery. Marcel Dekker, Inc., 335 p
11. Chen L, Ci T, Li T, Yu L, Ding J (2014) Effects of molecular weight distribution of amphiphilic block copolymers on their solubility, micellization, and temperature-induced sol-gel transition in water. Macromolecules 47(17):5895–5903
12. Abdelhamid D, Arslan H, Zhang Y, Uhrich KE (2014) Role of branching of hydrophilic domain on physicochemical properties of amphiphilic macromolecules. Polym Chem 5(4):1457–1462
13. Lu Y, Park K (2013) Polymeric micelles and alternative nanonized delivery vehicles for poorly soluble drugs. Int J Pharm 453(1):198–214
14. Letchford K, Liggins R, Burt H (2008) Solubilization of hydrophobic drugs by methoxy poly(ethylene glycol)-block-polycaprolactone diblock copolymer micelles: theoretical and experimental data and correlations. J Pharm Sci 97(3):1179–1190
15. Gelderblom H, Verweij J, Nooter K, Sparreboom A (2001) Cremophor EL: the drawbacks and advantages of vehicle selection for drug formulation. Eur J Cancer 37(13):1590–1598
16. Kwon G, Okano T (1996) Polymeric micelles as new drug carriers. Adv Drug Deliv Rev 21:107–116
17. Zhang Y, Huang Y, Li S (2014) Polymeric micelles: nanocarriers for cancer-targeted drug delivery. AAPS PharmSciTech 15(4):862–871
18. Opanasopit P, Yokoyama M, Watanabe M, Kawano K, Maitani Y, Okano T (2004) Block copolymer design for camptothecin incorporation into polymeric micelles for passive tumor targeting. Pharm Res 21(11):2001–2008
19. Kedar U, Phutane P, Shidhaye S, Kadam V (2010) Advances in polymeric micelles for drug delivery and tumor targeting. Nanomedicine 6(6):714–729
20. Le Garrec D, Ranger M, Leroux J-C (2004) Micelles in anticancer drug delivery. Am J Drug Deliv 2(1):15–42
21. Ohuchi M, Harada M, Amano Y, Kato Y, (2009) Physiologically active polypeptide- or protein-encapsulating polymer micelles, and method for production of the same (US 2009/0291130 A1)
22. Liaw J, Chang SF, Hsiao FC (2001) In vivo gene delivery into ocular tissues by eye drops of poly(ethylene oxide)-poly(propylene oxide)-poly(ethylene oxide) (PEO-PPO-PEO) polymeric micelles. Gene Ther 8(13):999–1004
23. Torchilin VP (2002) PEG-based micelles as carriers of contrast agents for different imaging modalities. Adv Drug Deliv Rev 54(2):235–252

24. Movassaghian S, Merkel OM, Torchilin VP (2017) Applications of polymer micelles for imaging and drug delivery. Wiley Interdisc Rev Nanomed Nanobiotechnol 7(5):691–707
25. Batrakova EV, Li S, Alakhov VY, Miller DW, Kabanov AV (2003) Optimal structure requirements for pluronic block copolymers in modifying P-glycoprotein drug efflux transporter activity in bovine brain microvessel endothelial cells. J Pharmacol Exp Ther 304(2):845–854
26. Chen L, Sha X, Jiang X, Chen Y, Ren Q, Fang X (2013) Pluronic P105/F127 mixed micelles for the delivery of docetaxel against Taxol-resistant non-small cell lung cancer: optimization and in vitro, in vivo evaluation. Int J Nanomed 8:73–84
27. Batrakova EV, Kelly DL, Li S, Li Y, Yang Z, Xiao L et al (2006) Alteration of genomic responses to doxorubicin and prevention of MDR in breast cancer cells by a polymer excipient: pluronic P85. Mol Pharm 3(2):113–123
28. Miller DW, Batrakova EV, Kabanov AV (1999) Inhibition of multidrug resistance-associated protein (MRP) functional activity with pluronic block copolymers. Pharm Res 16(3):396–401
29. Moghimi SM, Hunter AC (2000) Poloxamers and poloxamines in nanoparticle engineering and experimental medicine. Trends Biotechnol 18(10):412–420
30. Kabanov AV, Batrakova EV, Alakhov VY (2002) Pluronic block copolymers as novel polymer therapeutics for drug and gene delivery. J Controlled Release 82(2–3):189–212
31. Ho K, Li W, Wong C, Li P (2010) Amphiphilic polymeric particles with core-shell nanostructures: emulsion-based syntheses and potential applications. Colloid Polym Sci 288(16–17):1503–1523
32. Kwon GS (2006) Amphiphilic block copolymer micelles for nanoscale drug delivery. Drug Develop Res 67:15–22
33. Owens DE, Peppas NA (2006) Opsonization, biodistribution, and pharmacokinetics of polymeric nanoparticles. Int J Pharm 307(1):93–102
34. Moghimi SM, Szebeni J (2003) Stealth liposomes and long circulating nanoparticles: critical issues in pharmacokinetics, opsonization and protein-binding properties. Prog Lipid Res 42(6):463–478
35. Chaudhari KR, Ukawala M, Manjappa AS, Kumar A, Mundada PK, Mishra AK et al (2012) Opsonization, biodistribution, cellular uptake and apoptosis study of PEGylated PBCA nanoparticle as potential drug delivery carrier. Pharm Res 29(1):53–68
36. Rösler A, Vandermeulen GW, Klok HA (2001) Advanced drug delivery devices via self-assembly of amphiphilic block copolymers. Adv Drug Deliv Rev 53(1):95–108
37. Xu Q, Liu Y, Su S, Li W, Chen C, Wu Y (2012) Anti-tumor activity of paclitaxel through dual-targeting carrier of cyclic RGD and transferrin conjugated hyperbranched copolymer nanoparticles. Biomaterials 33(5):1627–1639
38. Shahin M, Ahmed S, Kaur K, Lavasanifar A (2011) Decoration of polymeric micelles with cancer-specific peptide ligands for active targeting of paclitaxel. Biomaterials 32(22):5123–5133
39. Kumar M, Kumar N, Domb A, Arora M (2002) Pharmaceutical polymeric controlled drug delivery systems. Filled Elastomers Drug Deliv Syst 160:45–117
40. Smith A, Xu X, McCormick C (2010) Stimuli-responsive amphiphilic (co)polymers via RAFT polymerization. Prog Polym Sci 35(1–2):45–93
41. Ward MA, Georgiou TK (2011) Thermoresponsive polymers for biomedical applications. Polymers 3:1215–1242
42. Motornov M, Roiter Y, Tokarev I, Minko S (2010) Stimuli-responsive nanoparticles, nanogels and capsules for integrated multifunctional intelligent systems. Prog Polym Sci 35(1–2):174–211
43. Hu Y, Litwin T, Nagaraja AR, Kwong B, Katz J, Watson N et al (2007) Cytosolic delivery of membrane-impermeable molecules in dendritic cells using pH-responsive core-shell nanoparticles. Nano Lett 7(10):3056–3064
44. Priya James H, John R, Alex A, Anoop KR (2014) Smart polymers for the controlled delivery of drugs—a concise overview. Acta Pharm Sin B 4(2):120–127
45. Guo W, Wang T, Tang X, Zhang Q, Yu F, Pei M (2014) Triple stimuli-responsive amphiphilic glycopolymer. J Polym Sci, Part A: Polym Chem 52(15):2131–2138

46. Ferlay J, Soerjomataram I, Dikshit R, Eser S, Mathers C, Rebelo M et al (2015) Cancer incidence and mortality worldwide: sources, methods and major patterns in GLOBOCAN 2012. Int J Cancer 136(5)
47. Seruga B, Ocana A, Tannock IF (2011) Drug resistance in metastatic castration-resistant prostate cancer. Nat Rev Clin Oncol 8(1):12–23
48. Plapied L, Duhem N, des Rieux A, Préat V (2011) Fate of polymeric nanocarriers for oral drug delivery. Curr Opin Colloid Interface Sci 16(3):228–237
49. Misra R, Acharya S, Sahoo SK (2010) Cancer nanotechnology: application of nanotechnology in cancer therapy. Drug Discovery Today 15(19):842–850
50. Csaba N, Garcia-Fuentes M, Alonso MJ (2006) The performance of nanocarriers for transmucosal drug delivery. Expert Opin Drug Delivery 3(4):463–478
51. Deshmukh AS, Chauhan PN, Noolvi MN, Chaturvedi K, Ganguly K, Shukla SS et al (2017) Polymeric micelles: basic research to clinical practice. Int J Pharm 532(1):249–268
52. Lv S, Tang Z, Li M, Lin J, Song W, Liu H et al (2014) Co-delivery of doxorubicin and paclitaxel by PEG-polypeptide nanovehicle for the treatment of non-small cell lung cancer. Biomaterials 35(23):6118–6129
53. Zhang W, Shi Y, Chen Y, Ye J, Sha X, Fang X (2011) Multifunctional Pluronic P123/F127 mixed polymeric micelles loaded with paclitaxel for the treatment of multidrug resistant tumors. Biomaterials 32(11):2894–2906
54. Bromberg L (2008) Polymeric micelles in oral chemotherapy. J Controlled Release 128(2):99–112
55. Mo R, Jin X, Li N, Ju C, Sun M, Zhang C et al (2011) The mechanism of enhancement on oral absorption of paclitaxel by *N*-octyl-*O*-sulfate chitosan micelles. Biomaterials 32(20):4609–4620
56. Li W, Li J, Gao J, Li B, Xia Y, Meng Y et al (2011) The fine-tuning of thermosensitive and degradable polymer micelles for enhancing intracellular uptake and drug release in tumors. Biomaterials 32(15):3832–3844
57. Li W, Zhao H, Qian W, Li H, Zhang L, Ye Z et al (2012) Chemotherapy for gastric cancer by finely tailoring anti-Her2 anchored dual targeting immunomicelles. Biomaterials 33(21):5349–5362
58. Némethy Á, Solti K, Kiss L, Gyarmati B, Deli MA, Csányi E et al (2013) pH-and temperature-responsive poly(aspartic acid)-*l*-poly(*N*-isopropylacrylamide) conetwork hydrogel. Eur Polymer J 49(9):2392–2403
59. Duan X, Xiao J, Yin Q, Zhang Z, Yu H, Mao S et al (2013) Smart pH-sensitive and temporal-controlled polymeric micelles for effective combination therapy of doxorubicin and disulfiram. ACS Nano 7(7):5858–5869
60. Guan J, Zhou Z-Q, Chen M-H, Li H-Y, Tong D-N, Yang J et al (2017) Folate-conjugated and pH-responsive polymeric micelles for target-cell-specific anticancer drug delivery. Acta Biomater 60:244–255
61. Hilgenbrink AR, Low PS (2005) Folate receptor-mediated drug targeting: from therapeutics to diagnostics. J Pharm Sci 94(10):2135–2146
62. Matsumura Y, Hamaguchi T, Ura T, Muro K, Yamada Y, Shimada Y et al (2004) Phase I clinical trial and pharmacokinetic evaluation of NK911, a micelle-encapsulated doxorubicin. Br J Cancer 91(10):1775–1781
63. Nakanishi T, Fukushima S, Okamoto K, Suzuki M, Matsumura Y, Yokoyama M et al (2001) Development of the polymer micelle carrier system for doxorubicin. J Controlled Release 74(1):295–302
64. Kim DW, Kim SY, Kim HK, Kim SW, Shin SW, Kim JS et al (2007) Multicenter phase II trial of Genexol-PM, a novel Cremophor-free, polymeric micelle formulation of paclitaxel, with cisplatin in patients with advanced non-small-cell lung cancer. Ann Oncol 18(12):2009–2014
65. Lee KS, Chung HC, Im SA, Park YH, Kim CS, Kim SB et al (2008) Multicenter phase II trial of Genexol-PM, a Cremophor-free, polymeric micelle formulation of paclitaxel, in patients with metastatic breast cancer. Breast Cancer Res Treat 108(2):241–250

66. Saif MW, Rubin MS, Figueroa JA, Kerr RO (2008) Multicenter phase II trial of Genexol-PM (GPM), a novel Cremophor-free, polymeric micelle formulation of paclitaxel in patients with advanced pancreatic cancer (APC): final results. In: Gastrointestinal cancers symposium; Orlando2008
67. Kim SC, Kim DW, Shim YH, Bang JS, Oh HS, Kim SW et al (2001) In vivo evaluation of polymeric micellar paclitaxel formulation: toxicity and efficacy. J Controlled Release 72(1):191–202
68. Valle JW, Armstrong A, Newman C, Alakhov V, Pietrzynski G, Brewer J et al (2011) A phase 2 study of SP1049C, doxorubicin in P-glycoprotein-targeting pluronics, in patients with advanced adenocarcinoma of the esophagus and gastroesophageal junction. Invest New Drugs 29(5):1029–1037
69. Hamaguchi T, Matsumura Y, Suzuki M, Shimizu K, Goda R, Nakamura I et al (2005) NK105, a paclitaxel-incorporating micellar nanoparticle formulation, can extend in vivo antitumour activity and reduce the neurotoxicity of paclitaxel. Br J Cancer 92(7):1240–1246
70. Kumar SR, Markusic DM, Biswas M, High KA, Herzog RW (2016) Clinical development of gene therapy: results and lessons from recent successes. Mol Ther Methods Clin Dev 3:16034
71. Kaufmann KB, Büning H, Galy A, Schambach A, Grez M (2013) Gene therapy on the move. EMBO Mol Med 5(11):1642–1661
72. Wang D, Gao G (2014) State-of-the-art human gene therapy: part I. gene delivery technologies. Discov Med 18(97):67–77
73. Videira M, Arranja A, Rafael D, Gaspar R (2014) Preclinical development of siRNA therapeutics: towards the match between fundamental science and engineered systems. Nanomedicine 10(4):689–702
74. EMA. Glybera. EPAR—product information. www.ema.europa.eu2012
75. Wasungu L, Hoekstra D (2006) Cationic lipids, lipoplexes and intracellular delivery of genes. J Control Release 116(2):255–264
76. Eliyahu H, Joseph A, Schillemans JP, Azzam T, Domb AJ, Barenholz Y (2007) Characterization and in vivo performance of dextran-spermine polyplexes and DOTAP/cholesterol lipoplexes administered locally and systemically. Biomaterials 28(14):2339–2349
77. Liu F, Huang L (2002) Development of non-viral vectors for systemic gene delivery. J Control Release 78(1–3):259–266
78. Zhang XX, McIntosh TJ, Grinstaff MW (2012) Functional lipids and lipoplexes for improved gene delivery. Biochimie 94(1):42–58
79. Zhang S, Xu Y, Wang B, Qiao W, Liu D, Li Z (2004) Cationic compounds used in lipoplexes and polyplexes for gene delivery. J Control Release 100(2):165–180
80. Scholz C, Wagner E (2012) Therapeutic plasmid DNA versus siRNA delivery: common and different tasks for synthetic carriers. J Control Release 161(2):554–565
81. Yang Z, Sahay G, Sriadibhatla S, Kabanov AV (2008) Amphiphilic block copolymers enhance cellular uptake and nuclear entry of polyplex-delivered DNA. Bioconjug Chem 19(10):1987–1994
82. Mishra S, Peddada LY, Devore DI, Roth CM (2012) Poly(alkylene oxide) Copolymers for Nucleic Acid Delivery. Acc Chem Res 45(7):1057–1066
83. Wang M, Wu B, Lu P, Tucker JD, Milazi S, Shah SN et al (2014) Pluronic-PEI copolymers enhance exon-skipping of 2′-O-methyl phosphorothioate oligonucleotide in cell culture and dystrophic mdx mice. Gene Ther 21(1):52–59
84. Davis ME (2009) The first targeted delivery of siRNA in humans via a self-assembling, cyclodextrin polymer-based nanoparticle: from concept to clinic. Mol Pharm 6(3):659–668
85. Oerlemans C, Bult W, Bos M, Storm G, Nijsen JF, Hennink WE (2010) Polymeric micelles in anticancer therapy: targeting, imaging and triggered release. Pharm Res 27(12):2569–2589
86. Mi P, Kokuryo D, Cabral H, Kumagai M, Nomoto T, Aoki I et al (2014) Hydrothermally synthesized PEGylated calcium phosphate nanoparticles incorporating Gd-DTPA for contrast enhanced MRI diagnosis of solid tumors. J Control Release 174:63–71
87. Kim KS, Park W, Hu J, Bae YH, Na K (2014) A cancer-recognizable MRI contrast agents using pH-responsive polymeric micelle. Biomaterials 35(1):337–343

88. Xiao Y, Lin ZT, Chen Y, Wang H, Deng YL, Le DE et al (2015) High molecular weight chitosan derivative polymeric micelles encapsulating superparamagnetic iron oxide for tumor-targeted magnetic resonance imaging. Int J Nanomed 10:1155–1172
89. Hong GB, Zhou JX, Yuan RX (2012) Folate-targeted polymeric micelles loaded with ultra-small superparamagnetic iron oxide: combined small size and high MRI sensitivity. Int J Nanomed 7:2863–2872
90. Kaida S, Cabral H, Kumagai M, Kishimura A, Terada Y, Sekino M et al (2010) Visible drug delivery by supramolecular nanocarriers directing to single-platformed diagnosis and therapy of pancreatic tumor model. Cancer Res 70(18):7031–7041
91. Lee SY, Yang CY, Peng CL, Wei MF, Chen KC, Yao CJ et al (2016) A theranostic micelleplex co-delivering SN-38 and VEGF siRNA for colorectal cancer therapy. Biomaterials 86:92–105
92. Guthi JS, Yang SG, Huang G, Li S, Khemtong C, Kessinger CW et al (2010) MRI-visible micellar nanomedicine for targeted drug delivery to lung cancer cells. Mol Pharm 7(1):32–40
93. Kessinger CW, Khemtong C, Togao O, Takahashi M, Sumer BD, Gao J (2010) In vivo angiogenesis imaging of solid tumors by $\alpha_v\beta_3$-targeted, dual-modality micellar nanoprobes. Exp Biol Med 235:957–965
94. Hoang B, Ekdawi SN, Reilly RM, Allen C (2013) Active targeting of block copolymer micelles with trastuzumab Fab fragments and nuclear localization signal leads to increased tumor uptake and nuclear localization in HER2-overexpressing xenografts. Mol Pharm 10(11):4229–4241
95. Sawant R, Jhaveri A (2014) Micellar nanopreparations for medicine. In: Torchilin V (ed) Handbook of nanobiomedical research: fundamentals, applications and recent developments. Frontiers in Nanobiomedical Research, vol 3. World Scientific, Singapore
96. Hammad A, Mosaad YM, Hammad EM, Elhanbly S, El-Bassiony SR, Al-Harrass MF et al (2016) Interleukin-17A rs2275913, Interleukin-17F rs763780 and rs2397084 gene polymorphisms as possible risk factors in Juvenile lupus and lupus related nephritis. Autoimmunity 49(1):31–40
97. Evaluate the efficacy and safety of genexol®-PM compared to genexol® in recurrent or metastatic breast cancer 2016 [Available from: https://clinicaltrials.gov/ct2/show/NCT00876486]
98. Lu HQ, Wang EQ, Zhang T, Chen YX (2016) Photodynamic therapy and anti-vascular endothelial growth factor for acute central serous chorioretinopathy: a systematic review and meta-analysis. Eye 30(1):15–22
99. A phase II trial of genexol-PM and gemcitabine in patients with advanced non-small-cell lung cancer 2016 [Available from: https://clinicaltrials.gov/ct2/show/NCT01770795]
100. Ronnekleiv-Kelly SM, Nukaya M, Diaz-Diaz CJ, Megna BW, Carney PR, Geiger PG et al (2016) Aryl hydrocarbon receptor-dependent apoptotic cell death induced by the flavonoid chrysin in human colorectal cancer cells. Cancer Lett 370(1):91–99
101. A study of NK012 in patients with relapsed small cell lung cancer 2016 [Available from: https://clinicaltrials.gov/ct2/show/NCT00951613]
102. Schneider M, Strobele S, Nonnenmacher L, Siegelin MD, Tepper M, Stroh S et al (2016) A paired comparison between glioblastoma "stem cells" and differentiated cells. Int J Cancer 138(7):1709–1718
103. A phase III study of NK105 in patients with breast cancer 2016 [Available from: https://clinicaltrials.gov/ct2/show/NCT01644890]
104. Matsumura Y, Hamaguchi T, Ura T, Muro K, Yamada Y, Shimada Y et al (2004) Phase I clinical trial and pharmacokinetic evaluation of NK911, a micelle-encapsulated doxorubicin. Br J Cancer 91(10):1775–1781
105. A phase 1 dose-escalation and pharmacokinetic study of NC-4016 in patients with advanced solid tumors or lymphoma 2016 [Available from: https://clinicaltrials.gov/ct2/show/NCT01999491]
106. Dose-escalation and expansion trial of NC-6300 in patients with advanced solid tumors or soft tissue sarcoma 2017 [Available from: https://clinicaltrials.gov/ct2/show/record/NCT03168061]

107. Combination therapy with NC-6004 and gemcitabine versus gemcitabine alone in pancreatic cancer 2016 [Available from: https://clinicaltrials.gov/ct2/show/NCT02043288]
108. Combination therapy with NC-6004 and gemcitabine in advanced solid tumors or non-small cell lung, biliary and bladder cancer 2016 [Available from: https://clinicaltrials.gov/ct2/show/NCT02240238]
109. Efficacy study of maintenance IT-101 therapy for ovarian cancer patients 2016 [Available from: https://clinicaltrials.gov/ct2/show/NCT00753740]
110. Study of CRLX101 (Formerly Named IT-101) in the treatment of advanced solid tumors 2016 [Available from: https://clinicaltrials.gov/ct2/show/NCT00333502]
111. BIND therapeutics. An open label, multicenter, Phase 2 study to determine the safety and efficacy of BIND-014 (docetaxel nanoparticles for injectable suspension) as a second-line therapy for patients with KRAS mutation positive or squamous cell non-small cell lung cancer. NCT02283320: https://clinicaltrials.gov
112. BIND therapeutics. A phase 1 open label, safety, pharmacokinetic and pharmacodynamic dose escalation study of BIND-014 (docetaxel nanoparticles for injectable suspension), given by intravenous infusion to patients with advanced or metastatic cancer. NCT01300533: https://clinicaltrials.gov
113. BIND therapeutics. An open label, multicenter, phase 2 study to determine the safety and efficacy of BIND-014 (docetaxel nanoparticles for injectable suspension), administered to patients with metastatic castration-resistant prostate cancer. NCT01812746: https://clinicaltrials.gov
114. Von Hoff DD, Mita MM, Ramanathan RK, Weiss GJ, Mita AC, LoRusso PM et al (2016) Phase I study of PSMA-targeted docetaxel-containing nanoparticle BIND-014 in patients with advanced solid tumors. Clin Cancer Res 22(13):3157–3163
115. Zhu Y, Sheng R, Luo T, Li H, Sun W, Li Y et al (2011) Amphiphilic cationic [dendritic poly(L-lysine)]-block-poly(L-lactide)-block-[dendritic poly(L-lysine)]s in aqueous solution: self-aggregation and interaction with DNA as gene delivery carriers. Macromol Biosci 11(2):174–186
116. Ulasov AV, Khramtsov YV, Trusov GA, Rosenkranz AA, Sverdlov ED, Sobolev AS (2011) Properties of PEI-based polyplex nanoparticles that correlate with their transfection efficacy. Mol Ther 19(1):103–112
117. Zhu C, Jung S, Luo S, Meng F, Zhu X, Park TG et al (2010) Co-delivery of siRNA and paclitaxel into cancer cells by biodegradable cationic micelles based on PDMAEMA-PCL-PDMAEMA triblock copolymers. Biomaterials 31(8):2408–2416
118. Lee SH, Lee JY, Kim JS, Park TG, Mok H (2017) Amphiphilic siRNA Conjugates for Co-Delivery of Nucleic Acids and Hydrophobic Drugs. Bioconjug Chem 28(8):2051–2061
119. Wang S, Zhang J, Wang Y, Chen M (2016) Hyaluronic acid-coated PEI-PLGA nanoparticles mediated co-delivery of doxorubicin and miR-542-3p for triple negative breast cancer therapy. Nanomed Nanotechnol Biol Med 12(2):411–420
120. Ebrahimian M, Taghavi S, Mokhtarzadeh A, Ramezani M, Hashemi M (2017) Co-delivery of doxorubicin encapsulated PLGA nanoparticles and Bcl-xL shRNA using alkyl-modified PEI into breast cancer cells. Appl Biochem Biotechnol
121. Devulapally R, Sekar NM, Sekar TV, Foygel K, Massoud TF, Willmann JK et al (2015) Polymer nanoparticles mediated codelivery of antimiR-10b and antimiR-21 for achieving triple negative breast cancer therapy. ACS Nano 9(3):2290–2302
122. Kang L, Gao Z, Huang W, Jin M, Wang Q (2015) Nanocarrier-mediated co-delivery of chemotherapeutic drugs and gene agents for cancer treatment. Acta Pharm Sinica B 5(3):169–175
123. Chen W, Yuan Y, Cheng D, Chen J, Wang L, Shuai X (2014) Co-delivery of doxorubicin and siRNA with reduction and pH dually sensitive nanocarrier for synergistic cancer therapy. Small 10(13):2678–2687
124. Tang S, Yin Q, Zhang Z, Gu W, Chen L, Yu H et al (2014) Co-delivery of doxorubicin and RNA using pH-sensitive poly(beta-amino ester) nanoparticles for reversal of multidrug resistance of breast cancer. Biomaterials 35(23):6047–6059

125. Li Y, Xu B, Bai T, Liu W (2015) Co-delivery of doxorubicin and tumor-suppressing p53 gene using a POSS-based star-shaped polymer for cancer therapy. Biomaterials 55:12–23
126. Itaka K, Osada K, Morii K, Kim P, Yun SH, Kataoka K (2010) Polyplex nanomicelle promotes hydrodynamic gene introduction to skeletal muscle. J Control Release 143(1):112–119
127. Uchida S, Itaka K, Chen Q, Osada K, Miyata K, Ishii T et al (2011) Combination of chondroitin sulfate and polyplex micelles from Poly(ethylene glycol)-poly{*N*′-[*N*-(2-aminoethyl)-2-aminoethyl]aspartamide} block copolymer for prolonged in vivo gene transfection with reduced toxicity. J Control Release 155(2):296–302
128. Chen Q, Osada K, Ishii T, Oba M, Uchida S, Tockary TA et al (2012) Homo-catiomer integration into PEGylated polyplex micelle from block-catiomer for systemic anti-angiogenic gene therapy for fibrotic pancreatic tumors. Biomaterials 33(18):4722–4730
129. Chen YC, Jiang LP, Liu NX, Ding L, Liu XL, Wang ZH et al (2011) Enhanced gene transduction into skeletal muscle of mice in vivo with pluronic block copolymers and ultrasound exposure. Cell Biochem Biophys 60(3):267–273
130. Davis ME, Zuckerman JE, Choi CH, Seligson D, Tolcher A, Alabi CA et al (2010) Evidence of RNAi in humans from systemically administered siRNA via targeted nanoparticles. Nature 464(7291):1067–1070
131. Zuckerman JE, Hsueh T, Koya RC, Davis ME, Ribas A (2011) siRNA knockdown of ribonucleotide reductase inhibits melanoma cell line proliferation alone or synergistically with temozolomide. J Invest Dermatol 131(2):453–460
132. Gaspar VM, Goncalves C, de Melo-Diogo D, Costa EC, Queiroz JA, Pichon C et al (2014) Poly(2-ethyl-2-oxazoline)-PLA-g-PEI amphiphilic triblock micelles for co-delivery of minicircle DNA and chemotherapeutics. J Controlled Release: Official J Controlled Release Soc 189:90–104
133. Bao X, Wang W, Wang C, Wang Y, Zhou J, Ding Y et al (2014) A chitosan-graft-PEI-candesartan conjugate for targeted co-delivery of drug and gene in anti-angiogenesis cancer therapy. Biomaterials 35(29):8450–8466
134. Yin T, Wang L, Yin L, Zhou J, Huo M (2015) Co-delivery of hydrophobic paclitaxel and hydrophilic AURKA specific siRNA by redox-sensitive micelles for effective treatment of breast cancer. Biomaterials 61:10–25
135. Loh XJ, Ong SJ, Tung YT, Choo HT (2013) Co-delivery of drug and DNA from cationic dual-responsive micelles derived from poly(DMAEMA-co-PPGMA). Mater Sci Eng C, Mater Bio Appl 33(8):4545–4550
136. Shi S, Shi K, Tan L, Qu Y, Shen G, Chu B et al (2014) The use of cationic MPEG-PCL-g-PEI micelles for co-delivery of Msurvivin T34A gene and doxorubicin. Biomaterials 35(15):4536–4547
137. Qian X, Long L, Shi Z, Liu C, Qiu M, Sheng J et al (2014) Star-branched amphiphilic PLA-b-PDMAEMA copolymers for co-delivery of miR-21 inhibitor and doxorubicin to treat glioma. Biomaterials 35(7):2322–2335
138. Davoodi P, Srinivasan MP, Wang CH (2016) Synthesis of intracellular reduction-sensitive amphiphilic polyethyleneimine and poly(epsilon-caprolactone) graft copolymer for on-demand release of doxorubicin and p53 plasmid DNA. Acta Biomater 39:79–93
139. Aji Alex MR, Nehate C, Veeranarayanan S, Kumar DS, Kulshreshtha R, Koul V (2017) Self assembled dual responsive micelles stabilized with protein for co-delivery of drug and siRNA in cancer therapy. Biomaterials 133:94–106
140. Zhu L, Perche F, Wang T, Torchilin VP (2014) Matrix metalloproteinase 2-sensitive multifunctional polymeric micelles for tumor-specific co-delivery of siRNA and hydrophobic drugs. Biomaterials 35(13):4213–4222
141. Xiao Y, Hong H, Javadi A, Engle J, Xu W, Yang Y et al (2012) Multifunctional unimolecular micelles for cancer-targeted drug delivery and positron emission tomography imaging—ScienceDirect. Biomaterials 33(11):3071–3082
142. Torchilin VP, Frank-Kamenetsky MD, Wolf GL (1999) CT visualization of blood pool in rats by using long-circulating, iodine-containing micelles. Acad Radiol 6(1):61–65

143. Guo J, Hong H, Chen G, Shi S, Zheng Q, Zhang Y et al (2013) Image-guided and tumor-targeted drug delivery with radiolabeled unimolecular micelles. Biomaterials 34(33):8323–8332
144. Patil RR, Yu J, Banerjee SR, Ren Y, Leong D, Jiang X et al (2011) Probing in vivo trafficking of polymer/DNA micellar nanoparticles using SPECT/CT imaging. Mol Ther 19(9):1626–1635
145. Li X, Li H, Yi W, Chen J, Liang B (2013) Acid-triggered core cross-linked nanomicelles for targeted drug delivery and magnetic resonance imaging in liver cancer cells. Int J Nanomed 8:3019–3031
146. Asem H, Zhao Y, Ye F, Barrefelt A, Abedi-Valugerdi M, El-Sayed R et al (2016) Biodistribution of biodegradable polymeric nano-carriers loaded with busulphan and designed for multimodal imaging. J Nanobiotechnol 14(1):82
147. Hong G, Yuan R, Liang B, Shen J, Yang X, Shuai X (2008) Folate-functionalized polymeric micelle as hepatic carcinoma-targeted, MRI-ultrasensitive delivery system of antitumor drugs. Biomed Microdevices 10(5):693–700
148. Liao C, Sun Q, Liang B, Shen J, Shuai X (2011) Targeting EGFR-overexpressing tumor cells using Cetuximab-immunomicelles loaded with doxorubicin and superparamagnetic iron oxide. Eur J Radiol 80(3):699–705
149. Guo J, Hong H, Chen G, Shi S, Nayak TR, Theuer CP et al (2014) Theranostic unimolecular micelles based on brush-shaped amphiphilic block copolymers for tumor-targeted drug delivery and positron emission tomography imaging. ACS Appl Mater Interfaces 6(24):21769–21779

Magnetic Nanoparticles for Cancer Therapy and Bioimaging

Ester Polo, Pablo del Pino, Alberto Pardo, Pablo Taboada and Beatriz Pelaz

Abstract Cancer is the second leading cause of death globally and was responsible for 8.8 million deaths in 2015. Nearly, 1 in 6 deaths worldwide is due to cancer. An ancient Egyptian textbook on trauma surgery is the oldest (3000 BC) description of cancer treatment, which mentioned a tool called the fire drill to remove tumors. Hand in hand with the development of modern medicine and physiology, other approaches have appeared although surgery remains the most popular "solution". During the twentieth century, our knowledge in cancer was greatly increased. Today, most common treatments are based on chemotherapy, which may be combined with surgery and radiation therapy. Despite the extended use of chemotherapeutic drugs, they lack good selectivity and cause manifold off-target effects. These issues combined with the fact that drug resistances are increasingly more common indicate the need for more selective drugs with better targeting capabilities. Some of the most promising candidates to develop these more effective drugs are based on the use of nanomaterials, in particular nanoparticles (NPs). Among the NPs, inorganic NPs are very interesting because they exhibit unique properties derived from their nanometric size. In the literature, there are manifold examples about how to take advantage of the properties of NPs to develop more effective therapies, imaging agents, or biosensors [1]. Since the beginning interest about how nanomaterials interact with living matter, magnetic NPs (MNPs) attracted a huge interest. This attention is related to their unique physicochemical properties. The different responses to external magnetic fields (alternating and non-alternating) or gradients have been applied for different purposes, including therapy, diagnosis, and theranostics. In this book chapter, synthetic fine tailoring of MNPs required to work as therapeutic and/or imaging agents will be discussed in the context of cancer. Most relevant MNP's applications in therapy and imaging will be described.

E. Polo · P. del Pino · B. Pelaz (✉)
Centro Singular de Investigación en Química Biolóxica e Materiais Moleculares (CIQUS), Universidade de Santiago de Compostela, 15782 Santiago de Compostela, Spain
e-mail: beatriz.pelaz@usc.es

E. Polo · P. del Pino · A. Pardo · P. Taboada · B. Pelaz
Departamento de Física de Partículas, Universidade de Santiago de Compostela, 15782 Santiago de Compostela, Spain

G. Gonçalves and G. Tobias (eds.), *Nanooncology*, Nanomedicine and Nanotoxicology, https://doi.org/10.1007/978-3-319-89878-0_7

Keywords Magnetic Nanoparticles · Cancer · Therapy · Imaging · Theranostics · Surface engineering

1 General Considerations About MNPs: Synthesis and Magnetic Properties

As mentioned above, the unique features of MNPs explain why these NPs have been used to such an extent in nanomedicine. To understand better how these properties are related to the material composition, structure, and size, their general properties and tuning parameters will be briefly explained in the following. A magnetic material is defined by few parameters [2]:

- Coercive Field (H_c): this parameter refers to the minimum energy to reverse the material net magnetization back to zero, and is related to the magnetic anisotropy (K_a).
- K_a is defined by the NP size and shape, and by the symmetry of the crystal lattice. This property determines the barrier energy to reverse the direction of the magnetic dipoles.
- M_S, saturation magnetization, is the maximum value of net magnetization under the influence of a magnetic field.
- M_R is the remanent magnetization and represents the residual magnetization with no applied field.

These parameters can be obtained from the hysteresis loop (i.e., magnetization vs. applied field), typically obtained by measurements in a superconducting quantum interference device (SQUID). All these parameters depend directly on the size (ratio of atoms on the NP surface vs. total atoms number), shape, crystalline structure, and composition of the materials. Fine control over these magnetic features can be achieved by controlling the synthetic method for their fabrication.

Superparamagnetism is a specific magnetic behavior which only occurs at the nanometric scale. In general, for spherical MNPs, a specific radius exists, below which MNPs become superparamagnetic, that is, they are monodomain. This means that the MNP's energy due to temperature is able to exceed the energy of magnetic stabilization of the MNP. Hence, this type of MNPs in the absence of external magnetic fields has M_R and H_c equal to 0. In principle, this fact makes superparamagnetic MNPs ideal candidates for bioapplications. The absence of permanent magnetic moment prevents from their aggregation, which facilitate their potential applicability in vivo.

To control and obtain the best response when designing MNPs, the selection of the right synthetic methodology is of utmost importance. In the last decades, the control over size, shape, and composition of NPs has been greatly enhanced. Typically, the synthetic methods might be divided into physical and chemical methods. In general, physical methods yield bigger amounts of material but with poor properties (e.g., homogeneity). Also, these methods normally require the use of expensive equipment.

On the other hand, chemical methods generate smaller product quantities but with more homogeneous properties and better quality. In general, there are many chemical methods; however, we will only discuss the most relevant ones:

(i) Sol–gel synthesis methodologies are used in which MNPs are produced by the condensation of the precursors upon their heating. Controlling the size and particle structure is possible using this methodology [3].
(ii) Co-precipitation is the most commonly used methodology to prepare MNPs within biomedical purposes. This method is based on the co-precipitation of ferrous and ferric salts; by modifying the stoichiometry of the salts, the ionic strength, the pH, the temperature, etc., the final size and magnetic behavior of the MNPs can be controlled. This method benefits from the use of cheap chemicals, mil reaction conditions, produces water-soluble MNPs that can be easily further modified, and it is easily scalable. Despite these advantages, this methodology lacks in good reproducibility, produces MNPs with a broad size distribution which needs from further purification to narrower their size distribution and exhibits an uncontrolled oxidation [4].
(iii) Hydrothermal syntheses take place in reactors or autoclaves under high-pressure conditions and temperatures; factors such as temperature, time, solvents, precursors nature and stoichiometry, etc. will have an influence in the structure, size, and shape of the final MNPs. As benefits, these methods produce also water-soluble MNPs, with a better size distribution compared to the co-precipitation method, but the use of high temperatures is required [5].
(iv) Thermal decomposition consists in the reduction of the metallic precursors under controlled conditions and at high temperatures (>200 °C), yielding highly homogeneous MNPs with an exceptional control over its composition and size. This methodology can be scaled up for the production of grams of NPs. Due to the high control over the crystalline structure and size, the magnetic properties are highly tunable. However, the as-synthesized MNPs are not soluble in water, and therefore, a phase transfer step is needed. In addition, some of the chemicals needed are toxic and, as mentioned previously, it requires the use of high temperatures [6].

As stated above, all these methodologies possess advantages and drawbacks. However, when designing a "smart" nanomaterial, the control over the size, shape, and composition is key factors, which define the final behavior and response of the material. So, the use of scalable methods that produce homogeneous NPs is highly recommended. Under these considerations, the best methodology among the previously mentioned is the thermal decomposition. In Fig. 1, a schematic representation of this method procedure is presented [7].

There are many magnetic materials; however, the preferred materials for bioapplications are the iron oxide based NPs. These MNPs are typically made of maghemite (Fe_2O_3) or magnetite (Fe_3O_4). Iron oxide materials are biocompatible, and they are known to be biodegradable and exhibit low toxicity. The US Food and Drug Administration (FDA) and the European Medicines Agency (EMA) have approved the use of magnetic oxides NPs including magnetite [4]. They have good magnetic capabil-

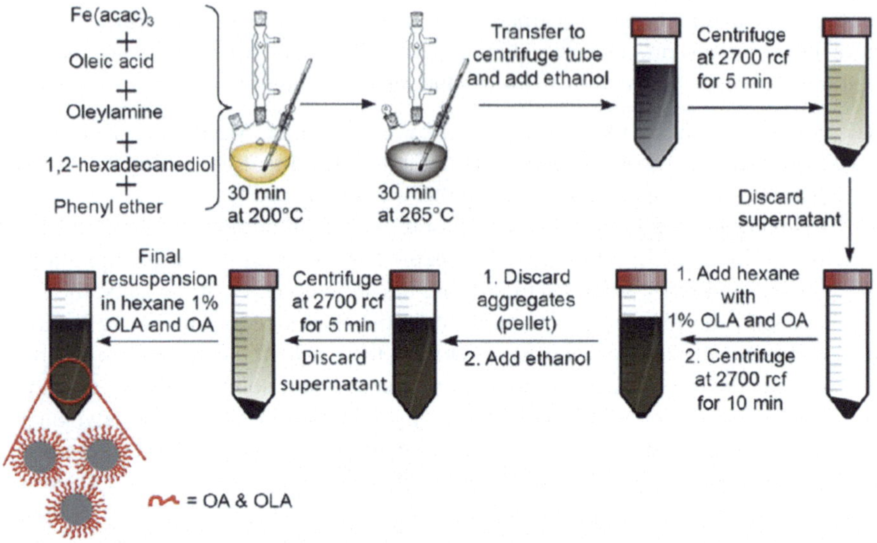

Fig. 1 Schematic of the process by thermal decomposition to produce iron oxide NPs. Figure reproduced from Hühn et al. [7]

ities [e.g., magnetic saturation values, or specific absorption rate (SAR)]. However, in some cases, these magnetic properties need to be improved. The most general method is the doping of these materials with other transitions metals, which result in better magnetic performances. As example, iron cations can be replaced by others such as cobalt, manganese, zinc, nickel, magnesium, etc. (c.f. Fig. 2a, d). The inclusion of a small percentage of these other cations drastically affects the final magnetic properties of the NPs [8] by the alteration of the magnetic saturation, as well as the magnetic anisotropy. The effect of the alteration of the composition by doping the NPs during the synthesis can be combined with the production of core–shell structures (core@shell) of soft and hard magnetic materials, aiming to tune the magnetic anisotropy by the synergy of the properties (c.f. Fig. 2b, c) [9, 10].

However, the synthesis of anisotropic ferrite-based materials with high quality remains still a challenge. Controlling the size of spherical and cubic NPs is possible but the existing methods to produce elongated, branched, or flat NPs [11] are still lacking good reproducibility and yield. There are still few examples showing this control over the shape [12]. In Table 1, the most common parameters for different types of MNPs are summarized.

Another alternative to modulate the magnetic behavior consists of changing the material composition; an example is the preparation of iron carbide NPs (c.f. Fig. 2e) which consist of iron (0) and carbon. These NPs exhibit good biocompatibility, stability, and magnetic response [13, 14]. Typically, the M_s values are higher while their coercivity is lower than iron oxide NPs, thanks to the Fe (0) core. They are

Table 1 Summary of the most relevant physicochemical characteristics and magnetic behavior of MNPs with different sizes, compositions, and shapes

Shape	Composition	Coating	d_c [nm]	M_S [emu/g]	SAR [W/$g_{magnetic}$]	r_1 [$mM^{-1}s^{-1}$]	r_2 [$mM^{-1}s^{-1}$]	References
Sphere	Fe_3O_4	OA	16	60.5 (5 K)	–	6.6 ± 0.3 (0.5T)	125.7 ± 5.6 (0.5T)	[12]
Plate	Fe_3O_4	OA	20	44.9 (5 K)	–	16.6 ± 0.4 (0.5T)	86.5 ± 2.3 (0.5T)	[12]
Truncated Octahedra	Fe_3O_4	OA	14	57.1 (5 K)	–	22.7 ± 1.3 (0.5T)	239.2 ± 7.55 (0.5T)	[12]
Tetrahedra	Fe_3O_4	OA	8.7	52.7 (5 K)	–	19.9 ± 0.7 (0.5T)	176.5 ± 5.22 (0.5T)	[12]
Cube	Fe_3O_4	OA	6.1	47.3 (5 K)	–	25.8 ± 1.1(0.5T)	76.4 ± 0.54 (0.5T)	[12]
Cube	Fe_3O_4	OA	12.9	54.1 (5 K)	–	19.3 ± 0.1 (0.5T)	80.84 ± 1.37 (0.5T)	[12]
Cube	Fe_3O_4	OA	18.2	56.1 (5 K)	–	27.9 ± 1.6 (0.5T)	298.02 ± 5.8 (0.5T)	[12]
Cube	Fe_3O_4	OA	25.9	44.7 (5 K)	–	13.7 ± 0.1 (0.5T)	215.6 ± 2.81 (0.5T)	[12]
Concave cube	Fe_3O_4	OA	22.5	42.9 (5 K)	–	8.3 ± 0.3 (0.5T)	224.4 ± 4.6 (0.5T)	[12]
Branched	Fe_3O_4	OA	82.5	39.2 (5 K)	–	4.9 ± 0.1 (0.5T)	103.9 ± 2.2 (0.5T)	[12]
Co rod@Au	Co@SnPtAu	HDA	75 × 6	158 (4 K)	–	–	–	[22]
Sphere	$MnFe_2O_3$	PMA	14	90.5 (r.t.)	184 ± 2 (water) (24 kA/m; 412.5 kHz)	–	–	[9]

(continued)

Table 1 (continued)

Shape	Composition	Coating	d_c [nm]	M_S [emu/g]	SAR [W/$g_{magnetic}$]	r_1 [$mM^{-1}s^{-1}$]	r_2 [$mM^{-1}s^{-1}$]	References
core@shell sphere	$CoFe_2O_3$@$MnFe_2O_3$	PMA	13@7	87 (r.t.)	553 ± 10(water) (24 kA/m; 412.5 kHz)	–	–	[9]
core@shell sphere	$MnFe_2O_3$@$CoFe_2O_3$	PMA	14@7	84 (r.t.)	302 ± 15(water) (24 kA/m; 412.5 kHz)	–	–	[9]
Sphere	$CoFe_2O_4$	PMA	15	80 (r.t.)	316 ± 9 (water) (24 kA/m; 412.5 kHz)	–	–	[9]
core@shell sphere	Fe@Fe_2C/Fe_5C_2	HDA	8.6@4.8	202(2 K)/ 195(300)	258.7 (20 mT/Mesitylene)	–	–	[13]
Endorem®	Fe_3O_4	Dextran	6	–	<0.1 (water; 6.5 KA/m; 300	24 (0.47T)	107 (0.47T)	[99]
Janus spheres	Au@Fe_2C	OLA	12	52.8 (r.t.)		–	210.6 (3T)	[23]
Cube	Fe_3O_4	PMAO	20	80 (5 K)	(25 mT; 520 kHz)	–	–	[11]
Cube	$Co_{0.5}Fe_{2.5}O_4$	PMAO	20	60 (298 K)	430 (water, 32 kA/m; 105 kHz)	–	950 (0.5 T)	[8]
Spherical NP aggregates	$MnFe_2O_4$	PAA	20	28 (310 K)	110 (water, 24 KA/m; 105 kHz)	27.5 (0.5 T)	571 (0.5 T)	[15]
Raspberry-like Cluster (CLIO)	γ-Fe_2O_3	PEG	13 (100)	80	–	–	450 (1.4 T)	[138]

(continued)

Table 1 (continued)

Shape	Composition	Coating	d_c [nm]	M_S [emu/g]	SAR [$W/g_{magnetic}$]	r_1 [$mM^{-1}s^{-1}$]	r_2 [$mM^{-1}s^{-1}$]	References
Sphere (Zn-doped)	$Zn_{0.4}Mn_{0.6}Fe_2O_4$	DMSA	15	170(300)	432 (water, 3.7 KA/m; 500 kHz)	–	860 (4.5 T)	[139]
Sphere (Mn-MEIO)	$MnFe_2O_4$	DMSA	12	110	–	–	358 (1.5 T)	[109]
Cube (FION)	Fe_3O_4	PEG-phospholipids	42 (378)	100 (r.t.)	–	–	343 (3 T)	[140]
Sphere (WSION)	Fe_3O_4	DMSA	9	100	–	–	130 (1.5 T)	[86]
Sphere (ESION)	γ-Fe_2O_4	PEG-bisphosphonate	2.2	50 (5 K)	–	4.8 (3 T)	17.5 (3 T)	[87]
Nanoplate	Fe_3O_4	DMSA	30 × 5	57.6 (5 K)	–	43.2 (0.5 T)	182.2 (0.5 T)	[128]
Sphere (WSION)	^{64}Cu-NOTA-SPIO	PEG/NOTA	11	–	–	–	101.9 (4.7)	[124]
Sphere	Cy5.5-CLIO	Epichlorohydrin ammonia	32	–	–	22 (8.5T)	72 (8.5 T)	[121]
Sphere	NaYF4:Yb^{3+}, Tm^{3+}@Fe_xO_y	Dopamine	25@5	12	–	–	190 (3T)	[122]

d_c = inorganic core diameter, @ = indicates the thickness of each part in core@shell structures, the numbers in parenthesis indicate the size of aggregates built by smaller NPs; x = indicates the different dimensions in anisotropic NPs; M_s = magnetic saturation; SAR = specific absorption rate; OA = oleic acid; HDA = hexadecylamine; PMA = poly(maleic-alt-anhydride)-grafted-dodecylamine; OLA = oleylamine; PMAO = poly(maleic anhydride)-1-octadecene; PAA = poly(acrylic acid); PEG = polyethyleneglycol; DMSA = dimercaptosuccinic acid; NOTA = 1,4,7-triazacyclononane-*N*, *N*, *N*-triacetic acid

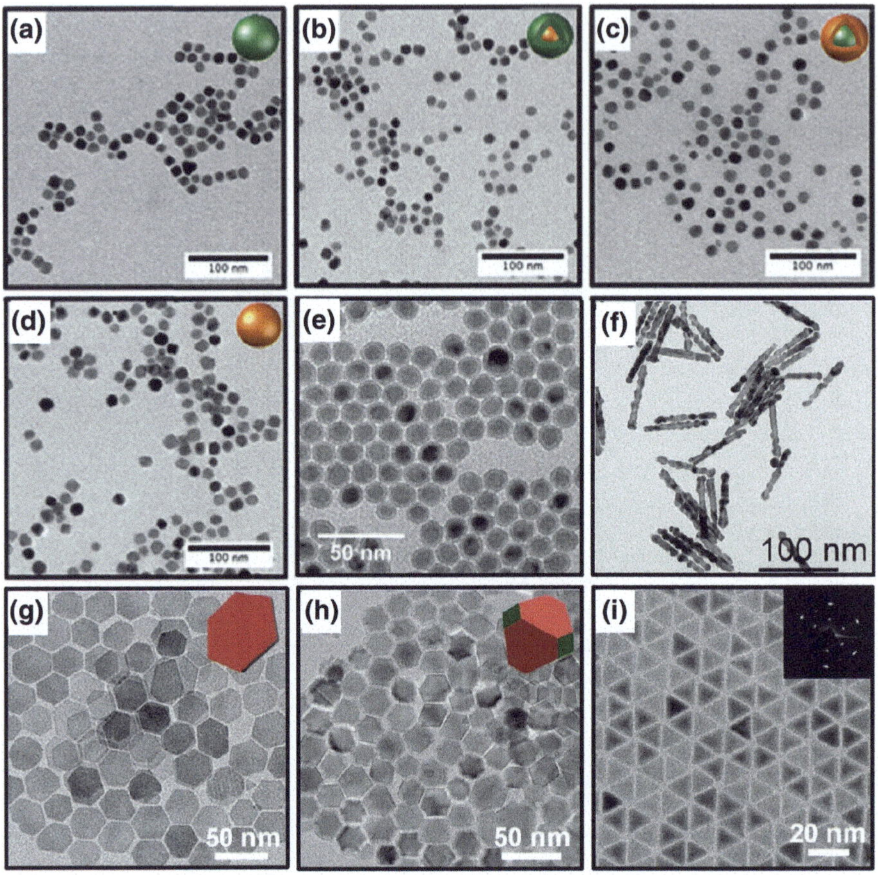

Fig. 2 Transmission electron micrographs of different MNPs **a** ca. 14 nm $MnFe_2O_4$; **b** ca. 14 nm $CoFe_2O_4$@$MnFe_2O_4$ core@shell; **c** ca. 14 nm $MnFe_2O_4$@$CoFe_2O_4$ core@shell; **d** ca. 14 nm $CoFe_2O_4$ [9]; **e** Iron carbide NPs [13]; **f** Co nanorods coated with Sn, Pt, and Au [22]; anisotropic magnetite NPs; **g** hexagonal plates; **h** truncated cubes; and **i** tetrahedra [12]

preferred against just iron (0) NPs due to their better stability to corrosion provided by the presence of carbon in the outer layer.

Not only purely magnetic materials have been used to fight against cancer but also many efforts have been done in seeking highly active materials. Thus, MNPs have been combined with both inorganic and organic materials to get interesting nanostructures and more efficient and versatile nanoplatforms. For example, the formation of controlled aggregates of MNPs using different polymers has been used to enhance their magnetic response, being a good strategy for the development of responsive materials. With this approach is possible not only getting higher saturation magnetization while preserving the superparamagnetic nature of the individual NPs but also generating new optomagnetic features [15]. The controlled aggregates formation can also be an interesting approach in order to understand and mimic what happens inside lysosomes after the NPs' internalization inside cells. After endocytosis, NPs are tight closely, and therefore strong interactions among them will occur [16]. Depending on the nature of the polymers used, this NP's arrangements can respond to external stimuli such as pH changes [17]. The formation of NPs' aggregates can also be done using mesoporous silica NPs as templates to deposit MNPs (e.g., ca. 8 nm Fe_3O_4 NPs to form ca. 70 nm aggregates) [18].

The controlled depositions in polymeric capsules using the layer-by-layer technique allow for the combination of MNPs in bigger structures (nano- and micrometric). Gathering together NPs can help to greatly increase the magnetic responsiveness to external magnetic gradients in order to extract or accumulate locally the capsules, which might be filled with a drug [19].

The preparation of magnetic liposomes has been indeed one of the most successful strategies to develop magnetic drug delivery systems (DDSs). This might be because liposomes are the leading DDSs for clinical translation. Currently, there are some liposomes (not magnetic), which are approved by the FDA to treat cancer. However, all of them have some limitations such as low capacity for lipophilic drugs, leaking and stability issues, and the fact that their preparation involves the use of organic solvents, which might impair the bioactivity of sensitive molecules such as enzymes [1, 20].

On the other hand, many highly magnetic materials lack biocompatibility. This fact can be minimized coating them with noble materials such as gold, creating core@shell nanocomposites. This strategy has been widely employed to coat iron oxide [21] NPs or cobalt nanorods (c.f. Fig. 2f) [22]. The Au coating provides the MNPs with resistance to air and water, preventing their corrosion, and therefore the potential leakage of toxic ions [22]. Moreover, Au has remarkable optical properties which combined with the magnetic properties of the cores can yield an extremely versatile platform for theranostic applications.

In the same direction of combining properties of different nanomaterials, examples of combinations of MNPs with semiconductor NPs to produce fluorescent and MNPs have been reported. Currently, the combination of materials in the form of Janus NPs, in which both materials are located in opposed positions of the NPs, is gaining importance due to a better control in the synthetic processes [23]. Another additional strategy to protect MNPs and help for the NP stabilization relies on coating with a silica shell, [24] providing, at the same time, potential anchoring points for subsequent functionalization steps.

2 General Considerations to Design Suitable MNPs for Nanomedicine

The ideal NP for nanomedicine applications has to fulfill several requirements [25]: (i) to be colloidal stable in physiological media (e.g., water media with high ionic

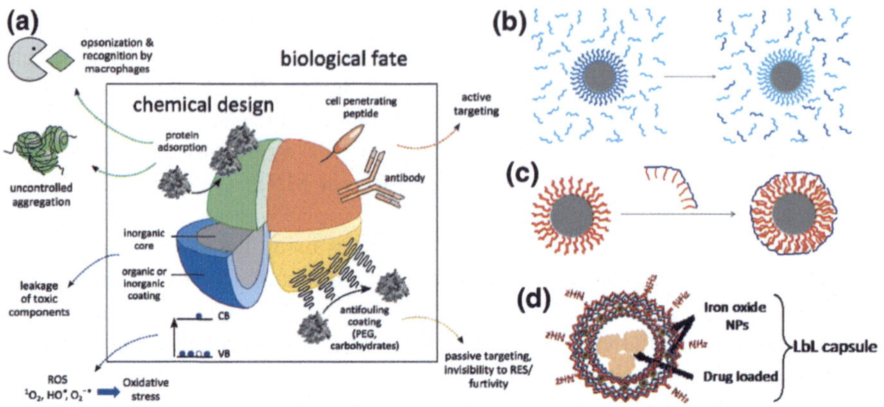

Fig. 3 **a** Different scenarios to represent the possible biological fates of inorganic NPs [25]. **b** Scheme representing the ligand exchange process on an NP surface [7], **c** the polymer coating of NPs [7], and **d** the structure of polymeric capsules with MNPs soaked on the polymeric walls

strength, high concentrations of proteins, and other biomolecules); (ii) to keep their magnetic capabilities or keep them unaltered along time in these media; (iii) to be nontoxic, in general this refers to avoid the leakage of toxic ions; and (iv) to be active to achieve an effective targeting (ideally active targeting). Therefore, this ideal nanovehicle will be a hybrid system composed of an organic and an inorganic domain.

Since the application of nanomaterials is to face medical problems, scientists have realized that many parameters are uncontrolled when NPs enter into contact with cells and living organisms. This is the main reason to explain why the expectations arisen by the nanomedicine field have not been accomplished yet [1, 26]. In the case of MNPs, the nature and quality of the organic layer are even more crucial to prevent NPs aggregation once an external magnetic gradient is applied. This aggregation induced by the NPs' magnetic moments is very dramatic for ferromagnetic NPs, which exhibit a permanent magnetic moment. This is why the use of these NPs is typically avoided. However, these MNPs are among the best candidates for some applications, as it will be discussed later.

Once NPs are immersed in biological media, proteins will be non-specifically adsorbed on top of them modifying their surface properties. If their colloidal stability is preserved, the capabilities provided by the inorganic core (magnetic in this case) will remain, although they might be affected by this adsorption due to changes in their hydrodynamic radius (c.f. Fig. 3a) [25]. Yet, the effect of this non-specific adsorption is not only limited to the colloidal stability, as what cells will "see" will be determined by this protein corona. Recently, the importance about the orientation of each particular protein on the NP surface and which epitopes are exposed to the media, and therefore to the recognition entities or cells (in vivo), is being largely explored [27].

As mentioned previously, best quality NPs are obtained using thermal decomposition methods; however, the resulting NPs are only stable in organic solvents. Therefore, engineering their surface to obtain "water-soluble" (i.e., colloidally stable) NPs is needed. In any case, the surfactants used typically in the synthesis do not provide the NPs with the best colloidal stability and do not allow for their further modification with molecules of biological relevance, e.g., drugs, antibodies, peptides, carbohydrates, etc.

One of the most general approaches to transfer the organic-soluble NPs to water media is the polymer coating technique. This methodology consists of the formation of a micelle surrounding the NP's core. The advantages of this method are several, for example, it can be applied for NPs of any material, shape, or size (c.f. Fig. 3c) [28, 29]. The only requirement is that the NP surface is coated with aliphatic chains, e.g., oleic acid or oleylamine, which are normally used as surfactants during the thermal decomposition. As the second advantage of this methodology, a variety of amphiphilic polymers can be used yielding NPs with different net charges or even zwitterionic NPs [30]. Finally, depending on the polymer nature, easy functionalization can be done either in the polymer before the NP coating process or once the NPs are transferred to water.

Another alternative to enhance the stability of the NPs is the ligand exchange, which consists of the ligand replacement by others that interact stronger with the NPs (c.f. Fig. 3b) [7]. To obtain the stability of iron oxide NPs, typically ligands containing several coordination groups are preferred such as dimercaptosuccinic acid (DMSA) [31, 32] or catechols (e.g., dopamine-based anchoring chains) [33].

Another strategy to water transfer NPs is to encapsulate them into an organic NP. This process can be used if the synthesis of these organic NPs takes place in a solvent in which the inorganic NPs are soluble [34]. If this is not possible, the application of one of the abovementioned processes can be first used to provide the inorganic NPs with suitable organic/aqueous solubility and afterward be trapped in between polymeric layers of, for example, layer-by-layer prepared polymeric capsules (c.f. Fig. 3d) [19].

A current problem to normalize the experiments carried out with different nanomaterials is the metrics used to appropriately compare the doses. This issue has been proven to be critical when comparing the effect of free and conjugated drugs as well as when comparing the effects of different nanomaterials [35, 36]. This is a general issue for any kind of nanomaterial that obviously affects to MNPs. When comparing the interaction in vitro or in vivo of different carriers based on NPs, the specific physical features of the NPs have to be considered. Typically, the dose has been presented by mass per volume (e.g., mg mL^{-1}). This is not a good metric when comparing, for example, NPs with the same inorganic core and different organic shells. As described above, inorganic NPs are actually built from organic and inorganic domains. This factor effect becomes more important when decreasing the NP size. For example, in NP of ca. 5 nm, the weight percentage provided by the organic layer is in the same order of magnitude than the inorganic core, yielding differences in NPs concentration even up to a 50% [35]. To solve this problem, a good solution is to determine the molecular weight of the NPs and convert the mass concentration into molar concentration or number of NPs per unit of volume. This is a good metric to understand the effect of individual NPs (c.f. Fig. 4). However, to calculate precisely, the molecular weight of some NPs is not always straightforward, especially when the shape becomes more exotic, e.g., branched NPs. The role of the surface is another factor to consider since the atoms located on the surface are more reactive. Therefore, if the NPs are not properly stabilized or they are metabolized by cells or living organisms [37]; these atoms will be more reactive with their surroundings.

As already discussed above, some iron oxide based drugs are approved by the FDA, and they are known to be degraded by living organisms. It was demonstrated that iron oxide NPs are degraded by heme oxygenase-1 to form blood hemoglobin [38]. A long-term (a year) study about the degradation of NPs in vivo (e.g., mice model) reported by Kolosnjaj-Tabi et al. [39] demonstrated that eventually the NPs are excreted from the body. In this study, they prepare hybrid NPs, with a gold core (ca. 5 nm) and a shell of magnetite with an overall NP size of ca. 13 nm. Two different coatings were selected, i.e., an amphiphilic polymer and a catechol-modified PEG. Results showed that the iron oxide layer was first degraded, and iron storage proteins were found in the vicinity of the NPs, suggesting their involvement in the NPs degradation. Gold NPs were degraded more slowly; however, in this timescale, degradation of these NPs was also observed. Finally, the degradation speed was affected by the surface coating of the NPs, confirming the important role of this parameter. Mazuel et al. [40] confirmed this fast degradation of iron oxide NPs using stem cells composing a tissue-like structure. They observed the degradation of the MNPs within 1 month. Remarkably, this degradation occurred without affecting the cellular iron homeostasis. Recently, Chiarelly et al. [41] studied the biodistribution of PEGylated iron oxide NPs (ca. 12 nm) in macaques and mice, finding that the distribution was similar in a majority of organs (e.g., liver, spleen, and muscle) and

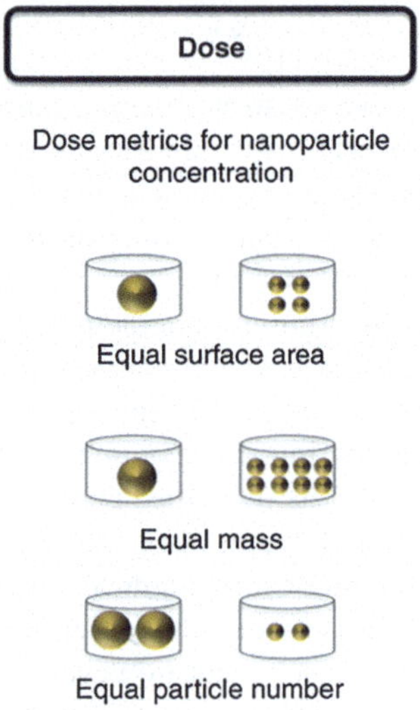

Fig. 4 Graphical representation of different metrics to compare the dose (from Ref. [35])

blood, and different in others such as kidneys, bone marrow, and brain. These results highlight the importance of the animal model to ensure the development of more efficient translational medicine.

3 Cancer Therapy: Drug Carriers; NPs Accumulation by External Magnets; Hyperthermia, Magnetogenetics

When the application of nanomaterials to treat diseases was envisaged, the major goal was the creation of ideal carriers that will specifically target the region of interest (ROI) aiming to reduce the drug dose required. This would reduce the off-target effects because only the "sick" cells would be treated. Therefore, the appearance of multidrug resistance (MDR) cells or tumors will be highly reduced. Currently, the use of nanomaterials is one of the stronger candidates to develop personalized therapies.

3.1 MNPs as Drug Carriers

The idea of highly selective drug nanocarriers has so far, however, failed. Revisiting which percentage of the dose was able to reach a solid tumor in the experimental works carried out during the last 10 years, in average only a ca. 0.7% of the NPs were accumulated in the tumor [42]. Despite this very low accumulation in the target tissue using the current strategies, researchers do not agree about whether this small accumulation would be enough to be effective. Probably, this dose would be enough to eliminate small tumors because the localized dose would be high enough. However, this dose would not be sufficient to treat bigger tumors [43]. Another important aspect to consider about these numbers is the fact that in many of these works an active targeting of the tumor was foreseen. Therefore, the current question is if active targeting really works in vivo and what is the real relevance of the enhanced permeability and retention (EPR) effect, which is normally used to promote the tumor targeting [43]. These are some of the current challenges that nanodrugs are facing.

In particular, for magnetic-based nanodrugs, the iron oxide drugs approved by the FDA are devoted for treating anemia related to chronic cardiac disease (Venofer®, Feraheme®, INFed®, Dexferrum®, and Ferrlecit®) or to be used as contrast agents (Feraheme®, Feridex®, and GastroMARK®). Currently, many drugs based on iron oxide particles are used in clinical trials such as Nanotherm™, which is currently in the late stage of clinical trials in the US to treat glioblastoma [44].

The most straightforward application of these NPs is their application as drug delivery carriers. Recently, the use of ferumoxytol to prevent tumor growth has been reported. It is known that NPs are internalized by tumor-associated macrophages (TAMs) [45]. Elemental iron plays an important role, as iron catalyzes the production of reactive oxidative species (ROS) through the Fenton reaction [46, 47]. This Fenton reaction is enhanced by the production of hydroxyl radicals by pro-inflammatory M1 macrophages. Daldrup-Link et al. described the increment of oxidative species in co-cultures of cancer cells and macrophages in the presence of ferumoxytol compared with the co-cultures alone. This leads to a higher mortality of the cancer cells, which under the light of their experiments was tightly related to M1 macrophages polarization. A complementary behavior was observed in vivo, with xenograft mammary tumors in female mice; in the mice treated with ferumoxytol, the tumor volume was reduced even to the half. A similar behavior was observed to prevent liver and lung metastases. These findings open the "off-label" therapeutic possibilities of a currently FDA-approved drug [48].

Recently, MacParland et al. demonstrated that using gold NPs as a model, M2 polarized macrophages, which are related to the cancer-associated inflammation, exhibited a preferential NP uptake [49]. These results are of interest when designing chemotherapeutic drugs for antitumor strategies. Considering their results, the presence of M2 macrophages in the tumoral microenvironment would decrease the capabilities of the drugs, due to their uptake by the macrophages before they have the chance to interact with the tumoral region. So, to consider the re-education of macrophages to an M1 phenotype would be a good strategy to increase the benefits of a high presence of M1 macrophages, while the nanodrugs would have more chances to reach and thereby treat the tumor [26].

3.2 MNPs for Magnetofection and Magnetic Accumulation

The use of MNPs as carriers has the advantage of the potential possibility of the application of an external gradient field to promote the spatial accumulation of the MNPs, which finally will lead to the NP's internalization by the process typically known as magnetofection. Magnetofection is currently used to transfect cells, and it is commercially available by several companies (e.g., Chemicell, Ozbiosciences) as an alternative to traditional transfection methods. This approach has several advantages such as the great reduction of the vector dose required, the shortening of the incubation time in order to achieve a high transfection efficiency, and the capability of transfection to non-permissive cells [50]. In the literature, many applications of magnetofection in vitro, in vivo, and ex vivo with different purposes have been described [51]. For example, lipospheres containing small interfering RNA (siRNA) were prepared by del Pino et al. as nonviral vectors to transfect cells. The lipospheres were internalized by the effect of an external magnetic field gradient even under flow conditions mimicking the bloodstream, and the siRNA was released into the cells [52]. The combination of cell-penetrating peptides (CPPs) and magnetofection to enhance the internalization of the NPs upon the exposure of an external gradient magnetic fields has been also reported in 2D [53] and 3D cell cultures [54] (Fig. 5).

The use of magnetofection in vivo has also found in commercial applications (e.g., Ozbiosciences) to target different organs such as brain, heart, or liver, as well as to target tumors. This technique is based on the localization of a magnet close to the treatment region; in some cases, this magnet needs to be implanted. The administration of the MNPs can be systemic or local, e.g., direct intratumoral administration [51]. A recent application of the magnetofection in vivo developed by Soto-Sánchez et al. consists of the transfection of DNA plasmids to promote the sensibilization of neurons ion channels to be controlled by light stimuli (e.g., optogenetic). The use of optogenetics allows activating selectively specific parts of the brain by light. The use of magnetofection compared with the transfection using viral vectors provides higher transfection rates without the drawbacks of viral vectors [55].

3.3 Magnetic Fluid Hyperthermia (MFH)

The use of alternating magnetic fields to induce local temperature increments is another major therapeutic application of MNPs. In general, hyperthermia refers to the treatment in which the target tissue's temperature is raised to temperatures in which it is destroyed (above 47 °C, thermal ablation) or the sensitivity of the cells in the tissue is increased for other treatments such as radiotherapy or chemotherapy (41–45 °C, thermal sensitization) [56]. The sensitization of the treated tissues normally is provoked by the increase of blood flow due to the enhanced vascular permeability induced by temperature, which finally leads to a decrease of hypoxia in the region and has been proven to promote the cytotoxic effects of many anticancer drugs [57]. This is also a consequence of the fact that tumoral cells are less capable to respond to increases of temperature [58]. At cellular level, the increment of temperature produces changes in protein folding, enzymatic activity, and cell growth and differentiation [59, 60]. Therefore, the combination of hyperthermia and

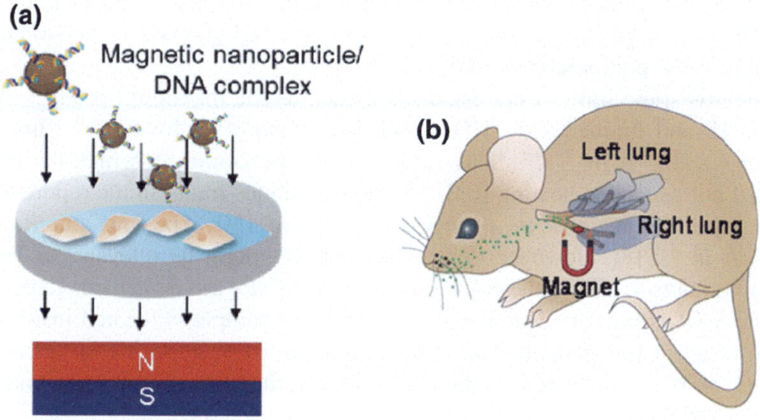

Fig. 5 Graphical representation of **a** in vitro and **b** in vivo magnetofection Nature Nanotechnology 2, 467–468 (2007). Adapted from Ref. [76]

radio or chemotherapy has been proven to be beneficial for patients by enhancing the survival rates in many types of cancers (e.g., cervix, bladder, brain, etc.) [56, 61]. The process of increasing temperature in a tissue can be achieved using an "out-inside" or "inside-out" approach; this is depending on where the heat source is located. In the case of nanomaterials which are either accumulated or injected in the tumor, the heat production is due to the NP heating which occurs inside the tumors (e.g., inside-out) [56]. The "out-inside" hyperthermia produced by radiofrequency, microwave, or ultrasound methods have several drawbacks such as the inhomogeneous heat/temperature distribution [62], which might create hot spots leading to healthy cell damage [56]. To obtain the required therapeutic temperatures in deep tumors is also a problem using this approach. The use of "inside-out" magnetic hyperthermia could also help to overcome this issue. The use of MNPs for hyperthermia typically is known as MFH and it is based on the capability of MNPs to absorb energy from an alternating magnetic field and convert it into heat.

In general, superparamagnetic NPs are preferred for MFH mainly because in the absence of an external magnetic field the NPs have not a magnetic moment. Briefly, depending on their physicochemical properties, MNPs will respond to an alternating external magnetic field by absorbing energy that will be converted by magnetic inversion to heat. The amount of heat produced by energy absorbed is expressed as the specific absorption rate (SAR), also known as specific loss power (SLP). SAR is defined as the absorbed power by mass of MNPs, under an AMF of a fixed frequency and intensity (Eq. 1) [10].

$$\mathrm{SAR} = \frac{\text{Absorbed power}}{\text{Mass of NPs}} \tag{1}$$

SAR is a magnitude which depends on many factors such as (i) the MNPs concentration which might be expressed related to the Fe (or magnetic) content as determined by ICP-MS, for example; or related to the total mass content of the NPs, that is, the magnetic content plus the oxygen contribution; (ii) the frequency and the intensity of the AMF; (iii) the physicochemical properties of the NPs; (iv) the dispersion media; and (v) the NPs' agglomeration state.

Clinically, some limits were established such as the maximum product of field strength, H, and frequency, f, (Hf), which Brezovich et al. found to be limited to $Hf < 4.85 \times 10^9$ A m^{-1} s^{-1}. This value was the maximum in which the patients exhibited not exceeded heating [63]. The typical values found in the reported studies employed frequencies in a range from 100 to 150 kHz, and amplitudes of 10–30 kA m^{-1} [38]. However, other combinations might be relevant. In Table 1, the conditions to measure the SAR for different MNPs are described. Experimentally, the SAR is measured in a setup in which the sample is located inside a coil Fig. 6a, which is refrigerated. The sample holder must be adiabatic to facilitate the calculations. If the setup fulfills the requirements, then the following formula may be applied [9, 64]:

$$SAR = \frac{C_{media} \cdot \rho_{media}}{\rho_{NPs}} \frac{dT}{dt} \tag{2}$$

$$SAR = \frac{C_{media} \cdot V}{m} \frac{dT}{dt} \tag{3}$$

where C_{media} corresponds to the heat capacity of the media where the sample is immersed (e.g., water = 4.2 J/g C), ρ_{media} refers to the media density (e.g., water = 1 g/cm^3), ρ_{NPs} is the density of the inorganic part of the MNPs in g/cm^3, and dT/dt is the slope of the curve obtained in the first 5–10 s of heating. V refers to the sample volume, and m to the mass of iron (or magnetic materials) in the sample.

Considering these equations, it is easy to understand why the dispersion media is affecting the SAR of the NPs. At this respect, Di Corato et al. studied the influence of the cellular location. To do so, they selected different types of MNPs synthesized by different routes and analyzed their magnetic properties. They found out that, in general, the MNPs lose heating capacity upon cellular internalization. This is explained by the physical restraint of the NPs inside the cells [64]. This is in line with other manuscripts in which the MNPs were physically restrained [9].

To elucidate the actual temperature that the MNPs are capable to produce upon excitation with an AMF, many approaches have been done. One of them, reported by Dias et al. [65], modified MNPs with oligonucleotides with different lengths in order to measure what was the release efficiency of each of these oligonucleotides upon the exposure to the AMF. The observed results indicated that the MNPs acted indeed as nanoheaters provoking the release of the complementary oligonucleotide strands even when the global temperature was 10 °C below the melting temperature. Using this methodology, the specific increments of temperature at a specific distance of the NP surface were determined. However, some questions about the release mechanisms remain unclear. These experiments were performed using NPs in solution, and the results were in line with others obtained by Riedinger et al. who used temperature-sensitive azo bonds to evaluate the local temperature increases [66].

Despite these issues, the use of MFH has been highly reported in vitro, in vivo, and ex vivo with promising results. One example is the efforts that have been done

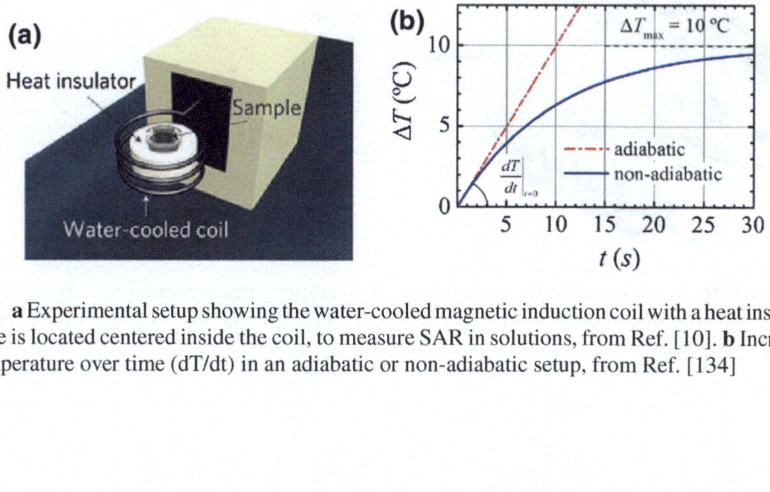

Fig. 6 **a** Experimental setup showing the water-cooled magnetic induction coil with a heat insulator; sample is located centered inside the coil, to measure SAR in solutions, from Ref. [10]. **b** Increment of temperature over time (dT/dt) in an adiabatic or non-adiabatic setup, from Ref. [134]

for glioma treatment [59]. Glioblastoma is the most malignant and frequent glioma. Although the incidence of the cancer originated by glial cells that are in the central nervous system is relatively low (ca. 1.79%), they are related to a high mortality and morbidity [67]. A review by Silva et al. published in 2011 revised the results obtained for glioma treatment obtained in 15 articles. For these experiments, in general, magnetite liposomes and superparamagnetic iron oxide NPs (SPIONs) coated with aminosilane were used as MNPs, and the experiments compiled data from in vitro, animal model studies with the implantation of tumors, and two clinical studies [67]. Jordan started the clinical studies of MFH for the gliomablastomas and prostate carcinoma [68]. In EU, Magforce (http://www.magforce.de/en/unternehmen/ueber-uns.html) has received the approval for treating patients with glioma. This company born in Germany is currently carrying out other clinical trials (e.g., prostate cancer), and six hospitals of Germany are offering their MFH treatment based on ca. 15 nm core iron oxide NPs.

Using more complex magnetic materials such as $CoFe_2O_4$@$MnFe_2O_4$ core@shell structures with a very high SAR value (ca. 2280 W g^{-1}), the remission of tumors in vivo has been described by thermal ablation [10] (c.f. Fig. 7a and b).

Despite these good results, any single treatment for cancer has been proven to be inefficient to eliminate the whole tumor, and especially to fight against metastasis. Therefore, combined therapy employing two or more forms of treatment modalities is being explored. This type of multimodal synergistic therapy consists of the cooperation of different treatments, which are integrated into a single nanoplatform [69]. For example, Kolosnjaj-Tabi et al. studied the effect of PEGylated iron oxide nanocubes (ca. 19 nm in diameter) loaded with doxorubicin to induce tumoral destruction (e.g., epidermoid carcinoma). They found out that after the direct intratumoral injection of the nanocubes and the application of MFH, the NPs were responsible for the tumoral

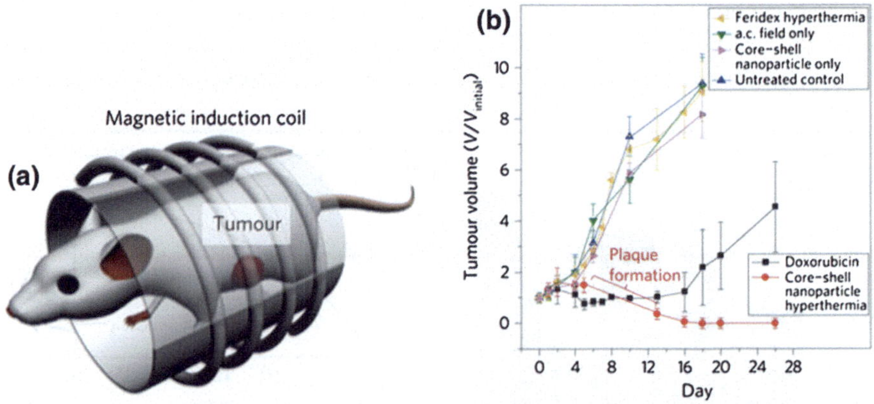

Fig. 7 **a** MFH setup for mice and **b** tumor volume of mice after different treatments, only AMF, after the treatment with doxorubicin, with only core@shell NPs and with the NPs and AMF treatment. Adapted from Ref. [10]

microenvironment destruction, which decreased the tumor growth and increased the sensitivity of the tumor to antitumoral drugs (e.g., doxorubicin) [70]. The combination of chemotherapy and MFH is currently providing very promising results. However, there is room for improvement, and more basic studies to learn about how these materials are affecting the tumoral microenvironment would help to improve the efficiency of this cancer therapy. Recently, Espinosa et al. [71] have explored the potential duality of iron oxide nanocubes to act as heating inductors excited by both and AMF and light. This might be the beginning of a dual hyperthermia therapy that would allow to minimize possible collateral damages. Similar results have been obtained using more complex systems, such as the formation of supramolecular MNPs loaded with doxorubicin. These NPs were formed by the agglomeration of 6 nm $Zn_{0.4}Fe_{2.6}O_4$ NPs in aggregates of ca. 160 nm containing doxorubicin. These assemblies showed exceptional heating capabilities upon AMF excitation, destroying malignant cells and slowing down in a very efficient manner the tumoral growing effect which was improved by the controlled release of doxorubicin [72].

Magnetogenetics is another recent, very promising, biomedical approach based on heating by MNPs [73, 74]. For instance, deciphering how neurocircuit activation causes particular behaviors requires modulating the activity of specific neurons. Pralle et al. demonstrated that magnetothermal genetic stimulation provides tetherless deep brain activation sufficient to evoke motor behavior in awake mice [73]. In this study, alternating magnetic fields were used to heat MNPs anchored on the neuronal membrane. Neurons, heat-sensitized by expressing thermosensitive channels, were locally activated with magnetic field application. This approach provides genetically and spatially targetable, repeatable, and temporarily precise activation of deep brain circuits without the need for surgical implantation of any device.

4 Imaging: Magnetic Resonance Imaging/Radiolabeling

MRI is the main anatomic tool used in clinical diagnosis including in cancer diagnosis, capable of reconstruction of atomic nuclear magnetization signal into 2D images. It is a noninvasive and radiation-free technique based on the same principle as *Nuclear Magnetic Resonance*, which presents enhanced soft tissue contrast permitting high spatial resolution [75].

The fundamental principle of MRI consists of acquiring the restored or residual magnetization after a specific time cycle by adjusting parameters in either the longitudinal direction or the transverse plane, showing bright and dark signal contrasts (T_1- and T_2-weighted MR images, respectively). The ^1H nuclei located at the specific ROI of a tissue is detected to construct a cross-sectional MR gray scale image. Under a strong magnetic field (B_0), the hydrogen nuclear spins are oriented in the same directions to the magnetic field (lower energy state) or in opposition to the magnetic field (higher energy state). The nucleus that acquired a higher energy state after being irradiated by a resonant radio frequency (1.5 and 3 T are typically used in clinics) return to the equilibrium state in a process referred as relaxation. Two different types of relaxation mechanisms can occur, longitudinal relaxation (spin–lattice relaxation) and transverse relaxation (spin–spin relaxation) [76]. The longitudinal relaxation time—T_1—represents the time required for longitudinal magnetization to recover (from zero to a value of 63% of the original state). Conversely, the transverse relaxation time—T_2—represents the time for the transverse magnetization to decrease from the maximum to a 37% of its excited state value (c.f. Fig. 8) [77]. In this manner, the faster the T_1 relaxation, the brighter the T_1 image; and the faster the T_2 relaxation, the darker the T_2 image; and vice versa. In general, T_1-weighted MRI is useful for assessing fat tissue or liquid retaining structures, while T_2-weighted MRI is suitable for assessing water-rich structures or local inflammation (c.f. Fig. 9).

The main drawback of MRI is that not always the contrast generated is enough for an accurate diagnosis. Despite the proton density of a tissue is fixed, modifying the relaxation time by administering a contrast agent can alter the magnetic characteristics of nearby water protons. The signal enhancement produced by a contrast agent depends on its longitudinal or transverse relaxivities (r_1 and r_2, respectively), which are defined as the increase of relaxation rates $R_1 = 1/T_1$ and $R_2 = 1/T_2$, respectively. A contrast agent can be categorized as T_1 or T_2 contrast agent based on their relaxation characteristics. Positive (or T_1) contrast agents increase the rate of longitudinal recovery thereby shortening the T_1 relaxation time and producing

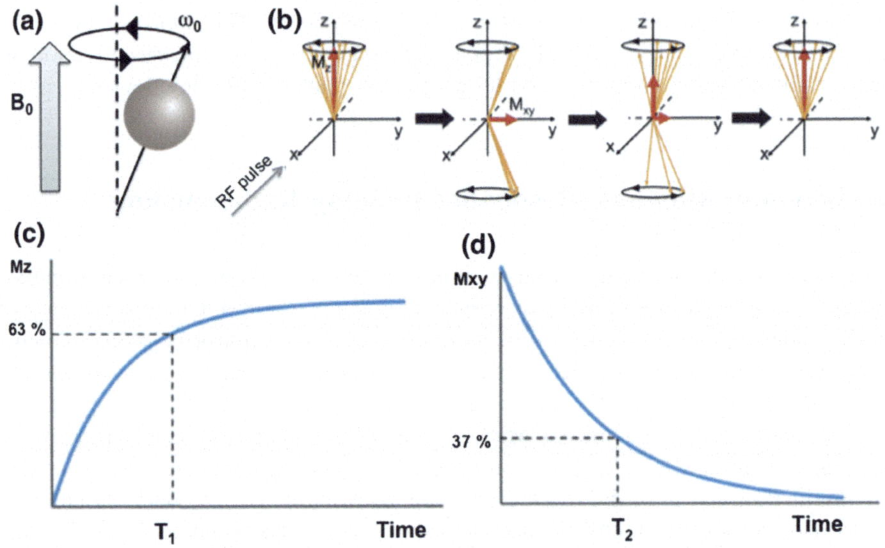

Fig. 8 **a** The hydrogen nuclear spins align parallel or antiparallel to the external magnetic field (B_0). **b** After resonant RF irradiation, the longitudinal magnetization (M_z) decrease and transverse magnetization (M_{xy}) is generated. **c** T_1 is the time required for longitudinal magnetization to recover to 63% of its equilibrium value. **d** T_2 is the time required for transverse magnetization to drop to 37% of its initial magnitude. Adapted from Ref. [77]

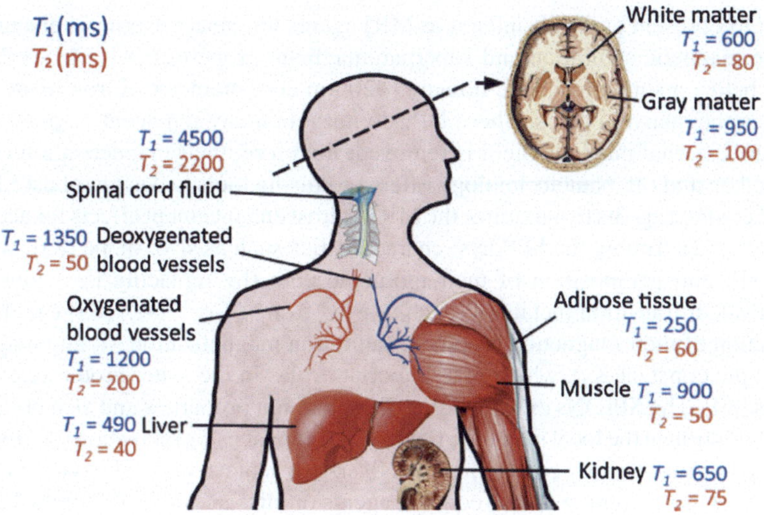

Fig. 9 Relaxation time constants of different tissues in the human body under a magnetic field of 1.5 T. Adapted from Ref. [114]

brighter signals. On the other hand, negative (or T_2) contrast agents increase the rate of transverse magnetization decays and shorten the T_2, generating a signal darkening on MRI. The signal produced by T_2 contrast agents can be confused sometimes with some endogenous conditions such as hemorrhage, air, or blood clots. The field inhomogeneity of the magnetic field led to a new relaxation rate designated as R_2* ($1/T_2$*) which is larger than R_2 value.

The efficacy of a contrast agent is evaluated by the following equation:

$$R_i = \left(1 / T_{i,0}\right) + r_i C \tag{4}$$

where R_i is the observed relaxation rate (s^{-1}); $T_{i,0}$ is the relaxation time before adding the contrast agent (s), r_i is the relaxivity coefficient ($mM^{-1}\ s^{-1}$), and C (mM) is the contrast agent concentration. The relaxivity coefficient (r_i) is the indicator of the contrast enhancement, and it is obtained as the gradient of the plot of R_i *versus* the molarity of the contrast agent [77].

The use of MRI contrast agents (paramagnetic or superparamagnetic metal ions), which influence the MRI signal properties of surrounding tissue, has made a significant impact in the use of MRI for clinical applications [78]. In particular, superparamagnetic iron oxide NPs (SPION) have been extensively studied as contrast agents for MRI in cancer, cardiovascular, and inflammation diseases [79]. They are promising candidates to substitute gadolinium-based agents (a typical SPION agent has a T_2 relaxivity of 100 $mM^{-1}\ s^{-1}$ and a T_1 relaxivity of 30 $mM^{-1}\ s^{-1}$, substantially larger than gadolinium-based contrast agents with T_2 relaxivity of 6 $mM^{-1}\ s^{-1}$ and T_1 relaxivity of 4 $mM^{-1}\ s^{-1}$, respectively) [80].

SPIONs present great advantages as MRI agents for cancer diagnosis due to their suitable magnetic saturation and superparamagnetic properties [81]. As explained above, below a single magnetic domain (~20 nm core diameter of iron oxide NPs) superparamagnetism occurs. These NPs do not retain any remanent magnetization when the external magnetic field is removed; therefore, their magnetization can be switched on and off. Nanotechnology offers promising tools to design robust SPION as MRI contrast agents to maximize the MR contrast enhancement effects for accurate diagnosis [82]. Tuning the SPIONs' characteristics such as size, shape [83], surface state [84], and composition of their inorganic core (by replacing Fe^{2+} ions with other divalent transition metal ions such as Mn^{2+}, Zn^{2+}, Co^{2+}, etc.) [85] enables the modulation of their magnetic properties (saturation magnetization M_s and magnetic anisotropic constant K_a), which play important roles in the water proton relaxation process [80]. The SPIONs exhibit strong T_1 relaxation properties and also produce a strong variation in the local magnetic field, which enhances T_2 relaxation (cf., Fig. 10) [86, 87]. For instance, by controlling the NP size to achieve high relaxivity, SPIONs can be used as biocompatible T_1 contrast agents or ultra-sensitive T_2 contrast agent.

As a consequence of their large size and magnetic moment, most of iron oxide based MRI contrast agents are detected as a signal loss on T_2-weighted MR images [86]. The high magnetic susceptibility of SPIONs enables them to be identified as darker signals on the MRI images being suitable T_2 contrast agents. The T_2 relaxivity for SPIONs is at least 5–10 fold higher than T_1 relaxivity [88]. An increase in the NP size leads to an increase in the T_2 relaxivity, and the r_2/r_1 ratio increases with increasing NP size [86]. Being R_2, the relaxation rate is defined by the following equation:

$$R_2 = \frac{1}{T_2} = \frac{256\pi^2\gamma^2}{405} M_s^2 V x \frac{r^2}{D\left(1 + \frac{L}{r}\right)} \tag{5}$$

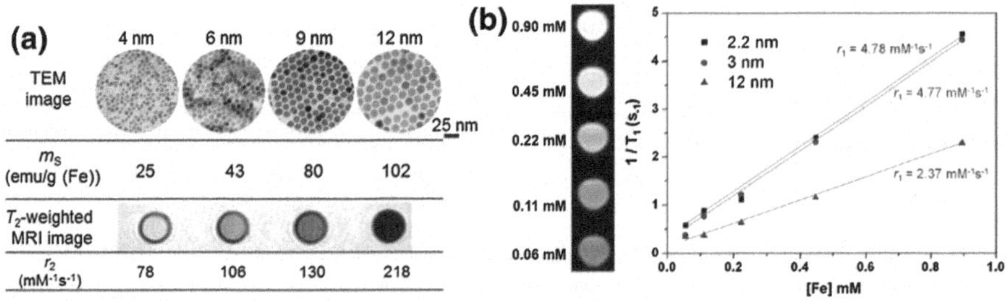

Fig. 10 Size effect on the magnetic properties of SPIONs as MRI contrast agents. **a** As the NP's size increases from 4 nm to 12 nm, M_s and r_2 increase. Adopted from Ref. [86]. **b** The T_1 contrast effects of extremely small SPION sizes: T_1-weighted MR images of 3 nm SPION and graphical representation of $1/T_1$ versus Fe concentration of SPION of 12, 3 and 2.2 nm in diameter. Adopted from Ref. [87]

where γ is the gyromagnetic ratio of the protons, M_s is the saturation magnetization of the NP, V is the NP volume fraction, r is the NP radius, D is the diffusion coefficient of water molecules, and L is the thickness of the surface coating.

Micrometer-sized iron oxide particles (MPIOs, with diameters of several microns), which can be detected by high magnetic field, have been also studied as T_2 contrast agents, being capable of visualizing individual cells [89, 90]. Moreover, other formulations such as monocrystalline iron oxide particles (MIONs), crosslinked iron oxide particles (CLIOs) [91], or assembly of SPION structures (e.g., nanocluster formation, controlled agglomerates of iron oxide NPs with polymers, by silica coating or by micelle encapsulation) [92, 93] can modulate their r_2/r_1 relaxivities to enhance the T_2 signal. However, smaller SPIONs are much better T_1 shortening agents and generate greater contrast enhancement due to the high surface-to-volume ratio. Ultrasmall superparamagnetic iron oxide particles (USPIONs) have also been reported with excellent T_1 enhancing properties [94, 95].

In addition to the size, shape, composition, etc., the design of their surface properties will influence their MNP colloidal stability, biocompatibility, and specific functionality, defining their nanointerface and thus their targeting capabilities and their synergistic effect for dual imaging modalities and therapeutic strategies [84, 96].

4.1 Applications of SPION as Contrast Agents for Cancer Diagnosis

The anatomical and structural differences between healthy and non-healthy tissues enable difference contrast between them [79]. These phenomena can be used in cancer diagnosis by MRI [97]. First generation of SPIO particles approved by the FDA as contrast agents for MRI consisted of big agglomerates of mono or polycrystalline iron oxide cores with diameter of 5–30 nm embedded in a polymer matrix (dextran, PEG, polysaccharide, etc.). They have been successfully applied as contrasts agent to detect tumor lesions in the liver (such as Feridex and Resovist, see Table 2) [98]. Due to their size (hydrodynamic diameter of 300 nm), these particles are taken up by macrophages and rapidly sequestered by the reticuloendothelial (RES) organs such as liver, spleen, etc. Phagocytized SPIO particles by Kupffer cells (macrophage cells of the liver) produce a shortening of the T_2 (due to their large r_2/r_1 ratio) in the liver tissue causing a reduction of the signal intensity. The signal lost is selective since the SPIO particles are not accumulated in the lesions due to the absence of Kupffer cells in the damaged regions, which facilitates the liver tumor detection in early stages and detection of metastases as small as 2–3 mm [99].

Solid tumors have a leaky and damaged vasculature that enables them to trap more SPIONs than healthy tissue, due to the enhanced permeation and retention (EPR) effect [100]. NP-based MR imaging to detect solid cancerous tissue at early stages tumor or metastasis has been reported. For instance, Combidex, dextran-coated

Table 2 Superparamagnetic iron oxide NP-based contrast agents. The r_1 and r_2 values are literature values and can be slightly variable depending on the magnetic field and MR pulse sequences

Name	Type	d_c (nm)	d_h (nm)	Coating	Properties (Relaxivity)	Applications	References
Feraheme (Ferumoxytol)	USPIO	6.7	17–31	Carbohydrate polymer	r_1 7.5 mM^{-1} s^{-1} r_2 92 mM^{-1} s^{-1}	Approved for iron deficiency anemia. Trial for undetectable cancer metastasis of lymph node. MR Angiography	[141]
Feridex US, Endorem EU (Ferumoxide)	SPIO	4.8–5.6	80–150	Dextran	r_1 23 mM^{-1} s^{-1} r_2 100 mM^{-1} s^{-1}	Approved for liver lesions detection. Trial for active cell tracking by MRI	[99]
Resovist (Ferucarbotran)	SPIO	4.2	45–60	Carboxy-dextran	r_1 19.4 mM^{-1} s^{-1} r_2 185.8 mM^{-1} s^{-1}	Approved for MRI of Liver lesions (Liver metastases detection) Organ-specific MRI contrast agent	[98]
Clariscan (PEG-Feron)	USPIO	4–7	20	Carbohydrate-PEG	r_1 3.1 mM^{-1} s^{-1} r_2 3.6 mM^{-1} s^{-1}	In clinical trials. MR angiography agent. Characterization of tumor microvasculature	[142]
Supravist	USPIO	3–5	21	Carboxy-dextran	r_1 7.3 mM^{-1} s^{-1} r_2 57 mM^{-1} s^{-1}	In clinical trials to detect metastatic in lymph nodes	[143]
Ferro pharma	USPIO	8.6	19	Citrate	r_1 8 mM^{-1} s^{-1} r_2 34 mM^{-1} s^{-1}	In clinical trials MR angiography	[94]
Lumirem EU, GastroMark US (Ferumoxsil)	SPIO	–	300	Siloxane	r_1 3.2 mM^{-1} s^{-1} r_2 72 mM^{-1} s^{-1}	To visualize abdominal structures	[144]
Combidex US, Sinerem EU (Ferumoxtran-10)	USPIO	4–6	30	Dextran + Citrate	r_1 10 mM^{-1} s^{-1} r_2 60 mM^{-1} s^{-1}	In clinical trials to detect metastatic in lymph nodes. Brain tumors	[145]

ferumoxtran-10, was employed as contrast agent for prostate cancer metastases detection.

As explained above, NP size can influence the magnetic properties of SPIONs. In addition, by decreasing the size of SPION, increasing T_1 contrast effect can be achieved. Second generation of SPION contrast agents is based on USPION (e.g., Clariscan, Supravist, Table 2) that are capable of escape phagocytosis and circulate longer times, so they can be used for T_1-weighted MRI for magnetic resonance angiography (MRA). Also, their small size enables them to reach the lymphatic system by crossing the capillary wall. USPIONs have potential to detect metastatic involvement of lymph nodes [101]. For instance, it has been reported the capability of Ferumoxtran-10 to detect and identify lymph node metastases in patients with prostate cancer [102].

Tumor acidic microenvironment is a characteristic of malignant tumor progression. Extremely small iron oxide NPs (ESION) with diameter below 3 nm exhibit weak magnetic properties. Their high r_1 and low r_2/r_1 made them suitable candidates for T_1 contrast agents [87]. Combination of two parameters such as size (by using ESION) and coating (using a polymer pH-sensitive) allows modulating their properties as contrast agents. ESION conjugated with a pH-sensitive polymer can target tumors via surface charge and switch triggered by the acidic environment [103]. Change in the pH of the surrounding medium led to change the size-clustering formation and, hence, to a different T_1 or T_2 contrast agent. When the pH decreases, the nanostructures are disassembled into the ESION that turns on T_1 contrast [104].

Finally, a new strategy takes advantages of the SPION phagocytosis by inflammation-related macrophages via endocytosis after intravenous injection [105]. The incorporation of these NPs in phagocytic monocytes and TAMs provides a valuable in vivo tool to locate lesions and predict damage. TAMs are usually associated with fast cancer progression and short patient survival period. MRI can be used to monitor involvement of macrophages in inflammatory processes and implement tumor location and prognostics information. In addition, the macrophage MRI detection with SPIONs of tumor morphology might facilitate the surgical biopsy of the tumor.

All these strategies described above are based on passive targeting, where SPIONs can reach interested regions with high angiogenesis (growing tumors) or large quantities of resident macrophages through the EPR effect. However, NPs should be delivered to specific regions to detect tumors present in organs, which lack phagocytic activity. Surface engineering of SPIONs with selective molecules for active targeting enables recognition of specific receptors highly expressed on tumor cells resulting in high NP accumulation. In addition, the cellular metabolism is altered during cancer growth and malignancy; and therefore, the altered metabolic profile of tumor (e.g., high increase of glucose uptake by tumor cells) can be also used as a target.

Next generation of iron oxide particles for MRI consists of small SPION (as T_2 contrast agent) which are modified with a polymer (such as PEG) [106] to prolong circulation time and conjugated with tumor targeting moieties (antibodies, aptamers, cell-penetrating peptides, carbohydrates, and other small molecules such as folic acid, etc.) [107, 108] in order to give selectivity to the contrast agent, acting as labels for in vivo targeting of tumors. A plethora of tumor biomarkers and corresponding antibodies have been described in the literature. Some biomarkers are specific of certain tumors, and some biomarkers exist in normal tissues, but they are overexpressed in the tumor. The selection of suitable tumor markers is crucial to achieve effective tumor targeting. For instance, using Mn-doped iron oxide NPs functionalized with Herceptin antibody (HER2/neu specific Ab) that can target breast cancer cells [which overexpress human epidermal growth factor receptor 2 (HER2/neu)] allows for in vivo detection of small tumors with high MR contrast effect [109, 110]. Aminosilane-coated SPION conjugated with EpCAM aptamers or USPIONs decorated with antibodies anti-α fetoprotein and anti-glypican 3 have been reported as successful examples of targeting of hepatocellular carcinoma cells. Other molecules such as prostate-specific membrane antigen, PSMA (expressed by the vascular endothelium of all solid sarcomas and carcinomas), and prostate stem cell antigen, PSCA, are common targets for prostate cancer detection. The utilization of targeted MRI contrast agents can help to improve the rate of cancer diagnosis in its earliest stages [111].

These tumor-selective targeting contrast agents are capable of real-time tracking of tumor angiogenesis and growth. The high number of NP accumulated in the target tissue allows substantial enhancing of MRI signal. Targeted SPIONs also can help to detect among different stages of cancer such as (i) detection of receptors overexpressed on the cancer cell surface, (ii) detection of unusual angiogenesis in the tumor microenvironment, (iii) detection of circulating tumor cells (CTCs); and (iv) detection of soluble tumor markers.

SPION-based MRI contrast agents not only provide an excellent imaging system for cancer but also can be employed for monitoring cell therapy and drug delivery [112]. They provide enough negative contrast for in vitro and in vivo cellular imaging, allowing for early discovery of diseases at the molecular and cellular levels. SPIONs and MRI have been described recently as a promising tool to perform long-term noninvasive imaging of transplanted cells in vivo to monitor their survival, migration, differentiation, and impact [113]. SPION can follow two different strategies for stem cell imaging and tracking. The first one consists of modifying SPIONs surface with a targeting ligand that recognizes a motif of the cellular surface. The second strategy is based on the SPION cell internalization. Some FDA-approved SPION formulations (such as Resovist, Feridex, Endorem) have been also studied as MRI contrast agents for stem cell labeling [114].

Although many SPION-based products approved by FDA are no longer used in clinics, many studies report their synergetic effect in combination with other molecules for multimodal imaging [115].

4.2 Multimodal Imaging

MRI has demonstrated great potential to obtain anatomic resolution and contrast; nevertheless, a single imaging technique cannot provide complete information. Moreover, the elimination of artifacts from MR images is crucial for a precise interpretation and accurate diagnosis. Different information about a specific ROI can be obtained by combination of several diagnostic-imaging techniques [116, 117]. Combination of MRI with other modalities such as optical fluorescence, computed tomography (CT), and positron emission tomography (PET), which provide functional information with high sensitivity for anatomical information, has shown enormous potential for accurate cancer diagnosis [115]. MNPs can be used as versatile platforms to conjugate with another imaging components (such as fluorescent molecules and radioisotopes) in order to generate signals in different imaging modalities (such as PET or optical MRI). Their inherent magnetic properties can also be used as a source of multimodal imaging signal as, for instance, in magnetic particle imaging (MPI), magneto-photoacoustic imaging (MPA), and magneto-motive ultrasound imaging (MMUS) (cf. Fig. 11). Multimodal SPION-based probes for dual T_1–T_2 MR imaging, radionuclide–MR imaging, CT–MR imaging, optical–MR imaging, and ultrasound–MR imaging are described in detail in the following reviews [76, 115].

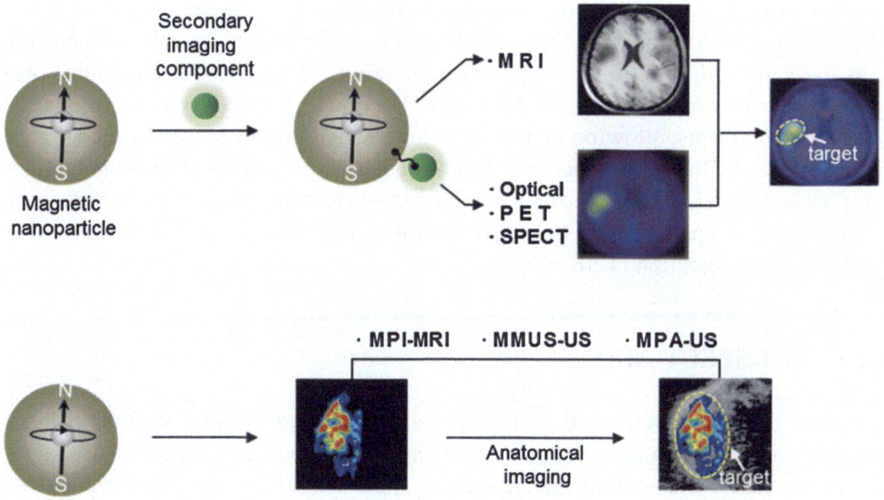

Fig. 11 MNP as versatile platform for multimodal imaging. Schematic representation of: **a** MNPs combination with another imaging component to generate multiple imaging signal; **b** MNPs are located with techniques such as MPI and MMUS. Adapted from Ref. [115]

4.2.1 CT–MRI

Computed tomography (CT) offers high spatial resolution and no tissue-penetrating limit, and is one of the most common imaging techniques used in clinics. CT is based on the different X-ray attenuations provoked by the various structures of the object according to their densities. The X-rays are detected from multiple angles and a 3D image of each slice of the object can be reconstructed. Bimodal CT–MRI probes can be prepared by combination of MNPs with a few elements such as iodine, gold, and lanthanides [118, 119]. These bimodal probes can provide information about tumor-associated vessels and detailed information of the tumor structure.

4.2.2 Optical–MRI

Many studies based on MRI–optical dual-mode imaging agents have been described by conjugated optically active compounds (such as organic fluorescent dyes) to SPIONs [120]. Although the optically active compounds present important advantages such as simplicity of use, generating activable signals and the possibility of multicolour analysis, the main limitation is the poor tissue penetration than can be achieved by fluorescence modality.

Crosslinked dextran-coated MNPs conjugated with a NIR fluorescent dye, (Cy5.5 dye) have been studied for brain tumor imaging (c.f. Fig. 12b) [121]. In addition, core-satellite structures based on a dye-encapsulated silica core conjugated with multiple SPIONs have been reported as successfully dual imaging probes capable of increasing sensitivity in both MRI and fluorescence imaging. More recently, quantum dots (QDs) and upconversion NPs (UCNPs) have been incorporated to iron oxide based MRI contrast agents, allowing high fluorescence sensitivity in deeply located organs (c.f Fig. 12c). NIR-to-NIR upconversion has been employed for in vivo imaging of lymphatic system [122]. In vivo dual-modal upconversion luminescence MR imaging of tumor and lymph node has been also demonstrated after intravenous injection or intracutaneous injection [120].

4.2.3 PET/SPECT–MRI

Recently, combination of radioisotope-based imaging techniques such as SPECT (photon emission computed tomography) and PET (positron emission tomography) with MRI offers great advantages of high sensitivity, quantitative, and no tissue-penetrating limit, providing highly resolved tomographic images. These imaging methodologies use a radioactive isotope incorporated in a metabolically active molecule. These radioactive tracers are injected into the circulating blood and the image collected represents the accumulation of the compound. Radiolabeling of SPION also allows for real-time noninvasive monitoring of drug delivery [123]. Iron oxide NPs coupled with radiometal chelates such as Cu-NOTA (chelating ligand, 1,4,7–triazacyclononane-*N*,*N*′, *N*″-triacetic acid) and a cRGD peptide have been investigated as PET–MRI contrast agents for targeted tumor imaging [124]. Instead of using chelators to incorporate the radiolabeling molecule, radioisotopes can be incorporated into the iron oxide core. PET/SPECT–MRI imaging agents have been applied for in vivo imaging of lymph nodes, cardiovascular organs, and the liver with high resolution (cf., Fig. 12d, e). Tri-modal imaging probes by attaching PET traces to dual MRI–optical probes enable monitoring tumor regions.

Some techniques for multimodal imaging such as PET–MRI present main drawbacks (such as radiation) and limitations, and still remain challenging and time-consuming. The multiple designs of MNPs make possible to develop multimodal imaging using only MNPs without the need to incorporate other types of materials (fluorescent molecules or radioisotopes). MNPs have become a versatile contrast agent platform for novel emerging imaging technologies such as MPI, MPA, and MMUS, enhancing the sensitivity, resolution, and accuracy of the image performance [115].

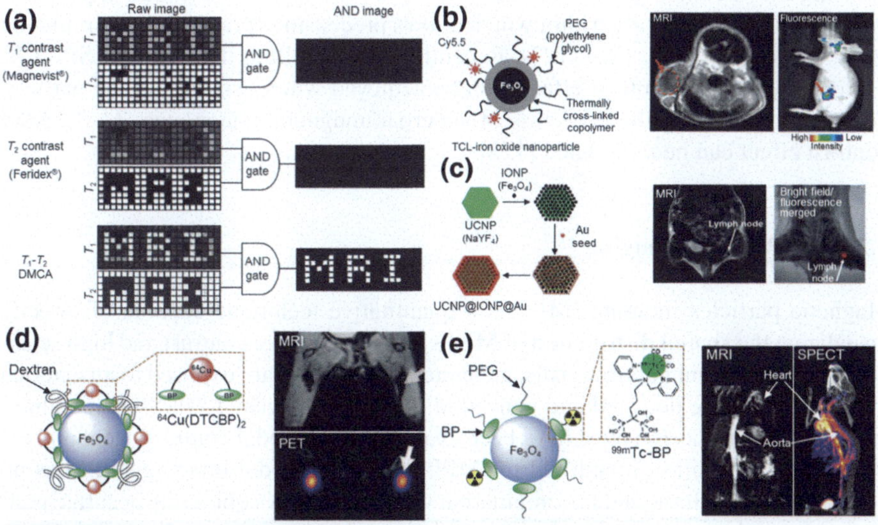

Fig. 12 **a** T_1- and T_2-weighted raw images of phantoms and final images using an AND logic algorithm of (i) Magnevist T_1 contrast agent, (ii) Feridex T_2 contrast agent, and (iii) T_1–T_2 DMCAs. Adapted from Ref. [127]. **b** Iron oxide NPs labeled with fluorescent dyes (Cy5.5) as MRI–optical dual-mode imaging agents enable visualization of tumors with high T_2 MRI signals and fluorescence signals. Adapted from Ref. [135]. **c** Hybrid nanostructures based on NaYF4 UCNP, iron oxide NP, and gold nanocluster are capable of labeling the lymph node of a mouse via both T_2 MRI and fluorescence imaging [136]. **d** MRI–PET dual-mode imaging agent based on iron oxide NPs, which are labeled with radioisotope (^{64}Cu) and chelator molecules (DTCBP) enable detection of lymph nodes with both MRI and PET signal [137]. **e** MRI–SPET dual-mode imaging agent based on iron oxide NP labeled with ^{99m}Tc enable visualization of cardiovascular organs such as the heart and aorta by MRI and SPECT imaging Ref. [94]. (Fig. 12c–e adapted from Ref. [115])

4.2.4 T_1–T_2 Dual-Modal MRI

T_1–T_2 dual-modal MRI has proved its capacity to improve the accuracy of MRI detection by enhancing bright MR signals in positive T_1 mode and dark MR signals in negative T_2 mode of the region of interest. It enables enhanced diagnostic accuracy and precision applying the same imaging technique [125]. Combination of paramagnetic metal ions (such as Gd^{3+} and Mn^{2+}) with SPIONs has been used in order to obtained complementary information on T_1-weighted and T_2-weighted MRI by positive and negative contrast simultaneous acquisition [126]. The paramagnetic ions and SPIONs are frequently separated by a polymer layer or incorporated in a silica coating in order to control the magnetic coupling.

Magnetically decoupled T_1–T_2 dual-mode contrast agents (DMCAs) present relaxivity coefficients approximately 2–3-fold higher than conventional T_1 and T_2 contrast agents. The perturbation of the T_1 contrast effects by the T_2 contrast agent is modulated by the separation control of the distance between the two materials (e.g., by using a silica shell spacer). In addition, T_1–T_2 DMCAs are capable of performing an AND logic gate algorithm which allows precise interpretation by eliminating MRI artifacts (c.f. Fig. 12a) [127]. Recently, by controlling the shape of SPIONs, a strong T_1 and T_2 contrast effects can be achieved when using the same particle. For instance, varying the thickness of superparamagnetic nanoplates, T_1–T_2 MRI contrast effect can be modulated [128].

4.2.5 MPI–MRI Imaging

Magnetic particles imaging (MPI) is a quantitative technique capable of directly visualizing the spatial distribution of MNPs with high image contrast and high sensitivity. The electromagnetic signals, generated after MNPs are exposed to an external magnetic field, are detected and converted into MPI images [129]. Some commercial SPIONs such as Resovist have been evaluated as bimodal contrast agent and the in vivo biodistribution monitored for MPI–MRI dual-modal imaging. MPI has no penetration depth limit and in combination with MRI allows obtaining accurate real-time information about NP location and high-resolution anatomical imaging [130]. Recently, cancer detection with MPI through passive targeting perfusion and EPR effect has been demonstrated by using long-circulating SPIO tracers [131].

4.2.6 Ultrasound MR Imaging

Ultrasound imaging is one of the most widely used biomedical imaging techniques in medicine. Despite this technique offers real advantages such as real-time imaging capability, high resolution, and portability, its sensitivity is very limited. The US sensitivity can be improved by the combination with MNPs as contrast agents for multimodal imaging modalities such as US–MRI, MMUS–US, and MPA–US, leading to visualize biological events at the cellular and molecular level in real time.

4.3 MMUS (Magneto-Motive Ultrasound Imaging)

A focused high intensity pulsed magnetic field is applied to the MNP, and their induced motion is monitored by US imaging. This technique allows determining the distribution of MNP in biological systems. Due to their large magnetic susceptibility, MNPs respond readily to magnetic fields and move toward the magnetic field. During the displacement, the NPs experience tissue resistance, against the magnetically induced displacement. Due to these two forces (magneto-motive and tissue resistance), MNPs induce vibrations that can be detected using US-based motion tracking technique. The combination of MMUS–US dual imaging allows monitoring the presence of MNPs along with the anatomical information. High-performance magneto-active MMUS contrast agents can be designed by synthesizing MNPs with higher magnetic susceptibilities. SPIONs have been applied to visualize sentinel lymph nodes by MPI [132]. Although it is still in the primitive stages, this technique has the potential to be a new imaging approach for cellular and molecular level dynamic imaging such as lymph node metastasis detection, stem cell tracking, and drug delivery monitoring.

4.4 MPA (Magneto-Photoacoustic Imaging)

Photoacoustic (PA) imaging is based on detecting ultrasound waves generated by tissues that absorb laser light and thermo-elastically expand. Using a transducer that converts acoustic waves into electric signals, the ultrasound wave is detected by a PA image. The combination with MNPs overcomes the limitation of the technique and enhances the quality of the images. When a pulsed external magnetic field is applied during PA acquisition, the MNPs create a vibrating motion that can be distinguished from the background signal and registered. Also, clustering MNPs enables higher PA signals. Designing multimodal MPA probes by combining MNP with NIR-absorbing nanomaterials led to powerful contrast agents for MRI and PA [133]. These probes are capable of capturing circulating tumor cells under flow conditions and detection by PA imaging. The detection of tumor cells circulating in vivo, which are some of the important biomarkers of cancer metastasis, tumor recurrence, and prognosis prediction, can be a powerful tool in cancer diagnosis.

5 Conclusions

The use of MNPs for cancer diagnosis and treatment has been proven to be a powerful biomedical tool. There are many promising applications that hopefully will translate to the clinic to increase the number of currently approved drugs based on these nanomaterials. However, there are still many challenges that need to be overcome, such as more efficient targeting in vivo. A better understanding of the underlying mechanisms in processes such as MFH, and the NPs degradation and interaction with the tumoral microenvironment will ensure the development of more effective therapies and diagnostic tools. Currently, the efforts of the scientist working in materials development are heading to expand the capabilities of the materials as dual platforms for combined capabilities in therapy and diagnosis (e.g., theranostics platforms) or developing multimodal materials that combine several therapies or imaging techniques just in one nanoplatform. We envisage that the following years will bring closer new approaches for clinical translational medicine.

References

1. Pelaz B, Alexiou C, Puebla RAA, Alves F, Andrews AM, Ashraf S, Balogh LP, Ballerini L, Bestetti A, Brendel C, Bosi S, Carril M, Chan WCW, Chen C, Chen X, Chen X, Cheng Z, Cui D, Du J, Dullin C, Escudero A, Feliu N, Gao M, George M, Gogotsi Y, Grünweller A, Gu Z, Halas N, Hampp N, Hartmann RK, Hersam MC, Hunziker P, Jian J, Jiang X, Jungebluth P, Kadhiresan P, Kataoka K, Khademhosseini A, Kopecek J, Kotov NA, Krug HF, Lee DS, Lehr C-M, Leong KW, Liang X-J, Lim ML, Marzán LML, Ma X, Macchiarini P, Meng H, Möhwald H, Mulvaney P, Nel AE, Nie S, Nordlander P, Okano T, Oliveira J, Park TH, Penner RM, Prato M, Puntes V, Rotello VM, Samarakoon A, Schaak RE, Shen Y, Sjöqvist S, Skirtach AG, Soliman MG, Stevens MM, Sung H-W, Tang BZ, Tietze R, Udugama BN, VanEpps JS, Weil T, Weiss PS, Willner I, Wu Y, Yang L, Yue Z, Zhang Q, Zhang Q, Zhang XE, Zhao Y, Zhou X, Parak WJ (2017) Diverse applications of nanomedicine. ACS Nano 11:2313–2381
2. del_Pino P, Pelaz B (2012) Hyperthermia using inorganic nanoparticles. In: Fuente JMdl, Grazu V (eds) Nanobiotechnology: inorganic nanoparticles versus organic nanoparticles: Frontiers of nanoscience, vol. 4. Elsevier, Amsterdam, The Netherlands, pp 309–335
3. Marciello M, Luengo Y, P. Morales M (2016) 24—Iron oxide nanoparticles for cancer diagnosis and therapy A2—Holban, Alina Maria. In: Grumezescu AM (ed) Nanoarchitectonics for smart delivery and drug targeting. William Andrew Publishing, pp 667–694. https://doi.org/10.1016/B978-0-323-47347-7.00024-0
4. Colombo M, Carregal-Romero S, Casula MF, Gutiérrez L, Morales MP, Böhm IB, Heverhagen JT, Prosperi D, Parak WJ (2012) Biological applications of magnetic nanoparticles. Chem Soc Rev 41:4306–4334
5. Lu AH, Salabas EL, Schuth F (2007) Magnetic nanoparticles: synthesis, protection, functionalization, and application. Angew Chem Int Ed 46(8):1222–1244
6. Ling D, Lee N, Hyeon T (2015) Chemical synthesis and assembly of uniformly sized iron oxide nanoparticles for medical applications. Acc Chem Res 48(5):1276–1285. https://doi.org/10.1021/acs.accounts.5b00038
7. Hühn J, Carrillo-Carrion C, Soliman MG, Pfeiffer C, Valdeperez D, Masood A, Chakraborty I, Zhu L, Gallego M, Zhao Y, Carril M, Feliu N, Escudero A, Alkilany AM, Pelaz B, Pino Pdel, Parak WJ (2017) Selected standard protocols for the synthesis, phase transfer, and characterization of inorganic colloidal nanoparticles. Chem Mater 29:399–461
8. Sathya A, Guardia P, Brescia R, Silvestri N, Pugliese G, Nitti S, Manna L, Pellegrino T (2016) $Co_xFe_{3-x}O_4$ Nanocubes for theranostic applications: effect of cobalt content and particle size. Chem Mater 28(6):1769–1780. https://doi.org/10.1021/acs.chemmater.5b04780
9. Zhang Q, Castellanos-Rubio I, Munshi R, Orue I, Pelaz B, Gries KI, Parak WJ, Pino Pdel, Pralle A (2015) Model driven optimization of magnetic anisotropy of exchange-coupled core-shell ferrite nanoparticles for maximal hysteretic loss. Chem Mater 27:7380–7387

10. Lee JH, Jang JT, Choi JS, Moon SH, Noh SH, Kim JW, Kim JG, Kim IS, Park KI, Cheon J (2011) Exchange-coupled magnetic nanoparticles for efficient heat induction. Nat Nanotechnol 6(7):418–422
11. Guardia P, Di Corato R, Lartigue L, Wilhelm C, Espinosa A, Garcia-Hernandez M, Gazeau F, Manna L, Pellegrino T (2012) Water-soluble iron oxide nanocubes with high values of specific absorption rate for cancer cell hyperthermia treatment. ACS Nano 6(4):3080–3091
12. Zhou Z, Zhu X, Wu D, Chen Q, Huang D, Sun C, Xin J, Ni K, Gao J (2015) Anisotropic shaped iron oxide nanostructures: controlled synthesis and proton relaxation shortening effects. Chem Mater 27(9):3505–3515. https://doi.org/10.1021/acs.chemmater.5b00944
13. Meffre A, Mehdaoui B, Kelsen V, Fazzini PF, Carrey J, Lachaize S, Respaud M, Chaudret B (2012) A simple chemical route toward monodisperse iron carbide nanoparticles displaying tunable magnetic and unprecedented hyperthermia properties. Nano Lett 12(9):4722–4728
14. Yu J, Chen F, Gao W, Ju Y, Chu X, Che S, Sheng F, Hou Y (2017) Iron carbide nanoparticles: an innovative nanoplatform for biomedical applications. Nanoscale Horizons 2(2):81–88. https://doi.org/10.1039/C6NH00173D
15. Casula MF, Conca E, Bakaimi I, Sathya A, Materia ME, Casu A, Falqui A, Sogne E, Pellegrino T, Kanaras AG (2016) Manganese doped-iron oxide nanoparticle clusters and their potential as agents for magnetic resonance imaging and hyperthermia. Phys Chem Chem Phys 18(25):16848–16855
16. Coral DF, Mendoza Zélis P, Marciello M, Morales MdP, Craievich A, Sánchez FH, Fernández van Raap MB (2016) Effect of nanoclustering and dipolar interactions in heat generation for magnetic hyperthermia. Langmuir 32(5):1201–1213. https://doi.org/10.1021/acs.langmuir.5b03559
17. Bigall NC, Curcio A, Leal MP, Falqui A, Palumberi D, Di Corato R, Albanesi E, Cingolani R, Pellegrino T (2011) Magnetic nanocarriers with tunable pH dependence for controlled loading and release of cationic and anionic payloads. Adv Mater 23(47):5645–5650. https://doi.org/10.1002/adma.201103505
18. Lee JE, Lee N, Kim H, Kim J, Choi SH, Kim JH, Kim T, Song IC, Park SP, Moon WK, Hyeon T (2010) Uniform mesoporous dye-doped silica nanoparticles decorated with multiple magnetite nanocrystals for simultaneous enhanced magnetic resonance imaging, fluorescence imaging, and drug delivery. J Am Chem Soc 132(2):552–557. https://doi.org/10.1021/ja905793q
19. Valdepérez D, Del Pino P, Sánchez L, Parak WJ, Pelaz B (2016) Highly active antibody-modified magnetic polyelectrolyte capsules. J Colloid and Interface Sci 474:1–8
20. Huynh NT, Passirani C, Saulnier P, Benoit JP (2009) Lipid nanocapsules: a new platform for nanomedicine. Int J Pharm 379(2):201–209
21. Salado J, Insausti M, Lezama L, de Muro I, Moros M, Pelaz B, Grazu V, de la Fuente J, Rojo T (2012) Functionalized Fe_3O_4@Au superparamagnetic nanoparticles: in vitro bioactivity. Nanotechnology 23(31)
22. Lentijo-Mozo S, Tan RP, Garcia-Marcelot C, Altantzis T, Fazzini P-F, Hungria T, Cormary B, Gallagher JR, Miller JT, Martinez H, Schrittwieser S, Schotter J, Respaud M, Bals S, Tendeloo GV, Gatel C, Soulantica K (2015) Air- and water-resistant noble metal coated ferromagnetic cobalt nanorods. ACS Nano 9(3):2792–2804. https://doi.org/10.1021/nn506709k
23. Ju Y, Zhang H, Yu J, Tong S, Tian N, Wang Z, Wang X, Su X, Chu X, Lin J, Ding Y, Li G, Sheng F, Hou Y (2017) Monodisperse Au–Fe_2C Janus nanoparticles: an attractive multifunctional material for triple-modal imaging-guided tumor photothermal therapy. ACS Nano 11(9):9239–9248. https://doi.org/10.1021/acsnano.7b04461
24. Wang Y, Gu HC (2015) Core-shell-type magnetic mesoporous silica nanocomposites for bioimaging and therapeutic agent delivery. Adv Mater 27(3):576–585
25. Pelaz B, Charron G, Pfeiffer C, Zhao YL, de la Fuente JM, Liang XJ, Parak WJ, del Pino P (2013) Interfacing engineered nanoparticles with biological systems: anticipating adverse nanobio interactions. Small 9(9–10):1573–1584
26. Polo E, Collado M, Pelaz B, del Pino P (2017) Advances toward more efficient targeted delivery of nanoparticles in vivo: understanding interactions between nanoparticles and cells. ACS Nano 11(3):2397–2402. https://doi.org/10.1021/acsnano.7b01197

27. Lo Giudice MC, Herda LM, Polo E, Dawson KA (2016) In situ characterization of nanoparticle biomolecular interactions in complex biological media by flow cytometry. Nat Commun 7:13475. https://doi.org/10.1038/ncomms13475, https://www.nature.com/articles/ncomms13475-supplementary-information
28. Soliman MG, B BP, Parak WJ, Pino Pd (2015) Phase transfer and polymer coating methods toward improving the stability of metallic nanoparticles for biological applications. Chem Mater 27:990–997
29. Pellegrino T, Manna L, Kudera S, Liedl T, Koktysh D, Rogach AL, Keller S, Rädler J, Natile G, Parak WJ (2004) Hydrophobic nanocrystals coated with an amphiphilic polymer shell: a general route to water soluble nanocrystals. Nano Lett 4(4):703–707
30. Hühn D, Kantner K, Geidel C, Brandholt S, De Cock I, Soenen SJH, Rivera Gil P, Montenegro J-M, Braeckmans K, Müllen K, Nienhaus GU, Klapper M, Parak WJ (2013) Polymer-coated nanoparticles interacting with proteins and cells: focusing on the sign of the net charge. ACS Nano 7(4):3253–3263. https://doi.org/10.1021/nn3059295
31. Salas G, Casado C, Teran FJ, Miranda R, Serna CJ, Morales MP (2012) Controlled synthesis of uniform magnetite nanocrystals with high-quality properties for biomedical applications. J Mater Chem 22(39):21065–21075. https://doi.org/10.1039/C2JM34402E
32. Joris F, Valdeperez D, Pelaz B, Soenen S, Manshian B, Parak W, De Smedt S, Raemdonck K (2016) The impact of species and cell type on the nanosafety profile of iron oxide nanoparticles in neural cells. J Nanobiotechnol 14:69
33. Amstad E, Gillich T, Bilecka I, Textor M, Reimhult E (2009) ultrastable iron oxide nanoparticle colloidal suspensions using dispersants with catechol-derived anchor groups. Nano Lett 9(12):4042–4048. https://doi.org/10.1021/nl902212q
34. Shi C, Thum C, Zhang Q, Tu W, Pelaz B, Parak WJ, Zhang Y, Schneider M (2016) Inhibition of the cancer-associated TASK 3 channels by magnetically induced thermal release of tetrandrine from a polymeric drug carrier. J Controlled Release 237:50–60
35. Feliu N, Pelaz B, Zhang Q, del_Pino P, Nyström A, Parak WJ (2016) Nanoparticle dosage-a nontrivial task of utmost importance for quantitative nanosafety research. Wiley Interdiscip Rev Nanomed Nanobiotechnol 8:479–492
36. Rivera Gil P, Jimenez de Aberasturi D, Wulf V, Pelaz B, del Pino P, Zhao Y, de la Fuente J, Ruiz de Larramendi I, Rojo T, Liang X-J, Parak WJ (2013) The challenge to relate the physicochemical properties of colloidal nanoparticles to their cytotoxicity. Acc Chem Res 46(3):743–749. https://doi.org/10.1021/ar300039j
37. Kreyling WG, Abdelmonem AM, Ali Z, Alves F, Geiser M, Haberl N, Hartmann R, Hirn S, de Aberasturi DJ, Kantner K, Khadem-Saba G, Montenegro JM, Rejman J, Rojo T, de Larramendi IR, Ufartes R, Wenk A, Parak WJ (2015) In vivo integrity of polymer-coated gold nanoparticles. Nat Nanotechnol 10(7):619–623
38. Deatsch AE, Evans BA (2014) Heating efficiency in magnetic nanoparticle hyperthermia. J Magn Magn Mater 354:163–172. https://doi.org/10.1016/j.jmmm.2013.11.006
39. Kolosnjaj-Tabi J, Javed Y, Lartigue L, Volatron J, Elgrabli D, Marangon I, Pugliese G, Caron B, Figuerola A, Luciani N, Pellegrino T, Alloyeau D, Gazeau F (2015) The one year fate of iron oxide coated gold nanoparticles in mice. ACS Nano 9(8):7925–7939
40. Mazuel F, Espinosa A, Luciani N, Reffay M, Le Borgne R, Motte L, Desboeufs K, Michel A, Pellegrino T, Lalatonne Y, Wilhelm C (2016) Massive intracellular biodegradation of iron oxide nanoparticles evidenced magnetically at single-endosome and tissue levels. ACS Nano 10(8):7627–7638. https://doi.org/10.1021/acsnano.6b02876
41. Chiarelli PA, Revia RA, Stephen ZR, Wang K, Jeon M, Nelson V, Kievit FM, Sham J, Ellenbogen RG, Kiem H-P, Zhang M (2017) Nanoparticle biokinetics in mice and nonhuman primates. ACS Nano 11(9):9514–9524. https://doi.org/10.1021/acsnano.7b05377
42. Wilhelm S, Tavares AJ, Dai Q, Ohta S, Audet J, Dvorak HF, Chan WCW (2016) Analysis of nanoparticle delivery to tumours. Nat Rev Mater 1:16014. https://doi.org/10.1038/natrevmats.2016.14, http://www.nature.com/articles/natrevmats201614-supplementary-information
43. Anchordoquy TJ, Barenholz Y, Boraschi D, Chorny M, Decuzzi P, Dobrovolskaia MA, Farhangrazi ZS, Farrell D, Gabizon A, Ghandehari H, Godin B, La-Beck NM, Ljubimova

J, Moghimi SM, Pagliaro L, Park J-H, Peer D, Ruoslahti E, Serkova NJ, Simberg D (2017) Mechanisms and barriers in cancer nanomedicine: addressing challenges, looking for solutions. ACS Nano 11(1):12–18. https://doi.org/10.1021/acsnano.6b08244
44. Bobo D, Robinson KJ, Islam J, Thurecht KJ, Corrie SR (2016) Nanoparticle-based medicines: a review of FDA-approved materials and clinical trials to date. Pharm Res 33(10):2373–2387. https://doi.org/10.1007/s11095-016-1958-5
45. Daldrup-Link HE, Golovko D, Ruffell B, DeNardo DG, Castaneda R, Ansari C, Rao J, Tikhomirov GA, Wendland M, Corot C, Coussens LM (2011) MR imaging of tumor associated macrophages with clinically-applicable iron oxide nanoparticles. Clin Cancer Res: An Official J Am Assoc Cancer Res 17(17):5695–5704. https://doi.org/10.1158/1078-0432.CCR-10-3420
46. Wlaschek M, Scharffetter-Kochanek K (2005) Oxidative stress in chronic venous leg ulcers. Wound Repair and Regeneration 13(5):452–461. https://doi.org/10.1111/j.1067-1927.2005.00065.x
47. Sindrilaru A, Peters T, Wieschalka S, Baican C, Baican A, Peter H, Hainzl A, Schatz S, Qi Y, Schlecht A, Weiss JM, Wlaschek M, Sunderkötter C, Scharffetter-Kochanek K (2011) An unrestrained proinflammatory M1 macrophage population induced by iron impairs wound healing in humans and mice. J Clin Investig 121(3):985–997. https://doi.org/10.1172/JCI44490
48. Zanganeh S, Hutter G, Spitler R, Lenkov O, Mahmoudi M, Shaw A, Pajarinen JS, Nejadnik H, Goodman S, Moseley M, Coussens LM, Daldrup-Link HE (2016) Iron oxide nanoparticles inhibit tumour growth by inducing pro-inflammatory macrophage polarization in tumour tissues. Nat Nanotechnol 11:986. https://doi.org/10.1038/nnano.2016.168, https://www.nature.com/articles/nnano.2016.168-supplementary-information
49. MacParland SA, Tsoi KM, Ouyang B, Ma X-Z, Manuel J, Fawaz A, Ostrowski MA, Alman BA, Zilman A, Chan WCW, McGilvray ID (2017) Phenotype determines nanoparticle uptake by human macrophages from liver and blood. ACS Nano 11(3):2428–2443. https://doi.org/10.1021/acsnano.6b06245
50. Scherer F, Anton M, Schillinger U, Henke J, Bergemann C, Kruger A, Gansbacher B, Plank C (2002) Magnetofection: enhancing and targeting gene delivery by magnetic force in vitro and in vivo. Gene Ther 9(2):102–109
51. Plank C, Zelphati O, Mykhaylyk O (2011) Magnetically enhanced nucleic acid delivery. Ten years of magnetofection—Progress and prospects. Adv Drug Deliv Rev 63(14):1300–1331. https://doi.org/10.1016/j.addr.2011.08.002
52. del Pino P, Munoz-Javier A, Vlaskou D, Rivera Gil P, Plank C, Parak WJ (2010) Gene silencing mediated by magnetic lipospheres tagged with small interfering RNA. Nano Lett 10(10):3914–3921
53. Smith CAM, de la Fuente J, Pelaz B, Furlani EP, Mullin M, Berry CC (2010) The effect of static magnetic fields and tat peptides on cellular and nuclear uptake of magnetic nanoparticles. Biomaterials 31(15):4392–4400
54. Child HW, Del Pino PA, De La Fuente JM, Hursthouse AS, Stirling D, Mullen M, McPhee GM, Nixon C, Jayawarna V, Berry CC (2011) Working together: the combined application of a magnetic field and penetratin for the delivery of magnetic nanoparticles to cells in 3D. ACS Nano 5(10):7910–7919
55. Soto-Sánchez C, Martínez-Navarrete G, Humphreys L, Puras G, Zarate J, Pedraz JL, Fernández E (2015) Enduring high-efficiency in vivo transfection of neurons with non-viral magnetoparticles in the rat visual cortex for optogenetic applications. Nanomed Nanotechnol Biol Med 11(4):835–843. https://doi.org/10.1016/j.nano.2015.01.012
56. Beik J, Abed Z, Ghoreishi FS, Hosseini-Nami S, Mehrzadi S, Shakeri-Zadeh A, Kamrava SK (2016) Nanotechnology in hyperthermia cancer therapy: from fundamental principles to advanced applications. J Controlled Release 235(Supplement C):205–221. https://doi.org/10.1016/j.jconrel.2016.05.062
57. Wust P, Hildebrandt B, Sreenivasa G, Rau B, Gellermann J, Riess H, Felix R, Schlag PM (2002) Hyperthermia in combined treatment of cancer. Lancet Oncology 3(8):487–497

58. Cavaliere R, Ciocatto EC, Giovanella BC, Heidelberger C, Johnson RO, Margottini M, Mondovi B, Moricca G, Rossi-Fanelli A (1967) Selective heat sensitivity of cancer cells. Biochemical and clinical studies. Cancer 20(9):1351–1381
59. Silva AC, Oliveira TR, Mamani JB, Malheiros SMF, Malavolta L, Pavon LF, Sibov TT, Amaro E, Tannús A, Vidoto ELG, Martins MJ, Santos RS, Gamarra LF (2011) Application of hyperthermia induced by superparamagnetic iron oxide nanoparticles in glioma treatment. Int J Nanomed 6:591–603. https://doi.org/10.2147/IJN.S14737
60. Lepock JR (2003) Cellular effects of hyperthermia: relevance to the minimum dose for thermal damage. Int J Hyperth 19(3):252–266
61. Huilgol N, Gupta S, Sridhar C (2010) Hyperthermia with radiation in the treatment of locally advanced head and neck cancer: A report of randomized trial. J Cancer Res Ther 6(4):492–496. https://doi.org/10.4103/0973-1482.77101
62. Jordan A, Scholz R, Maier-Hauff K, van Landeghem FK, Waldoefner N, Teichgraeber U, Pinkernelle J, Bruhn H, Neumann F, Thiesen B, von Deimling A, Felix R (2006) The effect of thermotherapy using magnetic nanoparticles on rat malignant glioma. J Neurooncol 78(1):7–14
63. Brezovich IA (1988) Low frequency hyperthermia: capacitive and ferromagnetic thermoseed methods. Med Phys Monogr 16:82–111
64. Di Corato R, Espinosa A, Lartigue L, Tharaud M, Chat S, Pellegrino T, Menager C, Gazeau F, Wilhelm C (2014) Magnetic hyperthermia efficiency in the cellular environment for different nanoparticle designs. Biomaterials 35(24):6400–6411
65. Dias JT, Moros M, del Pino P, Rivera S, Grazú V, de la Fuente JM (2013) DNA as a molecular local thermal probe for the analysis of magnetic hyperthermia. Angew Chem Int Ed Engl 52(44):11526–11529
66. Riedinger A, Guardia P, Curcio A, Garcia MA, Cingolani R, Manna L, Pellegrino T (2013) Subnanometer local temperature probing and remotely controlled drug release based on azo-functionalized iron oxide nanoparticles. Nano Lett 13(6):2399–2406. https://doi.org/10.1021/nl400188q
67. Fischer A, Sackmann E (1985) A charge-decoration technique for studying the heterogeneity of coexistent monolayer phases by electron microscopy. Nature 313:299–301
68. Jordan A, Scholz R, Maier-Hauff K, Johannsen M, Wust P, Nadobny J, Schirra H, Schmidt H, Deger S, Loening S, Lanksch W, Felix R (2001) Presentation of a new magnetic field therapy system for the treatment of human solid tumors with magnetic fluid hyperthermia. J Magn Magn Mater 225(1):118–126. https://doi.org/10.1016/S0304-8853(00)01239-7
69. Fan W, Yung B, Huang P, Chen X (2017) Nanotechnology for multimodal synergistic cancer therapy. Chem Rev 117(22):13566–13638. https://doi.org/10.1021/acs.chemrev.7b00258
70. Kolosnjaj-Tabi J, Di Corato R, Lartigue L, Marangon I, Guardia P, Silva AKA, Luciani N, Clement O, Flaud P, Singh JV, Decuzzi P, Pellegrino T, Wilhelm C, Gazeau F (2014) Heat-generating iron oxide nanocubes: subtle "Destructurators" of the tumoral microenvironment. ACS Nano 8(5):4268–4283
71. Espinosa A, Di Corato R, Kolosnjaj-Tabi J, Flaud P, Pellegrino T, Wilhelm C (2016) Duality of iron oxide nanoparticles in cancer therapy: amplification of heating efficiency by magnetic hyperthermia and photothermal bimodal treatment. ACS Nano 10(2):2436–2446. https://doi.org/10.1021/acsnano.5b07249
72. Lee JH, Chen KJ, Noh SH, Garcia MA, Wang H, Lin WY, Jeong H, Kong BJ, Stout DB, Cheon J, Tseng HR (2013) On-demand drug release system for in vivo cancer treatment through self-assembled magnetic nanoparticles. Angew Chem Int Ed Engl 52(16):4384–4388
73. Munshi R, Qadri SM, Zhang Q, Castellanos Rubio I, del Pino P, Pralle A (2017) Magnetothermal genetic deep brain stimulation of motor behaviors in awake, freely moving mice. eLife 6:e27069. https://doi.org/10.7554/elife.27069
74. Chen R, Romero G, Christiansen MG, Mohr A, Anikeeva P (2015) Wireless magnetothermal deep brain stimulation. Science 347(6229):1477–1480. https://doi.org/10.1126/science.1261821

75. Modo MMJ, Bulte JWM, Kim EE (2007) Molecular and cellular MR imaging. J Nucl Med 48(12):2087. https://doi.org/10.2967/jnumed.107.045369
76. Lee N, Yoo D, Ling D, Cho MH, Hyeon T, Cheon J (2015) Iron oxide based nanoparticles for multimodal imaging and magnetoresponsive therapy. Chem Rev 115(19):10637–10689. https://doi.org/10.1021/acs.chemrev.5b00112
77. Lee N, Hyeon T (2012) Designed synthesis of uniformly sized iron oxide nanoparticles for efficient magnetic resonance imaging contrast agents. Chem Soc Rev 41(7):2575–2589. https://doi.org/10.1039/C1CS15248C
78. Hingorani DV, Bernstein AS, Pagel MD (2015) A review of responsive MRI contrast agents: 2005–2014. Contrast Media Mol Imaging 10(4):245–265. https://doi.org/10.1002/cmmi.1629
79. Jin R, Lin B, Li D, Ai H (2014) Superparamagnetic iron oxide nanoparticles for MR imaging and therapy: design considerations and clinical applications. Curr Opin Pharmacol 18(Supplement C):18–27. https://doi.org/10.1016/j.coph.2014.08.002
80. Ni D, Bu W, Ehlerding EB, Cai W, Shi J (2017) Engineering of inorganic nanoparticles as magnetic resonance imaging contrast agents. Chem Soc Rev. https://doi.org/10.1039/C7CS00316A
81. Vuong QL, Gillis P, Roch A, Gossuin Y (2017) Magnetic resonance relaxation induced by superparamagnetic particles used as contrast agents in magnetic resonance imaging: a theoretical review. Wiley Interdisc Rev: Nanomed Nanobiotechnol 9(6):n/a–n/a. https://doi.org/10.1002/wnan.1468
82. Y-w Jun, Lee J-H, Cheon J (2008) Chemical design of nanoparticle probes for high-performance magnetic resonance imaging. Angew Chem Int Ed 47(28):5122–5135. https://doi.org/10.1002/anie.200701674
83. Zhao Z, Zhou Z, Bao J, Wang Z, Hu J, Chi X, Ni K, Wang R, Chen X, Chen Z, Gao J (2013) Octapod iron oxide nanoparticles as high-performance T_2 contrast agents for magnetic resonance imaging. Nat Commun 4:2266. https://doi.org/10.1038/ncomms3266 http://www.nature.com/articles/ncomms3266-supplementary-information
84. Kang T, Li F, Baik S, Shao W, Ling D, Hyeon T (2017) Surface design of magnetic nanoparticles for stimuli-responsive cancer imaging and therapy. Biomaterials 136(Supplement C):98–114. https://doi.org/10.1016/j.biomaterials.2017.05.013
85. Javed Y, Akhtar K, Anwar H, Jamil Y (2017) MRI based on iron oxide nanoparticles contrast agents: effect of oxidation state and architecture. J Nanopart Res 19(11):366. https://doi.org/10.1007/s11051-017-4045-x
86. Y-w Jun, Huh Y-M, J-s Choi, Lee J-H, Song H-T, KimKim Yoon S, Kim K-S, Shin J-S, Suh J-S, Cheon J (2005) nanoscale size effect of magnetic nanocrystals and their utilization for cancer diagnosis via magnetic resonance imaging. J Am Chem Soc 127(16):5732–5733. https://doi.org/10.1021/ja0422155
87. Kim BH, Lee N, Kim H, An K, Park YI, Choi Y, Shin K, Lee Y, Kwon SG, Na HB, Park J-G, Ahn T-Y, Kim Y-W, Moon WK, Choi SH, Hyeon T (2011) Large-scale synthesis of uniform and extremely small-sized iron oxide nanoparticles for High-resolution T_1 magnetic resonance imaging contrast agents. J Am Chem Soc 133(32):12624–12631. https://doi.org/10.1021/ja203340u
88. Ahmad T, Bae H, Rhee I, Chang Y, Lee J, Hong S (2012) Particle size dependence of relaxivity for silica-coated iron oxide nanoparticles. Curr Appl Phys 12(3):969–974. https://doi.org/10.1016/j.cap.2011.12.020
89. Shen T, Weissleder R, Papisov M, Bogdanov A, Brady TJ (1993) Monocrystalline iron oxide nanocompounds (MION): physicochemical properties. Magn Reson Med 29(5):599–604. https://doi.org/10.1002/mrm.1910290504
90. Shapiro EM, Skrtic S, Sharer K, Hill JM, Dunbar CE, Koretsky AP (2004) MRI detection of single particles for cellular imaging. Proc Natl Acad Sci USA 101(30):10901–10906. https://doi.org/10.1073/pnas.0403918101
91. Wunderbaldinger P, Josephson L, Weissleder R (2002) Crosslinked iron oxides (CLIO): a new platform for the development of targeted MR contrast agents. Acad Radiol 9(2, Supplement):S304–S306. https://doi.org/10.1016/S1076-6332(03)80210-6

92. Balasubramaniam S, Kayandan S, Lin Y-N, Kelly DF, House MJ, Woodward RC, St. Pierre TG, Riffle JS, Davis RM (2014) Toward design of magnetic nanoparticle clusters stabilized by biocompatible diblock copolymers for T_2-weighted MRI contrast. Langmuir 30(6):1580–1587. https://doi.org/10.1021/la403591z
93. Ai H, Flask C, Weinberg B, Shuai XT, Pagel MD, Farrell D, Duerk J, Gao J (2005) Magnetite-loaded polymeric micelles as ultrasensitive magnetic-resonance probes. Adv Mater 17(16):1949–1952. https://doi.org/10.1002/adma.200401904
94. Sandiford L, Phinikaridou A, Protti A, Meszaros LK, Cui X, Yan Y, Frodsham G, Williamson PA, Gaddum N, Botnar RM, Blower PJ, Green MA, de Rosales RTM (2013) Bisphosphonate-anchored PEGylation and radiolabeling of superparamagnetic iron oxide: long-circulating nanoparticles for in vivo multimodal (T_1 MRI-SPECT) imaging. ACS Nano 7(1):500–512. https://doi.org/10.1021/nn3046055
95. Zhou Z, Wang L, Chi X, Bao J, Yang L, Zhao W, Chen Z, Wang X, Chen X, Gao J (2013) Engineered iron-oxide-based nanoparticles as enhanced T_1 contrast agents for efficient tumor imaging. ACS Nano 7(4):3287–3296. https://doi.org/10.1021/nn305991e
96. Kievit FM, Stephen ZR, Veiseh O, Arami H, Wang T, Lai VP, Park JO, Ellenbogen RG, Disis ML, Zhang M (2012) Targeting of primary breast cancers and metastases in a transgenic mouse model using rationally designed multifunctional SPIONs. ACS Nano 6(3):2591–2601. https://doi.org/10.1021/nn205070h
97. Wang Y-XJ (2015) Current status of superparamagnetic iron oxide contrast agents for liver magnetic resonance imaging. World J Gastroenterol 21(47):13400–13402. https://doi.org/10.3748/wjg.v21.i47.13400
98. Reimer P, Balzer T (2003) Ferucarbotran (Resovist): a new clinically approved RES-specific contrast agent for contrast-enhanced MRI of the liver: properties, clinical development, and applications. Eur Radiol 13(6):1266–1276. https://doi.org/10.1007/s00330-002-1721-7
99. Tanimoto A, Kuribayashi S (2006) Application of superparamagnetic iron oxide to imaging of hepatocellular carcinoma. Eur J Radiol 58(2):200–216. https://doi.org/10.1016/j.ejrad.2005.11.040
100. Acharya S, Sahoo SK (2011) PLGA nanoparticles containing various anticancer agents and tumour delivery by EPR effect. Adv Drug Deliv Rev 63(3):170–183. https://doi.org/10.1016/j.addr.2010.10.008
101. Harisinghani MG, Barentsz J, Hahn PF, Deserno WM, Tabatabaei S, van de Kaa CH, de la Rosette J, Weissleder R (2003) Noninvasive detection of clinically occult lymph-node metastases in prostate cancer. N Engl J Med 348(25):2491–2499. https://doi.org/10.1056/NEJMoa022749
102. Heesakkers RAM, Jager GJ, Hövels AM, Hoop Bd, Bosch HCMvd, Raat F, Witjes JA, Mulders PFA, Kaa CHvd, Barentsz JO (2009) Prostate cancer: Detection of lymph node metastases outside the routine surgical area with ferumoxtran-10–enhanced MR imaging. Radiology 251(2):408–414. https://doi.org/10.1148/radiol.2512071018
103. Ling D, Park W, Park S-j LuY, Kim KS, Hackett MJ, Kim BH, Yim H, Jeon YS, Na K, Hyeon T (2014) Multifunctional tumor pH-sensitive self-assembled nanoparticles for bimodal imaging and treatment of resistant heterogeneous tumors. J Am Chem Soc 136(15):5647–5655. https://doi.org/10.1021/ja4108287
104. Wang L, Huang J, Chen H, Wu H, Xu Y, Li Y, Yi H, Wang YA, Yang L, Mao H (2017) Exerting enhanced permeability and retention effect driven delivery by ultrafine iron oxide nanoparticles with T_1–T_2 Switchable magnetic resonance imaging contrast. ACS Nano 11(5):4582–4592. https://doi.org/10.1021/acsnano.7b00038
105. Weissleder R, Nahrendorf M, Pittet MJ (2014) Imaging macrophages with nanoparticles. Nat Mater 13:125. https://doi.org/10.1038/nmat3780
106. Huang C, Neoh KG, Wang L, Kang E-T, Shuter B (2010) Magnetic nanoparticles for magnetic resonance imaging: modulation of macrophage uptake by controlled PEGylation of the surface coating. J Mater Chem 20(39):8512–8520. https://doi.org/10.1039/C0JM01526A
107. Shevtsov MA, Yakovleva LY, Nikolaev BP, Marchenko YY, Dobrodumov AV, Onokhin KV, Onokhina YS, Selkov SA, Mikhrina AL, Guzhova IV, Martynova MG, Bystrova OA, Ischenko

AM, Margulis BA (2014) Tumor targeting using magnetic nanoparticle Hsp70 conjugate in a model of C6 glioma. Neuro-Oncology 16(1):38–49. https://doi.org/10.1093/neuonc/not141
108. Shahbazi-Gahrouei D, Abdolahi M (2013) Detection of MUC1-expressing ovarian cancer by C595 monoclonal antibody-conjugated SPIONs using MR imaging. Sci World J 2013:7. https://doi.org/10.1155/2013/609151
109. Lee J-H, Huh Y-M, Jun Y-w, Seo J-w, Jang J-t, Song H-T, Kim S, Cho E-J, Yoon H-G, Suh J-S, Cheon J (2006) Artificially engineered magnetic nanoparticles for ultra-sensitive molecular imaging. Nat Med 13:95. https://doi.org/10.1038/nm1467, https://www.nature.com/articles/nm1467-supplementary-information
110. Huh Y-M, Y-w Jun, Song H-T, Kim S, J-s Choi, Lee J-H, Yoon S, Kim K-S, Shin J-S, Suh J-S, Cheon J (2005) In vivo magnetic resonance detection of cancer by using multifunctional magnetic nanocrystals. J Am Chem Soc 127(35):12387–12391. https://doi.org/10.1021/ja052337c
111. Bakhtiary Z, Saei AA, Hajipour MJ, Raoufi M, Vermesh O, Mahmoudi M (2016) Targeted superparamagnetic iron oxide nanoparticles for early detection of cancer: possibilities and challenges. Nanomed Nanotechnol Biol Med 12(2):287–307. https://doi.org/10.1016/j.nano.2015.10.019
112. Lee D-E, Koo H, Sun I-C, Ryu JH, Kim K, Kwon IC (2012) Multifunctional nanoparticles for multimodal imaging and theragnosis. Chem Soc Rev 41(7):2656–2672. https://doi.org/10.1039/C2CS15261D
113. Ferreira L, Karp JM, Nobre L, Langer R (2008) New opportunities: the use of nanotechnologies to manipulate and track stem cells. Cell Stem Cell 3(2):136–146. https://doi.org/10.1016/j.stem.2008.07.020
114. Mahmoudi M, Hosseinkhani H, Hosseinkhani M, Boutry S, Simchi A, Journeay WS, Subramani K, Laurent S (2011) Magnetic resonance imaging tracking of stem cells in vivo using iron oxide nanoparticles as a tool for the advancement of clinical regenerative medicine. Chem Rev 111(2):253–280. https://doi.org/10.1021/cr1001832
115. Shin T-H, Choi Y, Kim S, Cheon J (2015) Recent advances in magnetic nanoparticle-based multi-modal imaging. Chem Soc Rev 44(14):4501–4516. https://doi.org/10.1039/C4CS00345D
116. Condeelis J, Weissleder R (2010) In vivo imaging in cancer. Cold Spring Harb Perspect Biol 2(12):a003848. https://doi.org/10.1101/cshperspect.a003848
117. Wang X, Niu D, Li P, Wu Q, Bo X, Liu B, Bao S, Su T, Xu H, Wang Q (2015) Dual-enzyme-loaded multifunctional hybrid nanogel system for pathological responsive ultrasound imaging and T_2-weighted magnetic resonance imaging. ACS Nano 9(6):5646–5656. https://doi.org/10.1021/nn5068094
118. Dongkyu K, Mi Kyung Y, Tae Sup L, Jae Jun P, Yong Yeon J, Sangyong J (2011) Amphiphilic polymer-coated hybrid nanoparticles as CT/MRI dual contrast agents. Nanotechnology 22(15):155101
119. Oh MH, Lee N, Kim H, Park SP, Piao Y, Lee J, Jun SW, Moon WK, Choi SH, Hyeon T (2011) Large-scale synthesis of bioinert tantalum oxide nanoparticles for X-ray computed tomography imaging and bimodal image-guided sentinel lymph node mapping. J Am Chem Soc 133(14):5508–5515. https://doi.org/10.1021/ja200120k
120. Zhang L, Zhou H, Belzile O, Thorpe P, Zhao D (2014) Phosphatidylserine-targeted bimodal liposomal nanoparticles for in vivo imaging of breast cancer in mice. J Controlled Release 183(Supplement C):114–123. https://doi.org/10.1016/j.jconrel.2014.03.043
121. Kircher MF, Mahmood U, King RS, Weissleder R, Josephson L (2003) A multimodal nanoparticle for preoperative magnetic resonance imaging and intraoperative optical brain tumor delineation. Can Res 63(23):8122–8125
122. Xia A, Gao Y, Zhou J, Li C, Yang T, Wu D, Wu L, Li F (2011) Core–shell NaYF4:Yb3 + , Tm3 + @FexOy nanocrystals for dual-modality T_2-enhanced magnetic resonance and NIR-to-NIR upconversion luminescent imaging of small-animal lymphatic node. Biomaterials 32(29):7200–7208. https://doi.org/10.1016/j.biomaterials.2011.05.094

123. Chen F, Ellison PA, Lewis CM, Hong H, Zhang Y, Shi S, Hernandez R, Meyerand ME, Barnhart TE, Cai W (2013) Chelator-free synthesis of a dual-modality PET/MRI agent. Angew Chem Int Ed 52(50):13319–13323. https://doi.org/10.1002/anie.201306306
124. Yang X, Hong H, Grailer JJ, Rowland IJ, Javadi A, Hurley SA, Xiao Y, Yang Y, Zhang Y, Nickles RJ, Cai W, Steeber DA, Gong S (2011) cRGD-functionalized, DOX-conjugated, and 64Cu-labeled superparamagnetic iron oxide nanoparticles for targeted anticancer drug delivery and PET/MR imaging. Biomaterials 32(17):4151
125. Zhou Z, Bai R, Munasinghe J, Shen Z, Nie L, Chen X (2017) T_1–T_2 dual-modal magnetic resonance imaging: from molecular basis to contrast agents. ACS Nano 11(6):5227–5232. https://doi.org/10.1021/acsnano.7b03075
126. Bae KH, Kim YB, Lee Y, Hwang J, Park H, Park TG (2010) Bioinspired synthesis and characterization of gadolinium-labeled magnetite nanoparticles for dual contrast T_1- and T_2-weighted magnetic resonance imaging. Bioconjug Chem 21(3):505–512. https://doi.org/10.1021/bc900424u
127. Shin T-H, J-s Choi, Yun S, Kim I-S, Song H-T, Kim Y, Park KI, Cheon J (2014) T_1 and T_2 dual-mode MRI contrast agent for enhancing accuracy by engineered nanomaterials. ACS Nano 8(4):3393–3401. https://doi.org/10.1021/nn405977t
128. Zhou Z, Zhao Z, Zhang H, Wang Z, Chen X, Wang R, Chen Z, Gao J (2014) Interplay between longitudinal and transverse contrasts in Fe_3O_4 nanoplates with (111) exposed surfaces. ACS Nano 8(8):7976–7985. https://doi.org/10.1021/nn5038652
129. Goodwill PW, Saritas EU, Croft LR, Kim TN, Krishnan KM, Schaffer DV, Conolly SM (2012) X-space MPI: magnetic nanoparticles for safe medical imaging. Adv Mater 24(28):3870–3877. https://doi.org/10.1002/adma.201200221
130. Pablico-Lansigan MH, Situ SF, Samia ACS (2013) Magnetic particle imaging: advancements and perspectives for real-time in vivo monitoring and image-guided therapy. Nanoscale 5(10):4040–4055. https://doi.org/10.1039/C3NR00544E
131. Yu EY, Bishop M, Zheng B, Ferguson RM, Khandhar AP, Kemp SJ, Krishnan KM, Goodwill PW, Conolly SM (2017) Magnetic particle imaging: a novel in vivo imaging platform for cancer detection. Nano Lett 17(3):1648–1654. https://doi.org/10.1021/acs.nanolett.6b04865
132. Evertsson M, Kjellman P, Cinthio M, Fredriksson S, Rit Zandt, Persson HW, Jansson T (2014) Multimodal detection of iron oxide nanoparticles in rat lymph nodes using magnetomotive ultrasound imaging and magnetic resonance imaging. IEEE Trans Ultrason Ferroelectr Freq Control 61(8):1276–1283. https://doi.org/10.1109/TUFFC.2014.3034
133. Hu X, Wei C-W, Xia J, Pelivanov I, O'Donnell M, Gao X (2013) Trapping and photoacoustic detection of CTCs at the single cell per milliliter level with magneto-optical coupled nanoparticles. Small 9(12):2046–2052. https://doi.org/10.1002/smll.201202085
134. Périgo EA, Hemery G, Sandre O, Ortega D, Garaio E, Plazaola F, Teran FJ (2015) Fundamentals and advances in magnetic hyperthermia. Appl Phys Rev 2(4):041302. https://doi.org/10.1063/1.4935688
135. Lee H, Yu MK, Park S, Moon S, Min JJ, Jeong YY, Kang H-W, Jon S (2007) Thermally cross-linked superparamagnetic iron oxide nanoparticles: synthesis and application as a dual imaging probe for cancer in vivo. J Am Chem Soc 129(42):12739–12745. https://doi.org/10.1021/ja072210i
136. Cheng L, Yang K, Li Y, Chen J, Wang C, Shao M, Lee S-T, Liu Z (2011) Facile preparation of multifunctional upconversion nanoprobes for multimodal imaging and dual-targeted photothermal therapy. Angew Chem Int Ed 50(32):7385–7390. https://doi.org/10.1002/anie.201101447
137. Torres Martin de Rosales R, Tavaré R, Paul RL, Jauregui-Osoro M, Protti A, Glaria A, Varma G, Szanda I, Blower PJ (2011) Synthesis of 64CuII–Bis(dithiocarbamatebisphosphonate) and its conjugation with superparamagnetic iron oxide Nanoparticles: in vivo evaluation as dual-modality PET–MRI agent. Angew Chem Int Ed 50(24):5509–5513. https://doi.org/10.1002/anie.201007894
138. Pöselt E, Kloust H, Tromsdorf U, Janschel M, Hahn C, Maßlo C, Weller H (2012) Relaxivity optimization of a PEGylated iron-oxide-based negative magnetic resonance contrast

agent for T_2-weighted spin-echo imaging. ACS Nano 6(2):1619–1624. https://doi.org/10.1021/nn204591r
139. J-t Jang, Nah H, Lee J-H, Moon SH, Kim MG, Cheon J (2009) Critical enhancements of MRI contrast and hyperthermic effects by dopant-controlled magnetic nanoparticles. Angew Chem Int Ed 48(7):1234–1238. https://doi.org/10.1002/anie.200805149
140. Lee N, Choi Y, Lee Y, Park M, Moon WK, Choi SH, Hyeon T (2012) Water-dispersible ferrimagnetic iron oxide nanocubes with extremely high r_2 relaxivity for highly sensitive in vivo MRI of tumors. Nano Lett 12(6):3127–3131. https://doi.org/10.1021/nl3010308
141. Li W, Tutton S, Vu AT, Pierchala L, Li BSY, Lewis JM, Prasad PV, Edelman RR (2005) First-pass contrast-enhanced magnetic resonance angiography in humans using ferumoxytol, a novel ultrasmall superparamagnetic iron oxide (USPIO)-based blood pool agent. J Magn Reson Imaging 21(1):46–52. https://doi.org/10.1002/jmri.20235
142. Neuwelt EA, Hamilton BE, Varallyay CG, Rooney WR, Edelman RD, Jacobs PM, Watnick SG (2009) Ultrasmall superparamagnetic iron oxides (USPIOs): a future alternative magnetic resonance (MR) contrast agent for patients at risk for nephrogenic systemic fibrosis (NSF)? Kidney Int 75(5):465–474. https://doi.org/10.1038/ki.2008.496
143. Chan N, Laprise-Pelletier M, Chevallier P, Bianchi A, Fortin M-A, Oh JK (2014) Multidentate block-copolymer-stabilized ultrasmall superparamagnetic iron oxide nanoparticles with enhanced colloidal stability for magnetic resonance imaging. Biomacromol 15(6):2146–2156. https://doi.org/10.1021/bm500311k
144. Jung CW, Jacobs P (1995) Physical and chemical properties of superparamagnetic iron oxide MR contrast agents: Ferumoxides, ferumoxtran, ferumoxsil. Magn Reson Imaging 13(5):661–674. https://doi.org/10.1016/0730-725X(95)00024-B
145. Casula MF, Floris P, Innocenti C, Lascialfari A, Marinone M, Corti M, Sperling RA, Parak WJ, Sangregorio C (2010) Magnetic resonance imaging contrast agents based on iron oxide superparamagnetic ferrofluids. Chem Mater 22(5):1739–1748. https://doi.org/10.1021/cm9031557

Dendrimers-Based Nanoparticles for Cancer Therapy and Bioimaging

Jørn B. Christensen

Abstract Using dendrimers for drug delivery and diagnostics was one of the earliest potential applications. The field has evolved tremendously since and has now matured to a state, where real applications are getting close. This chapter reviews the progress in the use of dendrimer-based nanoparticles for cancer therapy and for bioimaging, and presents selected results published during the last 10 years. Topics include the EPR-effect, RNA/DNA-therapy, dendrimers as drug delivery devices for chemotherapy, dendrimers as nanodrugs and dendrimer-based imaging reagents and theranostics.

Keywords Dendrimers · Nanoparticles · Drug delivery · Nanodrugs · Imaging · Theranostics

1 Introduction

Dendrimers are molecules built by repetitive branching from a core creating structure built layer wise. They grow rapidly in size and for a larger (the size is usually given by the generation) dendrimers they have a surface and an interior (Fig. 1).

Dendrimers have many advantages compared to other types of nanoparticles; they are essentially monodisperse, the size can be controlled by synthesis and an infinite number of structures are possible combining different A_xB_y building blocks. Dendrimers grow rapidly in size (diameter) and have a large number of surface groups, which can be utilized for multivalent presentation of ligands bound to the surface. Depending on the branch-cell unit, a dendrimer can have cavities inside the molecule, which can accommodate small molecules, but small molecules can also be non-covalently bound to the surface groups of the dendrimer giving rise to endo- or exo-complexation depending on where the guest molecule is located. So

J. B. Christensen (✉)
Department of Chemistry, University of Copenhagen, Thorvaldsensvej 40, 1871 Frederiksberg C, Denmark
e-mail: jbc@chem.ku.dk

G. Gonçalves and G. Tobias (eds.), *Nanooncology*, Nanomedicine and Nanotoxicology, https://doi.org/10.1007/978-3-319-89878-0_8

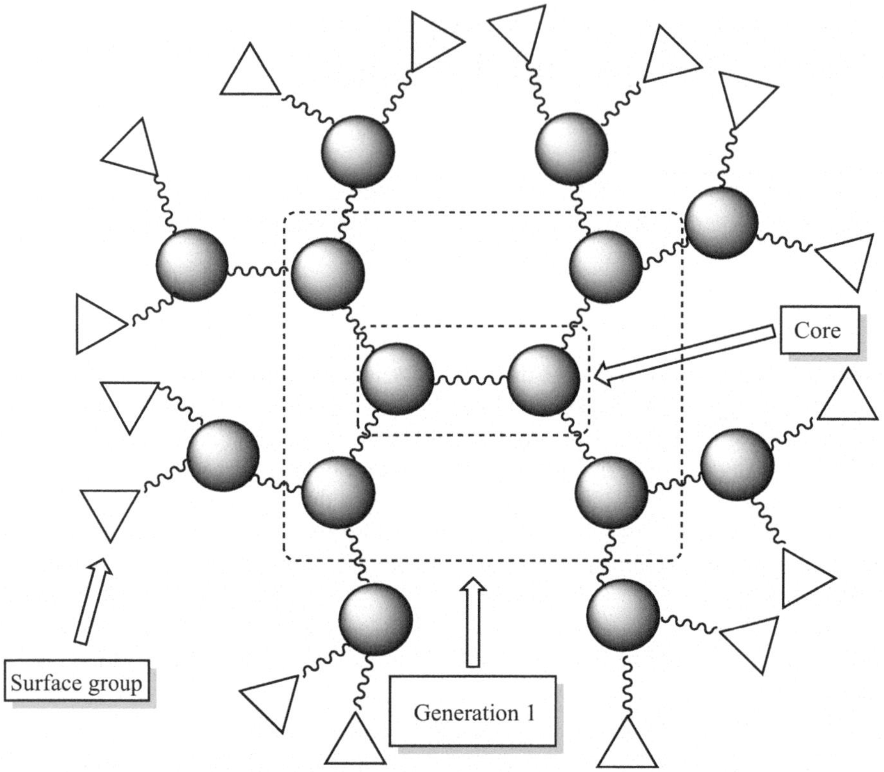

Fig. 1 Hypothetical generation 2 dendrimer based on an AB_4-core with CD_2 branch units

dendrimers can be used as drugs (Vivagel® from Starpharma Ltd) or as covalently bound drug-delivery systems.

Dendrimers have become increasingly popular in the last decade as the next big thing for the treatment, imaging, and analysis of cancer, which is quite logical considering the huge success with liposome-based formulations such as Doxil®. There have been published more than 1800 papers since 2005! This is a huge amount activity that reflects that dendrimers have aged from being chemical curiosities to slowly becoming a platform for nanomedicine. This also means that the present chapter cannot give a complete coverage of all the work done, but gives the major lines of research, that have been pursued.

2 The EPR-Effect

It is an experimental fact that nanoparticles tend to be concentrated in cancerous tissue instead of healthy tissue. The phenomenon has been explained as the enhanced permeation and retention effect (EPR), which should be due to the presence of leaky blood vessels in the cancerous tissue due to the fast growth of cancer cells and a lack of lymphatic drainage. The EPR-effect has, however, been questioned and although there could be other explanations of the phenomenon, it is still a good way of achieving targeted delivery without the need of targeting groups on the nanoparticles or dendrimers as seen in drugs like Doxil®.

3 RNA and DNA-Therapy Based on Dendritic Carriers

Small interfering RNA is a sequence of double-stranded RNA that interferes with the expression of specific genes preventing their translation into proteins. In cases where specific proteins are involved in a disease like cancer, it is possible to shut down their synthesis and if the correct protein is targeted, the state of disease. siRNA sequences are short only around 20 base pairs compared to DNA and siRNA does not change the genome, so it is a very attractive strategy provided that the target proteins can be identified.

siRNA cannot be administered directly due to the presence of exonucleases but has to be protected by a carrier, that can also assist in the cellular uptake. Due to the presence of phosphate groups in the backbone of siRNA and DNA, both can form complexes with positively charged dendrimers and these complexes (dendriplexes) are taken up by human cells. The dendriplexes are located in endosomes and the release of the siRNA/DNA from the dendrimer depends on protonation of the dendrimer increasing its hydrodynamic ratio due to Coulombic repulsion (also known as the umbrella or proton sponge model [1] leading to shredding off the coat of siRNA/DNA as shown in Fig. 2.

Designing the ideal dendritic carrier is a delicate balance, because the dendriplexes need to be stable enough to reach their target but they should still be able to dissociate from the siRNA/DNA once inside the cell. Another important aspect is the toxicity of the carrier due to the charge.

Cenã and co-workers [2] reported the use of an amino-terminated G1 PAMAM-dendrimer (commercial material with four amino groups on the surface; size of the core not disclosed) as a vehicle for targeting p42 MAPK in prostate cancer. Good stability and silencing effect was observed compared to the commercial products Lipofectamine and HiPerFect, which is interesting because the toxicity of amino-terminated PAMAM-dendrimers increases with size due to a higher number of amino groups. Huang et al. [3] synthesized and studied a G5 PAMAM-dendrimer (EDA-core), where the surface had been converted into 3,5-diamino-triazines (62 out of 128 surface groups; synthesis shown in Fig. 3. This dendrimer formed siRNA-complexes that showed excellent knockdown both in vitro and in vivo of the expression of the MDM2-gene, which is involved in non-small cell lung cancer. The in vivo results were obtained in a PC9-luc tumor xenograft mouse model (Fig. 4).

Carbosilane dendrimers have an internal structure based on carbon–silicon bonds which have a very flexible structure. Adding quaternary ammonium groups to the

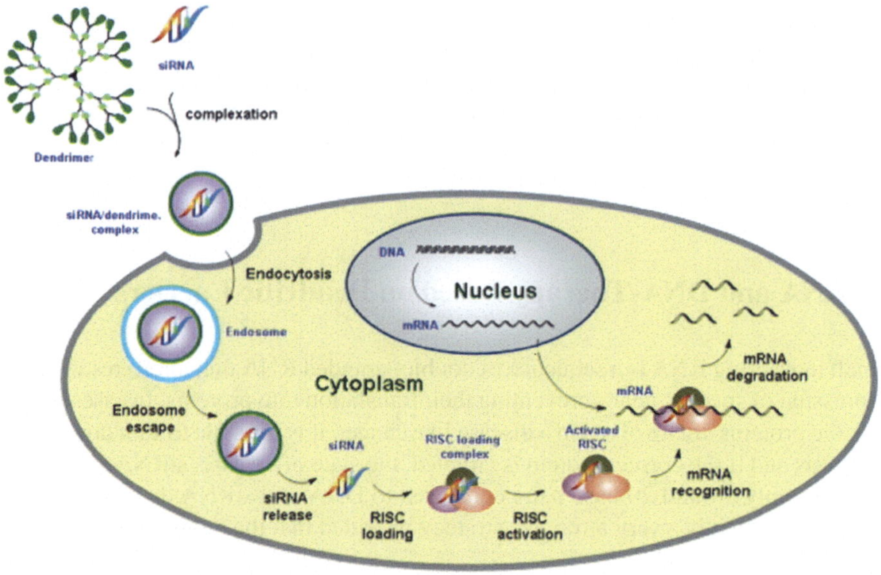

Fig. 2 Mechanism of dendrimer-mediated siRNA uptake and release [1]

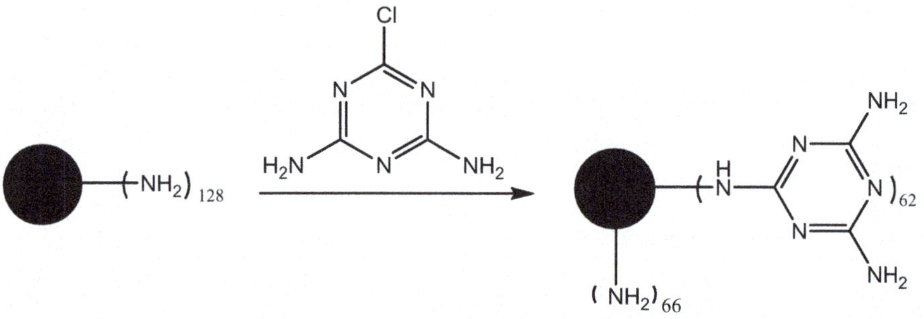

Fig. 3 The partially triazine-modified G5-PAMAM used by Huang et al. [3]

surface gives compounds that have been tested for in vitro siRNA-delivery against HIV and hepatocarcinoma [4–6] (Fig. 5).

The flexibility of the dendrimer is important for the stability of siRNA-dendrimer complexes, because a tight fit of the dendrimer to the phosphate backbone leads to a higher binding constant and stability of the complex. The interaction between RNA/DNA and different types of dendrimers have been studied by computational modeling as exemplified by the work of the groups of Danani [7, 8] and Pricl [7]. Pavan and co-workers [9] did molecular simulations to study the interactions between different generations of PAMAM-dendrimers and siRNA and have suggested a parameter to correlate the balance between enthalpy and entropy, which has been named the energetic flexibility index EF and is defined as $\mathrm{EF} = \frac{\Delta H}{T\Delta S}$, the

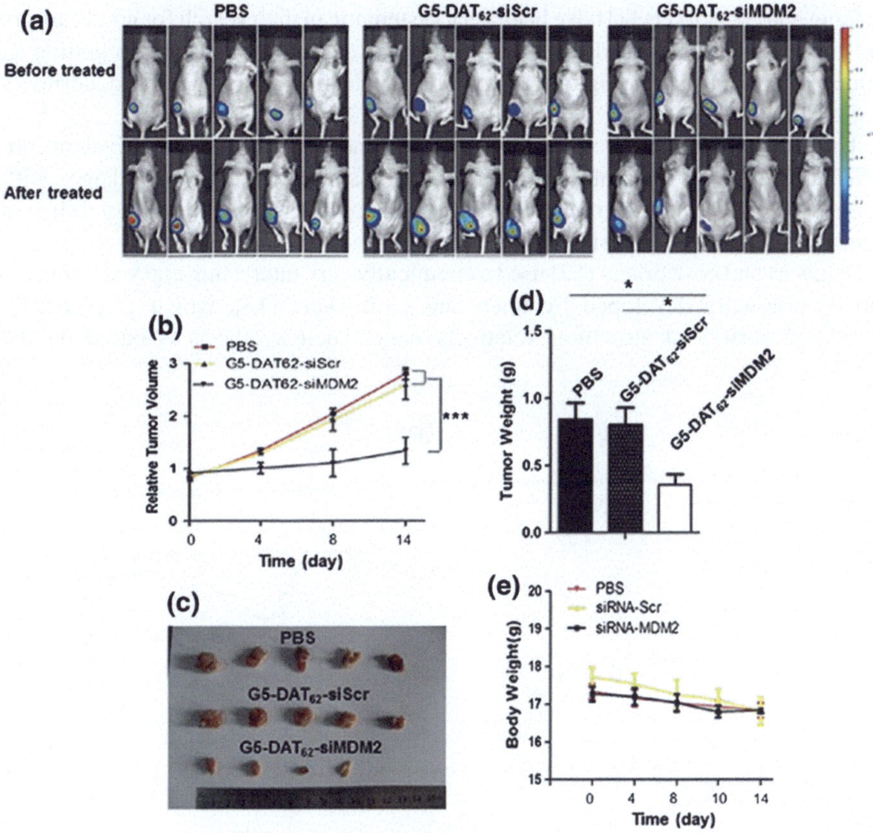

Fig. 4 **a** In vivo luminescence images of mice bearing PC-9-luc tumors before and after treatment with PBS, G5-DAT 62/siScr or G5-DAT 62/siMDM2. **b** The evolution of tumor volumes during the therapeutic period. **c** The photograph of the excised tumors after treatment. **d** The weight of the excised tumors after treatment. **e** Body weights of the mice during the treatment. $^{*}p < 0.05$ by student's t-test ($n = 5$) [3]

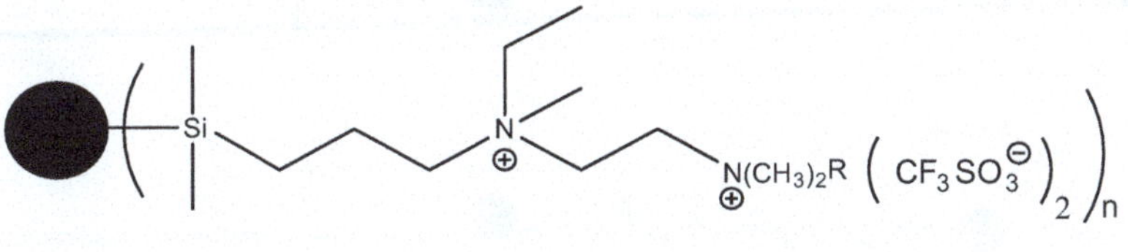

Fig. 5 Carbosilane dendrimer having 2 quaternary ammonium groups at the surface

parameters ΔH, and ΔS can be determined by binding studies and can be a guide to improving the design of a carrier. A high value of EF indicates the flexibility of the dendrimer and good binding, while a low value indicates a high entropic penalty and basically a misfit between the dendrimer and the siRNA.

Peng and co-workers [1] have been very systematic in their search for good carriers and have been focusing on dendrimers with a high degree of flexibility for getting an induced fit to the short and rigid RNA-strands. A selection of their flexible dendrimers is shown in Fig. 6.

The triethanolamine-cored dendrimers are characterized by having a trivalent core instead of the usual tetravalent and by replacing some of the amide linkages with ethers [10] or esters [11] greater flexibility is achieved while keeping solubility in water and a biomimetic design.

Siegwart and co-workers [12] used a chemically very interesting approach (Figs. 7 and 8) originally developed by Shen and co-workers [13], which gives access to a large variety of structures relatively easy. Their synthesis is based on the

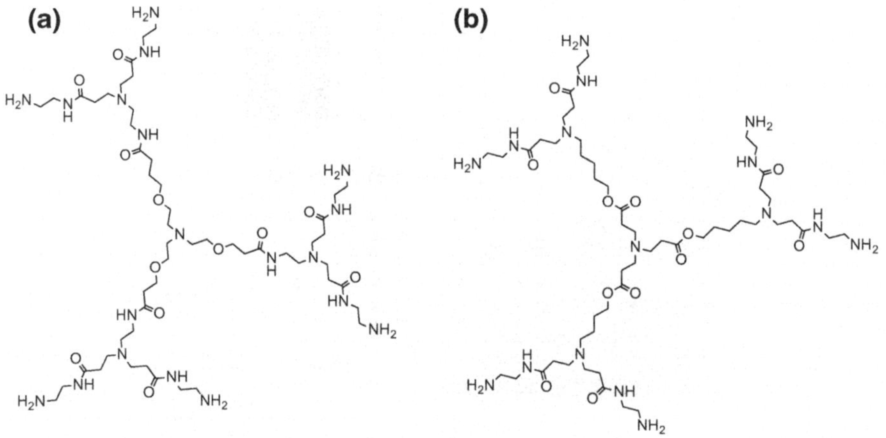

Fig. 6 **a** Triethanolamine-ether core-based dendrimer. **b** Nitrilotriacetic acid ester cored dendrimer

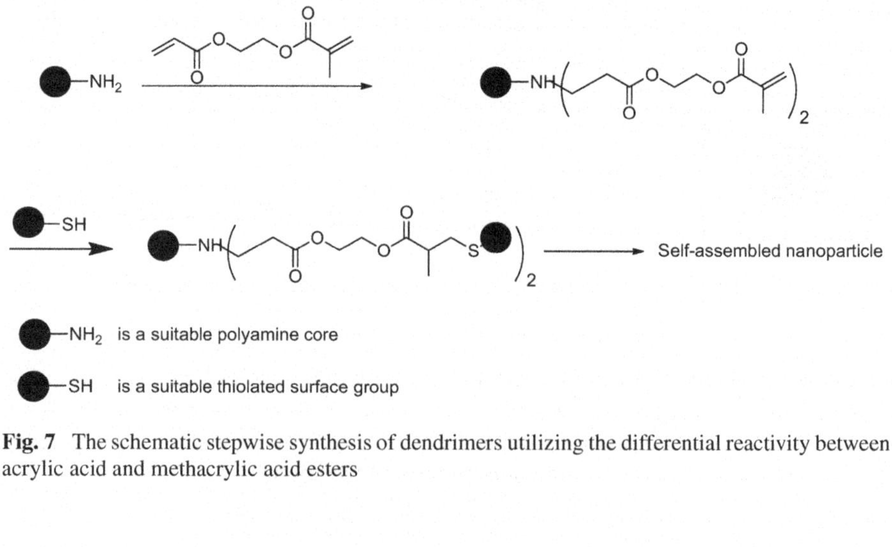

Fig. 7 The schematic stepwise synthesis of dendrimers utilizing the differential reactivity between acrylic acid and methacrylic acid esters

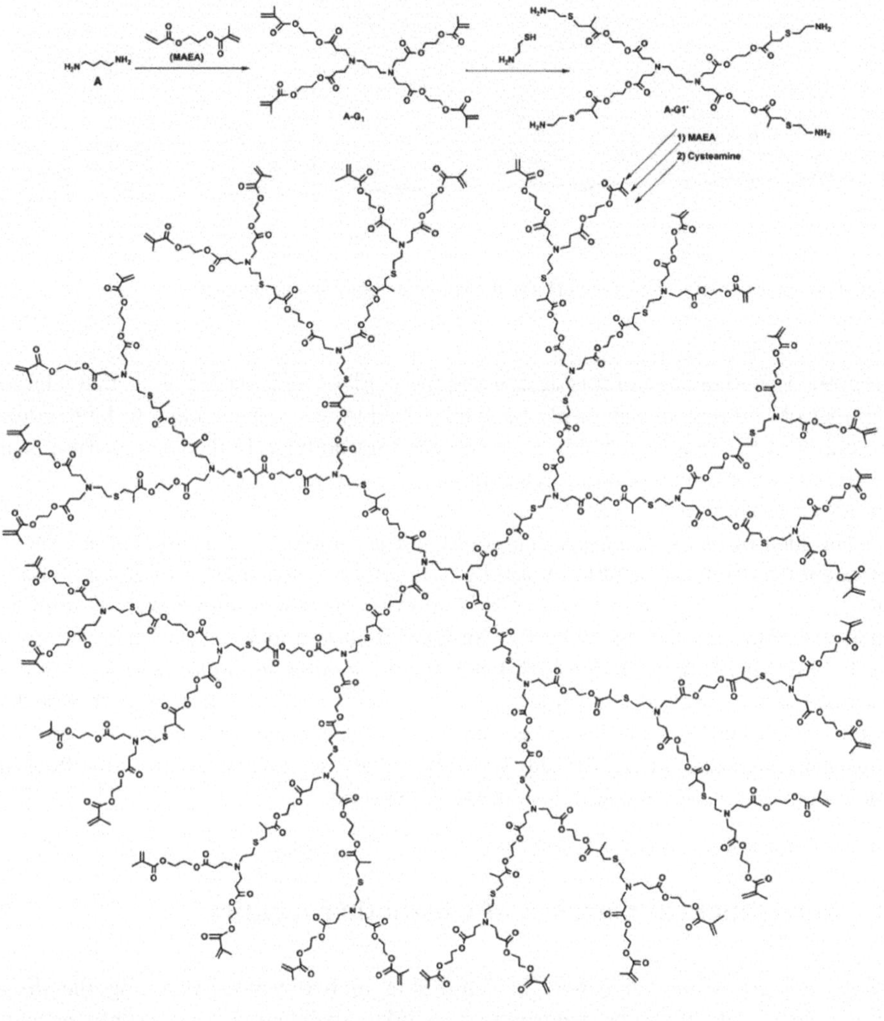

Fig. 8 Synthesis of a diaminobutane-cored dendrimer using the iterative method of Shen and co-workers [13]

differential reactivity of acrylates versus methacrylates in Michael additions. 2-(Acryloyloxy)ethyl methacrylate is reacted with the desired amine core leaving the methacrylic ester groups intact. They subsequently react with functionalized thiols using the "thiol-click" reaction to build the dendrimers, which were tested against liver cancer in mice showing promising properties.

Anderson and co-workers [14] synthesized a series of amphiphilic dendrimers from EDA-cored PAMAM- or PPI-dendrimers by reaction with different chain lengths of 2-alkyloxiranes giving the corresponding hydroxyalkylated dendrimers

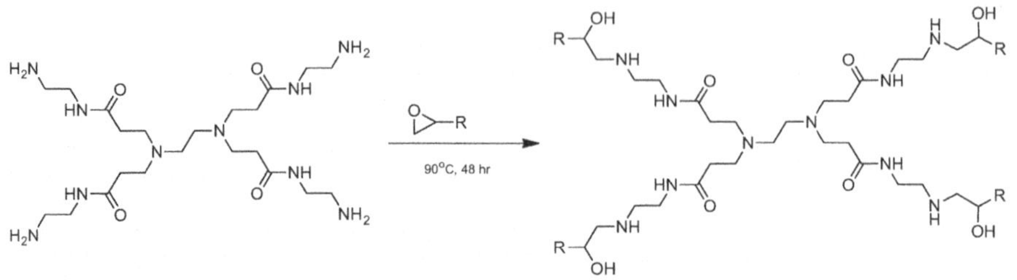

Fig. 9 Conversion of a EDA-core G1-PAMAM into an amphiphilic dendrimer

(Fig. 9). The advantage of this design and the work of Siegward et al. is, that a large number of compounds can easily be synthesized in few steps leading to large compound libraries, that then needs to be screened for activity. In the case of Anderson et al. [14], knockdown of alpha-fetoprotein in hepatocellular carcinoma cells in vivo in mice was shown.

The idea of using dendrons that could self-assemble to dendrimerisomes (non-covalent dendrimers) capable of siRNA-delivery has been studied by Peng and co-workers [1, 15]. The advantages of this approach is that it eliminates any conformational restrictions induced by the core and it should make elimination from the body easier because the individual dendrons are smaller and more flexible than the corresponding covalent dendrimer. There are, however, some potential drawbacks such as the stability in the bloodstream; a noncovalent system will have a critical micellar concentration and the entity will be subject to shear stress from the flow in the bloodstream, which could destabilize the system.

4 Dendrimers as Carriers of Chemotherapeutics

Drug delivery with dendrimers is achievable in two ways: Attaching the drug molecules covalently to the dendrimer via a linker, that degrade in vivo or by relying on non-covalently bound guest–host complexes of the drug molecule and a suitable dendrimer. Both types of systems have been investigated and special cases of the covalent systems are the self-immolative dendrimers, which degrade completely upon breaking a critical bond.

To get a covalently bound dendrimer-drug system, that can release the bound drug in vivo, it is necessary to have a cleavable linker. Derivatives of cis-aconitinic acid [10, 15–17], hydrazone-linkers [9, 10, 17–22], both are acid labile and disulfides [23, 17, 24, 25], that are cleaved in a reductive environment, for example, by reaction with glutathione inside cells, have all been used (Fig. 10).

Self-immolative dendrimers are dendrimers that undergo disintegration upon cleavage of a critical bond. The concept was first demonstrated by de Groot [26, 27], McGrath [28, 29], and Shabat [30], who independently developed dendrimers

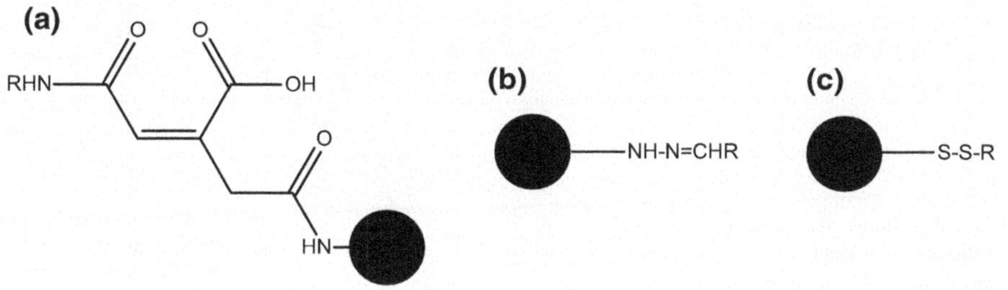

Fig. 10 Different types of biologically cleavable linkers

that easily underwent retro-Michael additions to form smaller molecules upon cleavage of a strategic bond.

Most of the systems described rely on the facile formation of quinone methides by elimination from suitably substituted benzylic amines or phenols and this might impose some limitations due to solubility issues of the products formed from the quinone methides, which are reactive intermediates that can polymerize.

In the non-covalent systems, two types of complexes exist: Exo-complexes, where the guest is bound to the surface and endo-complexes, where the guest is bound to the interior of the dendrimer. Guest–host chemistry in dendrimers and their use as drug-delivery systems have been extensively studied and for some general references see [32, 32, 33, 34].

Platinum(II) complexes have a long history in chemotherapy; they are highly cytotoxic acting by cross-linking DNA. Unfortunately, these compounds also cause permanent damage to the nervous tissue in the peripheral system on longer use. Many of the complexes like cisplatin (Fig. 11) are uncharged and since Pt(II) is a d^8-electron system, they have a square planar configuration with very slow ligand exchange. Guest–host complexes with dendrimers have been known for a long time and some examples are [35, 32, 36–52]. Changing oxidation state from Pt(II) to Pt(IV) gives octahedral d^6-complexes, which are also kinetically robust. Covalent binding of Pt(IV)-complex to a dendrimer has also been studied [53]. A very interesting study was done by Kulhari et al. [49], who took a critical look at formulations of carboxy-terminated PAMAM-dendrimers with cisplatin from a drug-delivery point of view; they studied the degree of loading, the influence of the size of the core of the dendrimer (1,2-diaminoethane vs. 1,4-butanediamine vs. 1,12-diaminododecane); batch-to-batch variability, which is critical for any practical use; release profile under different conditions and stability—unfortunately all are in vitro studies. The stability studies showed a change in release profile after storage at room temperature for 70 days where only 15% of the drug were releasable again. Taking the slow ligand exchange in Pt(II)-complexes into account, this could very likely the result of the formation of a Pt(II)-complex with the dendrimer.

Given the success with cisplatin, other transition metal complexes have also been tested for anticancer properties [25, 54] as, for example, the complexes between PPI-dendrimers and Ruthenocene(II) by Rodrigues and co-workers [55], which are

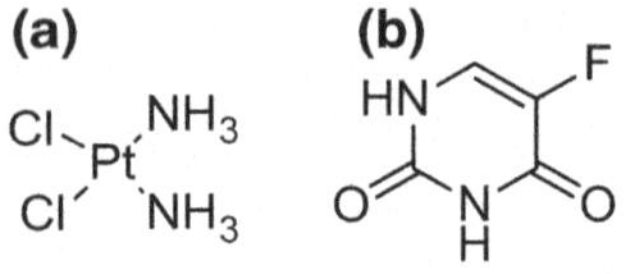

Fig. 11 Chemical structure of **a** cisplatin. **b** 5-Fluorouracil

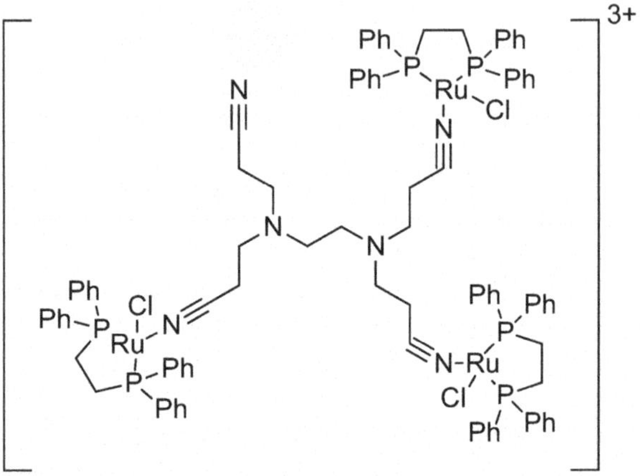

Fig. 12 Chemical structure of the Ruthenium(II) complex of a EDA-core PPI-dendrimer

interesting because the Ruthenium is complexed to the –C≡N nitrogen of a half generation PPI-nitrile (Fig. 12).

Guest–host systems with other chemotherapeutics have also been reported for example 5-Fluorouracil (Fig. 11) and peptide dendrimers [20] form guest–host complexes, which showed good skin permeating ability. These systems have subsequently been analyzed by molecular dynamics to give insights into the complexation and the interesting conclusion is, that a neutral dendritic host gives better binding than charged systems in the case of 5-fluorouracil [56]. Different surface modified PAMAM-dendrimers has also been studied as pH-responsive systems aiming at releasing the drugs inside the tumor due to a lower pH [49, 57–59].

Doxorubicin (Fig. 13) is a very popular drug for studying drug-delivery systems for treatment of cancer; it is highly cyto- and cardiotoxic and is not excreted from the human organism limiting the total amount of drug that can be given to a patient. It was also the first real success for nanomedicine in the form of Doxil®, which is a micellar formulation of doxorubicin that allows the use of smaller amounts of drug to get the desired effect. Some examples of doxorubicin complexed in or to dendrimers are [9, 16, 18, 60–63].

5 Dendrimers as Nanodrugs

In the majority of cases, the dendrimers are vehicles for transporting drugs to a target, but in recent years, examples have started to appear, where the dendrimer

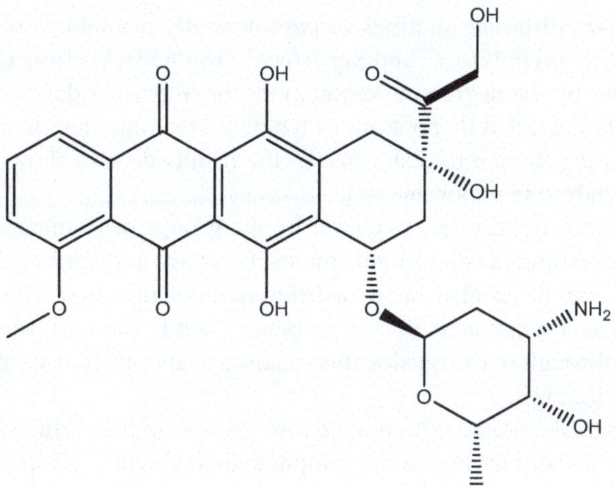

Fig. 13 Molecular structure of doxorubicin

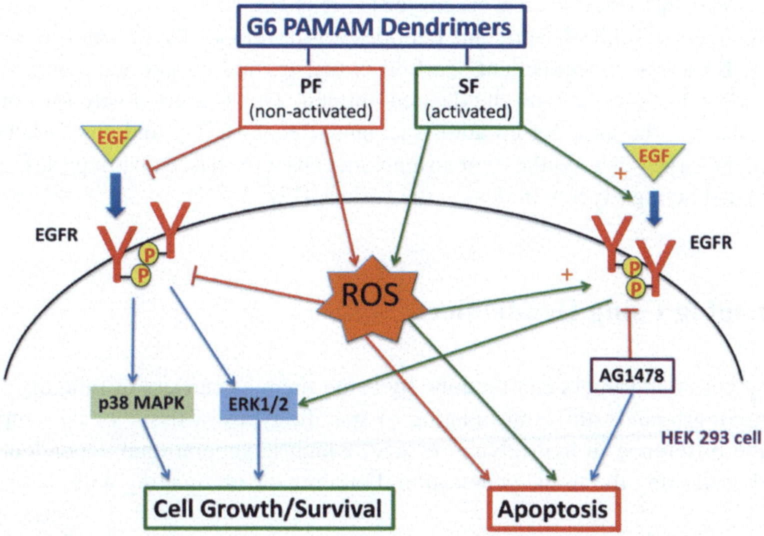

Fig. 14 Cellular uptake and fate of a spherical (PF) G6-PAMAM-dendrimer and a degraded (SF) G6-PAMAM-dendrimer

itself acts as a drug. Amino-terminated PAMAM-dendrimers are in general toxic due to the highly cationic surface, but a recent work by Akhtar et al. [64] have shown that they also affect EGFR signaling in vitro and in vivo, and might as such be useful for cancer therapy (Fig. 14). A very interesting aspect of this study is, that they are using two different qualities of commercially available amino-terminated PAMAM-dendrimers (Polyfect® and Superfect® G6-PAMAMs from Qiagen). These dendrimers differ by the degree of degradation; the Superfect-dendrimers that have been deliberately degraded by cooking in 1-butanol making them less spherical and probably converting them into dendrons, so the results do also show that the shape of dendrimer/dendron is important.

The phosphorous dendrimers explored by the groups of Caminade and Majoral (Fig. 15) are interesting; the dendrimers themselves have anticancer properties but so does the Cu(II)-complexes also and by a different mode of action. The phosphorous-dendrimer acts as a weak activator of caspase-3, while the corresponding Cu(II)-complex works through BAX translocation leading to apoptosis as shown by Mignani and co-workers [65, 66].

PPI-dendrimers decorated with maltotriose-groups on the surface (Fig. 16) show apoptosis-inducing activity in chronic lymphocytic leukemia cells [67], that is comparable with that of the drug fludarabine. This is a very good illustration of the use of the multivalency of dendrimers to achieve a biological effect.

Amphiphilic lysine-dendrons developed for siRNA-therapy was also found to have activity against glioblastoma cells without affecting neurons or astrocytes as demonstrated by Ceña and co-workers [68] (Fig. 17).

The concept of multiple antigenic peptides was developed by Tam and co-workers [69–71]. Basically, a multivalent scaffold is constructed by peptide synthesis and decorated on the surface with the desired antigen. This is a good way for boosting activity that can be used for creating vaccines. Recently Bay and co-workers [72] reported the large-scale synthesis of an anticancer vaccine based on a peptide dendron scaffold displaying glycopeptides on the surface (Fig. 18).

6 Imaging Using Dendrimers

Imaging with dendrimers can be done by using them as carriers of imaging agents and targeting ligands allowing imaging of specific types of tissue or by simply by using the difference in hydrodynamic size, which is generational dependent, as a method of limiting the tissue penetration. For a review on imaging with dendrimers see [73].

Imaging with X-ray techniques such as CT relies on the attenuation of X-rays, and tissue containing heavier elements such as calcium in bones have a higher natural contrast than soft tissue. Therefore, contrast agents are needed in order to examine soft tissue and organs. A contrast agent for X-ray-based techniques needs to incorporate heavier elements that should be in a nontoxic form. Iodine has been very popular and a number of contrast agents for CT scanning are poly(iodinated) compounds. Metal nanoparticles of metals such as gold are also useful, because they are inert and the only potential issue is excretion from the body. Shi and co-workers [74] reported recently that a G5 PAMAM-dendrimer, where the surface was modified with carboxybetaine acrylamide units, could bind gold nanoparticles (Au-DENs)

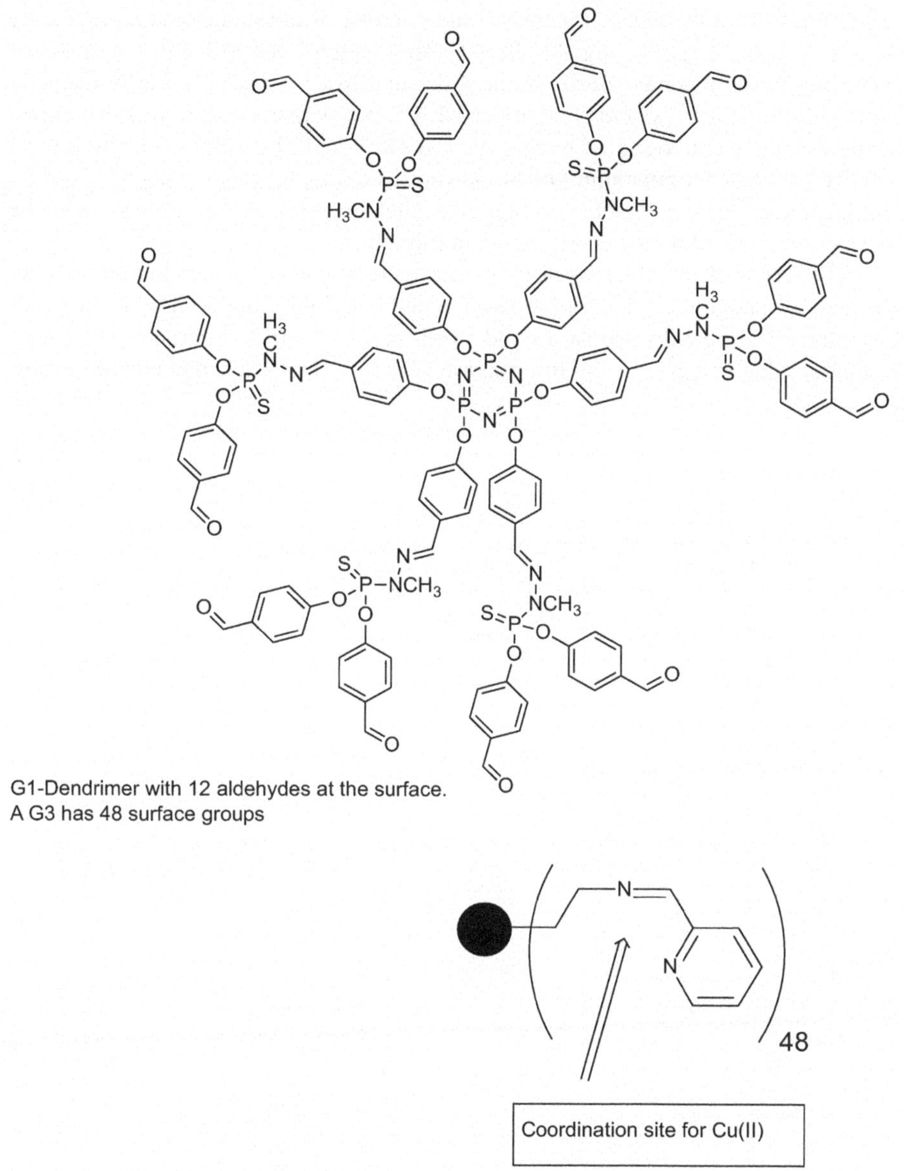

Fig. 15 A G1 Caminade-Majoral dendrimer and the G3-dendrimer studies in Refs. [65, 66]

to the surface and used them subsequently for lymphatic imaging by CT in mice (Fig. 19).

MR-imaging relies on mapping the proton density in the tissue by NMR. The relaxation of the [1] H-nuclei determines the contrast of the image. Shortening the relaxation times (T1 and T2) increases the contrast of the image and shortens the time necessary for collecting the data. Contrast agents for MR can be classified according to the mechanism of action: Gd^{3+} and Mn^{3+} affect T1, while magnetic iron oxides (magnetites) and Dy^{3+} affect T2. The problem encountered with contrast agents for MRI is toxicity of metals such as Gd^{3+} and Dy^{3+} as well as biological activity (and risk for poisoning) of Mn in various oxidation states. This is solved by administrating them as stable complexes or chelates that can be excreted from the body. A few typical reagents are shown in Fig. 20.

Methods based on radioactive isotopes have the advantage of very high sensitivity but require special facilities for preparation and handling. Single Photon Emission-Computed Tomography (SPECT) and Positron Emission Tomography (PET) are techniques that can give useful information [75, 76]. The radioactive nuclide is usu-

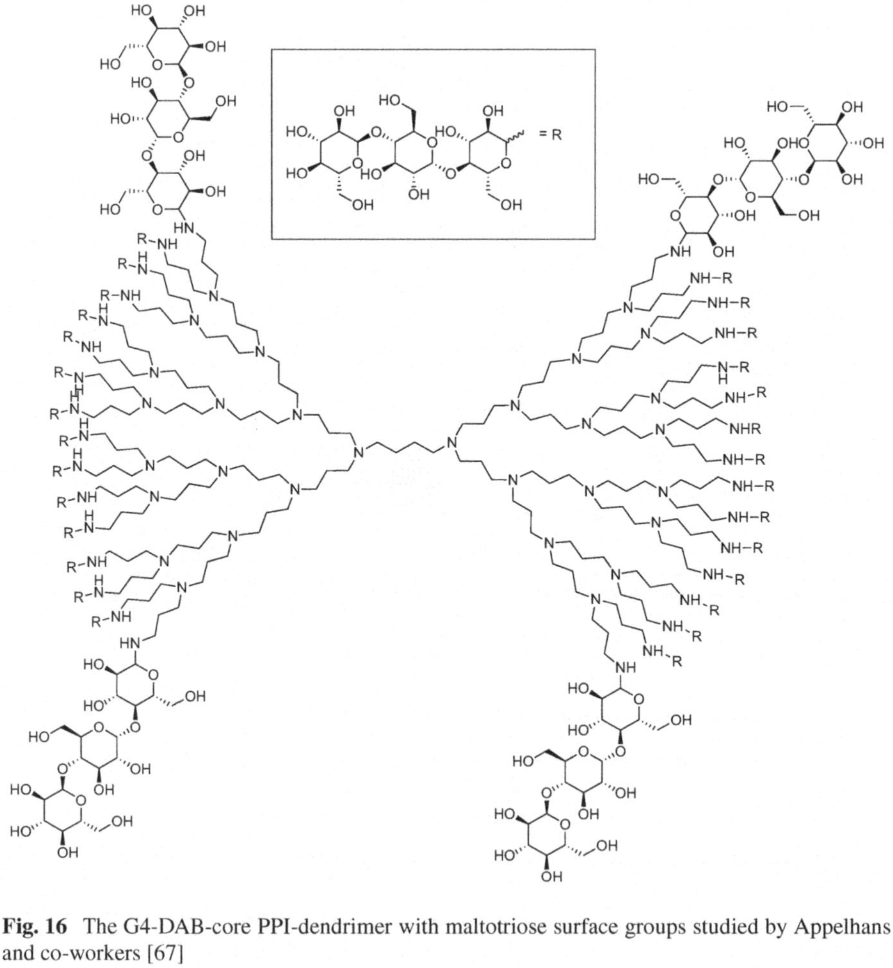

Fig. 16 The G4-DAB-core PPI-dendrimer with maltotriose surface groups studied by Appelhans and co-workers [67]

ally bound to a suitable ligand, but in the case of PET the half-life of the nuclides are usually very short minimizing the exposure to radiation for the patient, but on the other hand also putting restraints on the number of synthetic steps, that can be performed on the material containing the radioactive nuclide.

The multivalent surface of a dendrimer allows the preparation of "molecular Xmas-trees", where a dendrimer is decorated with different types of modalities. Tsien and co-workers [77, 78] made constructs carrying a cell-penetrating peptide masked by a peptide sequence specifically cleavable by tumor-specific proteases as

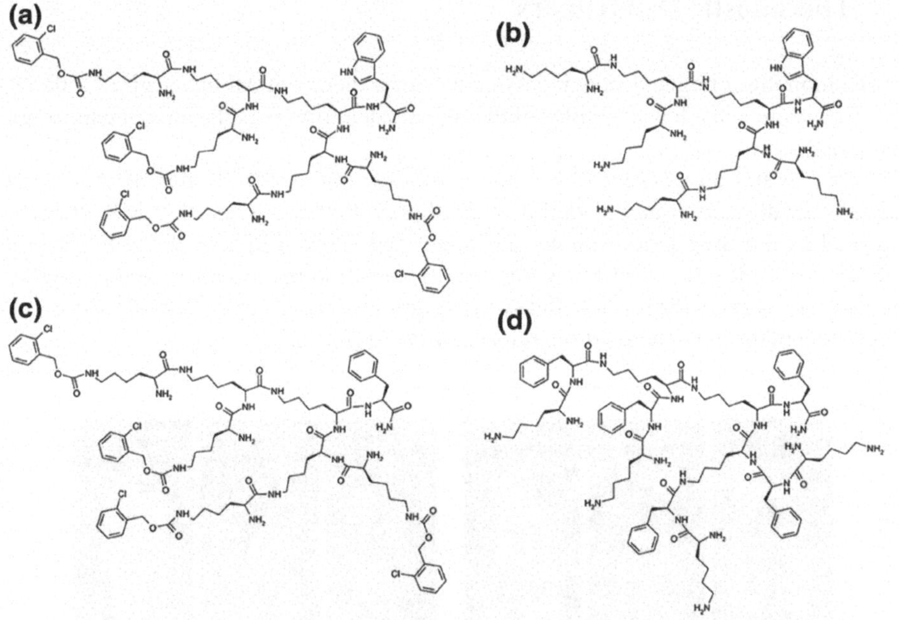

Fig. 17 Four lysin-based dendrons investigated by Ceña and co-workers [68]

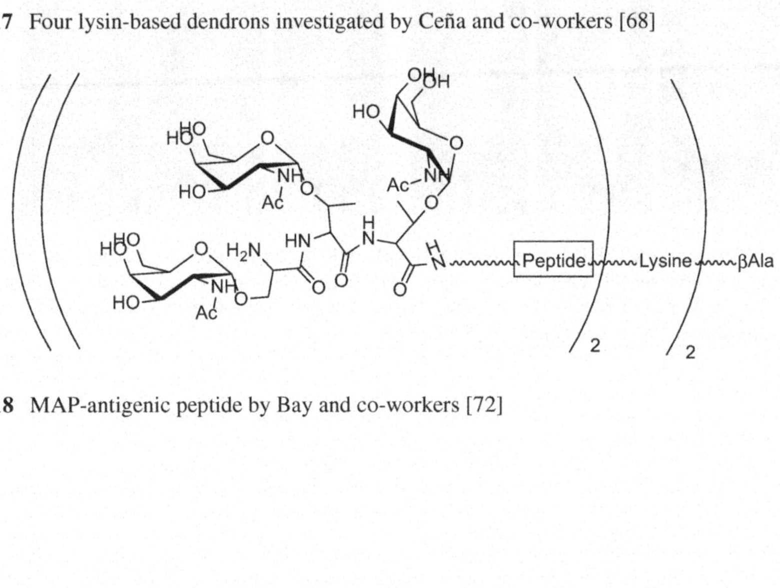

Fig. 18 MAP-antigenic peptide by Bay and co-workers [72]

well as fluorescence- and or MR-probes allowing the location of the tumor and its boundaries to the surrounding healthy tissue (Fig. 21).

Wang and co-workers reported constructs consisting of a G5-PAMAM-dendrimer, an MR-imaging agent, a tumor-specific antibody against HER-2 and hosting a gold nanoparticle in the interior of the dendrimer, which acts as a contrast agent for CT scanning [79] and a trimodal having a folic acid targeting ligand, a gadolium complex for MR, an interior gold nanoparticle for CT and Cy5.5 for NIR-fluorescence [23].

7 Theranostic Dendrimers

Combining a targeted imaging system with a therapeutical agent gives a theranostic system and a few examples of theranostic dendrimers in the area of cancer are next discussed.

Terreno and co-workers [80] made a self-assembling dendrimer from a small Janus face dendrimer and loaded it with a lipophilic Gd-complex for MR-imaging as well as the drug prednisolone phosphate and showed uptake and activity in a melanoma tumor model in mice. Another self-assembling system was reported by Gu and co-workers [81], where the system combined detection by NIR-fluorescence with dual pH/redox-triggered the release of Pt(II) (Fig. 22).

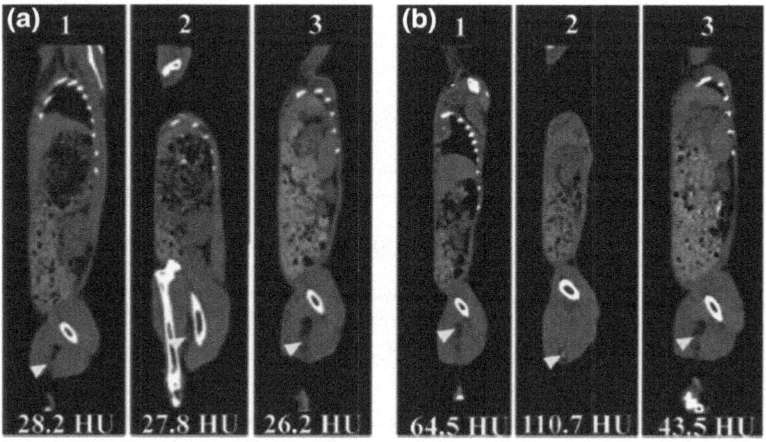

Fig. 19 CT images and CT values of a rabbit lymph node before (**a**) and at 2 h **b** post hoc injection of [(Au0)$_{100}$-G5·NHAc-CBAA$_{20}$] (1), [(Au0)$_{100}$-G5·NHAc-CBAA$_{80}$] (2) or [(Au0)$_{100}$-G5·NHAc-mPEG$_{20}$] (3) ([Au] = 0.1 M, in 500 μL saline) (n = 3). Reproduced from Shi and co-workers [74] with permission of The Royal Society of Chemistry

8 Outlook

The area of using dendrimers for diagnostics and treatment of cancer has been growing exponentially in recent years, and an interesting development has been the self-assembling dendrimers, which offers a simple shortcut from smaller dendrons to dendrimers avoiding the synthetic difficulties. It will be interesting to see how these systems translate into clinical use. Dendrimers having different surface groups arranged in a well-defined pattern allowing systematic studies of topics such as uptake mechanisms, therapeutical effect, theranostics, or biodistribution in tumors are a goal that requires a control in dendrimer synthesis that we do not have at the present. Understanding the interaction between dendrimers and the immune system is very important, because some types of cancer seems to be able to avoid discovery by the immune system, and at present it is known that dendrimers can be anti-, pro-, and non-inflammatory [82] and this might be utilized for developing more dendrimer-based drugs or theranostics against cancer. Another and more troublesome problem was presented in a recent commentary on the status of nanomedicine and cancer entitled "Just so stories: The random acts of anticancer nanomedicine performance"

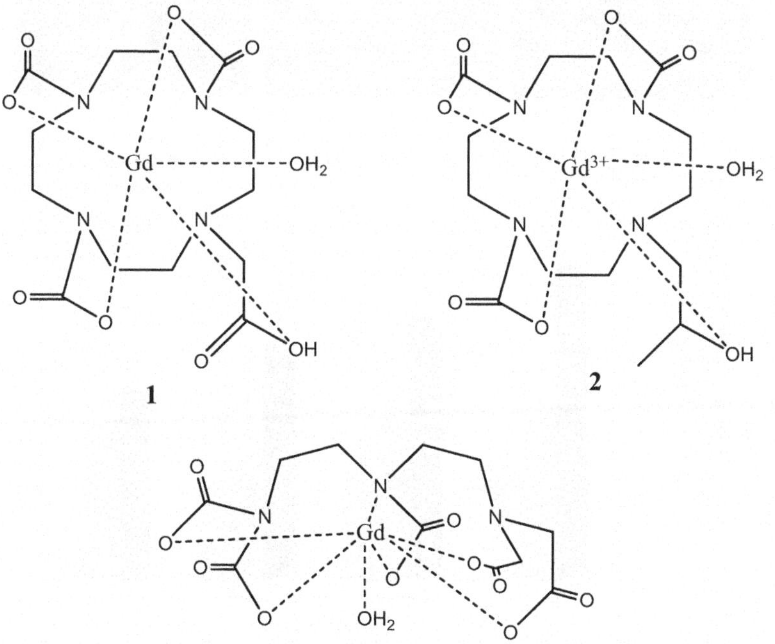

1 2 3

Fig. 20 Three commonly used chelates in MR-imaging, shown with Gd^{3+} as the metal and an exchangeable water molecule coordinated to the metal. The ligands are: 1: DOTA; 2: DO3A; 3: DTPA

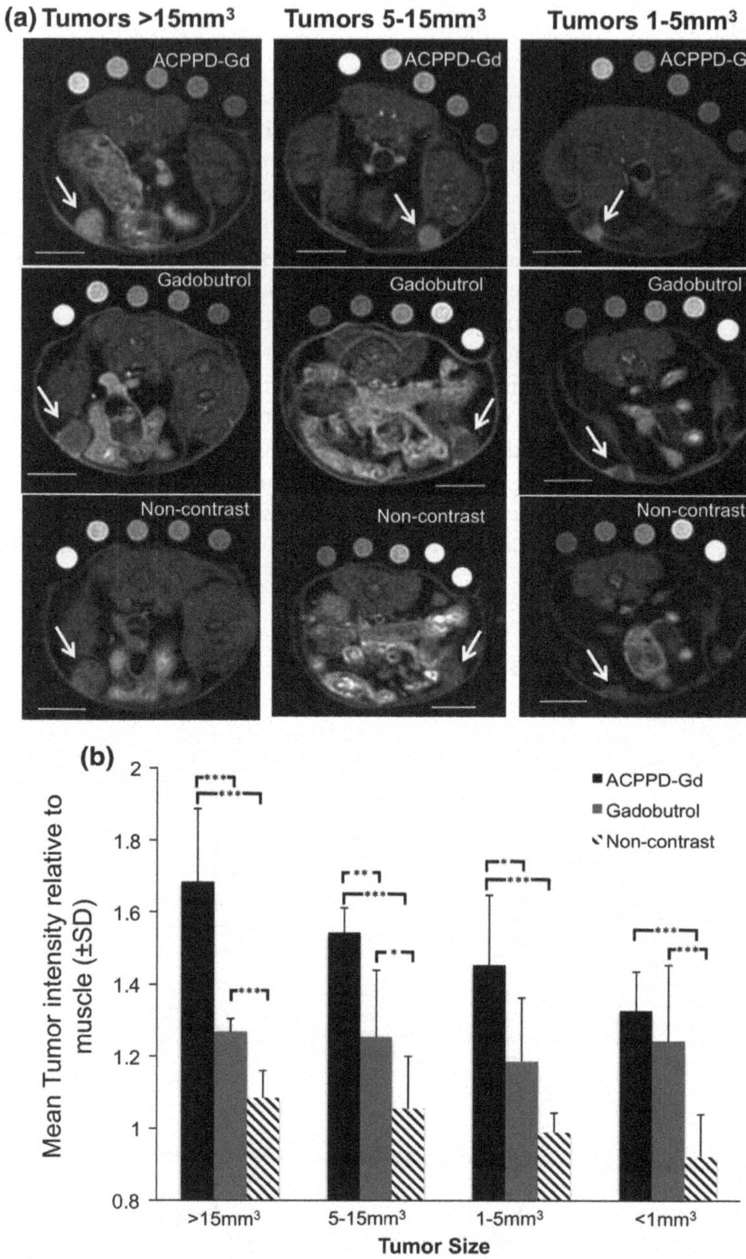

Fig. 21 Degree and pattern of enhancement as a function of tumor size. **a** Representative axial fat-saturated T1w MR images of mice given ACPPD-Gd (top row) and others imaged before (bottom row) and after gadobutrol (middle row) with comparable size tumors shown with the same window and level. Note that while ACPPD-Gd produced homogenous diffuse tumor enhancement (arrows), gadobutrol enhanced predominantly the tumor rim. Scale bar in each image = 5 mm. b, Bar graph shows mean tumor enhancement ± SD for each contrast group as a function of tumor size groups. * indicates $p < 0.05$, ** $p < 0.01$, *** $p < 0.001$

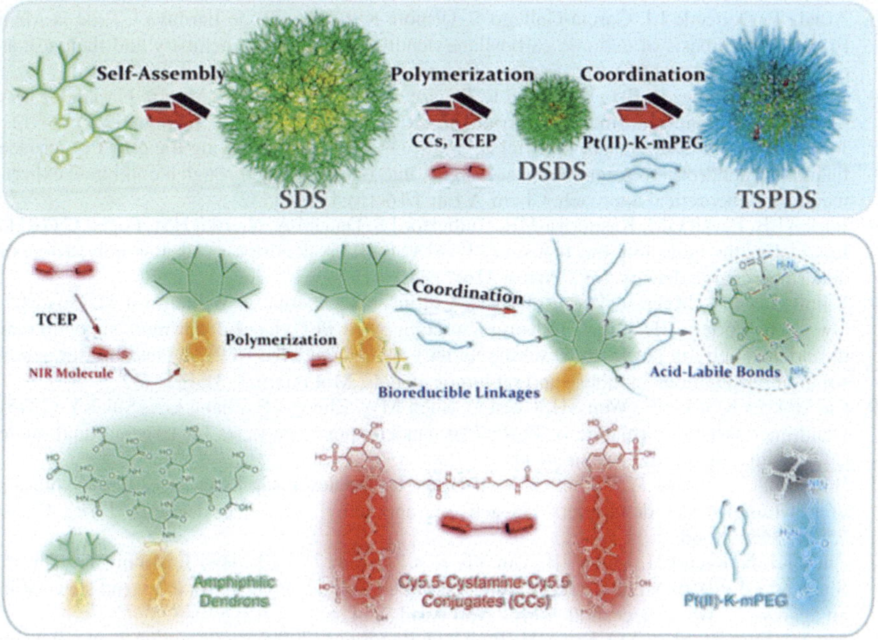

Fig. 22 Synthesis of a self-assembling theranostic dendrimer as described by Gu and co-workers [81]

[83]. Many nanomaterials show excellent results in vitro or in mouse models, but fail or show unpredictable activity and serious side effects, when put to the test in humans. There can be many reasons for this and it is a challenge for the field of dendrimers in nanomedicine that will generate a lot of good research in the future.

References

1. Liu XX, Liu C, Catapano CV, Peng L, Zhou JH, Rocchi P (2014) Structurally flexible triethanolamine-core poly(amidoamine) dendrimers as effective nanovectors to deliver RNAi-based therapeutics. Biotechnol Adv 32(4):844–852
2. Monteagudo S, Perez-Martinez FC, Perez-Carrion MD, Guerra J, Merino S, Sanchez-Verdu MP, Cena V (2012) Inhibition of p42 MAPK using a nonviral vector-delivered siRNA potentiates the anti-tumor effect of metformin in prostate cancer cells. Nanomedicine 7(4):493–506
3. Huang Q, Li L, Li L, Chen H, Dang YY, Zhang JS, Shao NM, Chang H, Zhou ZJ, Liu CY, He BW, Wei HF, Xiao JR (2016) MDM2 knockdown mediated by a triazine-modified dendrimer in the treatment of non-small cell lung cancer. Oncotarget 7(28):44013–44022
4. de las Cuevas N, Garcia-Gallego S, Rasines B, de la Mata FJ, Guijarro LG, Munoz-Fernandez MA, Gomez R (2012) In vitro studies of water-stable cationic carbosilane dendrimers as delivery vehicles for gene therapy against HIV and hepatocarcinoma. Curr Med Chem 19(29):5052–5061

5. Arnaiz E, Doucede LI, Garcia-Gallego S, Urbiola K, Gomez R, de Ilarduya CT, de la Mata FJ (2012) Synthesis of cationic carbosilane dendrimers via click chemistry and their use as effective carriers for DNA transfection into cancerous cells. Mol Pharm 9(3):433–447
6. Jimenez JL, Gomez R, Briz V, Madrid R, Bryszewsk M, de la Mata FJ, Munoz-Fernandez MA (2012) Carbosilane dendrimers as carriers of siRNA. J Drug Deliv Sci Technol 22(1):75–82
7. Jones SP, Pavan GM, Danani A, Pricl S, Smith DK (2010) Quantifying the effect of surface ligands on dendron-DNA interactions: insights into multivalency through a combined experimental and theoretical approach. Chem-A Eur J 16(15):4519–4532
8. Jensen LB, Pavan GM, Kasimova MR, Rutherford S, Danani A, Nielsen HM, Foged C (2011) Elucidating the molecular mechanism of PAMAM-siRNA dendriplex self-assembly: effect of dendrimer charge density. Int J Pharm 416(2):410–418
9. Kaminskas LM, McLeod VM, Kelly BD, Cullinane C, Sberna G, Williamson M, Boyd BJ, Owen DJ, Porter CJH (2012) Doxorubicin-conjugated PEGylated dendrimers show similar tumoricidal activity but lower systemic toxicity when compared to PEGylated liposome and solution formulations in mouse and rat tumor models. Mol Pharm 9(3):422–432
10. Cai HD, Li KG, Li JC, Wen SH, Chen Q, Shen MW, Zheng LF, Zhang GX, Shi XY (2015) Dendrimer-assisted formation of Fe_3O_4/Au nanocomposite particles for targeted dual mode CT/MR imaging of tumors. Small 11(35):4584–4593
11. Bouillon C, Tintaru A, Monnier V, Charles L, Quelever G, Peng L (2010) Synthesis of poly(amino)ester dendrimers via active cyanomethyl ester intermediates. J Org Chem 75(24):8685–8688
12. Zhou KJ, Nguyen LH, Miller JB, Yan YF, Kos P, Xiong H, Li L, Hao J, Minnig JT, Zhu H, Siegwart DJ (2016) Modular degradable dendrimers enable small RNAs to extend survival in an aggressive liver cancer model. Proc Natl Acad Sci U S A 113(3):520–525
13. Ma XP, Tang JB, Shen YQ, Fan MH, Tang HD, Radosz M (2009) Facile synthesis of polyester dendrimers from sequential click coupling of asymmetrical monomers. J Am Chem Soc 131(41):14795–14803
14. Khan OF, Zaia EW, Yin H, Bogorad RL, Pelet JM, Webber MJ, Zhuang I, Dahlman JE, Langer R, Anderson DG (2014) Ionizable Amphiphilic dendrimer-based nanomaterials with alkyl-chain-substituted amines for tunable siRNA delivery to the liver endothelium in vivo. Angew Chem-Int Ed 53(52):14397–14401
15. Medina SH, Tiruchinapally G, Chevliakov MV, Durmaz YY, Stender RN, Ensminger WD, Shewach DS, ElSayed MEH (2013) Targeting hepatic cancer cells with PEGylated dendrimers displaying N-Acetylgalactosamine and SP94 peptide ligands. Adv Healthc Mater 2(10):1337–1350
16. Yabbarov NG, Posypanova GA, Vorontsov EA, Popova ON, Severin ES (2013) Targeted delivery of doxorubicin: drug delivery system based on PAMAM dendrimers. Biochem 78(8):884–894
17. Werengowska-Ciecwierz K, Wisniewski M, Terzyk AP, Furmaniak S (2015) The chemistry of bioconjugation in nanoparticles-based drug delivery system. Adv Condens Matter Phys
18. Kaminskas LM, Kelly BD, McLeod VM, Sberna G, Owen DJ, Boyd BJ, Porter CJH (2011) Characterisation and tumour targeting of PEGylated polylysine dendrimers bearing doxorubicin via a pH labile linker. J Controlled Release 152(2):241–248
19. Machakanur SS, Patil BR, Badiger DS, Bakale RP, Gudasi KB, Bligh SWA (2012) Synthesis, characterization and anticancer evaluation of novel tri-arm star shaped 1,3,5-triazine hydrazones. J Mol Struct 1011:121–127
20. Mutalik S, Shetty PK, Kumar A, Kalra R, Parekh HS (2014) Enhancement in deposition and permeation of 5-fluorouracil through human epidermis assisted by peptide dendrimers. Drug Deliv 21(1):44–54
21. Hemmati M, Kazemi B, Najafi F, Zarebkohan A, Shirkoohi R (2016) Synthesis and evaluation of a glutamic acid-modified hPAMAM complex as a promising versatile gene carrier. J Drug Target 24(5):408–421
22. Lidicky O, Janouskova O, Strohalm J, Alam M, Klener P, Etrych T (2015) Anti-lymphoma efficacy comparison of anti-CD20 monoclonal antibody-targeted and non-targeted star-shaped polymer-prodrug conjugates. Molecules 20(11):19849–19864

23. Chen C, Posocco P, Liu XX, Cheng Q, Laurini E, Zhou JH, Liu C, Wang Y, Tang JJ, Dal Col V, Yu TZ, Giorgio S, Fermeglia M, Qu FQ, Liang ZC, Rossi JJ, Liu MH, Rocchi P, Pricl S, Peng L (2016) Mastering dendrimer self-assembly for efficient siRNA delivery: from conceptual design to in vivo efficient gene silencing. Small 12(27):3667–3676
24. Lee C, Lo ST, Lim J, da Costa VCP, Ramezani S, Oz OK, Pavan GM, Annunziata O, Sun XK, Simanek EE (2013) Design, synthesis and biological assessment of a triazine dendrimer with approximately 16 paclitaxel groups and 8 PEG groups. Mol Pharm 10(12):4452–4461
25. El Kazzouli S, El Brahmi N, Mignani S, Bousmina M, Zablocka M, Majoral JP (2012) From metallodrugs to metallodendrimers for nanotherapy in oncology: a concise overview. Curr Med Chem 19(29):4995–5010
26. de Groot FMH, Albrecht C, Koekkoek R, Beusker PH, Scheeren HW (2003) "Cascade-release dendrimers" liberate all end groups upon a single triggering event in the dendritic core. Angew Chem Int Ed 42(37):4490–4494
27. de Groot FMH, Broxterman HJ, Adams H, van Vliet A, Tesser GI, Elderkamp YW, Schraa AJ, Kok RJ, Molema G, Pinedo HM, Scheeren HW (2002) Design, synthesis, and biological evaluation of a dual tumor-specific motive containing integrin-targeted plasmin-cleavable doxorubicin prodrug. Mol Cancer Ther 1(11):901–911
28. McGrath DV (2005) Dendrimer disassembly as a new paradigm for the application of dendritic structures. Mol Pharm 2(4):253–263
29. Szalai ML, Kevwitch RM, McGrath DV (2003) Geometric disassembly of dendrimers: dendritic amplification. J Am Chem Soc 125(51):15688–15689
30. Amir RJ, Shabat D (2006) Domino dendrimers. In: SatchiFainaro R, Duncan R (eds) Polymer therapeutics I: polymers as drugs, conjugates and gene delivery systems. 192:59–94
31. Kannan RM, Nance E, Kannan S, Tomalia DA (2014) Emerging concepts in dendrimer-based nanomedicine: from design principles to clinical applications. J Int Med 276(6):579–617
32. Harper BW, Krause-Heuer AM, Grant MP, Manohar M, Garbutcheon-Singh KB, Aldrich-Wright JR (2010) Advances in platinum chemotherapeutics. Chem-A Eur J 16(24):7064–7077
33. Fox ME, Guillaudeu S, Frechet JMJ, Jerger K, Macaraeg N, Szoka FC (2009) Synthesis and in vivo antitumor efficacy of PEGylated Poly(L-lysine) dendrimer-camptothecin conjugates. Mol Pharm 6(5):1562–1572
34. Bharali DJ, Khalil M, Gurbuz M, Simone TM, Mousa SA (2009) Nanoparticles and cancer therapy: a concise review with emphasis on dendrimers. Int J Nanomed 4(1):1–7
35. Markelov DA, Falkovich SG, Neelov IM, Ilyash MY, Matveev VV, Lahderanta E, Ingman P, Darinskii AA (2015) Molecular dynamics simulation of spin-lattice NMR relaxation in poly-L-lysine dendrimers: manifestation of the semiflexibility effect. Phys Chem Chem Phys 17(5):3214–3226
36. Krause-Heuer AM, Grant MP, Orkey N, Aldrich-Wright JR (2008) Drug delivery devices and targeting agents for platinum(II) anticancer complexes. Aust J Chem 61(9):675–681
37. Haxton KJ, Burt HM (2009) Polymeric drug delivery of platinum-based anticancer agents. J Pharm Sci 98(7):2299–2316
38. Pisani MJ, Wheate NJ, Keene FR, Aldrich-Wright JR, Collins JG (2009) Anionic PAMAM dendrimers as drug delivery vehicles for transition metal-based anticancer drugs. J Inorg Biochem 103(3):373–380
39. Kapp T, Dullin A, Gust R (2010) Platinum(II)-dendrimer conjugates: synthesis and investigations on cytotoxicity, cellular distribution, platinum release, DNA, and protein binding. Bioconjug Chem 21(2):328–337
40. Cheng YY, Zhao LB, Li YW, Xu TW (2011) Design of biocompatible dendrimers for cancer diagnosis and therapy: current status and future perspectives. Chem Soc Rev 40(5):2673–2703
41. Haririan I, Alavidjeh MS, Khorramizadeh MR, Ardestani MS, Ghane ZZ, Namazi H (2010) Anionic linear-globular dendrimer-cis-platinum (II) conjugates promote cytotoxicity in vitro against different cancer cell lines. Int J Nanomed 5:63–75
42. Farrell NP (2011) Platinum formulations as anticancer drugs clinical and pre-clinical studies. Curr Top Med Chem 11(21):2623–2631

43. Kirkpatrick GJ, Plumb JA, Sutcliffe OB, Flint DJ, Wheate NJ (2011) Evaluation of anionic half generation 3.5–6.5 poly(amidoamine) dendrimers as delivery vehicles for the active component of the anticancer drug cisplatin. J Inorg Biochem 105(9):1115–1122
44. Yellepeddi VK, Vangara KK, Palakurthi S (2013) Poly(amido)amine (PAMAM) dendrimer-cisplatin complexes for chemotherapy of cisplatin-resistant ovarian cancer cells. J Nanopart Res 15(9)
45. Oberoi HS, Nukolova NV, Kabanov AV, Bronich TK (2013) Nanocarriers for delivery of platinum anticancer drugs. Adv Drug Deliv Rev 65(13–14):1667–1685
46. Zikos C, Evangelou A, Karachaliou CE, Gourma G, Blouchos P, Moschopoulou G, Yialouris C, Griffiths J, Johnson G, Petrou P, Kakabakos S, Kintzios S, Livaniou E (2015) Commercially available chemicals as immunizing haptens for the development of a polyclonal antibody recognizing carbendazim and other benzimidazole-type fungicides. Chemosphere 119:S16–S20
47. Kim JH, Yoon HJ, Sim J, Ju SY, Jang WD (2014) The effects of dendrimer size and central metal ions on photosensitizing properties of dendrimer porphyrins. J Drug Target 22(7):610–618
48. Kesavan A, Ilaiyaraja P, Beaula WS, Kumari VV, Lai JS, Arunkumar C, Anjana G, Srinivas S, Ramesh A, Rayala SK, Ponraju D, Venkatraman G (2015) Tumor targeting using polyamidoamine dendrimer-cisplatin nanoparticles functionalized with diglycolamic acid and herceptin. Eur J Pharm Biopharm 96:255–263
49. Kulhari H, Pooja D, Singh MK, Chauhan AS (2015) Optimization of carboxylate-terminated poly(amidoamine) dendrimer-mediated cisplatin formulation. Drug Dev Ind Pharm 41(2):232–238
50. Barth RF, Wu G, Meisen WH, Nakkula RJ, Yang W, Huo T, Kellough DA, Kaumaya P, Turro C, Agius LM, Kaur B (2016) Design, synthesis, and evaluation of cisplatin-containing EGFR targeting bioconjugates as potential therapeutic agents for brain tumors. Oncotargets Ther 9:2769–2781
51. England RM, Hare JI, Barnes J, Wilson J, Smith A, Strittmatter N, Kemmitt PD, Waring MJ, Barry ST, Alexander C, Ashford MB (2017) Tumour regression and improved gastrointestinal tolerability from controlled release of SN-38 from novel polyoxazoline-modified dendrimers. J Controlled Release 247:73–85
52. Parker JP, Ude Z, Marmion CJ (2016) Exploiting developments in nanotechnology for the preferential delivery of platinum-based anticancer agents to tumours: targeting some of the hallmarks of cancer. Metallomics 8(1):43–60
53. Sommerfeld NS, Hejl M, Klose MHM, Schreiber-Brynzak E, Bileck A, Meier SM, Gerner C, Jakupec MA, Galanski M, Keppler BK (2017) Low-generation polyamidoamine dendrimers as drug carriers for platinum(IV) complexes. Eur J Inorg Chem 12:1713–1720
54. Callari M, Aldrich-Wright JR, de Souza PL, Stenzel MH (2014) Polymers with platinum drugs and other macromolecular metal complexes for cancer treatment. Prog Polym Sci 39(9):1614–1643
55. Jardim MG, Rissanen K, Rodrigues J (2010) Preparation and characterization of novel poly(alkylidenamine) nitrile ruthenium metallodendrimers. Eur J Inorg Chem 11:1729–1735
56. De Luca S, Seal P, Ouyang DF, Parekh HS, Kannam SK, Smith SC (2016) Dynamical interactions of 5-fluorouracil drug with dendritic peptide vectors: the impact of dendrimer generation, charge, counterions, and structured water. J Phys Chem B 120(25):5732–5743
57. Buczkowski A, Waliszewski D, Urbaniak P, Palecz B (2016) Study of the interactions of PAMAM G(3)-NH_2 and G3-OH dendrimers with 5-fluorouracil in aqueous solutions. Int J Pharm 505(1–2):1–13
58. Rengaraj A, Subbiah B, Haldorai Y, Yesudhas D, Yun HJ, Kwon S, Choi S, Han YK, Kim ES, Shenpagam NH, Huh YS (2017) PAMAM/5-fluorouracil drug conjugate for targeting E6 and E7 oncoproteins in cervical cancer: a combined experimental/in silico approach. RSC Adv 7(9):5046–5054
59. Bai QK, Zhao ZG, Sui HJ, Chen J, Xie XH, Wen F (2015) The preparation and application of dendrimer modified CdTe/CdS near infrared quantum dots for brain cancer cells imaging. Appl Sci Basel 5(4):1076–1085

60. She WC, Pan DY, Luo K, He B, Cheng G, Zhang CY, Gu ZW (2015) PEGylated dendrimer-doxorubicin cojugates as pH-sensitive drug delivery systems: synthesis and in vitro characterization. J Biomed Nanotechnol 11(6):964–978
61. Niidome T, Yamauchi H, Takahashi K, Naoyama K, Watanabe K, Mori T, Katayama Y (2014) Hydrophobic cavity formed by oligopeptide for doxorubicin delivery based on dendritic poly(L-lysine). J Biomater Sci Polym Ed 25(13):1362–1373
62. Li XJ, Takashima M, Yuba E, Harada A, Kono K (2014) PEGylated PAMAM dendrimer-doxorubicin conjugate-hybridized gold nanorod for combined photothermal-chemotherapy. Biomaterials 35(24):6576–6584
63. Yabbarov NG, Posypanova GA, Vorontsov EA, Obydenny SI, Severin ES (2013) A new system for targeted delivery of doxorubicin into tumor cells. J Controlled Release 168(2):135–141
64. Akhtar S, Al-Zaid B, El-Hashim AZ, Chandrasekhar B, Attur S, Yousif MHM, Benter IF (2015) Cationic polyamidoamine dendrimers as modulators of EGFR signaling in vitro and in vivo. Plos One 10(7)
65. Mignani S, El Brahmi N, Eloy L, Poupon J, Nicolas V, Steinmetz A, El Kazzouli S, Bousmina MM, Blanchard-Desce M, Caminade AM, Majoral JP, Cresteil T (2017) Anticancer copper(II) phosphorus dendrimers are potent proapoptotic Bax activators. Eur J Med Chem 132:142–156
66. Ottaviani MF, El Brahmi N, Cangiotti M, Coppola C, Buccella F, Cresteil T, Mignani S, Caminade AM, Costes JP, Majoral JP (2014) Comparative EPR studies of Cu(II)-conjugated phosphorous-dendrimers in the absence and presence of normal and cancer cells. Rsc Adv 4(69):36573–36583
67. Franiak-Pietryga I, Ostrowska K, Maciejewski H, Appelhans D, Misiewicz M, Ziemba B, Bednarek M, Bryszewska M, Borowiec M (2017) PPI-G4 glycodendrimers upregulate TRAIL-induced apoptosis in chronic lymphocytic leukemia cells. Macromol Biosci 17(5)
68. Janiszewska J, Posadas I, Jativa P, Bugaj-Zarebska M, Urbanczyk-Lipkowska Z, Cena V (2016) Second generation amphiphilic poly-lysine dendrons inhibit glioblastoma cell proliferation without toxicity for neurons or astrocytes. Plos One 11(11)
69. Lu YA, Clavijo P, Galantino M, Shen ZY, Liu W, Tam JP (1991) Chemically unambiguous peptide immunogen—preparation, orientation and antigenicity of purified peptide conjugated to the multiple antigen peptide system. Mol Immunol 28(6):623–630
70. Defoort JP, Nardelli B, Huang W, Tam JP (1992) A rational design of synthetic peptide vaccine with a built-in adjuvant—a modular approach for unambiguity. Int J Pept Protein Res 40(3–4):214–221
71. Tam JP (1996) Recent advances in multiple antigen peptides. J Immunol Methods 196(1):17–32
72. Ganneau C, Simenel C, Emptas E, Courtiol T, Coic YM, Artaud C, Deriaud E, Bonhomme F, Delepierre M, Leclerc C, Lo-Man R, Bay S (2017) Large-scale synthesis and structural analysis of a synthetic glycopeptide dendrimer as an anti-cancer vaccine candidate. Org Biomol Chem 15(1):114–123
73. Qiao Z, Shi XY (2015) Dendrimer-based molecular imaging contrast agents. Prog Polym Sci 44:1–27
74. Xiong ZJ, Wang Y, Zhu JY, Li X, He Y, Qu J, Shen MW, Xia JD, Shi XY (2017) Dendrimers meet zwitterions: development of a unique antifouling nanoplatform for enhanced blood pool, lymph node and tumor CT imaging. Nanoscale 9(34):12295–12301
75. Ghobril C, Lamanna G, Kueny-Stotz M, Garofalo A, Billotey C, Felder-Flesch D (2012) Dendrimers in nuclear medical imaging. New J Chem 36(2):310–323
76. Kesharwani P, Xie LX, Banerjee S, Mao GZ, Padhye S, Sarkar FH, Iyer AK (2015) Hyaluronic acid-conjugated polyamidoamine dendrimers for targeted delivery of 3,4-difluorobenzylidene curcumin to CD44 overexpressing pancreatic cancer cells. Colloids Surf B 136:413–423
77. Malone CD, Olson ES, Mattrey RF, Jiang T, Tsien RY, Nguyen QT (2015) Tumor detection at 3 Tesla with an Activatable Cell Penetrating Peptide Dendrimer (ACPPD-Gd), alpha T1 Magnetic Resonance (MR) molecular imaging agent. Plos One 10(9)
78. Olson ES, Jiang T, Aguilera TA, Nguyen QT, Ellies LG, Scadeng M, Tsien RY (2010) Activatable cell penetrating peptides linked to nanoparticles as dual probes for in vivo fluorescence and MR imaging of proteases. Proc Natl Acad Sci U. S. A 107(9):4311–4316

79. Crotty S, Gerislioglu S, Endres KJ, Wesdemiotis C, Schubert US (2016) Polymer architectures via mass spectrometry and hyphenated techniques: a review. Anal Chim Acta 932:1–21
80. Filippi M, Catanzaro V, Patrucco D, Botta M, Tei L, Terreno E (2017) First in vivo MRI study on theranostic dendrimersomes. J Controlled Release 248:45–52
81. Li YK, Li YC, Zhang X, Xu XH, Zhang ZJ, Hu C, He YY, Gu ZW (2016) Supramolecular PEGylated dendritic systems as pH/Redox dual-responsive theranostic nanoplatforms for platinum drug delivery and NIR imaging. Theranostics 6(9):1293–1305
82. Janaszewska A, Gorzkiewicz M, Ficker M, Petersen JF, Paolucci V, Christensen JB, Klajnert-Maculewicz B (2018) Pyrrolidone modification prevents PAMAM dendrimers from activation of pro-inflammatory signaling pathways in human monocytes. Mol Pharm 15:12–20
83. Moghimi SM, Farhangrazi ZS (2014) Just so stories: the random acts of anti-cancer nanomedicine performance. Nanomed Nanotechnol Biol Med 10(8):1661–1666

Porous Silicon Particles for Cancer Therapy and Bioimaging

Steven J. P. McInnes, Abel Santos and Tushar Kumeria

Abstract Porous silicon (pSi) engineered by electrochemical etching of silicon has been explored as a drug delivery carrier with the aim of overcoming the limitations of traditional therapies and medical treatments. pSi is biodegradable, non-cytotoxic and has optoelectronic properties that make this platform material a unique candidate for developing biomaterials for drug delivery and theranostics therapies. pSi provides new opportunities to improve existing therapies in different areas, paving the way for developing advanced theranostic nanomedicines, incorporating payloads of therapeutics with imaging capabilities. However, despite these outstanding advances, more extensive in-vivo studies are needed to assess the feasibility and reliability of this technology for real clinical practice. In this Chapter, we present an updated overview about the recent therapeutic systems based on pSi, with a critical analysis on the problems and opportunities that this technology faces as well as highlighting the growing potential of pSi technolgy.

Keywords Porous silicon · Drug delivery · Theranostics · Therapy · Toxicity

S. J. P. McInnes
Division of Information Technology, Engineering and the Environment, The University of South Australia, Mawson Lakes, South Australia 5095, Australia

A. Santos (✉) · T. Kumeria
School of Chemical Engineering, The University of Adelaide, Adelaide 5005, Australia
e-mail: a1620797@adelaide.edu.au

A. Santos
Institute for Photonics and Advanced Sensing (IPAS), The University of Adelaide, Adelaide 5005, Australia

A. Santos
ARC Centre of Excellence for Nanoscale BioPhotonics (CNBP), The University of Adelaide, Adelaide 5005, Australia

T. Kumeria
School of Pharmacy, The University of Queensland, Brisbane, QLD, Australia

G. Gonçalves and G. Tobias (eds.), *Nanooncology*, Nanomedicine and Nanotoxicology, https://doi.org/10.1007/978-3-319-89878-0_9

1 Introduction

Since the beginning of the twentieth century when Paul Ehrlich postulated his visionary concept of "magic bullet" until the modern concept of nanomedicine, an intensive activity in fundamental research and medical translation has been carried out for the development of nanomedicines [1–3]. Nanomedicines are drug delivery systems, where therapeutics are encapsulated within nanocarriers and administered through different routes in order to release their therapeutic content over diseased cells, tissues, or organs [4, 5]. The aim of drug delivery systems based on nanocarriers is to selectively target disease-causing organisms, while keeping the rest of the host organism spared in order to minimize side effects and toxicity associated with drugs. Furthermore, these systems can also overcome other intrinsic limitations of conventional medicines by enhancing their biodistribution, solubility, selectivity, and pharmacokinetics [6, 7]. However, the modern concept of nanomedicine goes beyond simple drug delivery and more sophisticated forms of nanomedicines have been envisaged. Among them, theranostic agents combine both therapeutic and diagnostic entities into a single drug delivery carrier. This revolutionary concept opened a plethora of possibilities, not only to treat diseases but also to diagnose and understand the biological response and progression of living systems to different treatments [8, 9]. Drug nanocarriers with theranostic capabilities enable imaging of nanomedicines to localize and monitor drug release and assess the therapeutic efficacy by analyzing the response to a given treatment in situ, using numerous imaging modalities. Many different types of theranostic nanomedicines have been developed over the recent years [10, 11]. Typically, these nanomedicines are composed of nanocarriers such as polymers, liposomes, micelles, antibodies, and nanoparticles. In many cases, these vehicles are endowed with therapeutic and diagnosis capabilities by co-loading drugs and imaging agents. However, some of these vehicles can feature theranostic capabilities using the intrinsic properties of the nanocarrier. Among these, porous silicon (pSi) is considered an excellent theranostic platform due to its unique set of physical and chemical properties, which may be precisely tuned by engineering its porosity.

pSi fabrication via the electrochemical etching of silicon was discovered by the Uhlirs at Bell Laboratories in 1956 [12–15]. Years after, Gösele and Canham identified quantum confinement effects in pSi structures, opening new opportunities to expand the application of pSi in optoelectronics [16, 17]. However, while its unique optoelectronic properties make silicon and its porous forms the material of choice for microelectronic systems, the discovery of its biocompatibility and biodegradability by Canham in 1995 boosted and spread its use for drug delivery applications [18, 19]. Canham demonstrated for the first time the in vitro biocompatibility of pSi films featuring micro-, meso-, and macro-porosity. His pioneering studies revealed that the biodegradability and biointegration of pSi structures can be modulated by means of the level of porosity and the size of its pores. Canham's works boosted an increasing activity in the field of pSi for drug delivery applications and enabled new opportunities to explore the use of this promising material in medicine and related fields [20–22]. More recently, taking advantage of the optoelectronic properties of pSi, Sailor demonstrated for the first time that pSi-based nanoparticles can also act as theranostic platforms using a time-gated imaging approach [23–25]. pSi has lower toxicity than heavy metal-based quantum dots, offers better photostability than organic fluorophores, emits light in the NIR region (biological window 650–1300 nm), which is optimal to penetrate tissue, it is biodegradable with low cytotoxicity and can load high doses of therapeutic agents [25]. Furthermore, the combination of pSi nanoparticles with photosensitizers enables photodynamic therapies for localized treatments to reduce the side effects of anticancer medicines by the temporal and spatial control of the production of cytotoxic species in diseased cells, tissues, or organs [24]. As a result, pSi is considered a unique and promising vehicle platform for theranostic nanomedicines.

Despite these advantages, the bench-to-bedside translation of pSi-based nanomedicines faces both technical and commercial challenges [20]. Nanomedicines are increasingly reaching the investors' attention and it is expected that the investment in this area will grow strongly during the next decade [26]. However, clinical developments require considerable financial support from investors and the investors' enthusiasm in biotechnological companies is mainly limited to later-stage projects with demonstrated performance, which require extensive clinical trials and detailed long-term studies. Furthermore, clinical studies may undergo failure and delays and the unexpected long-term side effects associated with pSi are yet to be assessed in real clinical practice [27–34].

Herein, we summarize the most recent and significant advances in pSi-based technology for cancer therapy and bioimaging, providing a comprehensive overview of the different fabrication methods, properties, and applicability. The different drug delivery concepts based on pSi are reviewed with a critical overview of their advantages and intrinsic limitations. Furthermore, we provide detailed information about the recent progress in the clinical translation of pSi-based systems for different therapies. We conclude our chapter with an outlook about the future challenges and developments in this exciting research field.

2 Fabrication and Properties of Porous Silicon

2.1 *Fabrication Methods*

pSi structures can be produced by a series of methods, including both electrochemical and chemical etching processes. The most extensively used method is electrochemical etching of crystalline silicon wafers in hydrofluoric acid electrolytes based on organic solvents such as acetonitrile (CH_3CN) or dimethylformamide (C_3H_7NO) [13–15]. However, other fabrication methods such as metal-assisted etching and chemical etching of silicon in mixtures of HF and nitric acid (HNO_3) are gaining attention due to their versatility to produce pSi structures featuring complex geometrical features (e.g., nanopillars, nanowires, etc.) [35–40]. Furthermore, more recently, the conversion of natural or synthetic silicon dioxide micro/nanostructures into silicon replicas has been envisaged as an advantageous approach toward the development of innovative pSi structures by reduction of silicon dioxide into silicon [41–44].

2.1.1 Electrochemical Etching

Silicon wafers can be electrochemically etched via anodization [13–15]. In this electrochemical process, the silicon wafer acts as anode and, typically, a platinum wire is used as cathode. The application of a current bias between anode and cathode in a suitable electrolyte is translated into the selective dissolution of the crystalline structure of silicon into its porous form (pSi). The structure of pSi, reported for the first time by Watanabe and Sakai in 1971, can be readily engineered by the fabrication conditions (i.e., anodization current, type of silicon wafer, level of doping, illumination, patterning, electrolyte, temperature, etc.) [45]. Table 1 summarizes the most representative conditions and pore morphologies for pSi structures. During electrochemical etching, hydrogen-terminated silicon is electrochemically dissolved by the replacement of superficial H atoms by fluoride ions (F^-), which are present in the electrolyte solution (Eq. 1). Given that the surface of silicon is passivated with H atoms under open circuit conditions, the application of an anodic current bias is required to create an electronic hole (h^+) for the formation of a neutralized Si-F bond through polarization induced by the more electronegative F atom over the Si atom. In this mechanism, a new F atom bounds the Si atom at a different position, resulting in the generation of hydrogen molecules (H_2). The repetition of this process progressively weakens the Si–Si bonds in the silicon structure by means of the strong electronegativity of F atoms. This process, known as nucleophilic attack, results in the formation of pSi structures by electrochemical etching, where silicon atoms are etched away from the crystalline structure of silicon, engineering a porous structure from top to bottom.

$$Si + 4HF_2^- + h^+ \rightarrow SiF_6^{2-} + 2HF + H_2 + e^- \quad (1)$$

Although many experimental studies have tried to discern the formation mechanism of pSi, a theoretical model that entirely explains this phenomenon is yet to come. However, it is generally accepted that pores in pSi are a result of different mechanisms, depending on the etching conditions, including crystallographic face selectivity, enhanced electric field, tunneling, and quantum confinement.

2.1.2 Metal-Assisted Etching

pSi structures can be fabricated by metal-assisted chemical etching, technique pioneered by Dimova and co-workers in 1997, which enables the production of pSi

Table 1 Compilation of the most representative electrochemical etching conditions used to produce pSi structures and details of their morphological features

Silicon wafer	Current density (mA cm^{-2})		Doping density (cm^{-3})		
			1×10^{17}	1×10^{18}	1×10^{19}
p	300	Pore morphology and porosity	Sponge like 80%	Narrow long pores with dendritic branches 75%	Wide long pores with high wall roughness 80%
	30		Sponge like 70%	Sponge like 60%	Medium long pores with dendritic branches 50%
	3		Sponge like 60%	Sponge like 40%	Narrow long pores with dendritic branches 40%
n	300	Pore morphology and porosity	Wide long pores with low wall roughness 35%	Wide long pores with medium wall roughness 40%	Narrow long pores with dendritic branches 45%
	30		Medium long pores with dendritic branches 10%	Narrow long pores with dendritic branches 15%	Narrow long pores with dendritic branches 20%
	3		Sponge like 10%	Sponge like 20%	Sponge like 30%

structures with high surface-to-volume ratio, low level of defects, high crystalline quality and finely tuned geometric features [46–48]. The characteristics of the resulting pSi-based structures are not attainable by traditional electrochemical etching since metal-assisted etching of silicon enables the control over the etching direction on the (100) and non-(100) planes produce pSi with slanted pore growth [35–40]. The morphology of the resulting pSi structures can range from pores to wires or pillars, the feature size of which can be tuned from a few nanometers to micrometers. Metal-assisted etching of silicon occurs when a silicon substrate partially covered by a noble metal is immersed in an etchant electrolyte containing an oxidative agent, such as hydrogen peroxide, and hydrofluoric acid. In this process, the silicon beneath the metal is etched much faster than uncovered silicon and the noble metal film sinks progressively into the silicon substrate. The control over the metallic film features, which acts as an etching mask, makes it attainable the production of pores or pillars in the structure of silicon by this top-down approach. Several chemical and electrochemical reactions are involved in the formation of pSi by metal-assisted etching, which occurs at the interface between the noble metal and the silicon substrate when the system is immersed in the etchant solution [49–52]. In this system, the metal acts as the cathode by reducing hydrogen peroxide (Eqs. 2 and 3).

$$H_2O_2 + 2H^+ \rightarrow 2H_2O + 2H^+ \tag{2}$$

$$2H^+ \rightarrow H_2 \uparrow +2H^+ \tag{3}$$

The silicon wafer, which works as anode, is progressively oxidized during the metal-assisted etching. Several reaction models have been proposed to explain this phenomenon: namely; (i) silicon is oxidized to silicon dioxide, which is dissolved by HF (Eqs. 4 and 5), (ii) silicon is directly dissolved by HF in divalent state (Eq. 6) and (iii) silicon is dissolved by HF in tetravalent state (Eqs. 7 and 8).

$$Si + 2H_2O \rightarrow SiO_2 + 4H^+ + 4e^- \tag{4}$$

$$SiO_2 + 6HF \rightarrow H_2SiF_6 + 2H_2O \tag{5}$$

$$Si + 4HF_2^- \rightarrow SiF_6^{-2} + 2HF + H_2 \uparrow +2e^- \tag{6}$$

$$Si + 4HF_2^- + 4H^+ \rightarrow SiF_4 + 4H^+ \tag{7}$$

$$SiF_4 + 2HF \rightarrow H_2SiF_6 \tag{8}$$

During the chemical etching, a charge transfer phenomenon between the noble metal and the silicon occurs, initiating the oxidation and dissolution of silicon by hole injection mechanism [47, 53–58]. Electronic holes generated in the noble metal during the reduction of the oxidant are subsequently injected into the silicon, which is oxidized and dissolved by chemical reaction with HF. Therefore, this process is highly dependent on the doping level of silicon and the composition of the etchant electrolyte.

2.1.3 Thermal-Assisted Reduction

So far, different methods and approaches have been explored to convert silica (i.e., silicon dioxide) structures into silicon. Traditional carbothermal reduction of silica to form silicon occurs at temperatures above 2000 °C, limiting the application of this approach to convert 3D porous silica structures into their silicon replicas due to high temperatures [41–43]. Nevertheless, recent studies have demonstrated that when silica is thermally treated in the presence of magnesium at 650 °C (i.e., melting temperature of magnesium), the latter promotes the conversion of bulk silica structures into their silicon replicas (Eq. 9). The process is finished by a chemical etching to dissolve magnesium oxide from the crystalline structure of silicon [41]. The resulting silicon structures feature approximately the same structure than that of the original silica counterpart. These structures have been demonstrated to be optimal for the development of sensors, optical and optoelectronic devices, and batteries [59–62].

$$2Mg(g) + SiO_2(s) \rightarrow 2MgO(s) + Si(s) \quad (9)$$

2.2 Physical and Chemical Properties

pSi structures can be produced with a plethora of pore morphologies, sizes, and geometries. Pores in pSi can be engineered not only by means of the electrochemical/chemical etching conditions but also by pre- and post-treatments such as photolithography, wet chemical etching in alkaline etchants, thermal treatments, etc. The International Union of Pure and Applied Chemistry (IUPAC) establishes that pSi structures can be classified into three categories according to their pore size (d_p): namely; (i) micro-pSi (μpSi – $d_p < 2$ nm), meso-pSi (mpSi – 2 nm $< d_p < 50$ nm), and macro-pSi (MpSi – $d_p > 50$ nm). It is well known that pSi structures can feature a broad range of morphological details, although the most common structures are sponge-like pSi with randomly distributed branched pores and pSi with cylindrical pores of rough or smooth walls (Fig. 1) [63]. Typically, μpSi and mpSi present the former porous morphology while the latter is characteristic of MpSi structures. All these structures feature a random spatial distribution of pores across the surface of silicon. However, perfectly ordered porous structures based on MpSi can be produced if the surface of silicon is patterned by lithographic methods prior to chemical or electrochemical etching [52, 64, 65]. Furthermore, hierarchical porous structures combining μpSi, mpSi, and MpSi can be produced by different chemical and electrochemical approaches [66].

As-produced pSi structures, are hydride terminated (Si–H) [67], and are, therefore, prone to oxidize and degrade when exposed to environmental conditions or biological fluids [68]. While this can be an advantage for some applications such as drug delivery or tissue engineering, this can also prevent the use of pSi in some applications, where long-term chemical stability under different conditions is desired (e.g., in vivo sensors, theranostics, etc.). To overcome this limitation, researchers have

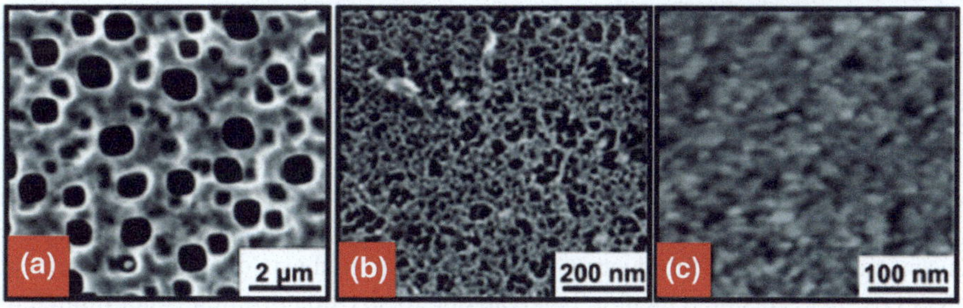

Fig. 1 Images of the surface morphology of the three forms of pSi **a** macropores with the opening of 1 μm (SEM), **b** mesopores with pore size>15 nm (SEM) and **c** nanopores with pore sized<15 nm (AFM) [63]

established several chemical and physical modification approaches to endow pSi structures with chemical stability. The pioneering approach by Linford and Chidsey made it possible the covalent functionalization of hydrogen-terminated silicon surfaces with alkenes and alkynes, opening new opportunities for passivating the surface of pSi-based structures [69]. For instance, functionalization of nonoxidized silicon surfaces by organic molecules enables the chemical passivation of pSi by bound layers such as Si–C, Si–O, and Si–N [70]. Thermal oxidation is a well-known approach used to passivate the surface of pSi structures by creating an oxide layer of controlled thickness. However, although this approach makes it possible to keep the optoelectronic properties of pSi while achieving partial passivation, thermally oxidized pSi is not stable under humidity [71]. Salonen and co-workers pioneered the carbonization (i.e., deposition of carbon) of pSi and envisaged this modification approach as a means of passivating the surface of pSi structures for expanding its applicability [71]. In this approach, acetylene molecules adsorbed onto the surface of pSi undergo dissociation at temperatures above 400 °C, upon which, hydrogen desorbs from the surface of hydrogen-terminated pSi, being replaced by carbon atoms, which bind to silicon to form a carbonized Si–C surface [71]. This modification minimizes the dissolution rate of pSi under physiological and harsh conditions. This property has been used to extend the drug release from pSi nanocarriers as the dissolution rate of these porous structures in biological fluids is significantly reduced [72]. However, it is worthwhile noting that the characteristic luminescence of pSi structures vanishes after carbonization treatment, which can prevent its applicability for other uses such optical sensing or theranostics [71]. The optoelectronic properties of pSi structures make it an attractive material for a broad range of applications. pSi was long envisaged as a platform to develop optoelectronic devices since the discovery of quantum confinement effects by Gösele and Canham [16, 17]. These properties boosted the use of pSi for a range of sensing systems based on changes in its optical (e.g., refractive index, photoluminescence, reflection, etc.) or electronic properties (e.g., resistivity, capacitance, etc.) [73]. MpSi has been demonstrated as an outstanding material for the development of photonic crystal and optical waveguides structures in the visible and NIR spectral regions [74, 75]. The geometric features of MpSi can be finely engineered by combining lithography, electrochemical etching and post-etching, resulting in MpSi structures with perfectly ordered cylindrical micro-pores, which can be grown from top to bottom with precision [76–80]. Pickering and co-workers reported for the first time the photoluminescence of pSi in the visible spectrum in 1984 [81]. A few years after, Gösele and Canham identified quantum confinement effects in μpSi structures, the size and geometry of which can be approximated to a bundle of silicon nanowires. The size of silicon crystals in these structures is large enough to have a rich electronic band and small enough to present an increment of the electron–hole pair (exciton) due to quantum confinement effect [16, 17]. This increment in bandgap due to size quantization is not limited to pSi but it is a characteristic of semiconductor materials. More recently, Sailor and co-workers made good use of this property and demonstrated pSi nanocarriers as theranostic platforms for imaging and photodynamic therapy [23–25].

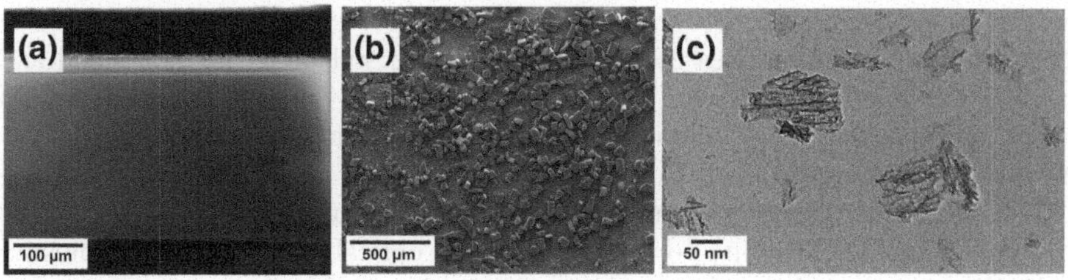

Fig. 2 Images of the morphology of the three types of pSi materials available after the pSi film is cleaved from the crystalline silicon wafer. **a** Membranes [82], **b** microparticles, and **c** nanoparticles [86]

2.3 Types of pSi Platforms

Once fabricated, pSi films can be further processed into the following 3 additional formats (Fig. 2):

- Membranes, which are intact films that have been detached from the crystalline silicon wafer via electropolishing [82];
- Microparticles (pSi MPs), fabricated from "free-standing" membranes by size reduction methods such as sonication [83] or other mechanical forces, such as ball milling [84] or controlled perforation of pSi film [85] and
- Nanoparticles (pSi NPs), produced via a combination of perforating the film structure on the nanoscale and subsequent mechanical force or sonication [86].

2.4 Common Surface Chemistry Modification of Porous Silicon

As-produced pSi structures are prone to oxidize and degrade when exposed to environmental conditions or biological fluids [68]. While this can be an advantage for some applications such as drug delivery or tissue engineering, this can prevent the use of pSi from some applications where long-term chemical stability under different conditions is desired (e.g., in vivo sensors, theranostics, etc.). To overcome this limitation, researchers have established several chemical and physical modification approaches to endow pSi structures with chemical stability [69, 70, 87]. Thermal oxidation is a well-known approach used to passivate the surface of pSi structures by creating an oxide layer of controlled thickness. However, although this approach makes it possible to keep the optoelectronic properties of pSi while achieving partial passivation, thermally oxidized pSi is not stable under humid conditions. Salonen and co-workers pioneered the carbonization (i.e., deposition of carbon) of pSi and envisaged this modification approach as a means of passivating the surface of pSi structures for expanding its applicability [71]. In this approach, acetylene molecules adsorbed onto the surface of pSi undergo dissociation at temperatures above 400 °C. This modification minimizes the dissolution rate of pSi under physiological and harsh conditions. This property has been used to extend the drug release from pSi nanocarriers as the dissolution rate of these porous structures in biological fluids is significantly reduced [74]. However, it is worthwhile noting that the characteristic luminescence of pSi structures vanishes after carbonization treatment, which can prevent its applicability for other uses such optical sensing and theranostics [73]. However, benefits include increased thermal and electrical conductivity, and enhanced mechanical strength and chemical stability [71].

Recent advances have also allowed for the dual functionalization of pSi with two different chemistries both laterally [88, 89] and vertically [90, 91]. To pattern laterally uses photolithographic techniques and has been used for both silanization [88] and hydrosilylation [89]. Vertical dual functionalization is performed by exploiting the wettability of the pSi when functionalizing [90, 91].

Surface passivation of fresh as-produced hydride-terminated (Si–H) pSi, which is unstable in both air and water, can be performed by the following techniques [92]:

- Ozonolysis; surfaces are exposed to a flow of O_3 and converted to Si–OH groups
- Thermal oxidation; surfaces are exposed to hight temperature under ambient conditions and are converted primarily Si–O–Si groups
- Thermal nitridization; surfaces are exposed to a nitrogen-gas-rich atmosphere and high temperature [93–95], generating a silicon nitride surface
- Thermal carbonization (TC) and thermal hydrocarbonization (THC); surfaces are exposed N_2 and acetylene at high temperatures [96, 97].
- Hydrosilylation; surfaces are exposed to alkene or alkyne groups [92], generating a Si–C bond, this reaction can be initiated by methods such as thermal [98], chemical [99, 100], photochemical [101, 102], electrochemical [103], and microwave assisted methods [104].
- Hydrosilanization; surfaces are exposed to a hydrosilane and a Lewis acid catalyst [105].

Further functionalization of already passivated surfaces can be performed using the following techniques:

- Silanisation; Si–OH- and Si–O–Si-terminated surfaces can be modified via compounds containing silane groups (Si–OH) or chlorosilanes (Si–Cl), these reactions can be performed in ambient conditions or at elevated temperatures [106].

Following surface passivation and functionalization via the methods listed above, various functional groups, such as, amines, isocyanates, methacrylates, and PEG moieties [88, 107–109] can be introduced via further modification by conventional organic synthesis routes [110]. This allows for the introduction of various functional groups and even biomolecules.

The chemistries introduced by silanization can then be used for further modification via commercial or custom made cross-linking agents, some even allow for the direct conjugation to various biomolecules, to promote cell adhesion, protein absorption, or even anti-fouling surfaces [110, 111].

pSi surfaces can also be modified through the incorporation of more complex organic and biomolecular species. There exist many different methods for covalently attaching molecules such as peptides [112, 113], proteins [114, 115], enzymes [116, 117], antibodies [118, 119], and DNA [120, 121] to pSi. These are typically employed after the pSi surface has been passivated via oxidation, nitridization, silanization, hydrosilylation, carbonization, hydrocarbonization, or electrografting. Generally, the moieties linked to the pSi surfaces react with either amine or thiol groups contained in the biomolecule being targeted. These species typically include isocyanates [88], epoxides [122], and N-hydroxysuccinimide [89, 102] while other examples include and sulfhydryl reactive groups [114, 123]. among many others [124]. The two key considerations when immobilizing biomolecules are surface coverage and retention of biological activity and properties for the desired application [110].

Subsequently, pSi can be modified with polymers [125]. Modification of pSi with polymers can enhance mechanical stability [126] while simultaneously improving the stability of the pSi in aqueous biological fluids [127]. The ability to cap with polymeric layers allows for the design of responsive devices that are sensitive to pH [128], temperature [109, 129], or specific ligands [130]. More recently, pSi surfaces have been modified with dual chemistries both laterally [88, 102, 131] and vertically [90, 91]. Lateral patterning relies on photolithographic techniques to the pattern during a silanization [88, 131] or hydrosilylation [102] reaction. Vertical patterning, uses liquid masking [91], exploits the wettability [90] or by carefully controlling the chemical reaction environment [132].

The opportunity to apply modified pSi structures spans areas such as microarrays, tissue culture and drug delivery, and theranostics.

3 Biocompatibility of Porous Silicon

3.1 Assessment of pSi as a Biocompatible Material

Biocompatibility of a biomaterial is of utmost importance for their safe and harmless use in humans due to their direct interaction of the biomaterials with human tissue. Hence, much work has been directed toward assessing and demonstrating the biocompatibility of pSi-based biomaterials. The biocompatibility of a biomaterial dictates it fate in the body, and interaction and responses elicited in the body. An ideal biomaterial should be compatible with living tissues and not generate toxicity, injury, or physiological reaction and immunological rejection. Hence, pSi has been subjected to stringent toxicity and biocompatibility through numerous in vitro and in vivo studies [111, 133–137]. pSi is not only biocompatible but also biodegradable, it has been demonstrated that pSi is non-cytotoxic in vitro and in vivo [133, 134], making it highly attractive for injectable drug delivery systems. Biocompatibility of pSi has been related to its structural (size and porosity) chemical properties (hydrophobicity and hydrophilicity), and local environment [135, 136, 138]. The biomedical applications of pSi picked pace after the first biocompatibility study in 1995, by Canham, which showed successful growth of hydroxyapatite onto the surface [18]. Soon after, degradation of pSi as silicic acid and its clearance via renal pathway was reported. These two discoveries triggered enormous interest in pSi as a biomaterial. The ability of pSi to biodegrade [18] is dependent on several factors including the acidity of the local environment and intrinsic properties such as size, porosity, and chemical functionality [111, 135–137, 139, 140].

In 1995, the first investigation of pSi biocompatibility showed the successful growth of hydroxyapatite on pSi substrates [18]. Soon after, Popplewell et al. [141] determined that orthosilicic acid, a product of pSi degradation in the body, is readily absorbed, gastrointestinally and subsequently excreted via the urinary system, additionally orthosilicic acid is linked to homeostasis and the regulation of key processes in the body [141, 142]. Recent in vitro studies of various pSi formulations have found effects such as: concentration-dependent cytotoxicity [143], particle size and concentration-dependent apoptosis and cell damage [138], enhanced cell proliferation and attachment [107], the ability to produce and suppress ROS formation [144], and the enhanced association of pSi and drug permeation to and across cell membranes [145, 146]. All these effects are caused by variations in the size, shape, or surface chemistry of the pSi; highlighting the need for robust in vitro assessment of the material before moving to further in vivo and clinical studies.

The performance of pSi as a biomaterial is dependent on being able to control and stabilize its surface chemistry [71]. Chemical instability of pSi is a big challenge in its biomedical applications, thus the surface chemistry of pSi is one of the most studied parameters in regard to biocompatibility and cytotoxicity [67, 147]. Inherently the surface of pSi is hydride terminated (Si–H), which is highly reactive and rapidly oxidizes in air to form a more stable silicon dioxide shell [67]. This reactive nature of pristine pSi has been associated with cytotoxicity due to the generation of reactive oxygen species that lead to hindered proliferation and cell death. To promote a controlled surface chemistry with less variability, pSi surfaces can be intentionally oxidized with ozonolysis or thermal oxidation to result in stable Si–OH or Si–O–Si groups on the pSi surface [67, 92]. Low et al. reported that incubation of Human lens epithelial cells (SRA 01/04) with pristine pSi particles resulted in complete cell death, whereas oxidized pSi did not elicit any cell death [144]. Using a fluorescent probe (2′,7′-dichlorofluorescin), they proved generation of ROS in the cell culture medium as the reason for observed cytotoxicity. They further extended this study by implanting thermally oxidized pSi in rabbit eye, which showed no toxicity over a period of 9 weeks. In vitro biocompatibility of pSi has been extensively studied with other cell lines like Chinese hamster ovary and rat neuronal B50 cells by Bayliss et al. in 1999 in two studies, who showed pSi is biocompatible and non-cytotoxic [148, 149]. Santos et al. reported a systematic study demonstrating the effect of pSi particle size (1.2–75 μm), particle concentration (0.2–4 mg/mL), and surface chemistry (thermally oxidized, thermally hydrocarbonized, and carbonized) [138]. They discovered concentration-dependent effect of particle size and surface chemistry on the viability of CaCo-2 cells. Thermally oxidized particles were the least toxic with threshhold concentration (<4 mg/mL) compared to the other functionalization (<2 mg/mL) of particles larger than 25 μm. A number of other studies have evaluated in vitro cytotoxicity and biocompatibility of pSi based on various other chemical modifications including chloro or alkoxysilanes [88, 150] with amines, isocyanates, methacrylates, and PEG moieties, oxide layer, polymer grafting, and metal nanoparticles. These in vitro biocompatibility studies have provided us a wealth of information about the interaction of cells with pSi in terms of cell toxicity [143], cell proliferation, attachment [107], differentiation, ability to produce ROS [144], and drug permeation [151].

pSi can be intrinsically luminescent [152], conjugated with dyes [153], or other contrast agents [66, 153–155] and these features can be exploited to image target tissues or organs either in whole animals or after resection from the animal [156]. However, most ex vivo applications of pSi have been studied for tissue engineering [89] or sensing of biomolecules [152]. The ability of pSi to act as a support for cell growth [102] lends itself towards the ex vivo regeneration of tissues and eventually organs [157] with the potential to be used to grow a patient's own skin for later autografting or a donor's cells for allografting onto burns or other serious wounds. Ex vivo applications with pSi can benefit from manipulating the excellent optical properties of the underlying pSi for reporting the health of the cells [158], and for the reporting of the delivery of drugs from within the pSi scaffold [159]. Various biomolecules and even their metabolites can be detected using ultra-sensitive mass spectroscopy techniques [110].

Based on the biocompatibility of pSi, established by in vitro and ex vivo studies, in vivo applications have been found in the treatment of a variety of diseases such as cancer, diabetes, lung infection, eye diseases, and many more [147]. The performance of pSi in vitro and ex vivo is completely different to in vivo applications due to the complex, dynamic, and responsive nature of a living animal model. In vivo administration of pSi may elicit an immune response, typically via the mononuclear phagocyte system (MPS), fibrotic capsulation, enzymatic degradation, and the particles will encounter cell membranes, interstitial pressure, and efflux pumps [160, 161]. There is also the potential for any degradation by-products to cause secondary reactions [162]. While implantation of pSi in vivo for extended periods can result in inflammation around the implant and may also cause the formation of fibrotic capsules, meanwhile, the by-products of modified pSi degradation may also cause cascades of unpredictable events [160, 163]. In vivo studies using pSi-based microparticles, nanoparticles, and films have provided us a plethora of information about their biodistribution, resorption rate, and chronic/acute toxicity [161, 162]. Bowditch et al., in 1998, tested the biocompatibility of pSi in guinea pigs after subcutaneous injection over a period of 26 weeks, showing no toxicity [164]. Santos's group has studied in vivo resorption and biodistribution of ^{18}F-labeled thermally hydrocarbonized pSi (THCpSi) particles in rats after oral gavage and parenteral injection [165]. Oral administration of THCpSi leads to clearance via a renal pathway in 6 h and most of the administered particles accumulated in Ileum and Cecum, whereas intravenously administered THCpSi accumulated in the liver, lungs, and spleen and cleared by renal pathway. In a recent study, the same group tested THCpSi and TOpSi in vivo in mice with myocardial infarction. The particles showed minute toxicity and no impact on hematological cardiological function. One of the most important studies related to in vivo biocompatibility of pSi, by Park et al. demonstrated the degradability and in vivo imaging capability based on its inherent photoluminescence [68].

It is worth noting, that most biocompatibility studies so far using pSi have been limited to in vitro assessments and only recently in vivo evaluation of pSi have picked pace in last decade. Therefore, complete and comprehensive knowledge of in vivo interaction and accumulation of pSi with different pore sizes, porosity, surface chemistry, particle size, and payload is still missing. Nevertheless, the future of pSi as a drug carrier looks bright based on efforts devoted to understanding its biological properties in the past two decades.

pSi is regarded as one of the most prominent and promising new materials for biosensing, bioimaging, drug delivery, and other biomedical medical applications. However, before pSi can reach its true biomedical potential and be widely used it must pass rigorous in vitro, ex vivo, and in vivo tests, as well as human clinical trials, to be deemed safe within the human body.

3.2 Modes of In Vivo Porous Silicon Delivery

Each pSi administration route; intravenous, subcutaneous, oral, or other, will vary in efficiency and bioavailability. Hence, dramatically different biological responses could be observed when pSi is implanted/administered to different tissues, even for the exact same material [166].

Intravenous injection (IV) allows for the systemic administration of pSiNPs, lending itself toward targeted applications. However, IV requires the specific functionalization of the pSi surface with the correct targeting ligands to direct the pSi and the payload it is carrying to the desired organ or site of pathology [118, 157]. Once adequately functionalized with targeting molecules, pSi can adequately hone to specific sites such as the stroma of pancreatic tumor mouse functionalized with the Ly6C antibody [167]. Materials administered by IV will suffer opsonization to some extent. Opsonization occurs when the MPS entraps the material [168] and can be combatted by appropriately modifying the surface chemistry/properties and/or size of the particles [169].

Particles administered by subcutaneous (SC) injection are trafficked in a size-dependent fashion to the lymphatic system [170]. This size dependence occurs due to large particles being consumed by peripheral antigen-presenting cells (APCs), and smaller particles being internalized by resident APCs after cell-free trafficking [170, 171]. Hudson et al. [172] demonstrated little toxicity of mesoporous silicates following subcutaneous injection in rats. However, intraperitoneal and intravenous injections caused either death or the animals needed to be euthanized. They reported the cause of the death was due to the formation of thrombus [172]. Subcutaneous pSi delivery systems are a particularly attractive option for the delivery of newly developed peptide drugs, which possess relatively short half-lives [173]. Various pSi functionalizations (thermally oxidized pSi, undecylenic-acid-modified thermally hydrocarbonized pSi and thermally hydrocarbonized pSi) have been investigated for sustained subcutaneous peptide delivery [173].

Oral administration is the most common route of administration as it is convenient, safe, and inexpensive. The biggest drawback with oral administration, however, is the presence of digestive enzymes and the low pH in the stomach. These two factors in combination with the low permeability of the intestinal walls can lead to further delivery difficulties. Fortuitously, pSi particles possess more stability at acidic pH's than alkali; hence, pSi is somewhat resistant to the gastric fluids of the stomach. Furthermore, pSi provides a small pore size to confine the drug in a completely tunable fashion [21, 174, 175] and coating pSi with enteric polymers permits tunable release at the desired pH [94].

4 Cancer Biology

Cancer, uncontrolled division and spreading of body's cells, is a deadly disease that claimed over 9 million lives in 2016 [176, 177]. Cancer disease starts as a benign tumor with an uncontrolled growth of cells due to genetic changes in genes that regulate their growth and division due to inherent (hereditary) or exogenous (environmental, habits—food, smoking, etc.) causes. Cancer cells are highly resourceful in recruiting other cells, molecules, and tissues to supply the tumor with oxygen and nutrients by building new blood vessels and extracellular matrix with dense stroma in some cases, and evade immune cells. Among the different types of tumors, malignant tumors are the most dangerous since these can spread into or invade other nearby tissues by a biological process called "metastasis". This phenomenon occurs when benign tumor cells break off from the solid tumor mass and travel to distant tissues in the body through the blood or the lymphatic system to form new tumors away from the original tumor site. Metastatic tumors cause severe damage to the patient, and most cancer patients die of cancer due to metastatic disease.

In general, cancers are categorized based on the type of cells from which they originate. Some representative examples are: (i) *Carcinoma:* originated by epithelial cells that form the body skin and, hence is the most common type of cancer (e.g., skin cancer); (ii) *Sarcoma:* produced by bone and soft tissues including muscle, tendons, blood vessels, lymph vessels, etc. (e.g., osteosarcoma); (iii) *Leukemia:* originated from abnormal white blood cells, which do not form a solid tumor (e.g., blood cancer); (iv) *Lymphoma:* cancer with lymph origin (lymphocytes—T cells and B cells) which has two main types—Hodgkin lymphoma and non-Hodgkin lymphoma; (v) *Myeloma:* produced by plasma cells in the bone marrow that form tumors in bones all through the body; (vi) *Melanoma:* cancer originated from the pigment-producing cells (melanocytes), most of which form on the skin; and (vii) *Glaucoma:* brain tumor.

The multivalent and highly complex nature of cancer makes this disease very difficult to treat and cure [178, 179]. Current treatments and therapeutics are generally insoluble, highly toxic, nonselective, and cleared rapidly. Therefore, it is imperative to develop novel targeted therapies that selectively reach the cancer site and deliver a powerful and effective treatment for personalized medicine. With the advances in our understanding of cancer biology over the past decade, a number of cancer cell surface receptors have been identified, which can be specifically targeted to deliver the therapeutic agent to the target tumor passively or actively [180, 181].

Passive cancer tissue targeting relies on the leaky nature of the tumor blood vessels, which is a signature of the nutrient-deprived core of fast-growing tumor. Nanoparticle-based cancer therapies accumulate preferentially at the tumor site utilizing leaky vessels that allow influx through their enlarged gap junctions. These enlarged gap junctions can range in size from 100 nm to 2 μm, for various tumor types. This phenomenon is known as "enhanced permeation and retention" effect (EPR). A number of EPR effect-based passive targeting drugs like Doxil (polyethylene glycol-coated doxorubicin liposomes) and Abraxane (albumin-bound paclitaxel nanoparticles) are already approved by the USFDA [176]. The advantage of passive accumulation is that it increases the half-life of the drug and the efficiency of the treatment up to several orders of magnitude. While limitations include the inability to specifically target cancer cells in all tumor types, due to the variation in tumor vascularization and porosity of the tumor vessels.

Active targeting relies on specific moieties such as antibodies or antibody fragments, peptides, and aptamers, among others, which endow therapeutics with selectivity toward cancer cells. Although very efficient and selective, active targeting has had limited clinical success due to the resource and cost intensiveness associated with the discovery and design of the target proteins for cancer cell surfaces. In addition, active targeting involves high cost and multistep processes to prepare the active cancer-targeted formulations. Therefore, active cancer targeting formulations are yet to be commercialized.

To understand the complex cancer biology and its response to medical treatment is a key aspect to develop advanced therapeutic agents to treat this resilient and deadly disease. The optoelectronic properties of pSi make it an excellent platform to develop theranostic agents combining both diagnostics and therapeutic capabilities in a single nanocarrier entity. pSi can be produced in different forms (micro and nanoporous particles, films, etc.), the physical (pore size, particle size, optical signals, etc.) and chemical (surface chemistry) properties of which can be precisely engineered for specific therapeutic applications. pSi offers a plethora of opportunities to investigate and treat cancer in situ, using both passive and active approaches. However, further fundamental and applied investigations of pSi using in vitro, ex vivo, and in vivo models are required to make this technology feasible for real clinical practice.

5 Porous Silicon for Cancer Therapy

pSi in the form of nano- and microparticles have been shown to be highly efficient in the treatment of a variety of cancers. This is mainly due to control over the size, surface chemistry, biocompatibility, biodegradability of pSi, and the ability to load and deliver a range of therapeutics. The pSi nanoparticles of controlled size (smaller than 300 nm) and a narrow size distribution can take the advantage of EPR effect in cancer and passively accumulate at the tumor site. Whereas, larger pSi particles can be administered locally at the tumor site for local therapy for reduced off-target effects of the loaded anticancer therapeutics. However, most of the work with pSi has been carried out using particles less than 2 μm and modified with a specific targeting moiety.

One of the first steps toward the application of a bioactive material as a therapeutic agent or carrier for cancer treatment is the demonstration of its ability to load and release anticancer drugs. In this regards, pSi particles have been successfully loaded and shown to provide a sustained release of a number of anti-tumor drugs including doxorubicin (DOX), Daunorubicin (DNR), and platinum-based drugs like cisplatin, carboplatin, and $Pt(En)Cl_2$. Vaccri et al., in 2006 demonstrated that pSi can successfully load doxorubicin and provide a sustained release over a period of 72 h, while the in vitro cytotoxicity assay with LoVo cells showed a significant decrease in cell survival in 6 h on incubation with DOX-loaded pSi particles (2 μm in size) [182]. Gu et al., on the other hand, used pSi particles impregnated iron oxide nanoparticles of 5 nm to load with DOX and demonstrated that the effect of DOX can be spatially modulated on the application of magnetic field. In one of their experiments, a permanent magnet is placed at the edge of a Petri dish containing HeLa cells and on incubation of the cells with DOX-loaded magnetic-pSi particles enhanced cell death is observed at the edge [183]. This is due to the magnetic field modulated migration of magnetic-pSi particles, which accumulate in large concentrations at the edge providing higher local DOX release, whereas the DOX amount in the center of the Petri dish is depleted. Also, they show that pSis are able to provide a sustained release of DOX over 8 days. Daunorubicin, another antimitotic cancer drug, has been widely studied for its loading and release behavior from pSi micro-and nanoparticles. However, most of the systems combining pSi micro- and nanoparticles with daunorubicin (DNR) have been focused on the treatment of ocular diseases, since DNR acts as an anti-proliferative drug to treat proliferative diseases of the eye (Proliferative vitreoretinopathy, PVR). Most of the works in this area has been carried out by Prof. Sailor's group at University of California, San Diego. The most interesting set of results on loading of DNR onto pSi is degradation of DNR due to redox activity of pSi, which can be avoided by thermally oxidizing the pSi particles [184]. Thus, most of the pSi-based DNR loading and release has been on oxidized or functionalized pSi or a composite system consisting of polymer and oxidized pSi delivering DNR to the rabbit eye (in vivo) [31, 33, 185–188]. Platinum-based cancer drugs tend to be highly cytotoxic like the other anticancer drugs and this it is important to formulate them in a system that is capable of passive and active targeting, and sustained release. Therefore, platinum-based anticancer drugs have been formulated by loading onto pSi by various groups [189–191]. Li et al., packed three platinum-based drugs into calcium phosphate layers on pSi for sustained release The cisplatin loading was found to be dependent on the concentration of the loading solution and surface chemistry of pSi, which was similar for the other platinum-based drugs as well. An interesting study on controlled cisplatin and transplatin release from Park et al. demonstrated an ingenious strategy to cap the pores of cisplatin-loaded pSi particles [191]. They first loaded the pSi particles with cisplatin, followed by electroless plating with platinum. The release, in this case, is a function of degradation of the SiO_2 matrix and opening of the pores capped by Pt particles. They were able to get a sustained release of the two drugs for 16 h and higher targeted toxicity toward human ovarian cancer cells under in vitro testing conditions. In terms of IC 50 values pSi particles showed the higher toxicity and IC50 values of 0.004, 0.006, and 0.018 μg (based on mass of free drug loaded) for the three pSi-based systems (oxidized pSi, undecylenic-acid-functionalized pSi, and 1-Dodecene-functionalized pSi particles, respectively), which is 60- and 700-fold lower relative to the free drug (IC 50 cisplatin: 2.9 μg and transplatin 6 μg).

Direct application of pSi as an anticancer therapeutic agent using its ability to generate reactive oxygen species (ROS) have also been demonstrated. Pristine pSi is known to generate ROS, which was shown to be usable as a therapeutic for cancer by Xiao et al. [192]. The production of ROS was initiated by irradiation with a suitable wavelength of light (458 nm), Photodynamic therapy. They showed that the rate of ROS production was much higher in presence of 458 nm trigger as compared to the samples incubated in the dark. In vitro experiments with HeLa and NIH-3T3 cells demonstrated 20–40% enhanced cell death on exposure to 458 nm light (at 1.5 or 5 J/cm^2) when incubated with pSi, in relation to the cells incubated in dark. While others have also the photothermal capability of pSi as a treatment for cancer [193, 194]. Lee et al., showed a drastic increase in local temperature when pSi suspension is irradiated with NIR (780–1400 nm) light, sufficient to ablate cancer cells.

Ferrari's group has focused their efforts on using precisely shaped disk-like pSi particles for cancer treatment (Fig. 3). The discoidal pSi particles are made by a combination of lithography and electrochemical etching process with the capability to control the size and shape of the particles with very high precision [195]. It is already known that shape of nanoparticles has a big impact on their biological properties. Therefore, they studied the biodistribution of the discoidal pSi particles in vivo [196]. Discoidal pSi particles with a diameter between 500 and 2600 nm, height ranging between 200 and 700 nm, and pore sizes from 5 to 150 nm were tested. The particles of diameter 600 and 1000 nm were tested for their rate of degradation in the serum, interaction with immune and endothelial cells, and biodistribution in mice bearing breast tumors. Both the type of particles degraded within 24 h in serum and did not elicit any immune response with macrophage cells. In terms of accumulation, the 600 nm pSi particles passively targeted the tumor better than their spherical counterparts, up to 5 times higher accumulation with 10.2% ID/g reaching tumor. They have also shown actively targeted discoidal pSi particles systems for treatment of pancreatic cancer and breast cancer. For pancreatic cancer treatment, pSi particles were modified with a Ly6C antibody that specifically targets Ly6C proteins expressed on the pancreatic cell surface [167]. The Ly6C-modified pSi particles showed almost 20 times better accumulation (~9.8% of the injected dose), relative to unmodified pSi particles (~0.5% of the injected dose). For breast cancer, they demonstrated a multistage targeted delivery system in which paclitaxel-loaded liposomes (PEG-PCL based) were packed into discoidal pSi particles [197]. As pSi

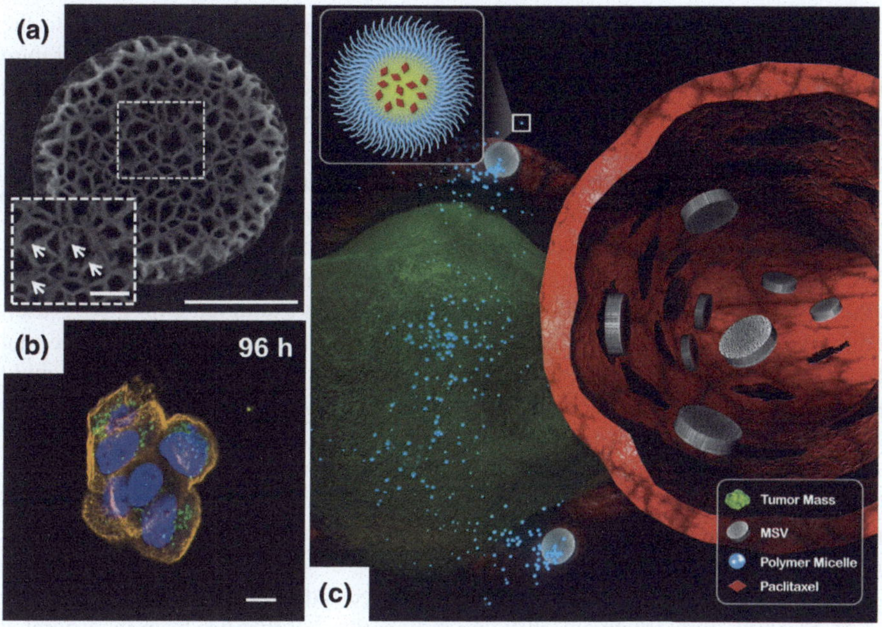

Fig. 3 Ferrari's group have fabricated precisely shaped disk-like pSi particles (designated as MSVs) for cancer treatment. Image and caption adapted from Blanco et al. [197]. **a** SEM micrograph showing loading of micelles within pores of MSVs. The scale bar represents 500 nm. The inset represents a magnification of the highlighted area of the original. Arrows point to micelle clusters. The scale bar in the inset represents 100 nm. **b** In vitro uptake and efficacy examination of MSVs in breast cancer cells. Confocal microscopy image of MCF-7 cells incubated with MSVs loaded with BODIPY-containing micelles 96 h after administration. The scale bar represents 5 microns. **c** Proposed mechanism of multistage delivery of micellar nanotherapeutics to tumors. Following navigation in blood vessels, MSVs marginate in tumor vasculature, accumulating in smaller capillaries. MSVs then become intravascular drug delivery depots, releasing paclitaxel-containing polymer micelles in a sustained fashion over time

particles reach and accumulate at the tumor site, either actively or passively, they locally release the drug-loaded liposomes which tend to be more effective.

A number of studies have demonstrated this ability of pSi. In addition, some studies even claim and demonstrate pSi as a cancer theranostic agent, which is a multifunctional system not only applicable in therapeutics but also cancer diagnostics and imaging. This has been made possible by the inherent photoluminescence (PL) of pSi that can be used for in vitro and in vivo imaging. In this regard, a landmark study by Park et al. paved the way, which was followed by a number of other studies [68]. Gu et al., demonstrated that pSi shows long PL lifetime (>100 μsec), which can be employed for time-gated in vivo imaging of cancer [23]. The long lifetime of pSi makes it possible to suppress the background autofluorescence that is a major challenge for in vivo PL imaging. Wang et al. used iRGD-modified pSi particles for active targeting tumor and monitored the in vivo biodistribution by SPECT in an ectopic PC3-MM2 mouse xenograft model [198]. Actively targeting pSi to the cancer with a number of targeting moieties have been demonstrated in the literature that includes homing with tumor-homing peptides like mammary-derived growth inhibitor, hyaluronic acid to target CD-44, Ly6C antibodies, anti-CD326 antibody, and RGD and iRGD peptides [199–202].

pSi, and other materials, can also be used for thermal therapy of cancer by inducing cell death via the local heating of NPs accumulated in the cancerous tissue [203]. The local heating can be initiated by microwave, ultrasound, radio frequency or laser stimuli [204, 205]. Photothermal therapy has also been demonstrated with pSi in colon carcinoma (CT-26) via both in vitro and in vivo excitation with a NIR laser, reducing viability to just 6.7% [193, 206]. With continued research photothermal therapy based on pSi could be used to treat deeper non-resectable tumors and could also be used as two-component treatment systems [207, 208].

Photodynamic therapy is performed using cytotoxic singlet oxygen and ROS generated from photosensitizers [209]. pSi NPs have been used in a similar manner using a two-photon method to induce the production of cytotoxic species from porphyrin modified pSi. This induced cell death in human breast cancer cells while allowing for targeting as well [24].

Brachytherapy treats tumors via locally emitted radiation. BrachySil^{TM} is a pSi-based brachytherapy treatment, utilizing radioactive ^{32}P that has been incorporated into pSi. The material is introduced to the tumors via intratumoral injection, subsequently, a high dose of radiation is delivered locally to the tumor site. The local effect of delivery relies on the fact that the ^{32}P emits β-radiation which penetrates just 8 mm of tissue [210, 211]. BrachySil^{TM} underwent clinical trials on hepatocellular carcinomas [212] and reduced the tumors over 24 weeks.

pSi is an ideal candidate for drug delivery based on its key properties, however, a lot of information is still missing. One of the biggest challenges with pSi is its stability, which needs to be addressed as it degrades very quickly under physiological conditions that make it very difficult to attach the targeting moieties. In addition, a therapeutic should be stable for the period of time it is stored, which is a challenge at the moment. Also, there is a growing trend toward more personalized (RNA, DNA, gene editing, etc.) and narrow spectrum therapeutics (proteins, enzymes, etc.), however, there are very little investigations into the use of pSi for proteins and nucleotide-based therapeutics for cancer treatment.

6 Porous Silicon for Bioimaging

6.1 Properties of Porous Silicon for Bioimaging

The optoelectronic properties of pSi structures make it an attractive material for a broad range of applications, the porous structure of pSi allows for the direct interrogation of the release of therapeutics via EOT monitoring or the detection of a simple color change. pSi was long envisaged as a platform to develop optoelectronic devices since the discovery of quantum confinement effects in its structure by Gösele and Canham [16, 17]. These studies enabled the use of pSi for a range of sensing systems based on changes in its optical (e.g., refractive index, photoluminescence, reflection) and electronic properties (e.g., resistivity, capacitance) [74]. pSi has been demonstrated as an outstanding material for the development of photonic crystal and optical waveguides structures in the visible and NIR range of the spectrum [75, 76]. The geometric features of pSi can be finely engineered by combining lithography, electrochemical etching, and post-etching, resulting in pSi structures with perfectly ordered cylindrical micro-pores, which can be grown precisely grown from top to bottom [80]. Pickering and co-workers reported for the first time the photoluminescence of pSi in the visible spectrum in 1984 [81], and a few short years after, Gösele and Canham identified quantum confinement effects in pSi structures, the size and geometry of which can be approximated to a bundle of silicon nanowires. The size of silicon crystals in these structures is large enough to have a rich electronic band and small enough to present an increment of the electron–hole pair (exciton) due to the quantum confinement effect [16, 17]. Recently, Sailor and co-workers made good use of this property and demonstrated pSi nanocarriers as theranostic agents for imaging and photodynamic therapy and that pSi-based nanoparticles (pSiNPs) can also act as imaging agents using a time-gated imaging approach [23–25].

The reflectance of white light from pSi surfaces results in interference patterns, these patterns can be monitored to elucidate changes in effective optical thickness (EOT). EOT is also influenced by the influx or efflux of material within the pores [213]. The capture of specific target molecules will give a positive EOT shift, while negative shifts are often molecules being released from the surface, or other processes such as oxidation and degradation [214]. This allows for pSi substrates to be used in applications such as biosensing [215–217], self-reporting drug delivery [31, 34] and imaging [68, 218, 219].

As a result of taking advantage of these optoelectronic properties, pSi is considered a unique and promising vehicle platform for theranostic applications and, therefore, considered as a unique vehicle platform for theranostic nanomedicines. Furthermore, given its optoelectronic properties, pSi technology can be easily integrated into implantable and biodegradable electronics, which is perfectly aligned with existing bionic devices currently used to treat medical conditions.

6.2 pSi Modifications for Bioimaging

In vivo imaging with pSi has been performed in various modes, such as luminescence [68], magnetic resonance imaging (MRI) [154, 155, 220], various computed tomography (CT) modes [161, 207, 211, 221, 222], and intravital microscopy [223, 224]. pSi NPs are highly advantageous in noninvasive imaging due to their ability to carry and concentrate contrast agents. The ability of pSi to be used for imaging will depend on the ability to correctly modify both surface properties and surface chemistry.

In vivo imaging using luminescent pSi NPs has been demonstrated [68, 219]. Luminescent pSi NPs with modified surfaces, with species such as PEG [225], have even demonstrated improved photoluminescence stability and increased circulation lifetimes. As maximum tissue penetration occurs in the NIR range due to minimal absorption by species such as hemoglobin, it is best to tune the luminescent pSi to absorb in the NIR range to allow for the imaging of deep tissue [226]. Luminescent pSi NPs have been coated with Dextran and subsequently injected into nude mice with tumors, these particles passively accumulated in the tumor and delivered doxorubicin, an antitumor agent [68]. Other amine terminated luminescent pSi NPs which are nontoxic, stable over the pH range of 4–14 and resistant to aging/degradation over several weeks have been developed [219]. The conjugation of fluorescent dyes to pSi can also be achieved [218] and these materials can be further enhanced via loading of materials like quantum dots [227, 228] or silver nanoparticles [229]. Intravitreal microscopy is used to help visualize biological processes that are occurring in live animals and is becoming more common [230]. The most common use of pSi for intravital microscopy is to assess the biodistribution of the fluorescent particles [167, 223, 224].

MRI contrast is able to be enhanced using of gadolinium (Gd)-loaded pSi, due to the Gd agent enhancing MRI sensitivity via improved longitudinal relaxivity [231]. However, Gizzatov et al. [154] found no enhancement contrast using Gd-loaded pSi particles to treat ovarian tumor cells, despite their accumulation within the cells. MRI contrast can also be enhanced by superparamagnetic iron oxide NPs (SPIONs) loaded into pSi MPs by precipitation [232] or entrapment by crosslinked aminosilanised [155]. SPION-loaded pSi MPs enhance imaging by inducing a shorter relaxation time, this effect was found to be concentration dependent. The use of amino-functionalized pSi subsequently allows for the attachment of targeting molecules such as vascular endothelial growth factor receptor 2 (VEGFR-2) and platelet endothelial cell adhesion molecule (PECAM) [155].

There are two common modes for nuclear imaging with pSi structures these are positron emission tomography (PET) and single-photon emission computed tomography (SPECT) [222, 233–235]. A radioactive label, commonly ^{13}N, ^{11}C, ^{68}Ga, ^{18}F, ^{64}Cu, and ^{124}I for PET modes and ^{99m}Tc, ^{111}In and ^{125}I for SPECT modes [226] is required for imaging. ^{18}F has been added to pSi via the direct substitution of Si–H or oxygen in Si–O–Si bonds resulting in Si–F bonds [156]. THCpSi NPs labeled with ^{18}F has been studied as a tracer in mice through the gastrointestinal tract for 6 h at nanogram quantities [236]. PEGylated pSi NPs have been used for SPECT imaging after labeling with ^{125}I. These particles were studied for their biodistribution and it was found that they were very quickly taken into the spleen and liver of healthy mice. Huhtala et al. [237] also tracked THC pSi labeled with ^{125}I and IGF-1 loaded, they observed that the IGF-1 remained in the liver to produce a sustained release that was more advantageous than IGF-1 administered on its own.

The ability to perform dual imaging has also been developed. Gu et al. [238] achieved this using a combination of fluorescence and magnetic properties. The authors used boron-activated luminescent pSi MPs loaded with SPIONs. The SPIONs were trapped in the pSi pores after performing a thermal oxidation, which adds oxide to the surface and narrows the pores. A small drug payload could be subsequently loaded into the narrowed pores. This combination of materials allowed for a local site to be targeted magnetically and the pSi to be tracked fluorescently. Other similar systems for dual functional pSi include cobalt NPs in luminescent pSi, opening the possibility for deep tissue imaging [239].

6.3 Advantages and Disadvantages of Porous Silicon for Bioimaging

As fabricated pSi needs to be functionalized appropriately in order to facilitate the correct amount of bioavailability and biostability. Without appropriate chemical modification, pSi will rapidly dissolve in the aqueous environments present in the body. The ability to readily modify pSi with so many different chemistries, however, does lead to a broad range of applications, while intrinsic optoelectronic properties can be exploited for imaging or other applications. Furthermore, pSi has lower toxicity than heavy metal-based quantum dots, offers better photostability than organic fluorophores, emits light in the NIR region, which is optimal to penetrate tissue, it is biodegradable with low cytotoxicity and can load high doses of therapeutic agents [25]. Furthermore, the combination of pSiNPs with photosensitizers enables photodynamic therapies, which pave the way for localized treatments to reduce the side effects of anticancer medicines by the temporal and spatial control of the production of cytotoxic species in diseased cells, tissues, or organs [24].

pSi technology brings new opportunities with the real potential to be translated into clinical therapies due to its unique properties. pSi can be engineered with versatility throughout well-established fabrication and modification methods. These methods also make it possible to engineer the physical and chemical properties of pSi, such as (e.g., high porosity, high loading capacity, controlled release, biodegradability, biocompatibility, biointegration, and self-reporting and imaging). pSi structures have intrinsic optoelectronic properties, which in combination with existing therapeutic agents can result in advanced nanomedicines for imaging and photodynamic therapies. This is a significant advantage as compared to other types of nanocarriers and drug delivery platforms such as liposomes or polymers. These factors make pSi a promising candidate for a new generation of nanomedicines.

7 Conclusions and Future Perspectives

As the studies reviewed and discussed throughout this chapter demonstrate, pSi is a promising platform for the development of a broad variety of nanomedicines and drug delivery systems, which could address some of the limitations of conventional therapies for cancer treatment. pSi structures can be produced by well-established fabrication process developed in the course of the past decades, and a multitude of pSi-based structures can be fabricated with high precision and tuned chemical and physical properties. Furthermore, although the long-term toxicity of pSi in real clinical practice is yet to be established, the significant number of in vitro and in vivo studies on pSi structures are a promising evidence of the biocompatibility of this material. pSi has a unique set of physical and chemical properties, including high porosity, controllable dimensions, tuneable surface chemistry, high loading capacity, biodegradability, biocompatibility, and biointegration. These properties make pSi a perfect platform to develop advanced and highly versatile drug nanocarriers and delivery systems, where the release of therapeutics can be engineered with precision to fulfill the demands of specific medical treatments. Furthermore, the optoelectronic properties of pSi make this material an excellent candidate for theranostic and photodynamic therapies, which is an advantage over systems based on the combination of imaging and therapeutic agents.

Despite these promising and outstanding advances in pSi technology for drug delivery and medical applications, it is worthwhile stressing that this technology still faces both technical and commercial challenges for its final clinical translation from bench to bedside. Therefore, more extensive fundamental research assessing the toxicity and side effects associated with pSi systems and their performance in terms of clinical efficiency will be needed before this technology becomes feasible and reliable. However, based on the evidence shown in the studies reviewed here, it is reasonable to conclude that there is considerable potential for pSi to become an alternative nanomedicine platform for some therapies such as cancer.

References

1. Rosen H, Abribat T (2005) The rise and rise of drug delivery. Nat Rev Drug Discov 4:381–385
2. Hoffman AS (2008) The origins and evolution of "controlled" drug delivery systems. J Controlled Release 132:153–163
3. Park K (2014) Controlled drug delivery systems: past forward and future back. J Controlled Release 190:3–8
4. Mitragotri S, Burke PA, Langer R (2014) Overcoming the challenges in administering biopharmaceuticals: formulation and delivery strategies. Nat Rev Drug Discov 13:655–672
5. LaVan DA, Lynn DM, Langer R (2002) Moving smaller in drug discovery and delivery. Nat Rev Drug Discov 1:77–84
6. Ferrari M (2005) Cancer nanotechnology: opportunities and challenges. Nat Rev Cancer 5:161–171
7. Santini JTJ, Richards AC, Scheidt R, Cima MJ, Langer R (2000) Microchips as controlled drug-delivery devices. Angew Chem Int Ed 39:2396–2407

8. Kelkar SS, Reineke TM (2011) Theranostics: combining imaging and therapy. Bioconjug Chem 22:1879–1903
9. Mann AP, Scodeller P, Hussain S. Joo J, Kwon E, Braun GB, Mölder T, She ZG, Ramana Kotamraju V, Ranscht B, Krajewski S, Teesalu T, Bhatia S, Sailor MJ, Ruoslahti E (2016) A peptide for targeted, systemic delivery of imaging and therapeutic compounds into acute brain injuries. Nat Comm 7:11980-1–11980-11
10. Lanza GM (2015) From micro to nano in seconds. Nat Nanotechnol 10:301–302
11. Svenson S (2013) Theranostics: are we there yet? Mol Pharm 10:848–856
12. Uhlir A (1956) Electrolytic shaping of germanium and silicon. Bell Syst Tech J 35(2):333–347
13. Lehmann V (2002) Electrochemistry of silicon. Wiley-VCH
14. Sailor MJ (2012) Porous silicon in practice: preparation, characterization and applications. Wiley-VCH
15. Canham LT (2014) Handbook of porous silicon. Springer International Publishing, Berlin
16. Canham LT (1990) Silicon quantum wire array fabrication by electrochemical and chemical dissolution of wafers. Appl Phys Lett 57(10):1046–1048
17. Lehmann V, Gösele U (1991) Porous silicon formation: a quantum wire effect. Appl Phys Lett 58:856–858
18. Canham LT (1995) Bioactive silicon structure fabrication through nanoetching techniques. Adv Mater 7(12):1033–1037
19. Canham LT, Reeves CL, King DO, Branfield PJ, Crabb JG, Ward MC (1996) Bioactive polycrystalline silicon. Adv Mater 8(10):850–852
20. Prestidge CA, Barnes TJ, Lau CH, Barnett C, Loni A, Canham LT (2007) Mesoporous silicon: a platform for the delivery of therapeutics. Expert Opin Drug Deliv 4:101–110
21. Salonen J, Kaukonen AM, Hirvonen J, Lehto VP (2008) Mesoporous silicon in drug delivery applications. J Pharm Sci 97(2):632–653
22. Anglin EJ, Cheng L, Freeman WR, Sailor MJ (2008) Porous silicon in drug delivery devices and materials. Adv Drug Deliv Rev 60:1266–1277
23. Gu L, Hall DJ, Qin Z, Anglin E, Joo J, Mooney DJ, Howell SB, Sailor MJ (2013) In vivo time-gated fluorescence imaging with biodegradable luminescent porous silicon nanoparticles. Nat Comm 4:2326
24. Secret E, Maynadier M, Gallud A, Chaix A, Bouffard E, Gary-Bobo M, Marcotte N, Mongin O, El Cheikh K, Hugues V, Auffan M, Frochot C, Morère A, Maillard P, Blanchard-Desce M, Sailor MJ, Garcia M, Durand JO, Cunin F (2014) Two-photon excitation of porphyrin-functionalized porous silicon nanoparticles for photodynamic therapy. Adv Mater 26:7643–7648
25. Joo J, Liu X, Ramana Kotamraju V, Ruoslahti E, Nam Y, Sailor MJ (2015) Gated luminescence imaging of silicon nanoparticles. ACS Nano 9:6233–6241
26. Ledford H (2016) Bankruptcy of nanomedicine firm worries drug developers. Nature 533:304–305
27. Santini JTJ, Cima MJ, Langer R (1999) A controlled-release microchip. Nature 397:335–338
28. Staples M, Daniel K, Cima MJ, Langer R (2006) Application of micro- and nano-electromechanical devices to drug delivery. Pharm Res 23:847–863
29. Nan K, Ma F, Hou H, Freeman WR, Sailor MJ, Cheng L (2014) Porous silicon oxide-plga composite microspheres for sustained ocular delivery of daunorubicin. Acta Biomater 10:3505–3512
30. Nieto A, Hou H, Sailor MJ, Freeman WR, Cheng L (2013) Ocular silicon distribution and clearance following intravitreal injection of porous silicon microparticles. Exp Eye Res 116:161–168
31. Wu EC, Andrew JS, Cheng L, Freeman WR, Pearson L, Sailor MJ (2011) Real-time monitoring of sustained drug release using the optical properties of porous silicon photonic crystal particles. Biomaterials 32:1957–1966
32. Chhablani J, Nieto A, Hou HY, Wu EC, Freeman WR, Sailor MJ, Cheng L (2013) Oxidized porous silicon particles covalently grafted with daunorubicin as a sustained intraocular drug delivery system. Invest Ophthalmol Vis Sci 54:1268–1279

33. Hartmann KI, Nieto A, Wu EC, Freeman WR, Kim JS, Chhablani J, Sailor MJ, Cheng L (2013) Hydrosilylated porous silicon particles function as an intravitreal drug delivery system for daunorubicin. J Ocul Pharmacol Ther 29:493–500
34. Cheng L, Anglin EJ, Cunin F, Kim D, Sailor MJ, Falkenstein I, Tammewar A, Freeman WR (2008) Intravitreal properties of porous silicon photonic crystals: a potential self-reporting intraocular drug-delivery vehicle. Br J Ophthalmol 92:705–711
35. Huang Z, Geyer N, Werner P, de Boor J, Gösele U (2011) Metal-assisted chemical etching of silicon: a review. Adv Mater 23:285–308
36. Huang Z, Shimizu T, Senz S, Zhang Z, Geyer N, Gösele U (2010) Oxidation rate effect on the direction of metal-assisted chemical and electrochemical etching of silicon. J Phys Chem C 114:10683–10690
37. Huang ZP, Shimizu T, Senz S, Zhang Z, Zhang XX, Lee W, Geyer N, Gösele U (2009) Ordered arrays of vertically aligned [110] silicon nanowires by suppressing the crystallographically preferred <100> etching directions. Nano Lett 9:2519–2525
38. Schmidt V, Senz S, Gösele U (2005) Diameter-dependent growth direction of epitaxial silicon nanowires. Nano Lett 5:931–935
39. Choi WK, Liew TH, Dawood MK (2008) Synthesis of silicon nanowires and nanofin arrays using interference lithography and catalytic etching. Nano Lett 8:3799–3802
40. de Boor J, Geyer N, Wittemann JV, Gösele U, Schmidt V (2010) Sub-100 Nm silicon nanowires by laser interference lithography and metal-assisted etching. Nanotechnology 21:095302–1–095302-5
41. Myrhaug EH, Tveit H (2000) Material balances of trace elements in the ferrosilicon and silicon processes. In: Electric furnace conference, Warrendale
42. Lui N, Huo K, MacDowell MT, Zhao J, Cui Y (2013) Rice husks as a sustainable source of nanostructured silicon for high performance Li-Ion battery anodes. Sci Rep 3:1919-1–1919-7
43. Kim KH, Lee DJ, Cho KM, Kim SJ, Park JK, Jung HT (2015) Complete magnesiothermic reduction reaction of vertically aligned mesoporous silica channels to form pure silicon nanoparticles. Sci Rep 5:9014-1–9014-7
44. Bao Z, Weatherspoon MR, Shian S, Cai Y, Graham PD, Allan SM, Ahmad G, Dickerson MB, Church BC, Kang Z, Abernathy HW, Summers CJ, Liu M, Sandhage KH (2007) Chemical reduction of three-dimensional silica micro-assemblies into microporous silicon replicas. Nature 446:172–175
45. Watanabe Y, Sakai T (1971) Application of a thick anode film to semiconductor devices. Rev Electr Commun Lab 19:899–903
46. Dimova-Malinovska D, Sendova-Vassileva M, Tzenov N, Kamenova M (1997) Preparation of thin porous silicon layers by stain etching. Thin Solid Films 297:9–12
47. Li X, Bohn PW (2000) Metal-assisted chemical etching in HF/H_2O_2 produces porous silicon. Appl Phys Lett 77:2572–2574
48. Tsujino K, Matsumura M (2005) Helical nanoholes bored in silicon by wet chemical etching using platinum nanoparticles as catalyst. Electrochem Solid State Lett 8:C193–C195
49. Chartier C, Bastide S, Levy-Clement C (2008) Metal-assisted chemical etching of silicon in HF-H_2O. Electrochim. Acta 53:5509–5516
50. Peng K, Lu A, Zhang R, Lee ST (2008) Motility of metal nanoparticles in silicon and induced anisotropic silicon etching. Adv Funct Mater 18:3026–3035
51. Zhang ML, Peng KQ, Fan X, Jie JS, Zhang RQ, Lee ST, Wong NB (2008) Preparation of large-area uniform silicon nanowires arrays through metal-assisted chemical etching. J Phys Chem C 112:4444–4450
52. Harada Y, Li XL, Bohn PW, Nuzzo RG (2001) Catalytic amplification of the soft lithographic patterning of Si. Nonelectrochemical orthogonal fabrication of photoluminescent porous Si pixel arrays. J Am Chem Soc 123:8709–8717
53. Peng KQ, Yan YJ, Gao SP, Zhu J (2003) Dendrite-assisted growth of silicon nanowires in electroless metal deposition. Adv Funct Mater 13:127–132
54. Kooij ES, Butter K, Kelly JJ (1999) Silicon Etching in HNO_3/Hf solution: charge balance for the oxidation reaction. Electrochem Solid State Lett 2:178–180

55. Turner DR (1960) On the mechanism of chemically etching germanium and silicon. J Electrochem Soc 107:810–816
56. Hadjersi T, Gabouze N, Kooij ES, Zinine A, Ababou A, Chergui W, Cheraga H, Belhousse S, Djeghri A (2004) Metal-assisted chemical etching in HF/$Na_2S_2O_8$ or HF/$KMnO_4$ produces porous silicon. Thin Solid Films 459:271–275
57. Cruz S, Honig-dOrville A, Muller J (2005) Fabrication and optimization of porous silicon substrates for diffusion membrane applications. J Electrochem Soc 152:C418–C424
58. Tsujino K, Matsumura M (2005) Boring deep cylindrical nanoholes in silicon using silver nanoparticles as a catalyst. Adv Mater 17:1045–1047
59. Yoo JK, Kim J, Jung YS, Kang K (2012) Scalable fabrication of silicon nanotubes and their application to energy storage. Adv Mater 24:5452–5456
60. Yi R, Dai F, Gordin ML, Chen S, Wang D (2013) Micro-sized Si-C composite with interconnected nanoscale building blocks as high-performance anodes for practical application in lithium-ion batteries. Adv Energy Mater 3:295–300
61. Magasinski A, Dixon P, Hertzberg B, Kvit A, Ayala J, Yushin G (2010) High-performance lithium-ion anodes using a hierarchical bottom-up approach. Nat Mater 9:353–358
62. Maher S, Alsawat M, Kumeria T, Fathalla D, Gihan F, Santos A, Fawzia H, Losic D (2015) Luminescent silicon diatom replicas: self-reporting and degradable drug carriers with biologically derived shape for sustained delivery of therapeutics. Adv Funct Mater 25:5107–5116
63. Sun W, Puzas JE, Sheu T, Liu X, Fauchet PM (2007) Nano- to microscale porous silicon as a cell interface for bone-tissue engineering. Adv Mater 19:921–924
64. Asoh H, Sakamoto S, Ono S (2007) Metal patterning on silicon surface by site-selective electroless deposition through colloidal crystal templating. J Colloid Interface Sci 316:547–552
65. Ono S, Oide A, Asoh H (2007) Nanopatterning of silicon with use of self-organized porous alumina and colloidal crystals as mask. Electrochim Acta 52:2898–2904
66. Cozzi C, Polito G, Strambinin LM, Barillaro G (2016) Electrochemical preparation of in-silicon hierarchical networks of regular out-of-plane macropores interconnected by secondary in-plane pores through controlled inhibition of breakdown effects. Electrochim Acta 187:552–559
67. Shahbazi M-A, Hamidi M, Mäkilä EM, Zhang H, Almeida PV, Kaasalainen M, Salonen JJ, Hirvonen JT, Santos HA (2013) The mechanisms of surface chemistry effects of mesoporous silicon nanoparticles on immunotoxicity and biocompatibility. Biomaterials 34(31):7776–7789
68. Park J-H, Gu L, Von Maltzahn G, Ruoslahti E, Bhatia SN, Sailor MJ (2009) Biodegradable luminescent Porous silicon nanoparticles for in vivo applications. Nat Mater 8(4):331–336
69. Linford MR, Chidsey CED (1993) Alkyl monolayers covalently bonded to silicon surfaces. J Am Chem Soc 115:12631–12632
70. Ciampi S, Harper JB, Gooding JJ (2010) Wet chemical routes to the assembly of organic monolayers on silicon surfaces via the formation of Si–C bonds: surface preparation, passivation and functionalization. Chem Soc Rev 39:2158–2183
71. Salonen J, Laine E, Niinisto L (2002) Thermal carbonization of porous silicon surface by acetylene. J Appl Phys 91:456–461
72. Jarvis KL, Barnes TJ, Prestidge CA (2012) Surface chemistry of porous silicon and implications for drug encapsulation and delivery applications. Adv Colloid Interface Sci 175:25–38
73. Harraz FA (2014) Porous silicon chemical sensors and biosensors: a review. Sensor Actuat B: Chem 202:897–912
74. Jamois C, Wehrspohn RB, Schilling J, Müller F, Hillebrand R, Hergert W (2002) Silicon-based photonic crystals slabs: two concepts. IEEE J Quantum Electron 38:805–810
75. Hillebrand R, Jamois C, Schilling J, Wehrspohn RB, Hergert W (2003) Computation of optical properties of Si-based photonic crystals with varying pore diameters. Phys Stat Solidi B 240:124–133
76. Müller F, Birner A, Gösele U, Lehmann V, Ottow S, Föll H (2000) Structuring of macroporous silicon for applications as photonic crystals. J Porous Mater 7:201–204

77. Schilling J, Wehrspohn RB, Birner A, Müller F, Hillebrand R, Gösele U, Leonard SW, Mondia JP, Genereux F, Van Driel HM, Kramper P, Sandoghdar V, Busch K (2001) A model system for two-dimensional and three-dimensional photonic crystals: macroporous silicon. J Opt A: Pure Appl Opt 3:S121–S132
78. Birner A, Wehrspohn RB, Gösele U, Bursch K (2001) Silicon-based photonic crystals. Adv Mater 13:377–388
79. Ottow S, Lehmann V, Föll H (1996) Processing of three-dimensional microstructures using macroporous N-type silicon. J Electrochem Soc 143:385–390
80. von Freymann G, Koch W, Meisel DC, Wegener M, Diem M, García-Martín A, Pereira S, Busch K, Schilling J, Wehrspohn RB, Gösele U (2003) Diffraction Properties of two-dimensional photonic crystals. Appl Phys Lett 83:614–616
81. Pickering C, Beale MIJ, Robbins DJ, Pearson PJ, Greef R (1984) Optical Studies of the Structure of porous silicon films formed in P-type degenerate and non-degenerate silicon. J Phys C 17:6535–6552
82. Irani YD, Klebe S, McInnes SJP, Jasieniak M, Voelcker NH, Williams KA (2017) Oral mucosal epithelial cells grown on porous silicon membrane for transfer to the rat eye. Sci Rep 7:10042-1–10042-11
83. Link J, Sailor M (2003) Smart dust: self-assembling, self-orienting photonic crystals of porous Si. PNAS 100:10607–10610
84. Donato MG, Monaca MA, Faggio G, Stefano LD, Jones PH, Gucciardi PG, Marago OM (2011) Optical trapping of porous silicon nanoparticles. Nanotechnology 22:505704–1–505704-8
85. McInnes SJP, Michl TD, Delalat B, Al-Bataineh SA, Coad BR, Vasilev K, Griesser HJ, Voelcker NH (2016) "Thunderstruck": plasma-polymer-coated porous silicon microparticles as a controlled drug delivery system. ACS Appl Mater Interfaces 8:4467–4476
86. Turner CT, McInnes SJP, Melville E, Cowin AJ, Voelcker NH (2016) Delivery of flightless I neutralizing antibody from porous silicon nanoparticles improves wound healing in diabetic mice. Adv Healthc Mater 6:1600707-1–1600707-13
87. Ciampi S, Böcking T, Kilian KA, Harper JB, Gooding JJ (2008) Click chemistry in mesoporous materials: functionalization of porous silicon rugate filters. Langmuir 24(11):5888–5892
88. Sweetman MJ, Shearer CJ, Shapter JG, Voelcker NH (2011) dual silane surface functionalization for the selective attachment of human neuronal cells to porous silicon. Langmuir 27(15):9497–9503
89. Coffer JL (2014) Porous silicon and related composites as functional tissue engineering scaffolds. In: Santos HA (ed) Porous silicon for biomedical applications. Woodhead, Cambridge
90. Kilian KA, Bocking T, Gaus K, Gooding JJ (2008) Introducing distinctly different chemical functionalities onto the internal and external surfaces of mesoporous materials. Angew Chem Int Ed 47:2697–2699
91. Wu C-C, Sailor MJ (2013) Selective functionalization of the internal and the external surfaces of mesoporous silicon by liquid masking. ACS Nano 7:3158–3167
92. Song J, Sailor MJ (1999) Chemical modification of crystalline porous silicon surfaces. Comment Inorg Chem 21:69–84
93. Lai M, Parish G, Dell J, Liu Y, Keating A (2011) Chemical resistance of porous silicon: photolithographic applications. Phys Stat Sol C 8:1847–1850
94. Lai M, Parish G, Liu Y, Dell JM, Keating AJ (2011) Development of an alkaline-compatible porous-silicon photolithographic process. J Microelectromech Syst 20:418–423
95. James TD, Keating A, Parish G, Musca CA (2009) Low temperature N_2-based passivation technique for porous silicon thin films. Solid State Commun 149:1322–1325
96. Salonen J, Bjorkqvist M, Laine E, Niinisto L (2004) Stabilization of porous silicon surface by thermal decomposition of acetylene. Appl Surf Sci 225:389–394
97. Salonen J, Lehto V-P, Bjorkqvist M, Laine E, Niinisto L (2000) Studies of thermally-carbonized porous silicon surfaces. Phys Stat Sol A 182:123–126
98. Bateman J, Eagling R, Worrall D, Horrocks B, Houlton A (1998) Alkylation of porous silicon by direct reaction with alkenes and alkynes. Angew Chem Int Ed 37:2638–2685

99. Kim N, Laibinis P (1998) Derivatization of porous silicon by grignard reagents at room temperature. J Am Chem Soc 120:4516–4517
100. Buriak JM, Stewart MP, Geders TW, Allen MJ, Choi HC, Smith J, Raftery D, Canham LT (1999) Lewis acid mediated hydrosilylation on porous silicon surfaces. J Am Chem Soc 121:11491–11502
101. Stewart MP, Buriak JM (1998) Photopatterned hydrosilylation on porous silicon. Angew Chem Int Ed 37:3257–3260
102. Sweetman MJ, Ronci M, Ghaemi SR, Craig JE, Voelcker NH (2012) porous silicon films micropatterned with bioelements as supports for mammalian cells. Adv Funct Mater 22:1158–1166
103. Lees I, Lin H, Canaria C, Gurtner C, Sailor M, Miskelly G (2003) Chemical stability of porous silicon surfaces electrochemically modified with functional alkyl species. Langmuir 19:9812–9817
104. Boukherroub R, Petit A, Loupy A, Chazalviel J-N, Ozanam F (2003) Microwave-Assisted chemical functionalization of hydrogen-terminated porous silicon surfaces. J Phys Chem B 107:13459–13462
105. Sweetman MJ, McInnes SJP, Vasani RB, Guinan T, Blencowe A, Voelcker NH (2015) Rapid, metal-free hydrosilanisation chemistry for porous silicon surface modification. Chem Commun 51:10640–10643
106. Stewart MP, Buriak JM (2000) Chemical and biological applications of porous silicon technology. Adv Mater 12:859–869
107. Low SP, Williams KA, Canham LT, Voelcker NH (2006) Evaluation of mammalian cell adhesion on surface-modified porous silicon. Biomaterials 27(26):4538–4546
108. Lowe RD, Szili EJ, Kirkbride P, Thissen H, Siuzdak G, Voelcker NH (2010) Combined immunocapture and laser desorption/ionization mass spectrometry on porous silicon. Anal Chem 82:4201–4208
109. Vasani RB, McInnes SJP, Cole MA, Jani AMM, Ellis AV, Voelcker NH (2011) Stimulus-Responsiveness and Drug Release from Porous Silicon Films ATRP-Grafted with Poly(N-Isopropylacrylamide). Langmuir 27:7843–7853
110. McInnes SJP, Lowe RD (2015) Biomedical uses of porous silicon. In: Losic D, Santos A (eds) Electrochemically engineered nanoporous materials. Springer, Switzerland 117–162 (Chapter 5)
111. Salonen J, Lehto V-P (2008) Fabrication and chemical surface modification of mesoporous silicon for biomedical applications. Chem Eng J 137:162–172
112. Clements LR, Wang P-Y, Harding F, Tsai W-B, Thissen H, Voelcker NH (2010) Mesenchymal stem cell attachment to peptide density gradients on porous silicon generated by electrografting. Phys Stat Solidi A 208:1440–1445
113. Clements LR, Wang PY, Harding F, Tsai WB, Thissen H, Voelcker NH (2011) Mesenchymal stem cell attachment to peptide density gradients on porous silicon generated by electrografting. Phys Stat Solidi A 208(6):1440–1445
114. Dancil K-PS, Greiner DP, Sailor MJ (1999) A porous silicon optical biosensor: detection of reversible binding of igg to a protein a-modified surface. J Am Chem Soc 121(34):7925–7930
115. Lin V, Motesharei K, Dancil K, Sailor MJ, Ghadiri MR (1997) A porous silicon-based optical interferometric biosensor. Science 278:840–843
116. Hart BR, Létant SE, Kane SR, Hadi MZ, Shields SJ, Reynolds JG (2003) New method for attachment of biomolecules to porous silicon. Chem Commun 3:322–323
117. Drott J, Lindstrom K, Rosengren L, Laurell T (1997) Porous silicon as the carrier matrix in microstructured enzyme reactors yielding high enzyme activities. J Micromech Microeng 7:14–23
118. Secret E, Smith K, Dubljevic V, Moore E, Macardle P, Delalat B, Rogers M-L, Johns TG, Durand J-O, Cunin F, Voelcker NH (2013) Antibody-functionalized porous silicon nanoparticles for vectorization of hydrophobic drugs. Adv Healthc Mater 2:718–727
119. Guan B, Magenau A, Ciampi S, Gaus K, Reece PJ, Gooding JJ (2014) Antibody modified porous silicon microparticles for the selective capture of CELLS. Bioconjug Chem 25:1282–1289

120. Holthausen D, Vasani RB, McInnes SJP, Ellis AV, Voelcker NH (2012) Polymerization-amplified optical DNA detection on porous silicon templates. ACS Macro Lett 1:919–921
121. McInnes SJP, Voelcker NH (2012) Porous silicon-based nanostructured microparticles as degradable supports for solid-phase synthesis and release of oligonucleotides. Nanoscale Res Lett 7:1–10
122. Shtenberg G, Massad-Ivanir N, Engin S, Sharon M, Fruk L, Segal E (2012) DNA-directed immobilization of horseradish peroxidase onto porous SiO_2 optical transducers. Nanoscale Res Lett 7:443–448
123. Letant SE, Hart BR, Kane SR, Hadi MZ, Shields SJ, Reynolds JG (2004) Enzyme immobilization on porous silicon surfaces. Adv Mater 16:689–693
124. Hermanson GT (2008) Bioconjugate techniques. Elsevier, San Fransisco
125. Pike A, Patole S, Murray N, Ilyas T, Connolly B, Horrocks B, Houlton A (2003) Covalent and non-covalent attachment and patterning of polypyrrole at silicon surfaces. Adv Mater 15:254–257
126. Yoon MS, Ahn KH, Cheung RW, Sohn H, Link JR, Cunin F, Sailor MJ (2003) Covalent crosslinking of 1-D photonic crystals of microporous Si by hydrosilylation and ring-opening metathesis polymerization. Chem Commun 6:680–681
127. McInnes SJP, Thissen H, Choudhury NR, Voelcker NH (2009) New Biodegradable materials produced by ring opening polymerisation of poly(L-lactide) on porous silicon substrates. J Colloid Interface Sci 332:336–344
128. McInnes SJP, Szili EJ, Al-Bataineh SA, Xu J, Alf ME, Gleason KK, Short RD, Voelcker NH (2012) Combination of iCVD and porous silicon for the development of a controlled drug delivery system. ACS Appl Mater Interfaces 4:3566–3574
129. McInnes SJP, Szili EJ, Al-Bataineh SA, Vasani RB, Xu J, Alf ME, Gleason KK, Short RD, Voelcker NH (2016) Fabrication and characterization of a porous silicon drug delivery system with an initiated chemical vapor deposition temperature-responsive coating. Langmuir 32:301–308
130. Climent E, Martínez-Máñez R, Maquieira Á, Sancenón F, Marcos MD, Brun EM, Soto J, Amorós P (2012) Antibody-capped mesoporous nanoscopic materials: design of a probe for the selective chromo-fluorogenic detection of finasteride. ChemistryOpen 1:251–259
131. Sweetman MJ, Voelcker NH (2012) Chemically patterned porous silicon photonic crystals towards internally referenced organic vapour sensors. RSC Adv 2:4620–4622
132. Guan B, Ciampi S, Le Saux G, Gaus K, Reece PJ, Gooding JJ (2011) Different functionalization of the internal and external surfaces in mesoporous materials for biosensing applications using "click" chemistry. Langmuir 27:328–334
133. Bowditch AP, Waters K, Gale H, Rice P, Scott EAM, Canham LT, Reeves CL, Loni A, Cox TI (1998) In-vivo assessment of tissue compatibility and calcification of bulk and porous silicon. MRS Proc 536:149–154
134. Rosengren A, Wallman L, Bengtsson M, Laurell T, Danielsen N, Bjursten L (2000) Tissue reactions to porous silicon: a comparative biomaterial study. Phys Stat Solidi A 182(1):527–531
135. Anderson SHC, Elliott H, Wallis D, Canham L, Powell J (2003) Dissolution of different forms of partially porous silicon wafers under simulated physiological conditions. Phys Stat Solidi A 197(2):331–335
136. Canham LT, Reeves C, Newey J, Houlton M, Cox T, Buriak J, Stewart M (1999) Derivatized mesoporous silicon with dramatically improved stability in simulated human blood plasma. Adv Mater 11(18):1505–1507
137. Godin B, Gu J, Serda RE, Ferrati S, Liu X, Chiappini C, Tanaka T, Decuzzi P, Ferrari M (2008) Multistage mesoporous silicon-based nanocarriers: biocompatibility with immune cells and controlled degradation in physiological fluids. Controlled Release Newslett 25(4):9–11
138. Santos HA, Riikonen J, Salonen J, Mäkilä E, Heikkilä T, Laaksonen T, Peltonen L, Lehto V-P, Hirvonen J (2010) In vitro cytotoxicity of porous silicon microparticles: effect of the particle concentration. Surf Chem Size Acta Biomater 6(7):2721–2731

139. Sailor MJ, Lee EJ (1997) Surface chemistry of luminescent silicon nanocrystallites. Adv Mater 9:783–793
140. Godin B, Gu J, Serda RE, Bhavane R, Tasciotti E, Chiappini C, Liu X, Tanaka T, Decuzzi P, Ferrari M (2010) Tailoring the degradation kinetics of mesoporous silicon structures through pegylation. J Biomed Mater Res A 94A(4):1236–1243
141. Popplewell JF, King S, Day J, Ackrill P, Fifield L, Cresswell R, di Tada M, Liu K (1998) Kinetics of uptake and elimination of silicic acid by a human subject: a novel application of ^{32}Si and accelerator mass spectrometry. J Inorg Biochem 69:177–180
142. Canham LT (2007) Nanoscale semiconducting silicon as a nutritional food additive. Nanotechnology 18(18):185704-1–185704-6
143. Bimbo LM, Mäkilä E, Laaksonen T, Lehto V-P, Salonen J, Hirvonen J, Santos HA (2011) Drug permeation across intestinal epithelial cells using porous silicon nanoparticles. Biomaterials 32(10):2625–2633
144. Low SP, Williams KA, Canham LT, Voelcker NH (2010) Generation of reactive oxygen species from porous silicon microparticles in cell culture medium. J Biomed Mater Res A 93(3):1124–1131
145. Bimbo LM, Mäkilä E, Raula J, Laaksonen T, Laaksonen P, Strommer K, Kauppinen EI, Salonen J, Linder MB, Hirvonen J, Santos HA (2011) Functional hydrophobin-coating of thermally hydrocarbonized porous silicon microparticles. Biomaterials 32:9089–9099
146. Bimbo LM, Sarparanta M, Mäkilä E, Laaksonen T, Laaksonen P, Salonen J, Linder MB, Hirvonen J, Airaksinen AJ, Santos HA (2012) Cellular interactions of surface modified nanoporous silicon particles. Nanoscale 4:3184–3192
147. Kumeria T, McInnes SJP, Maher S, Santos A (2017) Porous silicon for drug delivery applications and theranostics: recent advances, critical review and perspectives. Expert Opin Drug Deliv 14(12):1407–1422
148. Bayliss SC, Heald R, Fletcher DI, Buckberry LD (1999) The culture of mammalian cells on nanostructured silicon. Adv Mater 11:318–321 (SRC—GoogleScholar)
149. Mayne A, Bayliss S, Barr P, Tobin M, Buckberry L (2000) Biologically interfaced porous silicon devices. Phys Stat Solidi A 182(1):505–513
150. Mäkilä E, Bimbo LM, Kaasalainen M, Herranz B, Airaksinen AJ, Heinonen M, Kukk E, Hirvonen J, Santos HlA, Salonen J (2012) Amine modification of thermally carbonized porous silicon with silane coupling chemistry. Langmuir 28(39):14045–14054
151. Bimbo LM, Sarparanta M, Mäkilä E, Laaksonen T, Laaksonen P, Salonen J, Linder MB, Hirvonen J, Airaksinen AJ, Santos HA (2012) Cellular interactions of surface modified nanoporous silicon particles. Nanoscale 4(10):3184–3192
152. Kilian KA, Böcking T, Ilyas S, Gaus K, Jessup W, Gal M, Gooding JJ (2007) Forming antifouling organic multilayers on porous silicon rugate filters towards in vivo/ex vivo biophotonic devices. Adv Funct Mater 17:2884–2890
153. Wu EC, Park JH, Park J, Segal E, Cunin F, Sailor MJ (2008) Oxidation-triggered release of fluorescent molecules or drugs from mesoporous Si microparticles. ACS Nano 2(11):2401–2409
154. Gizzatov A, Stigliano C, Ananta JS, Sethi R, Xu R, Guven A, Ramirez M, Shen H, Sood A, Ferrari M, Wilson LJ, Liu X, Decuzzi P (2014) Geometrical confinement of Gd(Dota) molecules within mesoporous silicon nanoconstructs for MR imaging of cancer. Cancer Lett 352:1–5
155. Serda RE, Mack A, Pulikkathara M, Zaske AM, Chiappini C, Fakhoury JR, Webb D, Godin B, Conyers JL, Liu XW, Bankson JA, Ferrari M (2010) Cellular association and assembly of a multistage delivery system. Small 6:1329–1340
156. Sarparanta M, Heikkila T, Salonen J, Kukk E, Lehto V, Santos HA, Hirvonen J, Airaksinen AJ (2011) 1[8]F-labeled modified porous silicon particles for investigation of drug delivery carrier distribution in vivo with positron emission tomography. Mol Pharm 8:1799–1806
157. Schwartz MP, Derfus AM, Alvarez SD, Bhatia SN, Sailor MJ (2006) The smart petri dish: a nanostructured photonic crystal for real-time monitoring of living cells. Langmuir 22(16):7084–7090
158. Tanaka T, Godin B, Bhavane R, Nieves-Alicea R, Gu J, Liu X, Chiappini C, Fakhoury J, Amra S, Ewing A (2010) In vivo evaluation of safety of nanoporous silicon carriers following single and multiple dose intravenous administrations in mice. Int J Pharm 402(1):190–197

159. Koh Y, Jang S, Kim J, Kim S, Ko Y, Cho S, Sohn H (2008) Dbr pSi/Pmma composite materials for smart patch application. Colloids Surf A 313–314:328–331
160. Liu D, Shahbazi M-A, Bimbo LM, Hirvonen J, Santos HA (2014) Biocompatibility of porous silicon for biomedical applications. In: Santon HA (ed) Porous silicon for biomedical applications. Woodhead. pp 129–181
161. Serda RE, Ferrati S, Godin B, Tasciotti E, Liu X, Ferrari M (2009) Mitotic trafficking of silicon microparticles. Nanoscale 1:250–259
162. Liu D, Shahbazi M, Bimbo L, Hirvonen J, Santos H (2014) Biocompatibility of porous silicon for biomedical applications, porous silicon for biomedical applications. Woodhead Publishing, Cambridge, pp 129–181
163. Low SP, Voelcker NH, Canham LT, Williams KA (2009) The biocompatibility of porous silicon in tissues of the eye. Biomaterials 30:2873–2880
164. Bowditch A, Waters K, Gale H, Rice P, Scott E, Canham L, Reeves C, Loni A, Cox T (1998) In-vivo assessment of tissue compatibility and calcification of bulk and porous silicon. MRS Online Proc Libr Arch 536
165. Sarparanta M, Mäkilä E, Heikkilä T, Salonen J, Kukk E, Lehto V-P, Santos HA, Hirvonen J, Airaksinen AJ (2011) ^{18}f-labeled modified porous silicon particles for investigation of drug delivery carrier distribution in vivo with positron emission tomography. Mol Pharm 8(5):1799–1806
166. Dalilottojari A, Tong WY, McInnes SJP, Voelcker NH (2016) Biocompatible and bioactive porous silicon materials. In: Korotcenkov G (ed) Porous silicon: from formation to application, part 3 biomedical applications. Taylor & Francis: Boca Raton. pp 319–335 (Chapter 17)
167. Yokoi K, Godin B, Oborn CJ, Alexander JF, Liu X, Fidler IJ, Ferrari M (2013) Porous silicon nanocarriers for dual targeting tumor associated endothelial cells and macrophages in stroma of orthotopic human pancreatic cancers. Cancer Lett 334(2):319–327
168. Moghimi SM, Hunter AC, Murray JC (2001) Long-circulating and target-specific nanoparticles: theory to practice. Pharmacol Rev 53:283–318
169. Jiang W, Kim BY, Rutka JT, Chan WC (2008) Nanoparticle-mediated cellular response is size-dependent. Nat Nanotechnol 3:145–150
170. Lundquist CM, Loo C, Meraz IM, Cerda JDL, Liu X, Serda RE (2014) Characterization of free and porous silicon-encapsulated superparamagnetic iron oxide nanoparticles as platforms for the development of theranostic vaccines. Med Sci 2:51–69
171. Manolova V, Flace A, Bauer M, Schwarz K, Saudan P, Bachmann MF (2008) Nanoparticles target distinct dendritic cell populations according to their size. Eur J Immunol 38:1404–1413
172. Hudson SP, Padera RF, Langer R, Kohane DS (2008) The biocompatibility of mesoporous silicates. Biomaterials 29(30):4045–4055
173. Kovalainen M, Mönkäre J, Kaasalainen M, Riikonen J, Lehto V-P, Salonen J, Herzig K-H, Järvinen K (2013) Development of porous silicon nanocarriers for parenteral peptide delivery. Mol Pharm 10:353–359
174. Santos HA, Hirvonen J (2012) Nanostructured porous silicon materials: potential candidates for improving drug delivery. Nanomedicine 7(9):1281–1284
175. Santos HA, Salonen J, Bimbo LM, Lehto V-P, Peltonen L, Hirvonen J (2011) Mesoporous materials as controlled drug delivery formulations. J Drug Del Sci Tech 21(2):139–155
176. Institute NC What is cancer. https://www.cancer.gov/about-cancer/understanding/what-is-cancer. Accessed December 4
177. Organization WH Cancer. http://www.who.int/mediacentre/factsheets/fs297/en/. Accessed December 4
178. Yu MK, Park J, Jon S (2012) Targeting strategies for multifunctional nanoparticles in cancer imaging and therapy. Theranostics 2(1):3–44
179. Buchsbaum DJ, Rogers BE, Khazaeli MB, Mayo MS, Milenic DE, Kashmiri SV, Anderson CJ, Chappell LL, Brechbiel MW, Curiel DT (1999) Targeting strategies for cancer radiotherapy. Clin Cancer Res 5(10 Suppl):3048s–3055s
180. Trachootham D, Alexandre J, Huang P (2009) Targeting cancer cells by ROS-mediated mechanisms: a radical therapeutic approach? Nat Rev Drug Discov 8:579–591

181. Alexiou C, Schmid RJ, Jurgons R, Kremer M, Wanner G, Bergemann C, Huenges E, Nawroth T, Arnold W, Parak FG (2006) Targeting cancer cells: magnetic nanoparticles as drug carriers. Eur Biophys J 35(5):446–450
182. Vaccari L, Canton D, Zaffaroni N, Villa R, Tormen M, di Fabrizio E (2006) Porous silicon as drug carrier for controlled delivery of doxorubicin anticancer agent. Microelectron Eng 83(4):1598–1601
183. Gu L, Park J-H, Duong KH, Ruoslahti E, Sailor MJ (2010) Magnetic luminescent porous silicon microparticles for localized delivery of molecular drug payloads. Small 6(22):2546–2552
184. Chhablani J, Nieto A, Hou H, Wu EC, Freeman WR, Sailor MJ, Cheng L (2013) Oxidized porous silicon particles covalently grafted with daunorubicin as a sustained intraocular drug delivery system. Invest Ophthalmol Vis Sci 54(2):1268–1279
185. Hou H, Nieto A, Ma F, Freeman WR, Sailor MJ, Cheng L (2014) Tunable sustained intravitreal drug delivery system for daunorubicin using oxidized porous silicon. J Controlled Release 178(Supplement C):46–54
186. Nan K, Ma F, Hou H, Freeman WR, Sailor MJ, Cheng L (2014) Porous silicon oxide-plga composite microspheres for sustained ocular delivery of daunorubicin. Acta Biomater 10(8):3505–3512
187. Hou H, Huffman K, Rios S, Freeman WR, Sailor MJ, Cheng L (2015) A novel approach of daunorubicin application on formation of proliferative retinopathy using a porous silicon controlled delivery system: pharmacodynamics. Invest Ophthalmol Vis Sci 56(4):2755–2763
188. Wu EC, Andrew JS, Buyanin A, Kinsella JM, Sailor MJ (2011) Suitability of porous silicon microparticles for the long-term delivery of redox-active therapeutics. Chem Commun 47(20):5699–5701
189. Li X, St. John J, Coffer JL, Chen Y, Pinizzotto RF, Newey J, Reeves C, Canham LT (2000) Porosified silicon wafer structures impregnated with platinum anti-tumor compounds: fabrication, characterization, and diffusion studies. Biomed Microdevices 2(4):265–272
190. Coffer JL, Montchamp J-L, Aimone JB, Weis RP (2003) Routes to calcified porous silicon: implications for drug delivery and biosensing. Phys Stat Solidi A 197(2):336–339
191. Park JS, Kinsella JM, Jandial DD, Howell SB, Sailor MJ (2011) Cisplatin-loaded porous Si microparticles capped by electroless deposition of platinum. Small 7(14):2061–2069
192. Xiao L, Gu L, Howell SB, Sailor MJ (2011) Porous silicon nanoparticle photosensitizers for singlet oxygen and their phototoxicity against cancer cells. ACS Nano 5(5):3651–3659
193. Hong C, Lee J, Zheng H, Hong SS, Lee C (2011) Porous silicon nanoparticles for cancer photothermotherapy. Nanoscale Res Lett 6(1):321-1–321-8
194. Lee C, Kim H, Hong C, Kim M, Hong SS, Lee DH, Lee WI (2008) Porous silicon as an agent for cancer thermotherapy based on near-infrared light irradiation. J Mater Chem 18(40):4790–4795
195. Savage DJ, Liu X, Curley SA, Ferrari M, Serda RE (2013) Porous silicon advances in drug delivery and immunotherapy. Curr Opin Pharmacol 13(5):834–841
196. Godin B, Chiappini C, Srinivasan S, Alexander JF, Yokoi K, Ferrari M, Decuzzi P, Liu X (2012) Discoidal porous silicon particles: fabrication and biodistribution in breast cancer bearing mice. Adv Funct Mater 22(20):4225–4235
197. Blanco E, Sangai T, Hsiao A, Ferrati S, Bai L, Liu X, Meric-Bernstam F, Ferrari M (2013) Multistage delivery of chemotherapeutic nanoparticles for breast cancer treatment. Cancer Lett 334(2):245–252
198. Wang CF, Sarparanta MP, Mäkilä EM, Hyvönen MLK, Laakkonen PM, Salonen JJ, Hirvonen JT, Airaksinen AJ, Santos HA (2015) Multifunctional porous silicon nanoparticles for cancer theranostics. Biomaterials 48(Supplement C):108–118
199. Kinnari PJ, Hyvönen MLK, Mäkilä EM, Kaasalainen MH, Rivinoja A, Salonen JJ, Hirvonen JT, Laakkonen PM, Santos HA (2013) Tumour homing peptide-functionalized porous silicon nanovectors for cancer therapy. Biomaterials 34(36):9134–9141
200. Shahbazi M-A, Shrestha N, Mäkilä E, Araújo F, Correia A, Ramos T, Sarmento B, Salonen J, Hirvonen J, Santos HA (2015) A Prospective cancer chemo-immunotherapy approach mediated by synergistic Cd326 targeted porous silicon nanovectors. Nano Res 8(5):1505–1521

201. Almeida PV, Shahbazi M-A, Makila E, Kaasalainen M, Salonen J, Hirvonen J, Santos HA (2014) Amine-modified hyaluronic acid-functionalized porous silicon nanoparticles for targeting breast cancer tumors. Nanoscale 6(17):10377–10387
202. Kong F, Zhang X, Zhang H, Qu X, Chen D, Servos M, Mäkilä E, Salonen J, Santos HA, Hai M, Weitz DA (2015) Inhibition of multidrug resistance of cancer cells by co-delivery of DNA nanostructures and drugs using porous silicon Nanoparticles@Giant Liposomes. Adv Funct Mater 25(22):3330–3340
203. Puvanakrishnan P, Park J, Chatterjee D, Krishnan S, Tunnell JW (2012) In vivo tumor targeting of gold nanoparticles: effect of particle type and dosing strategy. Int J Nanomed 7:1251–1258 (SRC—GoogleScholar)
204. Kennedy LC, Bickford LR, Lewinski NA, Coughlin AJ, Hu Y, Day ES, West JL, Drezek RA (2011) A new era for cancer treatment: gold-nanoparticle-mediated thermal therapies. Small 7:169–183
205. Chatterjee DK, Diagaradjane P, Krishnan S (2011) Nanoparticle-mediated hyperthermia in cancer therapy. Ther Deliv 2:1001–1014
206. Hong C, Lee J, Son M, Hong SS, Lee C (2011) In-vivo cancer cell destruction using porous silicon nanoparticles. Anticancer Drugs 22:971–977
207. Shen H, You J, Zhang G, Ziemys A, Li Q, Bai L, Deng X, Erm DR, Liu X, Li C, Ferrari M (2012) Cooperative, nanoparticle-enabled thermal therapy of breast cancer. Adv Healthc Mater 1:84–89
208. Ma M, Chen H, Chen Y, Wang X, Chen F, Cui X, Shi J (2012) Au capped magnetic core/mesoporous silica shell nanoparticles for combined photothermo-/chemo-therapy and multimodal imaging. Biomaterials 33:989–998
209. Robertson CA, Evans DH, Abrahamse H Photodynamic therapy (Pdt): a short review on cellular mechanisms and cancer research applications for Pdt. J Photochem Photobiol B Biol 96:1–8
210. Canham LT, Ferguson F (2014) Porous silicon in branchytherapy. In: Canham LT (ed) Handbook of porous silicon. Springer International Publishing, Switzerland, pp 1−7
211. Zhang K, Loong SLE, Connor S, Sidney Yu SWK, Tan SY, Ng RTH, Lee KM, Canham LT, Chow PKH (2005) Complete tumor response following intratumoral 32P BioSilicon on human hepatocellular and pancreatic carcinoma xenografts in nude mice. Clin Cancer Res 11:7532–7537
212. Goh AS-W, Chung AY-F, Lo RH-G, Lau T-N, Yu SW-K, Chng M, Satchithanantham S, Loong SL-E, Ng DC-E, Lim B-C, Connor S, Chow PK-H (2007) A novel approach to brachytherapy in hepatocellular carcinoma using a Phosphorous32 (32P) brachytherapy delivery device—a first-in-man study. Int J Radiat Oncol Biol Phys 67:786–792
213. Bonanno LM, Kwong TC, DeLouise LA (2010) Label-free porous silicon immunosensor for broad detection of opiates in a blind clinical study and results comparison to commercial analytical chemistry techniques. Anal Chem 82:9711–9718
214. Bonanno LM, Segal E (2011) Nanostructured porous silicon-polymer-based hybrids: from biosensing to drug delivery. Nanomed 6:1755–1770
215. Pacholski C, Sartor M, Sailor MJ, Cunin F, Miskelly GM (2005) Biosensing using porous silicon double-layer interferometers: reflective interferometric fourier transform spectroscopy. J Am Chem Soc 127:11636–11645
216. Alvarez SD, Schwartz MP, Migliori B, Rang CU, Chao L, Sailor MJ (2007) Using a porous silicon photonic crystal for bacterial cell-based biosensing. Phys Stat Solidi A 204:1439–1443
217. Janshoff A, Dancil K-PS, Steinem C, Greiner DP, Lin VSY, Gurtner C, Motesharei K, Sailor MJ, Ghadiri MR (1998) Macroporous P-type silicon fabry−perot layers. Fabrication, characterization, and applications in biosensing. J Am Chem Soc 120:12108–12116
218. Tasciotti E, Godin B, Martinez JO, Chiappini C, Bhavane R, Liu X, Ferrari M (2011) Near-infrared imaging method for the in vivo assessment of the biodistribution of nanoporous silicon particles. Mol Imaging 10:56–58
219. Ahire JH, Wang Q, Coxon PR, Malhotra G, Brydson R, Chen R, Chao Y (2012) Highly luminescent and nontoxic amine-capped nanoparticles from porous silicon: synthesis and their use in biomedical imaging. ACS Appl Mater Interfaces 4:3285–3239

220. Sinha S, Tong WY, Williamson NH, McInnes SJP, Puttick S, Cifuentes-Rius A, Bhardwaj R, Plush SE, Voelcker NH (2017) Novel Gd-loaded silicon nanohybrid: a potential EGFR expressing cancer cell targeting MRI contrast agent. ACS Appl Mater Interfaces. Just Accepted Manuscript, https://doi.org/10.1021/acsami.7b14538
221. Kallinen AM, Sarparanta MP, Liu D, Salonen JJ, Hirvonen JT, Santos HA, Airaksinen AJ (2014) In vivo evaluation of porous silicon and porous silicon solid lipid nanocomposites for passive targeting and imaging. Mol Pharm 11:2876–2886
222. Sarparanta M, Bimbo LM, Laaksonen TJ, Laaksonen P, Nyman M, Salonen J, Linder MB, Hirvonen J, Santos HA, Airaksinen AJ, Plasma S (2012) Intravenous Delivery of hydrophobin-functionalized porous silicon nanoparticles: adsorption and biodistribution. Mol Pharm 9:654–663
223. Parodi A, Quattrocchi N, van de Ven AL, Chiappini C, Evangelopoulos M, Martinez JO, Brown BS, Khaled SZ, Yazdi IK, Enzo MV, Isenhart L, Ferrari M, Tasciotti E (2013) Synthetic nanoparticles functionalized with biomimetic leukocyte membranes possess cell-like functions. Nat Nanotechnol 8:61–68
224. van de Ven AL, Kim P, Haley O, Fakhoury JR, Adriani G, Schmulen J, Moloney P, Hussain F, Ferrari M, Liu X, Yun S, Decuzzi P (2012) Rapid tumoritropic accumulation of systemically injected plateloid particles and their biodistribution. J Control Release 158:148–155
225. Gallach D, Sanchez GR, Noval AM, Silvan MM, Ceccone G, Palma RJM, Costa VT, Duart JMM, Engineering B (2010) Materials science and mater. Sci Eng B 169:123–127
226. Santos HA (2014) Porous silicon for biomedical applications. Woodhead Publishing Limited
227. Tasciotti E, Liu X, Bhavane R, Plant K, Leonard AD, Price BK, Cheng MM-C, Decuzzi P, Tour JM, Robertson F, Ferrari M (2008) Mesoporous Silicon particles as a multistage delivery system for imaging and therapeutic applications. Nat Nanotechnol 3:151–157
228. Chiappini C, Tasciotti E, Fakhoury JR, Fine D, Pullan L, Wang Y-C, Fu L, Liu X, Ferrari M (2010) Tailored porous silicon microparticles: fabrication and properties. ChemPhysChem 11:1029–1035
229. Hernandez M, Recio G, Martin-Palma RJ, Garcia-Ramos JV, Domingo C, Sevilla P (2012) Surface enhanced fluorescence of anti-tumoral drug emodin adsorbed on silver nanoparticles and loaded on porous silicon. Nanoscale Res Lett 7:364–370
230. Kirui DK, Ferrari M (2015) Intravital microscopy imaging approaches for image-guided drug delivery systems. Curr Drug Targets 16(6):528–541
231. Ananta JS, Godin B, Sethi R, Moriggi L, Liu X, Serda RE, Krishnamurthy R, Muthupillai R, Bolskar RD, Helm L, Ferrari M, Wilson LJ, Decuzzi P (2010) Geometrical Confinement of gadolinium-based contrast agents in nanoporous particles enhances T1 contrast. Nature Nanotechnol 5:815–821
232. Nissinen T, Näkki S, Latikka M, Heinonen M, Liimatainen T, Xu W, Ras RHA, Gröhn O, Riikonen J, Lehto V-P (2014) Facile Synthesis of Biocompatible Superparamagnetic Mesoporous Nanoparticles for Imageable Drug Delivery. Micropor Mesopor Mat 195:2–8
233. Santos HA, Bimbo LM, Herranz B, Shahbazi M-A, Hirvonen J, Salonen J (2012) Nanostructured porous silicon in preclinical imaging: moving from bench to bedside. J Mater Res 28:152–164
234. Sarparanta MP, Bimbo LM, Mäkilä EM, Salonen JJ, Laaksonen PH, Helariutta AMK, Linder MB, Hirvonen JT, Laaksonen TJ, Santos HA, Airaksinen AJ (2012) The mucoadhesive and gastroretentive properties of hydrophobin-coated porous silicon nanoparticle oral drug delivery systems. Biomaterials 33(11):3353–3362
235. Rytkönen J, Miettinen R, Kaasalainen M, Lehto VP, Salonen J, Närvänen A (2012) Functionalization of mesoporous silicon nanoparticles for targeting and bioimaging purposes. J Nanomat 2012(Article ID 896562):1–9
236. Bimbo LM, Sarparanta M, Santos HA, Airaksinen AJ, Makila E, Laaksonen T, Peltonen L, Lehto V-P, Hirvonen J, Salonen J (2010) Biocompatibility of thermally hydrocarbonized porous silicon nanoparticles and their biodistribution in rats. ACS Nano 4(6):3023–3032
237. Huhtala T, Jalanko A, Kaasalainen M, Salonen J, Riikonen R, Complexed IGF (2012) Biodistribution and pharmacokinetics in infantile neuronal ceroid lipofuscinosis. J Drug Deliv 2012:626417

238. Gu L, Park J, Duong KH, Ruoslahti E, Sailor MJ (2010) Magnetic Luminescent porous silicon microparticles for localized delivery of molecular drug payloads. Small 6:2546–2552
239. Noval A, Vaquero V, Torres-Costa V, Gallach D, Ferro-Llanos V, Serrano JJ, Ruiz JP, Pozo F, Palma RJ (2011) Hybrid luminescent/magnetic nanostructured porous silicon particles for biomedical applications. J Biomed Opt 16:025002–025008

Metal/Metal Oxide Nanoparticles for Cancer Therapy

M. P. Vinardell and M. Mitjans

Abstract Cancer includes a large group of affections that are characterized by an uncontrolled cellular growth and the ability of these cells to spread to distant sites. The uncontrolled cellular growth in cancer cells is attributed to a failure in programmed cell death or apoptosis in which a possible role of oxidative stress has been described. Moreover, oxidative stress is involved in carcinogenesis by influencing intracellular signal transduction and transcription factors directly or indirectly via antioxidants. Metal and metal oxide nanoparticles (NPs) have emerged as innovative tools to address the development of new therapeutic anticancer agents alone or in combination with the classical treatments. The potential use of such NPs against cancer is based on the proapoptotic activity and autophagy, cell growth and metastasis inhibition, and generation of reactive oxygen species as well as radiosensitizing properties described in the studies included in this review. Moreover, we address the different strategies using metal/metal oxide NPs with and without conjugation alone or in combination with radiotherapy or chemotherapy as adjuvants or synergistic agents. NPs obtained from copper, cerium, and zinc were the most studied in various cancer types reported. In many cases, they have been preliminary studies and more research is necessary to increase the knowledge about the therapeutic activity of the metal oxide NPs studied. However, the possibilities that these NPs offer are both extensive and interesting.

Keywords Metal oxide nanoparticles · In vitro models · Animal models · Apoptosis · Oxidative stress

1 Introduction

Cancer is a major health problem around the world with an increasing number of cases diagnosed every year. There are several commonly used cancer therapies, including chemotherapy, surgery, and radiation therapy, but they all have disadvantages, such

M. P. Vinardell · M. Mitjans (✉)
Department of Biochemistry and Physiology, Faculty of Pharmacy and Food Sciences, Universitat de Barcelona, Av. Joan XXIII 27-31, 08028 Barcelona, Spain
e-mail: montsemitjans@ub.edu

G. Gonçalves and G. Tobias (eds.), *Nanooncology*, Nanomedicine and Nanotoxicology, https://doi.org/10.1007/978-3-319-89878-0_10

as limited efficiency and significant side effects. Thus, the search for new therapies is essential. One new therapy is based on the use of nanoparticles (NPs), which, due to their small size, can be used to target the cancer tissue more accurately. Inorganic NPs are usually the smallest, and are increasingly being studied for clinical applications as anticancer therapy drugs [55]. These inorganic NPs, including metal oxides, are promising materials for medical applications, such as cell imaging, drug delivery, and cancer therapy [36, 69]. Metal NPs exhibit unique properties in terms of optical, magnetic, and electrical activity and in particular, iron, nickel and cobalt NPs are applied in medical biotechnology because of their magnetic properties. Induction of oxidative stress, DNA damage, genotoxic effects, and anti-inflammatory responses are among the effects of metal/ metal oxide NPs that can contribute to their potential therapeutic anticancer use [53]. Moreover, metal oxide NPs have been used as carriers of anticancer drugs, allowing easier access of the drug to the cancer cells [19, 57]. Finally, synergistic effects of NPs with classical cancer treatments have also been reported [45, 73].

In this chapter, we focus on the use of metal and/or metal oxide NPs alone or in combination, exploring their mechanism of action and their application to different types of cancer. The in vitro methods used to evaluate their anticancer effects using different types of cancer cell lines are also explored.

2 Metal/Metal Oxide Nanoparticles for Brain Tumors: The Case of Glioma

Abnormal growth of brain tissue cells leads to the development of brain tumors that are classified as either primary (developing in the brain itself) or secondary (due to metastatic cells from other organs). Although primary malignant tumors are relatively infrequent, they are associated with substantial mortality and morbidity, causing approximately 25% of deaths among cancer cases in children [59].

Gliomas arise from constituent glial cells of the brain or their precursors and are the most common primary brain tumor, accounting for about 30% of all brain tumors and about 80% of all malignant brain tumors [15]. Difficulties identifying the specific disorder combined with the current treatment options lead to a median patient survival of 12–15 months after diagnosis [21] and undesirable side effects of treatment during that time. Nanomedicine offers a good opportunity to develop relevant therapeutic strategies to improve the inherent challenges of gliomas such as limitations to crossing the blood–brain barrier by improving its selective permeability, which can lead to a higher chance of survival. One strategy involves the use of metal and/or metal oxide NPs directly as therapeutic tools or as platforms for drug delivery.

In this context, a recent study demonstrated the beneficial effect of NPs on C6 glioma cells [74], although the use of a traditional anticancer drug such as doxorubicin was still necessary. To reduce the use of traditional anticancer drugs, other possibilities such as the use of copper oxide (CuO) NPs have been explored based on results on the uptake and toxicity of ionic copper by C6 glioma cells [5]. C6 glioma cells rapidly accumulate copper after exposure to cuprous oxide NPs in a time-, concentration- and temperature-dependent manner, and the increased copper content is accompanied by severe toxicity to C6 glioma cells [23].

Iron oxide NPs present numerous advantages in tumor treatment, however, several studies have already shown that the use of NPs has adverse effects on biological cells, and in particular the generation of reactive oxygen species (ROS) [31]. In contrast, certain NPs selectively induce autophagy in cancer cells but not in normal cells [25] and thus NPs can exhibit an intrinsic toxicity specifically toward cancer cells, which would lead to a high therapeutic effect [38].

The anti-oxidative properties of iron oxide NPs functionalized with caffeic acid have been investigated in vitro in U87-MG brain cancer cell lines. Moreover, the intravenous injection of NPs to mice intracranially xenografted with U87-MG tumors demonstrated that NPs circulate in the bloodstream to reach brain vessels, and passively accumulate in tumor tissue [43].

Other metal oxide NPs such as silicon oxide and zinc oxide have been studied in U373MG cells, a cell line obtained from human glioblastoma astrocytoma. The two metal oxides exhibited cytotoxic effects against these cells, at least partly via the induction of apoptosis involving caspase activation [27]. To determine whether SiO_2 or ZnO NPs induce apoptosis, U373MG cells were treated with silica oxide (SiO_2) or zinc oxide (ZnO) NPs, and caspase-3 activation was assessed by determining the proteolytic cleavage of poly-(adenosine diphosphate [ADP]-ribose) polymerase (PARP) from 0 to 9 h after treatment. The treatment of U373MG cells with SiO_2 NPs induced PARP cleavage at 1 h after treatment (Fig. 1).

Other studies have focused on the use of metal/metal oxide NPs as radiosensitizing agents to improve the therapeutic effect of radiotherapy in resistant tumor cells. In vitro studies with silver NPs (Ag NPs) in the U251 cell line indicated that they can be a good alternative to the traditional gold NPs [30]. The efficacy of Ag NPs was attributed to higher proapoptotic activity and autophagy than by gold NPs, leading to a better therapeutic outcome in the U251 cell line and also in an orthotopic mouse brain tumor model. Recently, the biocompatible bismuth oxide NPs, which have both medical and cosmetic applications, have also been assessed as potential nano-radiosensitizers in the radioresistant L9 gliosarcoma cell line [52]. The authors focused on this heavy metal because its high atomic number (Z) guarantees toxic effects. However, their results demonstrated that the effect of NPs not only depends on their Z but also on other properties such as morphology and aggregation, and thus further research on the subject should be conducted.

To sum up, the use of metal/metal oxide NPs alone or in combination with traditional cancer treatments (chemotherapy, radiotherapy) in brain tumors is still in the preliminary preclinical phase, with cell lines being the main tumor models. Further steps to demonstrate clinical relevance are mandatory.

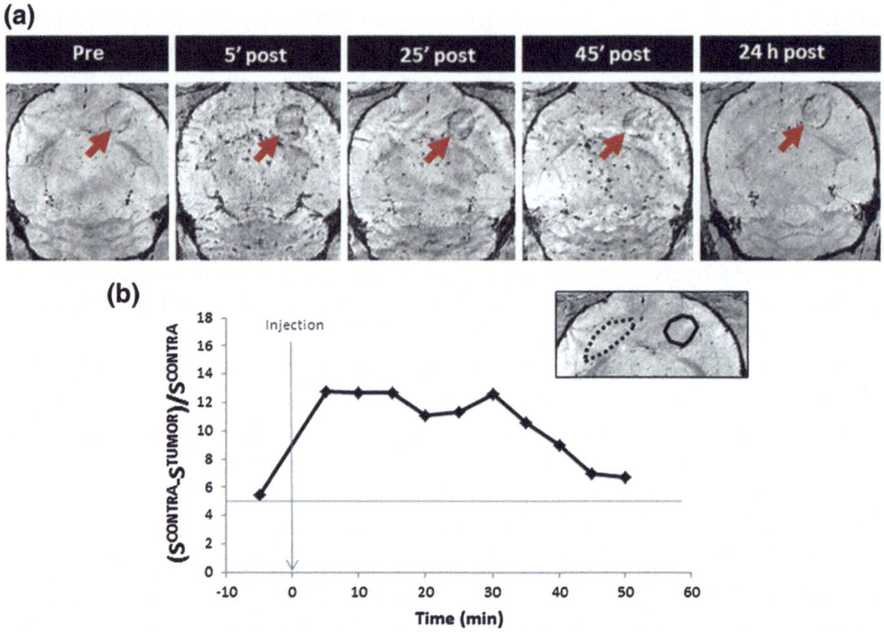

Fig. 1 Injection of γFe_2O_3@CA NP into a U87-MG mouse model of glioblastoma. T2*-weighted MR images of mouse brain pre- and at different times post-injection of γFe_2O_3@CA NP. **a** Tumor is indicated with a red arrow. **b** Contrast follow-up between tumor and contralateral healthy zone versus time. (Upper right corner of **b**). Example of manual segmentation of tumor and contralateral healthy zone on one brain slice (from Richard et al. [43] with permission)

3 Metal/Metal Oxide Nanoparticles for Melanoma

Melanoma is a highly aggressive cancer and the most lethal type of skin cancer due to its high metastatic potential, leaving untreated patients with a median survival of 12 months [58]. The development of new therapies is a crucial objective because melanoma has strong resistance to conventional treatment approaches, which include surgical intervention, radiation, and chemotherapy. Nanomedicines have been emerging as new treatment options, and include gold NPs, metallic multi-segments, silicon nanowire, iron core–gold shell NPs, and gold nanorods. CuO NPs have been considered as potential antitumor drugs because they have been demonstrated not only to induce apoptosis of tumor cells selectively in vitro but also to inhibit the growth and metastasis of melanoma with little hepatic and renal toxicity in mice [62].

CD271-positive ($CD271^+$) cells were demonstrated to be "genuine cancer stem cells" and studies have shown that $CD271^+$ cells can drive melanoma initiation, progression, immune evasion, and drug resistance [29]. For this reason, a good strategy to study the anti-melanoma effectiveness of new drugs is to use human melanoma cell lines containing a high percentage of $CD271^+$ cells. In this context, CuO NPs have

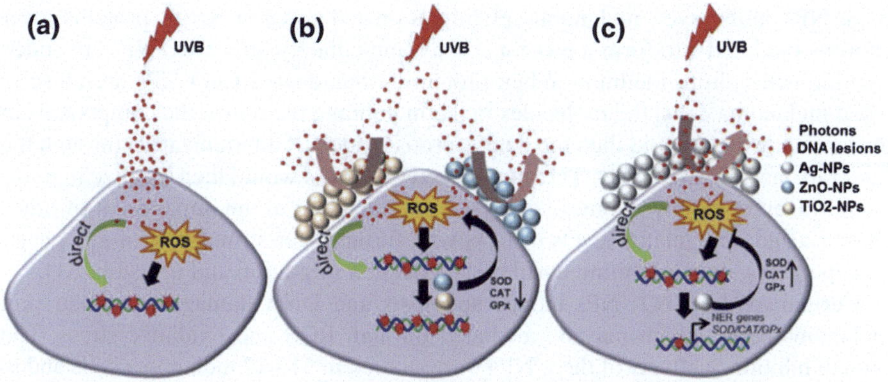

Fig. 2 Graphical representation of the UV-B-protective efficacy of silver, zinc oxide, and titanium dioxide nanoparticles (from Tyagi et al. [54] with permission)

been demonstrated to induce apoptosis in human melanoma stem cells (CD271^{+}/high cells) in A375 and WM266-4 melanoma cell lines. Moreover, they significantly suppressed the expression of SOX10 and CD271 (considered to be the genes involved in maintaining the stemness of melanoma stem cells) and downregulated the expression of MITF (which is involved in the tumorigenesis, growth, metastasis and drug resistance of melanoma). Cuprous oxide NPs also suppressed the growth of human melanoma in tumor-bearing non-obese diabetic-severe combined immunodeficiency (NOD-SCID) mice, which was accompanied by structural necrosis and fibrosis of the tumor (Fig. 2). All these results suggest that cuprous oxide NPs have significant potential for clinical therapy of melanoma [72].

Copper oxide NPs could selectively inhibit cancer cell growth at relatively low concentrations while having no effect on normal cells for various reasons. First, cancer cells are known to have a higher metabolism than normal cells, and thus more nanoparticles might be internalized along with growth factors, leading to high cytotoxicity in cancer cells. Second, normal cells excrete nanoparticles constantly, while cancer cells do not [61]. The same authors in other studies demonstrated that CuO NPs can selectively induce melanoma cancer cell apoptosis by targeting mitochondria, while being less toxic to normal cells [62, 64].

Uveal melanoma, which belongs to a rare subset of all melanomas, is the most common primary intraocular tumor in adults with approximately 50% patients developing the metastatic disease over 15 years. The median survival of uveal melanoma from the time of metastasis is approximately 6–12 months, and there is no effective therapy to improve overall survival to date [49]. Studies of both malignant and normal ocular cell lines have demonstrated that CuO NPs were able to impair uveal melanoma cell migration, invasion, and cytoskeletal assembly, suggesting that these NPs could selectively suppress uveal melanoma cell migration and invasion while having little effect on normal cells. CuO NPs were internalized mainly through lipid raft-mediated endocytosis. Song et al. [51] proposed a mechanism whereby cuprous oxide NPs inhibit uveal melanoma cell migration and invasion. Serum proteins cover cuprous oxide NPs to form a protein corona immediately after CuO NPs are added into the cell culture medium. When protein corona-coated CuO NPs are close to uveal melanoma cells, the molecules in the membrane recognize the components in the corona, which would then mediate cuprous oxide NP internalization through the lipid raft-mediated pathway. The internalized CuO NPs would then localize in mitochondria, autophagolysosomes, and lysosomes, leading to the impairment of mitochondria and upregulated levels of ROS with further over-stimulation of autophagy and apoptosis, thus inhibiting uveal melanoma cell migration and invasion [51].

Cerium oxide (CeO) NPs induce apoptosis and DNA damage in human skin melanoma cells, which may be mediated through ROS and oxidative stress. The growth inhibitory effects of these NPs were studied in 518A2 melanoma cells and in two normal human cell lines for comparative purposes. Normal cells were several-fold less sensitive than melanoma cells [37]. Similar results were observed by other authors, demonstrating that cerium oxide NPs induce apoptosis and DNA damage in human skin melanoma cells [1]. The combination of this class of NPs with the conventional cancer drug doxorubicin was also addressed in the study performed by Sack et al. [45], with interesting findings related to their effectiveness against tumor cells but also their protective capacity when exposed to human dermal fibroblasts (HDF). Cytotoxicity and ROS production in the human malignant melanoma cell line A375 was higher in the presence of both antitumor agents than with doxorubicin or CeO NPs alone, but the genotoxic effects of doxorubicin were also reduced. In contrast, when studied in HDF a protective effect of NPs against the cytotoxic effects of doxorubicin was observed. These findings open up the possibility of implementing novel therapeutic strategies based on the synergistic effects of doxorubicin and CeO NPs.

Exposure to ultraviolet (UV) radiation and more specifically exposure to UV-B is considered the most established etiological cause of skin carcinogenesis. UV-B irradiation harms skin by causing DNA damage, inflammation, cutaneous photoaging, oxidative stress, and immunosuppression, all of which may contribute to the process of skin carcinogenesis [42]. To limit the exposure of human skin to UV radiation, topical application of various commercial sunscreen creams is recommended. These sunscreen products contain ingredients that act as UV-B filters by forming a thin protective film on the skin surface. ZnO NPs and titanium dioxide (TiO_2) NPs are the most commonly used active ingredients in sunscreens due to their broad-spectrum protection against UV-B (absorption, reflection, and scattering) as compared to organic compounds that merely absorb UV-B radiation [32]. A recent study demonstrated the UV-B-protective effects of silver nanoparticles (Ag NPs) in human immortalized keratinocytes (HaCaT) [2]. The same authors compared the UV-B-protective efficacy of Ag NPs against more traditional active ingredients such as ZnO and TiO_2 NPs.

The results showed that Ag NPs were superior to TiO_2 and ZnO NPs in protecting human skin cells from the deleterious effects of UV-B. The authors proposed the mechanism of action of the three NPs (Fig. 2).

UV-B irradiation causes DNA damage by direct absorption, i.e., photo-oxidation, or indirectly by free radical generation (ROS). Although pretreatment with ZnO and TiO NPs provides physical protection (by absorbing or reflecting UV-B) it fails to provide any protection from DNA damage and even contributes to enhanced oxidative DNA damage via increased ROS production. However, Ag NPs protected HaCaT cells against UV-B-induced DNA damage by repairing bulky DNA lesions through the nucleotide excision repair mechanism. Moreover, Ag NPs are also effective at protecting HaCaT cells from UV-B-induced oxidative DNA damage by increasing the activity of the antioxidant enzymes SOD/CAT/GPx. These results suggest that Ag NPs could serve well as a replacement for current major active ingredients, i.e., ZnO and TiO_2 NPs, of commercially available sunscreen creams [54].

Nanodiamonds (NDs) have been proposed as alternatives to classical ZnO and TiO_2 NPs, due to their biocompatibility and low toxicity. Wu et al. [66] demonstrated for the first time that nanodiamonds attenuate UV-B and efficiently protect cells such as keratinocytes and fibroblasts from UV-B-induced damage. The 100-nm NDs exhibited superior UV-B attenuation compared with nanosized TiO_2, ZnO and 5-nm NDs in the HaCaT keratinocyte model, and both 5- and 100-nm NDs exhibited superior UV-B attenuation compared with nanosized ZnO in the two cell types. The protective efficiency of 100-nm NDs was comparable to that of nanosized rutile TiO_2 in an animal model. The animal model consisted of C57BL/6J mice subjected to 20-min UV-B irradiation six times daily for 3 consecutive days, with skin damage developing from the third day. An enzyme-linked immunosorbent assay (ELISA) detected the elevation of both tumor necrosis factor-α (TNF-α) and interleukin-1β (IL-1β), two proinflammatory cytokines. NDs protect C57BL/6J mouse skin from UV-B-induced inflammation by reducing the production of both cytokines. Additionally, NDs are safe materials as they do not stimulate the production of ROS during UV irradiation. These results collectively suggest that NDs could be a "diamond-class" sunscreen ingredient [66].

Indeed, none of these nanoparticles are used for melanoma treatment but they are all applied in the prevention of the development of this type of cancer.

4 Metal/Metal Oxide Nanoparticles for Lung Cancer

One of the limitations of the different therapeutic agents against cancer is the development of resistant cells in hypoxic conditions. Hypoxia is produced in tumors due to the high proliferation of cancer cells [11]. The non-small cell lung cancer, which represents about 85% of all lung tumors, predominantly affects patients with mutations in certain oncogenes and for these reasons, A549 cells overexpressing these oncogenes are a good representative of lung cancer biology and represent a good model for studying lung cancer treatments [44]. On the other hand, A549 cells have strong invasive capacity that can be enhanced under hypoxic conditions, thus increasing tumor progression [48].

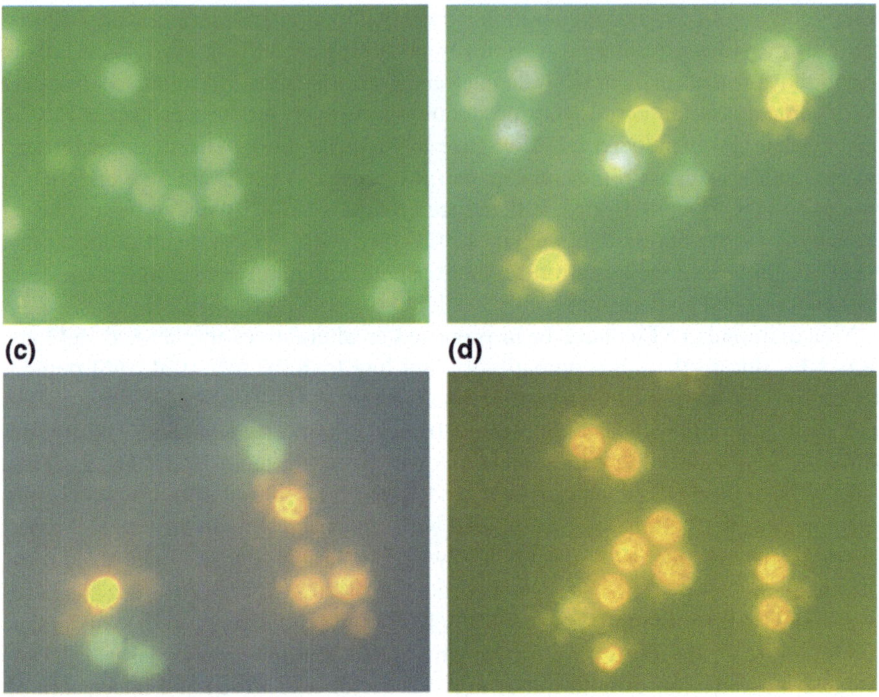

Fig. 3 ZnO nanoparticle-induced apoptosis in LTEP-a-2 cells after 4 h of exposure. Morphologic examination of LTEP-a-2 cells by AO/EB fluorescence staining: **a** control, **b** 0.05 μg/mL, **c** 0.1 μg/mL, and **d** 0.2 μg/mL (from Wang et al. [60] with permission)

In the search for new therapies that do not induce cell resistance new CuO, nickel oxide (NiO) and ferric oxide (Fe_2O_3) NPs were synthesized and studied for their in vitro effects on the growth of A549 cells, a human lung tumor cell line. All three metal oxide nanoparticles showed strong anticancer activity in media with normal levels of oxygen and with low levels or hypoxia, with CuO being the most active. In the case of CuO NPs, the mechanism of nanotoxicity appears to involve ROS generation, cell cycle arrest, and caspase-independent programmed necrotic death. The findings suggest that these metal oxide NPs may be used as effective therapeutic agents, especially in situations where hypoxia induces resistant cells [33].

Other NPs such as ZnO have been evaluated for their capacity for therapy of lung cancer in another human pulmonary adenocarcinoma cell line, LTEP-a (Fig. 3). The results show that ZnO NPs induce a decrease in cell viability with specific morphological changes and apoptosis. The proposed primary mechanism action of this NPs in LETP-a cells may be oxidative stress, based on the increase in ROS and depletion of intracellular glutathione in apoptotic cells [60].

Another strategy in lung cancer therapy is the use of green synthesized nanocomposites of silver-decorated highly reduced graphene oxide (PGE-HRG-Ag) [24]. One of the concerns related to the use of graphene is toxicity, because these materials are prepared by chemical methods and the reagents involved in their preparation are hazardous and harmful to both the environment and humans. The green synthesis of graphene-based materials reduces this toxicity because it involves biocompatible ingredients under physiological conditions. Anticancer evaluation studies of the PGE-HRG-Ag nanocomposites were performed against A549 human lung cancer cells using tamoxifen as the reference drug. Based on the studies performed, it can be concluded that the tested compound induced cell death by apoptosis. A correlation between the amount of Ag nanoparticles on the surface of highly reduced graphene oxide (HRGO) and the anticancer activity of the nanocomposite was observed, and thus an increase in the concentration of Ag nanoparticles on the surface of HRGO enhances the anticancer activity of the nanocomposite studied.

5 Metal/Metal Oxide Nanoparticles for Gastrointestinal Cancer

Gastrointestinal (GI) cancer is a term for the group of cancers that affect the gastrointestinal tract and accessory organs of the digestive system. This includes cancers of the esophagus, gallbladder, liver, pancreas, stomach, small intestine, bowel (large intestine or colon and rectum), and anus.

5.1 Esophageal Cancer

The development of novel metal or metal oxide-based nanoparticle therapies against esophageal cancer is still in the preliminary stages, because efforts have been focused on its early detection. However, the use of gold and magnetic nanodevices for this variety of cancer is more common. Jin et al. [22] reported the synthesis of small-scale core/shell nanoparticles containing tantalum oxide nanospheres (TaONS), doxorubicin, and polypyrrole conjugated with a fluorescent dye. The biocompatibility of TaONS was demonstrated in vitro using HUVEC cells while the photothermal activity was assessed in the human esophageal squamous adenocarcinoma cell line KYSE30. The in vivo model (Balb/c mice subcutaneously injected with KYSE30 cells) demonstrated that TaONS accumulated in the cancerous tumors, as observed using different imaging techniques (computed tomography, photoacoustic, and fluorescent imaging). Moreover, intravenous administration of TaONS plus near-infrared laser irradiation in mice when tumor sizes reached almost 100 mm^3 showed greater therapeutic efficacy than chemotherapy or photothermal treatment alone, reducing tumor volume and leading to complete eradication without recurrence. Finally, the local aerosol administration of TaNOS also showed good therapeutic efficacy and biocompatibility and thus indicated a better therapeutic strategy due to its lower side effects than intravenous injection.

Gallbladder and small intestinal cancers are relatively uncommon and rare diseases, and no interesting studies were found in our literature review.

5.2 *Gastric Cancer*

Gastric cancer has special relevance in Eastern Asia and currently, patient survival remains very low. The use of metal or metal oxide NPs has focused on the use of gold (Au) NPs, but Xiao et al. [70] looked at the potential use of cerium for its capacity to mimic some enzymatic activities. In vitro investigations in BGC823 and MKN28 gastric cancer cell lines showed the potential therapeutic activities of synthetic CeO NPs by reducing cell proliferation and migration. The intrinsic activity of CeO NPs can be attributed, among other potential cellular pathways, to upregulation of the DHX15 gene and activation of p38 MAPK. In vivo experiments corroborated the potential use of CeO NPs, showing a reduction in tumor cell growth and metastasis inhibition.

5.3 *Pancreatic Cancer*

Pancreatic cancer has an extremely poor prognosis, with 90% of pancreatic cancers being malignant and with a 5-year survival after diagnosis. Less than 20% of patients are candidates for surgical resection; therefore, chemotherapy and radiation therapy (RT) remain the only other treatment options. Unfortunately, RT induces side effects, including skin irritation, loss of appetite, fatigue, and nausea, as well as the pain associated with these conditions. Research to reduce the unwanted side effects of RT has yielded two categories of compounds: radiation protectants and radiation sensitizers. However, currently available protectants also induce side effects such as nausea, vomiting, and hypotension. Therefore, the creation and identification of novel compounds that improve the efficacy and therapeutic index of RT would directly improve cancer treatment. CeO NPs have been used as an adjuvant to improve RT in preclinical trials. The valence state and oxygen defects allow CeO NPs to act as auto-regenerative redox status modulators. Recently, CeO NPs have been shown to be capable of entering mammalian cells and have implications in biological systems.

One study examined the ability of CeO NPs to drive ROS accumulation, as well as the subsequent impact on pancreatic cancer cell survival in vitro and in vivo. The pro-oxidant activity of CeO NPs drives radiation-induced radical production selectively in pancreatic cancer cells resulting in radiation sensitization to apoptotic death and growth inhibition. These results identify CeO NPs as a potentially novel radiation sensitizer for the treatment of human pancreatic cancer [65].

5.4 Colorectal Cancer

Colorectal cancer together with breast and prostate cancer is the most commonly diagnosed and the fourth most common cause of cancer death. Patients suffering from inflammatory bowel disease are at an increased risk of developing colorectal cancer, with environmental factors and genetic susceptibility also playing a role in risk.

Nickel oxide NPs have anti-inflammatory properties and toxic effects against bacteria by inducing oxidative stress and releasing nickel ions (Ni^{2+}) inside the cell, and are currently being employed in the field of biomedicine. One possible application is in cancer therapy, especially in the treatment of colon cancer. For this reason, the in vitro cytotoxic activity of NiO NPs prepared using *Moringa oleifera* plant extract against HT-29 (colon carcinoma) cell lines was studied. NiO NPs showed effective cytotoxic activity against HT-29 with alteration of cell morphology and active generation of ROS, which led to the destruction of cell proteins, DNA, and cell membrane, resulting in cell death [13]. These are preliminary results that need to be studied in more depth.

Another preliminary study of ZnO NPs examined its effects on the LoVo human colon carcinoma cell line, the routes of entry into this cell, and their mechanism of action. The results demonstrated that the main cytotoxic effect of ZnO NPs on these cells is due partially to the entry of Zn^{2+} into the cells. ZnO NPs entered LoVo cells by either passive diffusion or endocytosis or both, and contact with the acid pH of lysosomes resulted in the release of Zn^{2+} ions. The simultaneous presence of ZnO NPs and Zn^{2+} ions in the LoVo cells led to the formation of ROS at the mitochondrial and nuclear level, inducing severe DNA damage and cell death [10]. Zijno et al. [75] using Caco-2 cells as in vitro model of human colon carcinoma, studied the genotoxic effects of ZnO and TiO_2 NPs. The study revealed that ZnO but not TiO_2 NPs were able to compromise the cell defense mechanisms and to provoke DNA damage. The genotoxic behavior of ZnO NPs was attributed to a significant and prolonged ROS production that caused direct and indirect genotoxicity. Thus, ROS could interact physically with DNA but also could affect the intracellular pathways impairing repair protein functions.

The cytotoxicity of CuO NPs was evaluated in HT-29 and SW620 human colon cancer cell lines. CuO NPs induced apoptosis in these cells at low concentrations by a mechanism based on the downregulation of Bcl-2 and Bcl-xL protein expression [26].

6 Metal/Metal Oxide Nanoparticles for Blood Cancer

Blood cancer is an umbrella term for cancers that affect the blood, bone marrow, and lymphatic system. There are three groups of blood cancer, according to the hematic cells affected: leukemia, lymphoma, and myeloma. The potential therapeutic application of different metal oxide nanoparticles on the different blood cancers has been investigated.

6.1 Leukemia

Leukemia is a group of cancers that usually begin in the bone marrow and result in high numbers of abnormal white blood cells. High levels of ROS are produced in these malignant disorders resulting in an imbalance in redox homeostasis [41] and a crucial signaling pathway for the initiation and execution of apoptosis.

The activation of ROS has some therapeutic benefits in arresting cancer cell growth. In this sense, cobalt oxide nanoparticles (CoO NPs) were found to induce oxidative stress in different human immortalized lines used as models of leukemia, such as Jurkat, K562 and KG-1A, relative to normal peripheral blood mononuclear cells [7]. The CoO NPs have been shown to have important anticancer properties. The cytotoxicity of CoO NPs in three leukemic cancer cell lines was measured by MTT assay, showing a similar dose-dependent effect in each line. The toxicity of cobalt oxide depends on the internalization of cobalt ions, which is higher in cancer cells than in normal cells. High levels of membrane phospholipids and the greater negative charge of cancer cells may enhance the attachment of CoO NPs to the cell membrane and release the cobalt ions and their subsequent uptake into cancer cells. CoO NPs induced cytotoxicity by activation of the stress-signaling pathway [8].

Besides CoO NPs, other metal/metal oxide nanoparticles have shown their potential beneficial therapeutic effects against leukemia. Thus, CONPs induce apoptosis in K562 cells through reactive oxygen species (ROS) generation [47] and ZnO NPs have the same effect in the HL60 leukemia cell line [39].

6.2 Lymphoma

Lymphoma is a cancer of the immune system and white blood cells. The two main types are Hodgkin lymphoma and non-Hodgkin lymphoma. Among the different metal nanoparticles studied for the treatment of such malignant disorders, platinum nanoparticles have been demonstrated to be effective at inducing cell death in the U937 human lymphoma cell line; nevertheless, the mechanism of cell killing remains unknown [20].

Moreover, (Ag NPs) have been found to prolong lifespan and reduce tumor volume and body weight in animals with Dalton's ascites lymphoma (DAL cells), a type of non-Hodgkin lymphoma. Ag NPs have potent antitumor activity, inducing apoptosis through mitochondrial-dependent and -independent pathways [18].

In order to minimize the side effects of metal-based drugs on normal cells a physiologically friendly approach must be used. In this respect, a nutraceutical involving selenium based-therapy seems to be a promising approach with fewer side effects than the traditional chemotherapeutic strategies. For this reason selenium nanoparticles have been proposed for the treatment of lymphoma using Dalton's lymphoma model [28]. The authors found that the size of the nanoparticles plays an important role in their biological activity, with smaller NPs being more cytotoxic to cancer cells than larger ones [28].

6.3 Myeloma

In the case of myeloma, there have been few studies of metal oxide nanoparticles in the treatment of multiple myeloma (MM). Despite an increase in the number of drugs used in the treatment of this type of blood cancer, multiple myeloma remains an incurable hematological disorder [73].

Nanoparticles have not been used alone to treat myeloma, but have shown good results in combination with other traditional treatments. Bortezomib is one of the drugs used in the treatment of myeloma but induces adverse effects due to the high dose required, and thus a lower dose of bortezomib in combination with a second or third agent could be a better strategy. Magnetic nanoparticles of Fe_3O_4 have been a major drug delivery system over recent years. For this reason, a nontoxic concentration of bortezomib with Gambogic acid-polymerized magnetic nanoparticles of Fe_3O_4 as a drug delivery system has been proposed for combination cancer chemotherapy, and is a good alternative to traditional chemotherapeutic agents for targeted therapeutic approaches to treat multiple myeloma. The apoptotic effect of treatment using this combined approach has been demonstrated in vitro in the human myeloma RPMI-8226 cell line, with greater effects than in cells treated with bortezomib alone by Zhang et al. [73]. In the subcutaneous RPMI-8226 tumor-bearing nude mouse model, the use of magnetic nanoparticles of Fe_3O_4 significantly reduced the tumor volume as reported by the same authors.

In another study, Fe_2O_3 NPs were used in combination with paclitaxel and anti-ABCG2 monoclonal antibody to inhibit the growth of cancer stem cells (CSCs) obtained from the human MM RPMI-8226 cell line in vitro and in their derived tumors in xenografted mice [71]. In a similar way, graphene oxide nanoparticles loaded with doxorubicin increased the effect of doxorubicin alone in human multiple myeloma cells (RPMI-8226) [67]. Much effort should still be invested to elucidate the clinical significance of these studies.

7 Metal/Metal Oxide Nanoparticles for Reproductive System Cancer

Reproductive cancers refer to those that develop in the organs related to reproduction located in the pelvis, but also in the mammary gland or breast. In women, these neoplastic diseases can affect the uterus (endometrial cancer being the most common), cervix, ovaries, vagina, and vulva. Breast cancer can also be considered as a female reproductive cancer. The most common cancers in men are located in the prostate, testicles, and penis.

7.1 Endometrial Cancer

Endometrial cancer is relatively frequent among postmenopausal women and treatment can include some form of surgery accompanied by chemotherapy, radiotherapy, or both. However, it is important to retain reproductive function when premenopausal women are diagnosed. Nanomedicine is a possible solution that might allow surgery to be avoided [17]. Nevertheless, current research efforts are focused on liposomal drugs and no novel metal or metal oxide-based therapies have been reported to date.

7.2 Cervical Cancer

Cervical cancer affects women of reproductive age. Cervical carcinoma is the most common gynecological malignancy and is one of the main causes of cancer-related deaths in women. The carcinogenic transformation from human papillomavirus (HPV) infection to invasive cancer takes a relatively long time. Thus, it is possible to detect the disease at a stage when treatment could be highly effective [3].

ZnO nanoparticles showed a significant cytotoxic effect in vitro on HeLa human cervical carcinoma cells, inducing cell apoptosis through increased intracellular ROS levels and the upregulation of apoptotic gene expression. These results imply the probable utility of ZnO nanoparticles in the treatment of cervical cancer [34]. The same authors demonstrated the efficacy of copper oxide nanorods at inducing apoptosis in human cervical carcinoma cells. Apoptotic-related gene expression increased following treatment, which further confirms the role of apoptosis in this strategy. All these data suggest the possible utility of CuO nanorods in cervical tumor treatment [35].

A recent study on cervical carcinoma cell lines and on cervical carcinoma xenograft nude mice demonstrated the potential therapeutic effects of CuO on cervical carcinoma therapy. CuO NPs strongly inhibit growth and induce apoptosis in both cultured cells and tumor-bearing mice. The cervical cancer nude mouse subcutaneous tumor model was established by transplanting MS751 cells behind the groin of nude mice. The diameter of the tumors in the CuO NPs-treated group was the same as in the cisplatin-treated group, but the animal's weight was not affected by the CuO treatment in contrast to the cisplatin treatment [68] (Fig. 4).

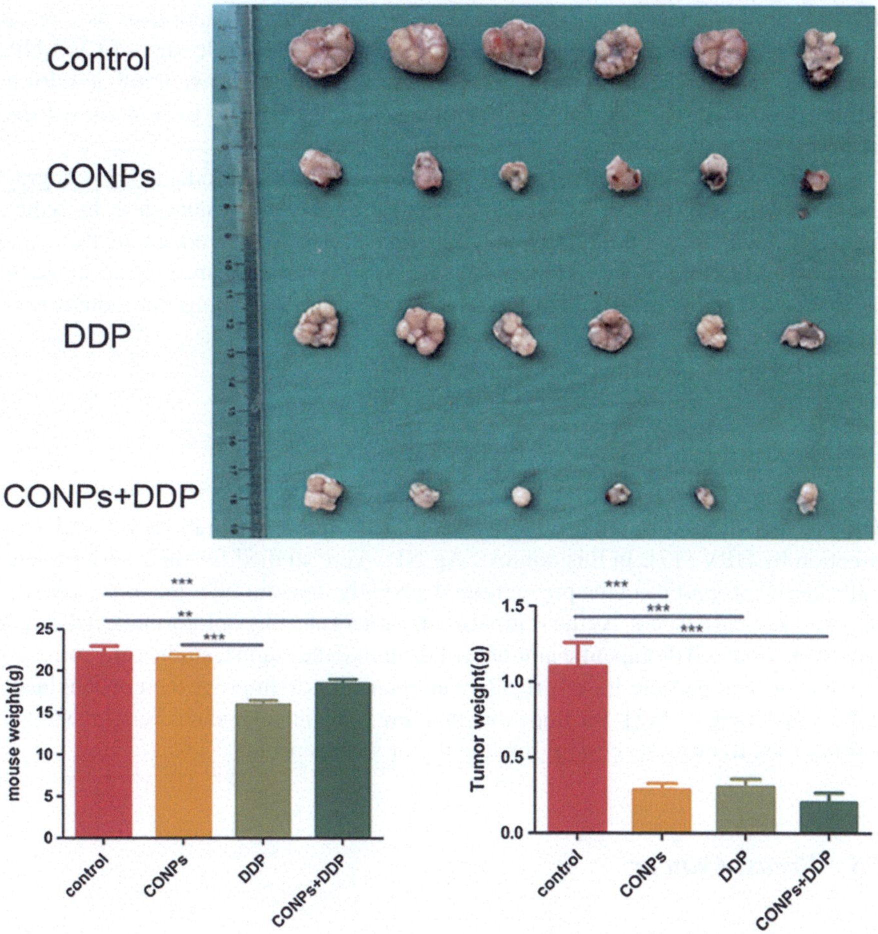

Fig. 4 Antitumor effects of CONPs. Representative pictures of the tumors in the different groups. The diameter of the tumors in the CONP group was equal to that in the DDP group receiving cisplatin but distinctly smaller than that in the glucose group. The smallest diameter was in the CuO NPs+DDP group. The weight of the cisplatin group was lower than that of the control group, while there was no difference in weight between the CONP group and the control group [68]

7.3 Ovarian Cancer

Ovarian cancer is also an important cause of cancer deaths in women because it is difficult to diagnose and shows chemoresistance and potential malignancy. In this type of cancer, the use of CuO NPs conjugated with folic acid (NCe-FA) has been investigated using in vitro and in vivo models [16]. The use of the cell lines A2780, C200, OVCAR3, and SKOV3 demonstrated higher accumulation of conjugated CuO NPs but significantly lower reduction of intracellular ROS. However, the conjugated

NPs reduced cell viability, increased cell death, and activated apoptosis as a result of higher NP uptake. A recent study concluded that the particle size of CuO NPs determines their uptake, intracellular distribution, and ROS scavenging activity in different cell lines [56] and that much effort should be made to transform these promising therapeutic strategies into reality.

Interestingly, the outcomes of an in vivo study in an ovarian model using xenograft nude mice [16] showed that NCe-FA prevents metastasis of ovarian cancer by reducing tumor size but without causing vital organ toxicity. Moreover, in the same xenograft model, the combination of NCe-FA with the anticancer drug cisplatin had better beneficial effects than the two agents alone, including the reduction of angiogenesis.

7.4 *Vaginal and Vulvar Cancer*

Vaginal and vulvar tumors are highly infrequent and sometimes associated with infection by HPV [17]. In this context, Ag NPs were studied for their antiviral and antibacterial properties in the prevention of sexually transmitted infections as causes of gynecological cancers. As the vaginal environment and physiology make it difficult to deliver sustained therapeutic amounts of drug into the vagina, studies are currently focusing on nanoparticle-based drug delivery platforms to improve their prophylactic and therapeutic uses [12]. We hope that in a few years novel and challenging studies will have led to more successful treatments for such cancers.

7.5 *Breast Cancer*

Breast cancer is one of the most common cancers affecting women worldwide. The majority of the deaths from breast cancer are due to its drug resistance and potential metastasis to distant organs. A recent paper reviewed the use of NPs carrying anticancer agents to target tumors for the treatment of breast cancer [50].

ZnO NPs have been studied in the breast cancer cell line MDA-MB-231, showing concentration-dependent cytotoxicity, whereas no positive correlation was found between cytotoxicity and increasing concentration of stress in normal cells [4].

In another study, the use of CeO NPs modified by the chemotherapeutic agent neogambogic acid (CNP-NGA) increased the effectiveness of radiotherapy in breast cancer [9]. The effects of such NPs in combination with radiotherapy increased the cytotoxic effects of the antitumoral treatments alone in the MCF-7 cell line by altering cell cycling and increasing autophagy and apoptosis. Based on these observations, the authors concluded that adjuvant therapies are good therapeutic strategies to diminish adverse effects by allowing a reduction in radiotherapy dose.

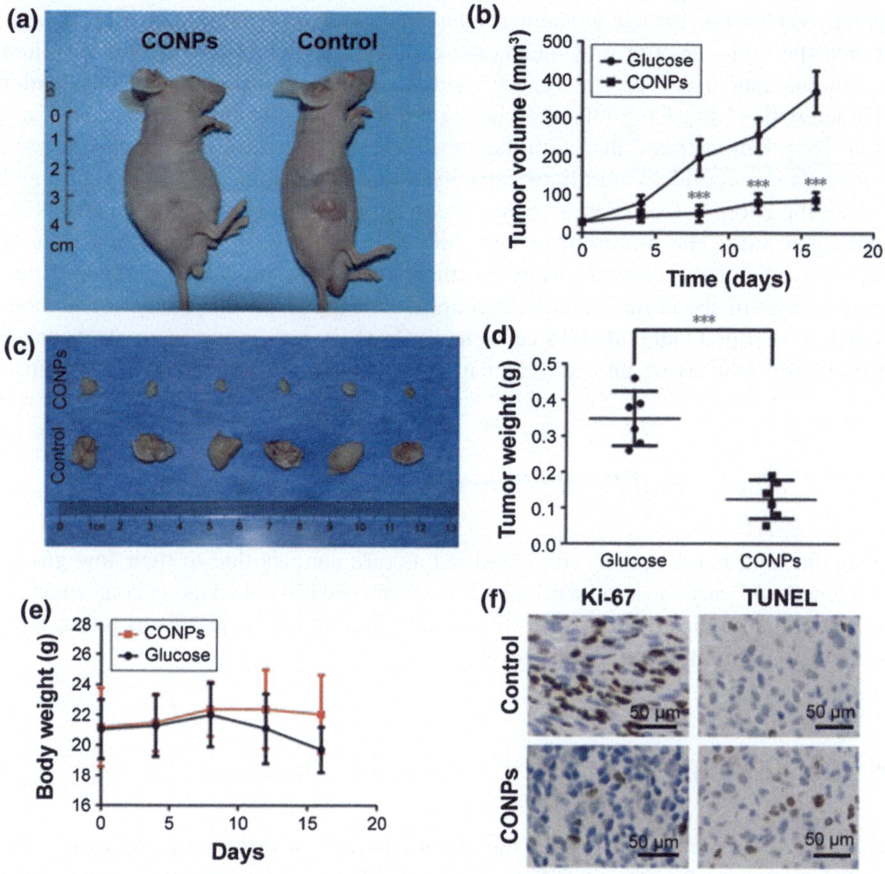

Fig. 5 Antitumor effects of cuprous oxide nanoparticles (CuO NPs therapy on subcutaneous prostate cancer). **a** Mice bearing subcutaneous prostate cancer on day 14 of treatment. **b** Plot of tumor volume versus time. **c** Images of excised subcutaneous tumors. **d** Statistical analysis of the tumor weight from the CuO NPs and control group after treatment for 16 days. **e** Body weight of mice in the CuO NPs and control group. **f** Ki-67 staining and TUNEL staining images (from Wang et al. [63] with permission)

7.6 Prostate Cancer

Prostate cancer is the most commonly diagnosed non-cutaneous malignancy and represents the second most common cause of cancer-related death among men and the most common cancer in Europe [6]. Androgen-deprivation therapy is the standard treatment for locally advanced or metastatic prostate cancer, but many patients develop castration resistance after approximately 2–3 years, with a poor prognosis. Thus, there is an urgent need for novel therapies for this type of cancer and a better understanding of the biology underlying treatment resistance [40]. One possible therapy is based on the use of nanoparticles. In this sense, some studies have demonstrated the high sensitivity of melanoma cells to CuO nanoparticles [62, 64], and melanoma cancer and prostate cancer present similar copper metabolism disorder characteristics [46]. CuO NPs may be a good treatment for prostate cancer. Wang et al. [63] demonstrated that with the subcutaneous prostate cancer mouse model, CuO NPs showed ideal antitumor capacity, and could inhibit the proliferation and induce the apoptosis of prostate cancer cells based on Ki-67 staining and TUNEL staining results. The tumors from the CuO NPs group were significantly smaller than those from the control group and there was no significant difference in the body weight of the control and treated animals (Fig. 5). In this study, the authors also demonstrated that CuO NPs can selectively induce apoptosis in prostate cancer cells, cause cycle arrest, and inhibit the migration and invasion of these cells in vitro.

7.7 Testicular and Penile Cancer

Testicular and penile cancer are classified as rare cancers due to their low global incidence, and they show good relative 5-year survival [14]. As a likely consequence, no important studies dealing with the use of metal/metal oxide NPs as therapeutic tools were found in our literature review.

8 Conclusions

There have been many studies on the potential effect of different metal oxide NPs in the treatment of cancer in different organs. These studies have been performed in vitro using cancer cell lines or in vivo with animal models as shown in Table 1. CuO NPs were the most studied metal oxide NPs, probably due to their ROS scavenging properties, although their potential therapeutic use has been predominantly described only in vitro. In a similar way, CeO and ZnO NPs were the focus of an important number of studies. Different strategies have been investigated, including the use of metal/metal oxide NPs with and without conjugation alone or in combination with radiotherapy or chemotherapy as adjuvants or synergistic agents. In many cases, they have been preliminary studies and more research is necessary to increase the knowledge about the therapeutic activity of the metal oxide NPs studied. However, the possibilities that these NPs offer are both extensive and interesting.

Table 1 Metal/metal oxide NPs with anticancer activity against different types of tumor

Tumor	Metal oxide	Model	Reference
Brain (glioma)	Cuprous oxide	In vitro rat glioma C6 cell line	Joshi et al. [23]
	Zinc oxide Silicon dioxide	In vitro human glioblastoma U373MG cell line	Kim et al. [27]
	Iron oxide	In vitro human glioblastoma U87-MG cell line In vivo orthotopic U87-MG tumor in nude mice	Richard et al. [43]
	Silver	In vitro rat glioma U251 cell line In vivo orthotopic U251 tumor in mice	Liu et al. [30]
	Bismuth oxide	In vitro rat malignant glioma 9L cell line	Stewart et al. [52]
Melanoma	Cerium oxide	In vitro human melanoma A375cell line	Ali et al. [1] Sack et al. [45]
		In vitro human melanoma 518A2 cell line	Pešić et al. [37]
	Cuprous oxide	In vitro human melanoma A375 and WM266-4 cell lines In vivo tumor-bearing non-obese diabetic-severe combined immunodeficiency (NOD-SCID) mice	Yu et al. [72]
Lung	Cuprous oxide Nickel oxide Ferric oxide	In vitro human lung carcinoma A549 cell line	Pandey et al. [33]
	Zinc oxide	In vitro human lung adenocarcinoma LTEP-a cell line	Wang et al. [60]
Esophageal	Tantalum oxide	In vitro human adenocarcinoma KYSE30 cell line	Jin et al. [22]
Gastric	Cerium oxide	In vitro human gastric adenocarcinoma BGC823 and MKN28 cell lines In vivo xenograft nude mice	Xiao et al. [70]
Pancreatic	Cerium oxide	In vitro human pancreatic cancer L3.6pl cell line In vivo orthotopic injection of pancreatic cancer cells in athymic nude mice	Wason and Zhao [65]
Colorectal	Zinc oxide	In vitro human colorectal adenocarcinoma LoVo cell line	Condello et al. [10]
	Nickel oxide	In vitro human colorectal adenocarcinoma HT-29 cell line	Ezhilarasi et al. [13]
	Cuprous oxide	In vitro human colorectal adenocarcinoma HT-29 and SW620 cell lines	Khan et al. [26]
	Zinc oxide Titanium oxide	In vitro human colorectal adenocarcinoma Caco-2 cell line	Zijno et al. [75]
Leukemia	Cobalt oxide	In vitro human leukemia Jurkat, K562 and KG-1A cell lines	Chattopadhyay et al. [7]
	Cuprous oxide	In vitro human leukemia K562 cell line	Chattopadhyay et al. [8]

(continued)

Table 1 (continued)

Tumor	Metal oxide	Model	Reference
Lymphoma	Platinum	In vitro human lymphoma U937 cell line	Jawaid et al. [20]
	Silver	In vivo mice injected with DAL cells	Jacob and Shanmugam [18]
	Selenium		Kumar et al. [28]
Myeloma	Iron oxide	In vitro human myeloma RPMI-8226 cell line	Zhang et al. [73]
	Ferric oxide		Yang et al. [71]
	Graphene oxide		Wu et al. [67]
Cervical	Zinc oxide	In vitro human cervical HeLa cell line	Parduragan et al. [34] Parduragan et al. [35]
	Cuprous oxide	In vitro human cervical carcinoma MS751 cell line In vivo M5751 xenograft nude mice	Xia et al. [68]
Ovarian	Cerium oxide	In vitro human ovarian cancer A2780, C200, OVCAR3 and SKOV3 cell lines In vivo A2780 xenografts nude mice	Hijaz et al. [16]
Breast	Zinc oxide	In vitro human breast MDA-MB-231 adenocarcinoma cell line	Biplab et al. [4]
	Cerium oxide	In vitro human breast cancer MCF-7 cell line	Chen et al. [9]
Prostate	Cuprous oxide	In vitro human prostate cancer PC-3 cell line	Wang et al. [63]

References

1. Ali D, Alarifi S, Alkahtani S et al (2015) Cerium oxide nanoparticles induce oxidative stress and genotoxicity in human skin melanoma cells. Cell Biochem Biophys 71(3):1643–1651
2. Arora S, Tyagi N, Bhardwaj A et al (2015) Silver nanoparticles protect human keratinocytes against UVB radiation-induced DNA damage and apoptosis: potential for prevention of skin carcinogenesis. Nanomed Nanotechnol 11:1265–1275
3. Basu P, Mittal S, Bhadra Vale D, Chami Kharaji Y (2018) Secondary prevention of cervical cancer. Best Pract Res Clin Obstet Gynaecol. https://doi.org/10.1016/j.bpobgyn.2017.08.012
4. Biplab KC, Paudel SN, Rayamajhi S et al (2016) Enhanced preferential cytotoxicity through surface modification: synthesis, characterization and comparative in vitro evaluation of TritonX-100 modified and unmodified zinc oxide nanoparticles in human breast cancer cell (MDA-MB-231). Chem Cent J 10:16. https://doi.org/10.1186/s13065-016-0162-3
5. Bulcke F, Santofimia-Castaño P, Gonzalez-Mateos A et al (2015) Modulation of copper accumulation and copper-induced toxicity by antioxidants and copper chelators in cultured primary brain astrocytes. J Trace Elem Med Biol 32:168–176
6. Center M, Jemal A, Lortet-Tieulent J (2012) International variation in prostate cancer incidence and mortality rates. Eur Urol 61:1079–1092

7. Chattopadhyay S, Dash SK, Tripathy S et al (2015) Cobalt oxide nanoparticles induced oxidative stress linked to activation of TNF-α/caspase-8/p38-MAPK signaling in human leukemia cells. J Appl Toxicol 35:603–613
8. Chattopadhyay S, Dash SK, Tripathy S et al (2015) Phosphonomethyl iminodiacetic acid-conjugated cobalt oxide nanoparticles liberate Co(++) ion-induced stress associated activation of TNF-α/p38 MAPK/caspase 8-caspase 3 signaling in human leukemia cells. J Biol Inorg Chem 20:123–141
9. Chen F, Zhang XH, Hu XD et al (2015) Enhancement of radiotherapy by ceria nanoparticles modified with neogambogic acid in breast cancer cells. Int J Nanomed 10:4957–4969
10. Condello M, De Berardis B, Ammendolia MG et al (2016) ZnO nanoparticle tracking from uptake to genotoxic damage in human colon carcinoma cells. Toxicol In Vitro 35:169–179
11. Cosse JP, Michiels C (2008) Tumour hypoxia affects the responsiveness of cancer cells to chemotherapy and promotes cancer progression. Anti-Cancer Agents Med 8(7):790–797
12. Ensign LM, Cone R, Hanes J (2014) Nanoparticle-based drug delivery to the vagina: a review. J Control Release 190:500–514
13. Ezhilarasi A, Vijaya JJ, Kaviyarasu K et al (2016) Green synthesis of NiO nanoparticles using *Moringa oleifera* extract and their biomedical applications: cytotoxicity effect of nanoparticles against HT-29 cancer cells. J Photochem Photobiol B 164:352–360
14. Gatta G, Trama A, Capocaccia R (in press) Epidemiology of rare cancers and inequalities in oncologic outcomes. EJSO. https://doi.org/10.1016/j.ejso.2017.08.018
15. Goodenberger ML, Jenkins RB (2012) Genetics of adult glioma. Cancer Genet 205:613–621
16. Hijaz M, Das S, Mert I et al (2016) Folic acid tagged nanoceria as a novel therapeutic agent in ovarian cancer. BMC Cancer 16:220. https://doi.org/10.1186/s12885-016-2206-4
17. Howard D, Garcia-Parra J, Healey GD et al (2016) Antibody–drug conjugates and other nanomedicines: the frontier of gynaecological cancer treatment. Interface Focus 6(6):20160054. https://doi.org/10.1098/rsfs.2016.0054
18. Jacob JA, Shanmugam A (2015) Silver nanoparticles provoke apoptosis of Dalton's ascites lymphoma in vivo by mitochondria dependent and independent pathways. Colloid Surf. B 136:1011–1016
19. Javed KR, Ahmad M, Ali S et al (2015) Comparison of doxorubicin anticancer drug loading on different metal oxide nanoparticles. Medicine 94(11):e617. https://doi.org/10.1097/MD.0000000000000617
20. Jawaid P, Rehman MU, Hassan MA et al (2016) Effect of platinum nanoparticles on cell death induced by ultrasound in human lymphoma U937 cells. Ultrason Sonochem, 206–215
21. Jhanwar-Uniyal M, Labagnara M, Friedman M et al (2015) Glioblastoma: molecular pathways, stem cells and therapeutic targets. Cancers 7(2):538–555. https://doi.org/10.3390/cancers7020538 (Basel)
22. Jin Y, Ma X, Zhang S et al (2017) A tantalum oxide-based core/shell nanoparticle for triple-modality image-guided chemo-thermal synergetic therapy of esophageal carcinoma. Cancer Lett 397:61–71
23. Joshi A, Rastedt W, Faber K et al (2016) Uptake and toxicity of copper oxide nanoparticles in C6 glioma cells. Neurochem Res 41(11):3004–3019
24. Khan M, Khan M, Al-Marri AH et al (2016) Apoptosis inducing ability of silver decorated highly reduced graphene oxide nanocomposites in A549 lung cancer. Int J Nanomed 11:873–883
25. Khan MI, Mohammad A, Patil G et al (2012) Induction of ROS, mitochondrial damage and autophagy in lung epithelial cancer cells by iron oxide nanoparticles. Biomaterials 33:1477–1488
26. Khan S, Ansari AA, Khan AA et al (2017) In vitro evaluation of cytotoxicity, possible alteration of apoptotic regulatory proteins, and antibacterial activity of synthesized copper oxide nanoparticles. Colloids Surf B 153:320–326
27. Kim JE, Kim H, Soo S et al (2014) In vitro cytotoxicity of SiO_2 or ZnO nanoparticles with different sizes and surface charges on U373MG human glioblastoma cells. Int J Nanomed 9(Suppl 2):235–241

28. Kumar S, Tomar MS, Acharya A (2015) Carboxylic group-induced synthesis and characterization of selenium nanoparticles and its anti-tumor potential on Dalton's lymphoma cells. Colloid Surf B 126:546–552
29. Lehraiki A, Cerezo M, Rouaud F et al (2015) Increased CD271 expression by the NF-kB pathway promotes melanoma cell survival and drives acquired resistance to BRAF inhibitor vemurafenib. Cell Discov 1:15030. https://doi.org/10.1038/celldisc.2015.30. eCollection 2015
30. Liu P, Jin H, Guo Z et al (2016) Silver nanoparticles outperform gold nanoparticles in radiosensitizing U251 cells in vitro and in an intracranial mouse model of glioma. Int J Nanomed 11:5003–5014
31. Manke A, Wang L, Rojanasakul Y (2013) Mechanisms of nanoparticle-induced oxidative stress and toxicity. BioMed Res Int 2013:942916. https://doi.org/10.1155/2013/942916
32. Monteiro-Riviere NA, Wiench K, Landsiedel R et al (2011) Safety evaluation of sunscreen formulations containing titanium dioxide and zinc oxide nanoparticles in UVB sunburned skin: an in vitro and in vivo study. Toxicol Sci 123:264–280
33. Pandey N, Dhiman S, Srivastava T et al (2016) Transition metal oxide nanoparticles are effective in inhibiting lung cancer cell survival in the hypoxic tumor microenvironment. Chem-Biol Interact 254:221–230
34. Pandurangan M, Enkhtaivan G, Kim DH (2016) Anticancer studies of synthesized ZnO nanoparticles against human cervical carcinoma cells. J Photochem Photobio B 158:206–211
35. Pandurangan M, Nagajyothi PC, Shim J, Kim DH (2016) Anti-proliferative effect of copper oxide nanorods against human cervical carcinoma cells. Biol Trace Elem Res 173:62–70
36. Papis E, Rossi F, Raspanti M et al (2009) Engineered cobalt oxide nanoparticles readily enter cells. Toxicol Lett 189:253–259
37. Pešić M, Podolski-Renić A, Stojković S et al (2015) Anti-cancer effects of cerium oxide nanoparticles and its intracellular redox activity. Chem-Biol Interact 232:85–93
38. Peynshaert K, Manshian BB, Joris F et al (2014) Exploiting intrinsic nanoparticle toxicity: the pros and cons of nanoparticle-induced autophagy in biomedical research. Chem Rev 114:7581–7609
39. Premanathan M, Karthikeyan K, Jeyasubramanian K et al (2011) Selective toxicity of ZnO nanoparticles toward gram-positive bacteria and cancer cells by apoptosis through lipid peroxidation. Nanomed-Nanotechnol 7:184–192
40. Rajan P, Sudbery IM, Eugenia M et al (2014) Next-generation sequencing of advanced prostate cancer treated with androgen-deprivation therapy. Eur Urol 66:32–39
41. Raza MH, Siraj S, Arshad A et al (2017) ROS-modulated therapeutic approaches in cancer treatment. J Cancer Res Clin 143(9):1789–1809
42. Reichrath J, Rass K (2014) Ultraviolet damage, DNA repair and vitamin D in nonmelanoma skin cancer and in malignant melanoma: an update. Adv Exp Med Biol 810:208–233
43. Richard S, Saric A, Boucher M (2016) Antioxidative theranostic iron oxide nanoparticles toward brain tumors imaging and ROS production. ACS Chem Biol 11(10):2812–2819
44. Rose-James A, Tt S (2012) Molecular markers with predictive and prognostic relevance in lung cancer. Lung Cancer Int 2012:729532. https://doi.org/10.1155/2012/729532
45. Sack M, Alili L, Karaman E et al (2014) Combination of conventional chemotherapeutics with redox-active cerium oxide nanoparticles—a novel aspect in cancer therapy. Mol Cancer Ther 13(7):1740–1749
46. Safi R, Nelson ER, Chitneni SK et al (2014) Copper signaling axis as a target for prostate cancer therapeutics. Cancer Res 74:5819–5831
47. Shafagh M, Rahmani F, Delirezh N (2015) CuO nanoparticles induce cytotoxicity and apoptosis in human K562 cancer cell line via mitochondrial pathway, through reactive oxygen species and P53. Iran J Basic Med Sci 18:993–1000
48. Simiantonaki N, Jayasinghe C, Michel-Schmidt R et al (2008) Hypoxia-induced epithelial VEGF-C/VEGFR-3 upregulation in carcinoma cell lines. Int J Oncol 32(3):585–592
49. Singh AD, Turell ME, Topham AK (2011) Uveal melanoma: trends in incidence, treatment, and survival. Ophthalmology 118(9):1881–1885

50. Singh SK, Singh S, Lillard JW Jr et al (2017) Drug delivery approaches for breast cancer. Int J Nanomed 12:6205–6218. https://doi.org/10.2147/IJN.S140325
51. Song H, Xu Q, Zhu Y et al (2015) Serum adsorption, cellular internalization and consequent impact of cuprous oxide nanoparticles on uveal melanoma cells: implications for cancer therapy. Nanomedicine 10(24):3547–3562
52. Stewart C, Konstantinov K, McKinnon S et al (2016) First proof of bismuth oxide nanoparticles as efficient radiosensitisers on highly radioresistant cancer cells. Phys Medica 32:1444–1452
53. Tuli HS, Kashyap D, Bedi SK et al (2015) Molecular aspects of metal oxide nanoparticle (MO-NPs) mediated pharmacological effects. Life Sci 143:71–79
54. Tyagi N, Srivastava SK, Arora S et al (2016) Comparative analysis of the relative potential of silver, zinc-oxide and titanium-dioxide nanoparticles against UVB-induced DNA damage for the prevention of skin carcinogenesis. Cancer Lett 383(1):53–61. https://doi.org/10.1016/j.canlet.2016.09.026
55. Urruticoechea A, Alemany R, Balart J et al (2010) Recent advances in cancer therapy: an overview. Curr Pharm Design 16(1):3–10
56. Vassie JA, Whitelock JM, Lord M (2017) Endocytosis of cerium oxide nanoparticles and modulation of reactive oxygen species in human ovarian and colon cancer cells. Acta Biomater 50:127–141
57. Vinardell MP, Mitjans M (2015) Antitumor activities of metal oxide nanoparticles. Nanomaterials 5:1004–1021 Basel
58. Vogelsang M, Wilson M, Kirchhoff T (2016) Germline determinants of clinical outcome of cutaneous melanoma. Pigm Cell Melanoma Res 29(1):15–26
59. Walsh KM, Ohgaki H, Wrensch MR (2016) Epidemiology. In: Berger MS and Weller M (ed) Handbook of clinical neurology. Elsevier BV, vol 134, pp 3–10
60. Wang C, Hu X, Gao Y, Ji Y (2015) ZnO nanoparticles treatment induces apoptosis by increasing intracellular ROS levels in LTEP-a-2 cells. Biomed Res Int 2015:423287. https://doi.org/10.1155/2015/423287
61. Wang L, Liu Y, Li W et al (2010) Selective targeting of gold nanorods at the mitochondria of cancer cells: implications for cancer therapy. Nano Lett. 11(2):772–780
62. Wang Y, Yang F, Zhang HX et al (2013) Cuprous oxide nanoparticles inhibit the growth and metastasis of melanoma by targeting mitochondria. Cell Death Dis 4:e783. https://doi.org/10.1038/cddis.2013.314
63. Wang Y, Yang Q-W, Yang Q et al (2017) Cuprous oxide nanoparticles inhibit prostate cancer by attenuating the stemness of cancer cells via inhibition of the Wnt signaling pathway. Int J Nanomed 12:2569–2579
64. Wang Y, Zi X-Y, Su J et al (2012) Cuprous oxide nanoparticles selectively induce apoptosis of tumor cells. Int J Nanomed 7:2641–2652
65. Wason MS, Zhao J (2013) Cerium oxide nanoparticles: potential applications for cancer and other diseases. Am J Transl Res 5:126–131
66. Wu M, Sun DS, Lin YC et al (2015) Nanodiamonds protect skin from ultraviolet B-induced damage in mice. J Nanobiotechnol 13:35. https://doi.org/10.1186/s12951-015-0094-4
67. Wu S, Zhao X, Cui Z et al (2014) Cytotoxicity of graphene oxide and graphene oxide loaded with doxorubicin on human multiple myeloma cells. Int J Nanomed 9:1413–1421
68. Xia L, Wang Y, Chen Y et al (2017) Cuprous oxide nanoparticles inhibit the growth of cervical carcinoma by inducing autophagy. Oncotarget 8:61083–61092
69. Xia T, Kovochich M, Liong M et al (2008) Comparison of the mechanism of toxicity of zinc oxide and cerium oxide nanoparticles based on dissolution and oxidative stress properties. ACS Nano 2:2121–2134
70. Xiao YF, Li JM, Wang SM et al (2016) Cerium oxide nanoparticles inhibit the migration and proliferation of gastric cancer by increasing DHX15 expression. Int J Nanomed 11:3023–3034
71. Yang C, He X, Song L et al (2014) Gamma-Fe_2O_3 nanoparticles increase therapeutic efficacy of combination with paclitaxel and anti-ABCG2 monoclonal antibody on multiple myeloma cancer stem cells in mouse model. J Biomed Nanotechnol 10:336–344

72. Yu B, Wang Y, Yu X (2017) Cuprous oxide nanoparticle-inhibited melanoma progress by targeting melanoma stem cells. Int J Nanomed 12:2553–2567
73. Zhang W, Qiao L, Wang X et al (2015) Inducing cell cycle arrest and apoptosis by dimercapto-succinic acid modified Fe_3O_4 magnetic nanoparticles combined with nontoxic concentration of bortezomib and gambogic acid in RPMI-8226 cells. Int J Nanomed 10:3275–3289
74. Zhang Y, Yu J, Zhang L et al (2016) Enhanced antitumor effects of doxorubicin on glioma by entrapping in polybutylcyanoacrylate nanoparticles. Tumor Biol 37:2703–2708
75. Zijno A, De Angelis I, De Berardis B et al (2015) Different mechanisms are involved in oxidative DNA damage and genotoxicity induction by ZnO and TiO_2 nanoparticles in human colon carcinoma cells. Toxicol In Vitro 29(7):1503–1512

Reconfigurable Nucleic Acid Materials for Cancer Therapy

Morgan Chandler, Weina Ke, Justin R. Halman, Martin Panigaj and Kirill A. Afonin

Abstract According to the WHO, we can expect the rise of around 24 million new cases of cancer per year by around 2030 worldwide, which is a 70% increase from 2012. The molecular key players leading to cancer are heterogeneous in respect to tissue origin and may vary from patient to patient, calling for an individualized approach. Nucleic acid biopolymers (DNA and RNA) lend themselves toward applications in personalized therapeutics with high programmability based on their primary structure of five building blocks as well as biocompatibility based on established roles and functional abilities in vivo. Based on the last decades of advances in synthetic methods and natural functions of RNA, various pathways for the regulation of gene expression and DNA/RNA protein binding pathways have been uncovered, leading to the development of novel nanoparticle formulations with governing design principles. For the translation of such approaches into the clinic, the immunogenicity, strategies for delivery, and integration of mechanisms for conditional activation of therapeutic nucleic acids are explored.

Keywords RNA nanotechnology · Immunology · Drug delivery · Dynamic RNA nanoparticles · RNA self-assembly

M. Chandler · W. Ke · J. R. Halman · K. A. Afonin (✉)
Nanoscale Science Program, Department of Chemistry, University of North Carolina at Charlotte, Charlotte, NC 28223, USA
e-mail: kafonin@uncc.edu

M. Panigaj
Institute of Biology and Ecology, Faculty of Science, Pavol Jozef Safarik University, Kosice 041 54, Slovak Republic

K. A. Afonin
The Center for Biomedical Engineering and Science, University of North Carolina at Charlotte, Charlotte, NC 28223, USA

G. Gonçalves and G. Tobias (eds.), *Nanooncology*, Nanomedicine and Nanotoxicology, https://doi.org/10.1007/978-3-319-89878-0_11

1 Therapeutic Nucleic Acids (TNAs) for Cancer Treatment

As an information storage medium, the genomic DNA plays a central role in the life of every cell. Inherited or acquired damage or mutations to encoded genes may lead to a lack of cell cycle control, leading to cancerogenesis. Despite the intensive research of genome editing techniques, therapy on the level of the gene sequence still lies in the future. However, targeting of genes at the transcriptional level, represented by messenger RNA, through antisense or splice-switching oligonucleotides provides a valuable therapeutic approach. In addition, a growing body of experimental evidence turns the image of RNA from passive rather to an active player in a broad spectrum of cellular processes [1]. The remarkable increase of our knowledge about the diverse functions of noncoding RNAs offers many avenues not only to exploit them as therapeutic targets but—together with advancements in RNA structure predictions—to also use nucleic acids as targeted therapeutics [2]. The concept of therapeutic nucleic acids is based on the application of functional RNA and/or DNA molecules (ribozymes, decoys, antisense oligonucleotides, mRNAs, antimiRs, aptamers, siRNAs, microRNAs, etc.) (Fig. 1) as a stand-alone drug or as a combination of various functional types of nucleic acids [3]. The relative simplicity of nucleic acid chemical structure allows for the simple linking of two or more functionally diverse nucleic acids into one chimeric molecule.

This rational design of novel nucleic acids has already shown potential and numerous advantages in their cell-specific delivery along with synergistic activities of formulations. Chimeric nucleic acids which contain an aptamer that can carry out the precise targeting of malfunctioned cells are of particular interest for biomedical applications. The nucleic acid aptamers are short single-stranded oligonucleotides (RNAs, DNAs, or their chemical analogs) that are selected to specifically bind to a target with an affinity comparable to monoclonal antibodies. Unlike antibodies, aptamers are selected in vitro from libraries of up to $\sim 10^{16}$ randomly synthesized sequences that are screened to recognize the specified target molecules under controlled conditions. The aptamer selection method termed systematic evolution of ligands by exponential enrichment (SELEX) [8, 9] allows for amplification of target binding sequences in each cycle until few or ideally one sequence with the highest binding affinity and specificity prevails in population. In comparison with antibodies, the synthesis of aptamers does not require living cells and is an easily scalable and time-saving process. The significant advantage of SELEX lies in the versatility of conditions under which we can perform aptamer selection. Specifically, a group of strategies based on living cell-SELEX, where aptamers are selected to bind certain cell types, has the potential to develop more efficient modes of cell targeting [10]. They offer the advantage of isolating aptamers which bind therapeutically relevant cell surface receptors in their native environment or distinguish between specific cell phenotypes even without prior detailed knowledge of cell surface protein population. Just to name a few successful selections, the cell-SELEX has been applied to discriminate between leukemic cell populations [11], small lung cancer cells versus large cell lung cancer [12], to detect renal cell carcinoma [13], or metastatic breast

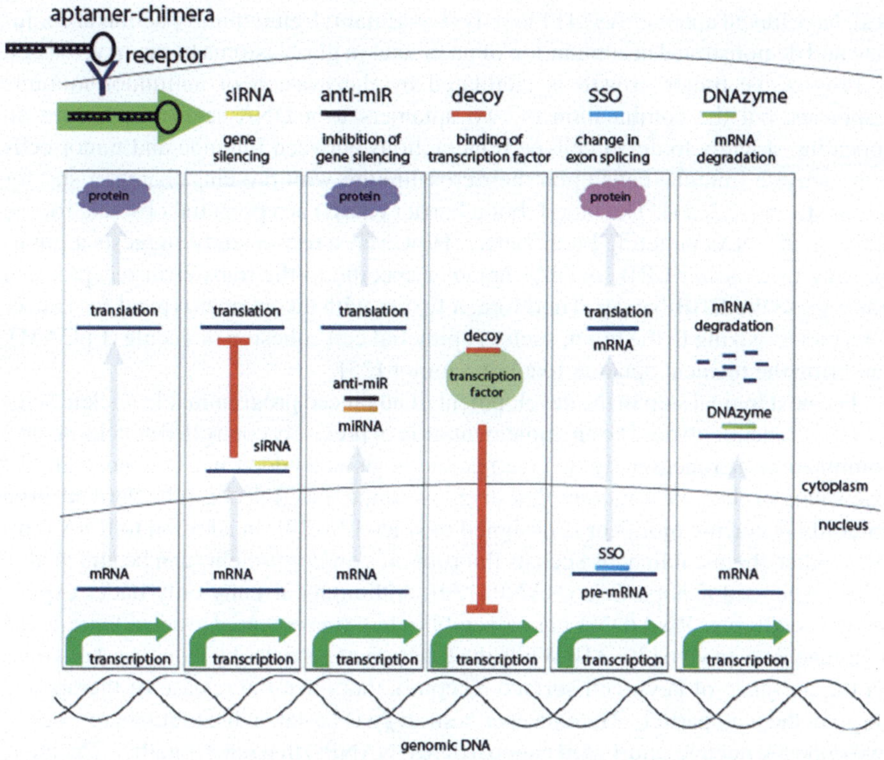

Fig. 1 The schematic illustration of several interactions of various aptamer-chimeras with gene expression. From left to right: normal expression leading to a protein production, delivery of siRNA (or miRNA) inhibits translation of corresponding mRNA while delivery of anti-miR blocks endogenous miRNA in binding to target mRNAs [4]. The result of anti-miR application is thus enhanced production of the specific protein. Decoys are oligonucleotide sequences intended to outcompete endogenous equivalent strands bound by transcription factor or another DNA/RNA binding protein [5]. Splice-switching oligonucleotides (SSOs) are short antisense nucleic acids base-pairing with a pre-mRNA. SSOs impair the normal splicing through blocking the RNA-RNA or protein–RNA binding interactions [6]. In vitro-selected DNAzymes are catalytically active DNA strands; in this context, cleaving a specific mRNA leads to downregulation of gene expression [7]

cancer cells [14]. One decade after the first description of a chimeric aptamer-siRNA application [15], aptamers have been tested to deliver various therapeutic cargos in vitro and in vivo as widely reviewed [3, 16, 17]. Importantly, in addition to receptor recognition, several aptamers can trigger antagonistic or agonistic responses via the target with the potential to enhance therapeutic effects [18, 19].

The significant potential in cancer therapy represents the connection of GC-rich double-stranded DNA (dsDNA) sequences to the aptamer. Single short double-stranded oligonucleotides or tethered in long train-like structures can bind chemotherapeutics such as doxorubicin and through the aptamer, specifically deliver it to cancer cells [20, 21]. While doxorubicin simply intercalates to carrier dsDNA, recently Luo et al. have linked aptamer AS1411 to poly (l-γ-glutamyl-glutamine)-paclitaxel conjugate and demonstrated accumulation of paclitaxel in glioblastoma tissue in vivo [22].

Progressive tumor growth is facilitated by the evasion of antitumor immune responses, but the combination of two aptamers to a DNA scaffold provides an appealing strategy to direct cell–cell interactions between immune and tumor cells or to activate immune cells upon their crosslinking with proteins characteristic for tumor stroma [23, 24]. The blood–brain barrier (BBB) is a potential obstacle for the delivery of TNAs targeting brain cancer. However, a recent study suggests a possible way of crossing BBB. A DNA aptamer specific to the transferrin receptor can transcytose the BBB barrier. Therefore, a fusion with an aptamer typical for cancer cells metastasizing to the brain, such as epithelial cell adhesion molecule (EpCAM), can bring the bivalent aptamer to a brain tumor [25].

The next logical step in the development of advanced programmable nucleic acid-based therapeutics would be an implementation of precisely controlled stoichiometric combination of functionally diverse nucleic acids embedded in 3D nanoscaffolds. The shape of the core structure that displays functional RNAs can be derived from naturally occurring motifs or is designed in silico [26–29]. In addition to RNA moieties, other chemical ligands such as fluorophores and/or proteins can be attached to the nucleic acid nanoparticle (NANP) [29]. Although currently only under experimental conditions, RNA/DNA nanoassemblies have the potential to act as logic-gated nanomachines in vivo [30–33]. While the function of particles by Douglas et al. relies on the presence of keys–cell surface receptors that open the release of therapeutic cargo in the nanoparticle, Halman et al.'s strategy is based on the mutual presence of two opposite nucleic acid-based nanoparticles (NANPs) that react together. The interaction leads to shape-switching which can be used for the production of RNA interface inducers, aptamers, a fluorescent signal, or templates for further transcription.

From the cancer therapy point of view, the introduction of double-stranded parts in a nanoparticle can serve as a delivery moiety for chemotherapeutics intercalated in the double helix [21]. Experimental evidence has shown the feasibility of RNA NANPs as a therapeutic agent in various models of cancer in vitro or in cancer xenografts. The field of RNA nanotechnology [28, 34, 35] applies the current knowledge related to the structure and function of natural RNAs to further address specific biomedical challenges. Also, the option of both canonical and noncanonical base pairings tremendously expands the diverse set of RNA structural motifs available as building blocks [36–38]. For example, by using tertiary RNA interactions, we demonstrated the computer-assisted design and formation of modular dimers [39–41], trimers [35], tetramers [42], pentamers [43], 2D and 3D hexamers and decamers, and long RNA fibers [29, 44–47]. We also designed and verified a series of RNA and RNA/DNA NANPs that can conditionally respond to the environment and through the subsequent conformational changes, activate different functionalities in vitro and in vivo [33, 48–50]. As another platform, a 3WJ motif derived from the DNA-packaging RNA (pRNA) motor of bacteriophage phi29 has been proven to be a versatile tool in RNA nanodesign for biomedical applications [51–53]. The central domain of monomer pRNA allows for the creation of the tetravalent X-motif where an increase of siRNA numbers from one up to four continuously enhances downregulation of gene expression. Replacement of one the four RNA strands with an RNA fragment linked to a fluorescent dye along with another with folate which served as a ligand for binding to the cancer cells revealed that the signal was localized only in the tumor, but not in other organs of the body [54]. Based on the same pRNA motif, Shu et al. designed, assembled, and characterized 14 different homogenous NANPs with a defined size, structure, and stoichiometry [55]. More recently, Shu et al. formulated a pRNA nanoparticle labeled with fluorescent dye for tracking purposes, RNA aptamer targeting epidermal growth factor receptor (EGFR), and anti-miR-21 as a therapeutic moiety. Systemic administration of designed NANPs in orthotopic breast cancer tumor-bearing mice inhibited tumor growth (Fig. 2) [52]. Although experimental data are very promising, the application of NANPs in therapy still may bring some complex problems as described later in the text.

The systemic delivery of naked therapeutic nucleic acids (TNAs) is very challenging. From the structural design, in vivo stability, immunogenicity, and cellular uptake to endosome escape, every mentioned feature affects the efficiency of DNA/RNA therapeutics. Specifically, RNAs are infamous for their sensitivity to nucleases, which complicate their in vivo application. However, experimental evidence suggests that incorporation of chemically modified nucleotides or internucleotide linkages extends

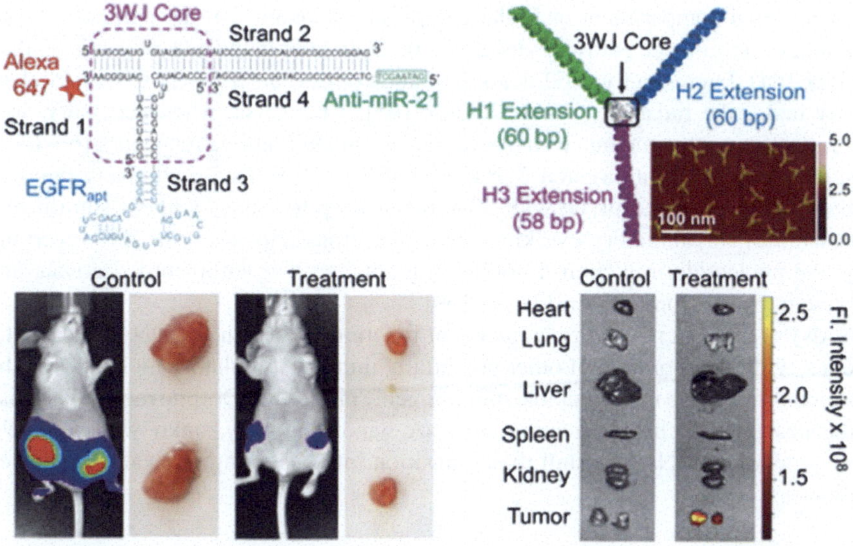

Fig. 2 Left upper corner—2D sequence of the nanoparticle equipped with: EGFR RNA aptamer for targeted delivery, anti-miR-21 for therapy, and Alexa-647 dye for imaging. Right upper corner—3D model of RNA nanoparticles with extensions using 3WJ core as scaffold and atomic force microscopy image of the nanoparticles. Below, from left to right—assessment of targeting and therapeutic effects of nanoparticles in mouse model over the course of five injections. Retention of targeted NANPs in tumors 8 h postinjection documented by fluorescence [52] (This figure was used with permission and can be found at http://pubs.acs.org/doi/10.1021/acsnano.5b02471. Further permissions related to this material should be directed to ACS.)

the half-life of short nucleic acids in the bloodstream [56, 57]. As for the RNA NANPs, it has also been recently confirmed that incorporation of locked nucleic acid nucleotides, alone or in combination with 2′F, prolongs the survival of NANPs in vivo [58].

Simple aptamer-chimeras or any advanced NANPs adopt a 3D structure that differs from naturally occurring nucleic acids [59]. It seems that stability/integrity of DNA nanostructures depend on their 3D structure [60]. However, still, it is hard to predict what form and what ratio of DNA versus RNA in a nanoparticle will be more or less stable. The size and structural diversity of nucleic acid nanostructures including the distribution of ligands on the DNA/RNA nanoparticle determine their renal clearance and size-dependent extravasation which, together with cellular factors, influence the rate of the nanoparticle by cellular uptake [61].

In the case of aptamer-driven targeting, binding to the corresponding receptor directs endocytosed payloads toward degradation in lysosomes, exporting out of the target cell, or in the best-case scenario, the nanoparticle escapes from the endosome. Accessing the cytoplasm is the most limiting step for naked nucleic acid therapeutics. The trafficking of aptamer-nucleic acid therapeutics depends on their concentration, targeted receptor, cell type, and its physiological status [62, 63]. On the cellular level, the trafficking of internalized NANPs comes jointly with defense activation.

The TNAs entering the cells can trigger an innate immune response either in the endolysosomal compartment or in the cytoplasm. Activation of the particular type of immune surveillance pathway, discussed in detail later in the text, depends on the cell type [64]. Interestingly, modification of nucleotides can also help RNA particles to stay under the radar of cellular defense [65]. Vice versa, when necessary, we can utilize immunostimulatory RNAs to induce innate immune responses [66–68]. Generally, it seems that targeted 2′-F modified RNA NANPs are relatively rapidly accumulated in targeted tissue with 50% survival up to 7 days without significant expression of inflammatory cytokines and interferons. However, the difference in only one nucleotide in a control nonbinding aptamer was sufficient to activate an increase in expression of both IFN-β/γ [69].

NANPs represent novel arrangements of functional RNA and/or DNA types that, together with incorporation of other potentially immunostimulatory molecules such as proteins, may have unpredictable immunogenic properties. Despite rapid advances in the design and characterization of TNA nanotechnology, unknown immunogenic properties of NANPs stall their transition into becoming clinically available medicines.

1.1 Immunogenic Responses to TNA Presence

Research over recent decades has demonstrated the immunogenicity of simple single- and double-stranded nucleic acids, yet there is little data reported on the immunogenicity of complex multistranded nucleic acid assemblies in general and NANPs in particular. Recently, there have been efforts to identify key physicochemical contributors to the immunogenic properties of nucleic acid-based nano-constructs. Our results demonstrate the link of physicochemical properties such as melting temperature and blood stability to their relative immunogenic properties [70, 71].

Cells sense and react to foreign nucleic acids through a number of different pattern recognition receptors (PRR). One specific class of PRRs are Toll-like receptors (TLRs) which are responsible for detecting extracellular and endosomal nucleic acids (as well as a variety of other pathogen-derived structures). For nucleic acids, TLRs 3, 7, 8, 9, and 13 (in mice) are known to bind specifically to double-stranded RNAs, single-stranded RNAs, single-stranded unmethylated CpG DNAs, and bacterial ribosomal RNAs, respectively [72–74]. Stimulation of corresponding TLRs can further activate the MyD88 pathway (or TRIF for TLR3), leading to the release of a deluge of cytokines and chemokines with pro-inflammatory properties [75]. Various signaling cascades have developed for cytosolic detection of nucleic acids as well. While there is a myriad of proteins which can act as sequence- and size-specific sensors for nucleic acids including Ku70 [76], DHX9 and DXH36 [77], and DAI [78], the best understood nucleic acid sensors in the cytosol are RIG-I and MDA5 [75, 79–81]. Each of these endolysosomal and cytosolic-based nucleic acid sensors can eventually activate the secretion of type I interferons and pro-inflammatory cytokines and chemokines through a series of common pathways (Fig. 3). Although sensors for simple single- and double-stranded nucleic acids have been identified, mechanistic activation of an innate immune response via the presence of larger multistranded NANPs remains to be elucidated.

Understanding the link between size, shape, diameter, melting temperature, blood stability, and content (regarding both to DNA vs. RNA as well as possible cargo) will prove invaluable when designing the next wave of novel non-immunogenic NANP-based TNAs as well as purposeful induction of an immune response for vaccine adjuvants. Recent research has begun to explore both the characteristics of NANPs that can modulate a target cell's immune response as well as functional conjugations that can modify how the nanoparticle is received by cells.

Understanding this link is especially important for TNAs as the benefit of their function must outweigh any possible immunogenic side effects. The design of nucleic acid-based cancer therapeutics is limited, in this fashion, to shorter nucleic acids to limit immune response as longer nucleic acids have been known to elicit greater immune responses [82, 83]. The most commonly used TNAs include aptamers, siRNAs, microRNAs, and shRNAs. One strategy for combining multiple TNAs for co-delivery is the use of NANPs [29, 34, 43, 45, 84]. We recently demonstrated that the tuning of the immune response of NANPs can be achieved by altering the ratio of DNA and RNA [32, 48, 70, 71]. Increasing the amount of RNA versus DNA in certain types of NANPs can enhance their immunogenicity in an additive fashion. This recently was demonstrated by our group by using NANPs of various shapes, sizes, and compositions (Fig. 4) [70]. Specifically, in a hexameric nucleic acid nanocube, it was demonstrated that replacing DNA strands with RNAs increased transfected cells' immune response across a panel of interferons and interleukins increasingly for each additional RNA [32]. Similarly, multistrand nucleic acid polygons demon-

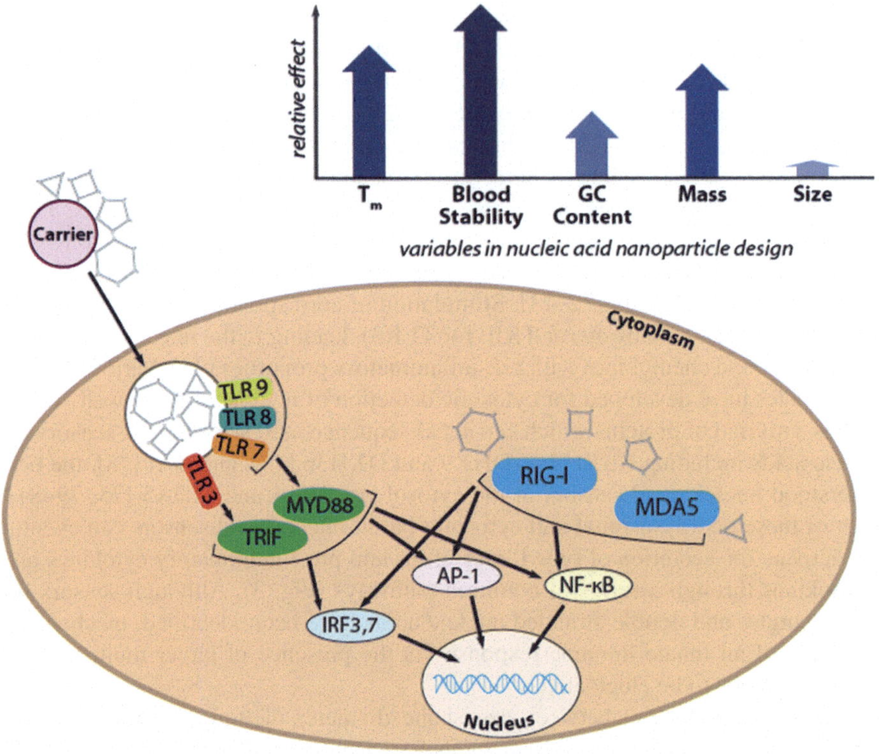

Fig. 3 Top: Graph representing how various physicochemical properties alter immunogenic responses. Bottom: General cellular response to nucleic acids of cells via both endosomal and cytosolic sensing. Downstream sensing leads to the induction and secretion of various cytokines and chemokines

strated the same properties, where integrating RNA into their structures increased their immunogenic properties proportional to the amount of RNA integrated [70].

Beyond alteration of the intrinsic nucleic acid nanostructure, conjugation of functional immunomodulators can act to either enhance or quench an immune response. CpG (an unmethylated single-stranded DNA) is an immune stimulator which is recognized by TLR9 and generates a robust immune response [85, 86]. A number of recent studies have shown that the integration of CpG motifs into established nucleic acid nanostructures can act to enhance the nanostructures' immunogenicity [68, 87–89]. This allows for a multivalent system in which the number of conjugated CpG motifs can be modified for the desired immune response. Additionally, GpG (a CpG analog with immune response reducing qualities) has been used to suppress the immune response of NANPs [90, 91].

Efforts to link various physicochemical properties to a given NANP's immune response have recently demonstrated distinct connections. Quantitative structure activity relationship (QSAR) modeling has been used by our group to try to better

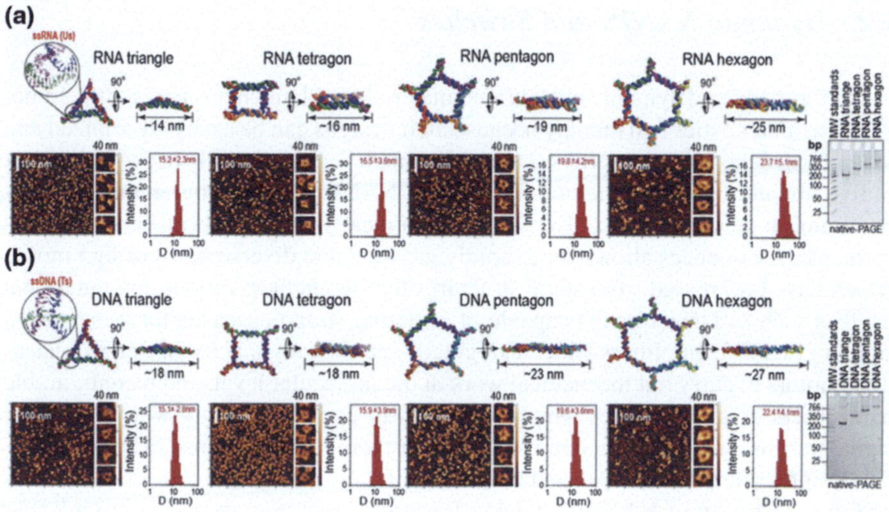

Fig. 4 **a** Programmable nucleic acid RNA and **b** DNA polygons. Each panel presents energy minimized 3D models of RNA and DNA nanoparticles (identical sequences are colored the same), with corresponding AFM images, hydrodynamic radii measured by DLS (presented as ±SEM), and ethidium bromide total staining native-PAGE results. “MW standards” denote the low molecular weight DNA ladder (NEB) used as the size marker [70] (This figure was used with permission and can be found at http://onlinelibrary.wiley.com/doi/10.1002/smll.201701255/abstract. Further permissions related to this material should be directed to *Small*.)

understand this association. This approach allows an input to be weighted comparatively to other inputs versus the magnitude of a specific response. In this fashion, one study has demonstrated the physicochemical properties of a nucleic acid nanoparticle including size, mass, diameter, melting temperature, and blood stability versus its specific immune response. In doing so, the authors elucidated key physicochemical contributors to a nucleic acid nanoparticle’s immunogenicity. Surprisingly, it was determined that the most immunomodulatory properties of these NANPs are their relative stabilities (both thermal and versus blood serum) and not as significantly their size nor shape [70]. Further evaluations using a wider range of NANPs will aid in clarifying additional features that modulate the immune response.

Although the immunogenicity of NANPs remains obfuscated, recent studies have demonstrated methods to modify their immunogenicity by altering either their structure or conjugating active agents. Furthermore, the design of dynamic NANPs, described in the next section, that are active only in the presence of selected stimuli has the potential to increase the safety and efficacy of nucleic acid-based therapy.

1.2 Dynamic NANPs and Switches

Reconfigurable and dynamic nucleic acid materials hold much promise for the future of cancer diagnostics and therapy because their designs can be easily customized and their physicochemical properties can be fine-tuned. With a naturally occurring toolset of five ubiquitous nucleotide building blocks, NANPs can be programmed to carry out a multitude of functions. The thermodynamically driven hybridization of complementary sequences allows for a rapidly growing and diverse set of design motifs which have progressed from static structures to now include various computational devices such as DNA circuits capable of initiating strand cascades for transmitting signals (Fig. 5a), amplifiers which can greatly increase signals for optimized detection, motors to carry out mechanical work at the molecular level, and reconfigurable nanostructures and switches which change form upon interacting with a target or signal to subsequently trigger an observable or therapeutic response [92]. The rational design principles incorporated into assembly improve both the reaction kinetics and the sensitivity of NANPs, allowing for targets such as cancer-causing mRNAs and long noncoding (lnc)RNA biomarkers to be detected from within complex samples and cellular environments [93, 94]. DNA structures have been programmed to respond to changes in pH, light, enzymes, or ion concentration as well as the addition of nucleic acid strands to activate multiple functionalities such as a switch between B- and Z-form DNA, RNA digestion, or release of cargo [95–100].

As a major step toward the development of dynamic NANPs, the reassociation of complementary hybrid strands can be used to activate functionalities such as RNA interference for specific gene silencing (Fig. 6) [33]. Building upon this, dynamic nucleic acid nanocubes composed of six individual DNA and/or RNA strands in varying ratios have been shown to interact as programmed with a cognate nucleic acid "anti-cube" for a shape switch into six double-stranded duplexes. Upon reassociating, the system can be programmed for further activation of multiple functionalities including the completion of a split RNA aptamer, energy transfer between attached chromophores, transcription initiation via the completion of T7 RNA polymerase promoters, and the release of Dicer Substrate RNA for triggering RNA interference (Fig. 5b) [32]. The programmability of DNA and RNA can also be utilized to direct the assembly of NANPs of other formulations. Proteins, metals, and other functional components can then be assembled using the same design strategies [101]. DNA templates have been used for the precise assembly of gold nanoparticles [102] as well as for control over the crystalline states of inorganic nanoparticles [103]. Programmability also allows for applications in biosensing by using a known sequence as a trigger for a measured response. Molecular beacons, for example, are simple probes consisting of a fluorophore and quencher moiety on opposite ends of an RNA strand which contains a sequence complementary to a target. Upon recognition of the target, the molecular beacon unfolds to bind it, thereby separating the fluorophore and quencher for the production of a fluorescent signal [104].

Precise control over programmability via the conditional activation of dynamic NANPs reinforces their potential in biomedicine [105]. By only activating functions

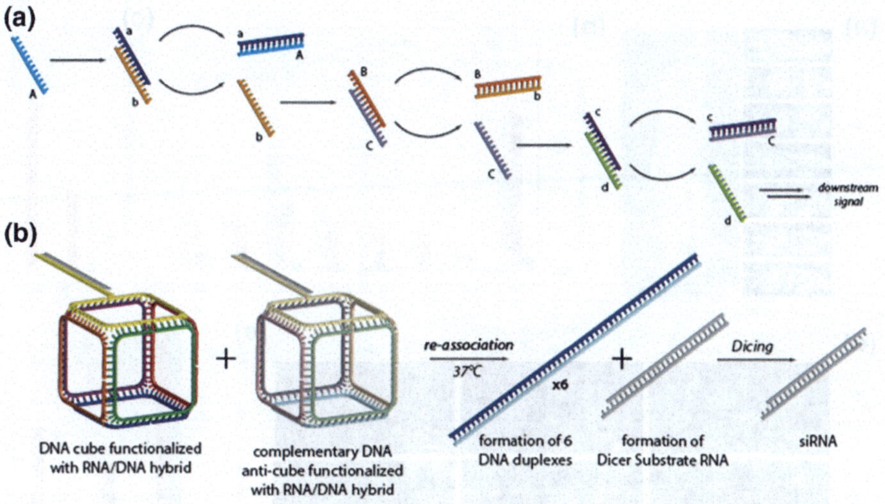

Fig. 5 **a** Strand displacement is commonly used to drive steps in dynamic nucleic acid constructs. As shown, strand A displaces strand b to bind to its complement, strand a. The released strand b displaces strand C to bind to its complement, strand B. The released strand C displaces strand d to bind to its complement, strand c. The final product, strand d, is free to propagate a signal downstream. **b** Dynamic nanoparticles such as three-dimensional cubes can carry functionalities for activation in vivo. As shown, a DNA cube carrying an RNA/DNA hybrid functionality reassociates with its cognate anti-cube at 37 °C. Upon interacting, the cubes switch into six double-stranded DNA duplexes plus one Dicer Substrate RNA duplex. In the intracellular environment, this duplex is cleaved by Dicer for the formation of siRNA and the silencing of targeted genes

upon the recognition of a target, NANPs are optimized for specific treatment to diseased cells which not only concentrates the response but also minimizes the impact on non-diseased cells in the environment. Furthermore, though early NANPs combined diagnostic and therapeutic steps into a single motion, more dynamicity through the customization of triggers and targets can be achieved if diagnostic and therapeutic responses are temporally separated [49]. For example, conditional switches often take advantage of RNAi by releasing a Dicer Substrate RNA upon interaction with a specific RNA biomarker [106–108]. Logic-gated assemblies such as self-cleaving ribozyme logic gates contain AND, OR, YES, and NOT functions for molecular circuits which can be implemented in combined biosensing and gene regulation [49, 109–111]. Upon their delivery, dynamic NANPs which are conditionally activated serve as promising cancer therapeutics.

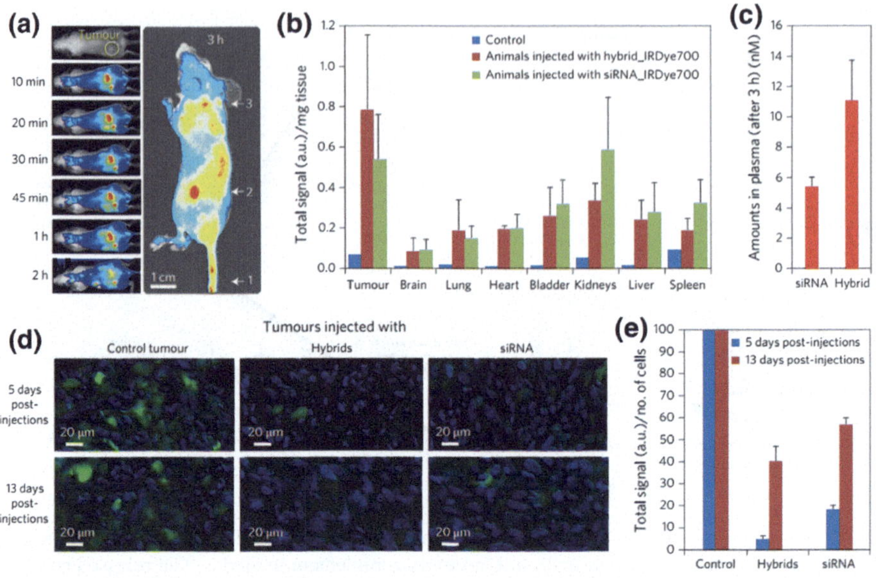

Fig. 6 In vivo and ex vivo studies of RNA–DNA hybrids in a tumor xenograft mouse model. **a** Time-dependent biodistribution imaging in vivo. At 3 h, whole mouse image, fluorescent maximums (in red) correspond to the places of injection (1), tumor (2), and blood withdrawal (3). **b** Relative organ uptakes of fluorescently labeled RNA–DNA hybrids and siRNAs in tumor-bearing mice 3 h post tail-vein injection. A relatively high level of hybrid accumulation occurs in tumor tissue. Error bars indicate standard deviation (s.d.); N ¼ 3. **c** Amounts of fluorescent probe (RNA–DNA hybrids and Dicer Substrate siRNAs labeled with IRDye700) in the mouse bloodstream were measured 3 h post injection. Error bars indicate s.d.; N ¼ 4. **d** Ex vivo fluorescent imaging of tumors 5 and 13 days post-injection in vivo demonstrate comparable levels of eGFP silencing caused by siRNA and RNA–DNA hybrids. **e** Ex vivo quantification (two animals per experiment) of eGFP expression 5 and 13 days post injection [33] (This figure was used with permission and can be found at https://www.nature.com/articles/nnano.2013.44. Further permissions related to this material should be directed to *Nature Nanotechnology*.)

1.3 Exploiting Human Cellular Vesicles-Based Delivery of TNAs

The delivery of TNAs in vivo remains one of the main challenges precluding their broad biomedical applications. Various carriers have been investigated by our group for their potential use for TNA delivery [112–114]. Here, we briefly discuss some possibilities of exosome-based TNA delivery.

Vesicles derived from human cell membranes represent an interesting carrier for the delivery of therapeutic agents. If prepared from a patient, they are non-immunogenic upon application back to the donor's body and therefore are suitable for repeated administration. We can consider these phospholipid particles as biological liposomes. Currently, two different types of vesicles—erythrocyte ghosts and exosomes—are investigated for their potential to serve as therapeutic cargo transportation systems [115, 116]. While ghosts are created artificially from erythrocytes by breaking their membranes and removing the cytoplasm, exosomes are natural extracellular vesicles (EV) of endocytic origin. They are secreted by most cells including tumor cells and they may have a role in intercellular communication [117].

The exosomes horizontally transfer miRNAs and mRNAs to the recipient cells where delivered RNA is still functional [118, 119]. The natural association of RNA with exosomes is very inspirational for exploiting exosomes for the delivery of TNAs. In addition, studies have shown that exosomes are capable of carrying small molecular drugs across the BBB with decreasing mononuclear phagocyte system (MPS) drug clearance [120], while 98% of potent central nervous system drugs cannot cross the BBB and their efficacy in clinical trials is not as successful as shown in laboratory conditions [121]. In the recent decade, the RNAi machinery has been extensively investigated due to its potential for gene modulation and treatment of various diseases through the introduction of exogenous synthetic siRNAs [29]. However, one of the biggest challenges for the application of siRNAs in therapy is the vulnerability of RNA to degradation by plasma and tissue nucleases [122, 123]. Several studies have been conducted to test the protection of exogenous RNA by using exosomes as therapeutic carriers [120]. The pioneering study by Alvarez-Erviti et al. has shown that exosomes delivered exogenous RNAi to targeted tissues more efficiently than liposomes and several viruses that are not of endogenous origin [124, 125]. Wahlgren et al. reported the successful transportation of exogenous siRNA to monocytes and lymphocytes using human plasma exosomes [126]. Shtam et al. provided evidence of successfully introducing the heterologous siRNAs to HeLa and fibrosarcoma cell lines (HT1080 cells) using exosomes [127]. Banizs et al. demonstrated that endothelial exosomes have the capability to accommodate and deliver short foreign nucleic acids into endothelial cells [128]. The use of exosomes fills in the gap of a customized therapeutic delivery vehicle due to their natural ability to deliver any suitable cargos from cell to cell with low immunogenicity to the host.

The detailed knowledge of exosome biogenesis can help us to hijack the pathway and modify it for our purposes. The formation of exosomes results from the inward budding of endosomal membranes; the resulting endosome and its content of intraluminal vesicles (ILVs) are referred as a multivesicular body (MVB) [129]. The process of forming MVBs is complicated and involves more than 20 proteins, most of which are categorized as ESCRT (endosomal sorting complex required for transport) complexes. There are ESCRT-0, -I, -II, and -III complexes. The ESCRT-0 complex initiates the formation of the MVB. It recognizes and binds to ubiquitinated membrane proteins on the endosomal membrane. Ubiquitin is critical in sorting membrane cargos into the MVB. The ESCRT-I complex has several roles such as transferring the ubiquitinated proteins between the ESCRT-0 and -II complexes, and it also recruits the ESCRT-III complex. ESCRT-I and -II induce a conformational change of the endosome membrane at the neck of the nascent vesicle. ESCRT-II and -III finally pinch off the neck of the nascent vesicle and release it into the MVB lumen [130]. ESCRT-III associates with an ATPase called Vps4p/SKD1, which participates in the recycling of the ESCRT complex and its ATPase activity provides the main energy source (Fig. 7) [131]. MVBs can fuse with the cell membrane to release its ILVs as exosomes or for degradation by the lysosome. The types of MVBs are various and their biogenesis and fusion with the lysosome or plasma membrane are still unclear [132, 133]. Under specific conditions such as starvation, rapamycin treatment [134], or ISGylation of MVB proteins [135], MVBs are promoted to fuse with the lysosome instead of fusing with the plasma membrane and releasing as exosomes. Therefore, the final fate of MVBs is not immutable but changes under different conditions. The intracellular trafficking and communication of exosomes can occur through several major processes. First, transmembrane proteins on the EV membrane can interact with receptors on the cell membrane. These receptor–ligand interactions can then initiate a signaling cascade. EVs can also enter the target cell through either direct fusion with the cell membrane or endocytosis to release their cargo, such as nucleic acids and proteins, which are released into the cytosol of the target cells (Fig. 7). The released mRNA can be translated into a protein while delivered miRNAs and exogenously modified siRNAs can inhibit mRNA translation and thus silence the specific genes. The cargo and function of exosomes depend on their producing cells demonstrating that intercellular trafficking and communication of exosomes is a dynamic system where producing cells determine the "message" that exosomes deliver [136].

Proteins, lipids, and nucleic acids are the basic biochemical composition of exosomes. The canonical exosomes are lipid-enclosed vesicles with transmembrane

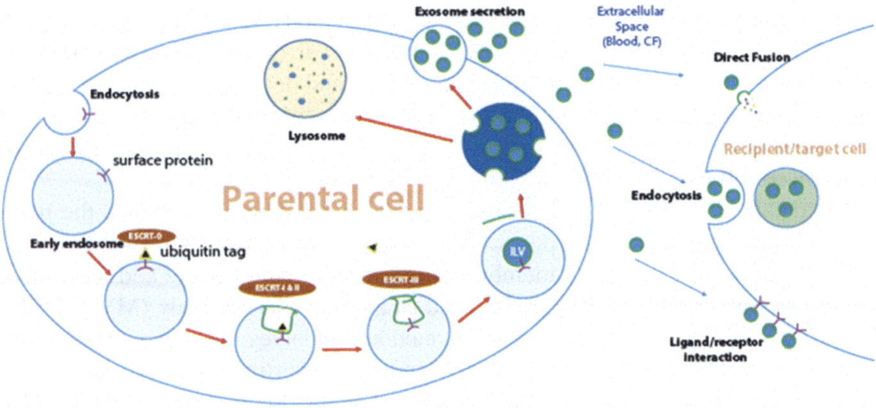

Fig. 7 After endocytosis, intraluminal vesicles (ILVs) are formed by the inward budding of the plasma membrane of maturing early endosomes (EE) to late endosomes (LE) or multivesicular bodies (MVB). The ESCRT-dependent mechanism involves ESCRT-0, -I, -II, and -III proteins together with an ATPase called Vps4p/SKD1 which aids in cargo sorting into ILVs. Some MVBs are destined for degradation in lysosomes, while other MVBs will fuse with the plasma membrane (PM) to release ILVs as exosomes. Once in the extracellular space, the secreted exosomes can travel through biological fluids and reach recipient/target cells, thus releasing their cargos. Such cell communication can be completed in several ways: (1) direct fusion with recipient/target cell membrane, (2) internalization through the endocytotic pathway, or (3) ligand/receptor interaction

proteins exposed at their surface (e.g., MHC class I and II, tetraspanins, integrins, TfR, flotillins). Inside the cytosol, there are various types of cytosolic proteins (e.g., ESCRT components, heterotrimeric G proteins, proteasome, actin, etc.) and nucleic acids (e.g., miRNA, mRNAs, and other noncoding RNAs) [137].

One of the challenges is to discriminate between exosomes and other MVs. The novel methods of isolation and purification of exosomes are being developed rapidly. The traditional methods are ultracentrifugation [138], size exclusion chromatography [139–142], and precipitation [143, 144]. The newer methods include affinity capture beads [145–148], size or immunoaffinity-based dynamic microfluidic systems [149–154], and time-affinity purification [155] and lipid-nanoprobe systems [156].

Due to the exosomes' capacity to induce cell signaling, it was under great investigation for its influence in tumor metastasis [157]. Tumor-derived exosomes are reported to form the pre-metastatic niche at distanced organs [158–160]. Therefore, the tumor-derived exosomes have the potential to predict future metastatic sites in a patient. Tumor-derived exosomes also can be used to distinguish different stages in cancer development. Melo et al. identified a cell surface proteoglycan glypican-1 which is abundant in tumor-derived exosomes; by using this marker, they were able to detect early- and late-stage pancreatic cancer [161]. Tumor-derived exosomes have not only the ability to predict developing cancer, but they also have therapeutic potential. It was reported that engineered dendritic cell-derived exosomes act as anticancer agents to power the immune cells to fight against cancer [129].

References

1. Cech TR, Steitz JA (2014) The noncoding RNA revolution-trashing old rules to forge new ones. Cell 157:77–94
2. Wu C et al (2013) Building a multifunctional aptamer-based DNA nanoassembly for targeted cancer therapy. J Am Chem Soc 135:18644–18650
3. Kruspe S, Giangrande PH (2017) Aptamer-siRNA chimeras: discovery, progress, and future prospects. Biomedicines 5(3):45
4. Catuogno S et al (2015) Selective delivery of therapeutic single strand antimiRs by aptamer-based conjugates. J Controlled Release: Official Journal of the Controlled Release Society
5. Porciani D et al (2015) Aptamer-mediated codelivery of doxorubicin and NF-κB decoy enhances chemosensitivity of pancreatic tumor cells. Mol Ther Nucleic Acids 4:e235
6. Havens MA, Hastings ML (2016) Splice-switching antisense oligonucleotides as therapeutic drugs. Nucleic Acids Res 44:6549–6563
7. Subramanian N et al (2015) Chimeric nucleolin aptamer with survivin DNAzyme for cancer cell targeted delivery. Chem Commun (Cambridge, England) 51:6940–6943
8. Tuerk C, Gold L (1990) Systematic evolution of ligands by exponential enrichment: RNA ligands to bacteriophage T4 DNA polymerase. Science (New York, N.Y.) 249:505–510
9. Ellington AD, Szostak JW (1990) In vitro selection of RNA molecules that bind specific ligands. Nature 346:818–822
10. Catuogno S, Esposito CL (2017) Aptamer cell-based selection: overview and advances. Biomedicines 5(3):49
11. Shangguan D et al (2006) Aptamers evolved from live cells as effective molecular probes for cancer study. Proc Natl Acad Sci USA 103:11838–11843

12. Chen HW et al (2008) Molecular recognition of small-cell lung cancer cells using aptamers. Chem Med Chem 3:991–1001
13. Wang J et al (2017) In vitro selection of DNA aptamers against renal cell carcinoma using living cell-SELEX. Talanta 175:235–242
14. Li X et al (2014) In vitro selection of DNA aptamers for metastatic breast cancer cell recognition and tissue imaging. Anal Chem 86:6596–6603
15. McNamara JO et al (2006) Cell type-specific delivery of siRNAs with aptamer-siRNA chimeras. Nat Biotechnol 24:1005–1015
16. Panigaj M, Reiser J (2016) Aptamer guided delivery of nucleic acid-based nanoparticles. DNA RNA Nanotechnol 2(1):42–52
17. Zhou J, Rossi J (2017) Aptamers as targeted therapeutics: current potential and challenges. Nat Rev Drug Discovery 16:181–202
18. Huang B-T et al (2017) A CTLA-4 antagonizing DNA aptamer with antitumor effect. Mol Ther Nucleic Acids 8:520–528
19. Yunn N-O et al (2015) Agonistic aptamer to the insulin receptor leads to biased signaling and functional selectivity through allosteric modulation. Nucleic Acids Res 43:7688–7701
20. Liu J et al (2016) Multifunctional aptamer-based nanoparticles for targeted drug delivery to circumvent cancer resistance. Biomaterials 91:44–56
21. Zhu G et al (2013) Self-assembled, aptamer-tethered DNA nanotrains for targeted transport of molecular drugs in cancer theranostics. Proc Natl Acad Sci USA 110:7998–8003
22. Luo Z et al (2017) Precise glioblastoma targeting by AS1411 aptamer-functionalized poly (l-γ-glutamylglutamine)-paclitaxel nanoconjugates. J Colloid Interface Sci 490:783–796
23. Liu X et al (2011) Targeted cell-cell interactions by DNA nanoscaffold-templated multivalent bispecific aptamers. Small (Weinheim an Der Bergstrasse, Germany) 7:1673–1682
24. Schrand B et al (2014) Targeting 4-1BB costimulation to the tumor stroma with bispecific aptamer conjugates enhances the therapeutic index of tumor immunotherapy. Cancer Immunol Res 2:867–877
25. Macdonald J et al (2017) Development of a bifunctional aptamer targeting the transferrin receptor and epithelial cell adhesion molecule (EpCAM) for the treatment of brain cancer metastases. ACS Chem Neurosci 8:777–784
26. Jaeger L, Leontis NB (2000) Tecto-RNA: One-Dimensional Self-Assembly through Tertiary Interactions. Angew Chem Int Ed Engl 39:2521–2524
27. Jaeger L, Westhof E, Leontis NB (2001) TectoRNA: modular assembly units for the construction of RNA nano-objects. Nucleic Acids Res 29:455–463
28. Guo P (2010) The emerging field of RNA nanotechnology. Nat Nanotechnol 5:833–842
29. Afonin KA et al (2014) Multifunctional RNA nanoparticles. Nano Lett 14:5662–5671
30. Afonin KA et al (2014) Co-transcriptional production of RNA-DNA hybrids for simultaneous release of multiple split functionalities. Nucleic Acids Res 42:2085–2097
31. Douglas SM, Bachelet I, Church GM (2012) A logic-gated nanorobot for targeted transport of molecular payloads. Science (New York, N.Y.) 335:831–834
32. Halman JR et al (2017) Functionally-interdependent shape-switching nanoparticles with controllable properties. Nucleic Acids Res 45(4):2210–2220
33. Afonin KA et al (2013) Activation of different split functionalities on re-association of RNA-DNA hybrids. Nat Nanotechnol 8:296–304
34. Afonin KA et al (2014) In silico design and enzymatic synthesis of functional RNA nanoparticles. Acc Chem Res 47:1731–1741
35. Afonin KA, Lindsay B, Shapiro BA (2013) Engineered RNA nanodesigns for applications in RNA nanotechnology. RNA Nanotechnol 1:1–15
36. Leontis NB, Westhof E (2003) Analysis of RNA motifs. Curr Opin Struct Biol 13:300–308
37. Leontis NB, Westhof E (2002) The annotation of RNA motifs. Comp Funct Genomics 3:518–524
38. Leontis NB, Lescoute A, Westhof E (2006) The building blocks and motifs of RNA architecture. Curr Opin Struct Biol 16:279–287

39. Afonin KA, Cieply DJ, Leontis NB (2008) Specific RNA self-assembly with minimal paranemic motifs. J Am Chem Soc 130:93–102
40. Afonin KA, Leontis NB (2006) Generating new specific RNA interaction interfaces using C-loops. J Am Chem Soc 128:16131–16137
41. Afonin KA et al (2012) Attenuation of loop-receptor interactions with pseudoknot formation. Nucleic Acids Res 40:2168–2180
42. Bindewald E et al (2011) Multistrand RNA secondary structure prediction and nanostructure design including pseudoknots. ACS Nano 5:9542–9551
43. Stewart JM et al (2016) Programmable RNA microstructures for coordinated delivery of siRNAs. Nanoscale 8:17542–17550
44. Afonin KA et al (2010) In vitro assembly of cubic RNA-based scaffolds designed in silico. Nat Nanotechnol 5:676–682
45. Afonin KA et al (2011) Design and self-assembly of siRNA-functionalized RNA nanoparticles for use in automated nanomedicine. Nat Protoc 6:2022–2034
46. Afonin KA et al (2016) The use of minimal RNA toeholds to trigger the activation of multiple functionalities. Nano Lett 16:1746–1753
47. Grabow WW et al (2011) Self-assembling RNA nanorings based on RNAI/II inverse kissing complexes. Nano Lett 11:878–887
48. Afonin KA et al (2015) Triggering of RNA interference with RNA-RNA, RNA-DNA, and DNA-RNA nanoparticles. ACS Nano 9:251–259
49. Bindewald E et al (2016) Multistrand structure prediction of nucleic acid assemblies and design of RNA switches. Nano Lett 16:1726–1735
50. Parlea L et al (2016) Cellular delivery of RNA nanoparticles. ACS Comb Sci 18:527–547
51. Binzel DW et al (2016) Specific delivery of miRNA for high efficient inhibition of prostate cancer by RNA nanotechnology. Mol Ther 24:1267–1277
52. Shu D et al (2015) Systemic delivery of anti-miRNA for suppression of triple negative breast cancer utilizing RNA nanotechnology. ACS Nano 9:9731–9740
53. Rychahou P et al (2015) Delivery of RNA nanoparticles into colorectal cancer metastases following systemic administration. ACS Nano 9:1108–1116
54. Haque F et al (2012) Ultrastable synergistic tetravalent RNA nanoparticles for targeting to cancers. Nano Today 7:245–257
55. Shu Y et al (2013) Fabrication of 14 different RNA nanoparticles for specific tumor targeting without accumulation in normal organs. RNA (New York, N.Y.) 19:767–777
56. Morrissey DV et al (2005) Activity of stabilized short interfering RNA in a mouse model of hepatitis B virus replication. Hepatology (Baltimore, Md.) 41:1349–1356
57. Keefe AD, Cload ST (2008) SELEX with modified nucleotides. Curr Opin Chem Biol 12:448–456
58. Piao X et al (2017) Assessment and comparison of thermal stability of phosphorothioate-DNA, DNA, RNA, 2′-F RNA and LNA in the context of Phi29 pRNA 3WJ. RNA (New York, N.Y.)
59. Li H, Labean TH, Leong KW (2011) Nucleic acid-based nanoengineering: novel structures for biomedical applications. Interface Focus 1:702–724
60. Keum J-W, Bermudez H (2009) Enhanced resistance of DNA nanostructures to enzymatic digestion. Chem Commun (Cambridge, England) 45:7036–7038
61. Toy R et al (2014) Shaping cancer nanomedicine: the effect of particle shape on the in vivo journey of nanoparticles. Nanomedicine (London, England) 9:121–134
62. Juliano R et al (2008) Mechanisms and strategies for effective delivery of antisense and siRNA oligonucleotides. Nucleic Acids Res 36:4158–4171
63. Roepstorff K et al (2009) Differential effects of EGFR ligands on endocytic sorting of the receptor. Traffic (Copenhagen, Denmark) 10:1115–1127
64. Wu J, Chen ZJ (2014) Innate immune sensing and signaling of cytosolic nucleic acids. Annu Rev Immunol 32:461–488
65. Robbins M, Judge A, MacLachlan I (2009) siRNA and innate immunity. Oligonucleotides 19:89–102

66. Bourquin C et al (2007) Immunostimulatory RNA oligonucleotides trigger an antigen-specific cytotoxic T-cell and IgG2a response. Blood 109:2953–2960
67. Radovic-Moreno AF et al (2015) Immunomodulatory spherical nucleic acids. Proc Natl Acad Sci USA 112:3892–3897
68. Khisamutdinov EF et al (2014) Enhancing immunomodulation on innate immunity by shape transition among RNA triangle, square and pentagon nanovehicles. Nucleic Acids Res 42:9996–10004
69. Dassie JP et al (2014) Targeted inhibition of prostate cancer metastases with an RNA aptamer to prostate-specific membrane antigen. Mol Ther: The Journal of the American Society of Gene Therapy 22:1910–1922
70. Johnson MB et al (2017) Programmable nucleic acid based polygons with controlled neuroimmunomodulatory properties for predictive QSAR modeling. Small 13:1701255
71. Bui MN et al (2017) Versatile RNA tetra-U helix linking motif as a toolkit for nucleic acid nanotechnology. Nanomed Nanotechnol Biol Med 13:1137–1146
72. Beutler B et al (2006) Genetic analysis of host resistance: Toll-like receptor signaling and immunity at large. Annu Rev Immunol 24:353–389
73. Moresco EMY, Beutler B (2011) Special delivery: granulin brings CpG DNA to Toll-like receptor 9. Immunity 34:453–455
74. Uematsu S, Akira S (2006) Toll-like receptors and innate immunity. J Mol Med (Berlin, Germany) 84:712–725
75. Motshwene PG et al (2009) An oligomeric signaling platform formed by the Toll-like receptor signal transducers MyD88 and IRAK-4. J Biol Chem 284:25404–25411
76. Zhang X et al (2011) Cutting edge: Ku70 is a novel cytosolic DNA sensor that induces type III rather than type I IFN. J Immunol (Baltimore, Md.: 1950) 186:4541–4545
77. Kim T et al (2010) Aspartate-glutamate-alanine-histidine box motif (DEAH)/RNA helicase A helicases sense microbial DNA in human plasmacytoid dendritic cells. Proc Natl Acad Sci USA 107:15181–15186
78. Takaoka A et al (2007) DAI (DLM-1/ZBP1) is a cytosolic DNA sensor and an activator of innate immune response. Nature 448:501–505
79. Pichlmair A et al (2006) RIG-I-mediated antiviral responses to single-stranded RNA bearing 5′-phosphates. Science (New York, N.Y.) 314:997–1001
80. Takeuchi O, Akira S (2008) MDA5/RIG-I and virus recognition. Curr Opin Immunol 20:17–22
81. Yoneyama M, Fujita T (2009) RNA recognition and signal transduction by RIG-I-like receptors. Immunol Rev 227:54–65
82. Unterholzner L et al (2010) IFI16 is an innate immune sensor for intracellular DNA. Nat Immunol 11:997–1004
83. Kato H et al (2008) Length-dependent recognition of double-stranded ribonucleic acids by retinoic acid-inducible gene-I and melanoma differentiation-associated gene 5. J Exp Med 205:1601–1610
84. Dao BN et al (2015) Triggering RNAi with multifunctional RNA nanoparticles and their delivery. DNA and RNA Nanotechnol 2:1–12
85. Bode C et al (2011) CpG DNA as a vaccine adjuvant. Expert Rev Vaccines 10:499–511
86. Shirota H, Klinman DM (2014) Recent progress concerning CpG DNA and its use as a vaccine adjuvant. Expert Rev Vaccines 13:299–312
87. Matsuoka N et al (2010) Structural and immunostimulatory properties of Y-shaped DNA consisting of phosphodiester and phosphorothioate oligodeoxynucleotides. J Control Release: Official Journal of the Controlled Release Society 148:311–316
88. Mohri K et al (2012) Design and development of nanosized DNA assemblies in polypod-like structures as efficient vehicles for immunostimulatory CpG motifs to immune cells. ACS Nano 6:5931–5940
89. Klinman DM, Sato T, Shimosato T (2016) Use of nanoparticles to deliver immunomodulatory oligonucleotides. Wiley Interdiscip Rev. Nanomed Nanobiotechnol 8:631–637
90. Ho PP et al (2003) An immunomodulatory GpG oligonucleotide for the treatment of autoimmunity via the innate and adaptive immune systems. J Immunol (Baltimore, Md.: 1950) 171:4920–4926

91. Hess KL et al (2017) Polyplexes assembled from self-peptides and regulatory nucleic acids blunt Toll-like receptor signaling to combat autoimmunity. Biomaterials 118:51–62
92. Zhang DY, Seelig G (2011) Dynamic DNA nanotechnology using strand-displacement reactions. Nat Chem 3:103–113
93. Li D, Song S, Fan C (2010) Target-responsive structural switching for nucleic acid-based sensors. Acc Chem Res 43(5):631–641
94. Kagan D et al (2011) Functionalized micromachines for selective and rapid isolation of nucleic acid targets from complex samples. Nano Lett 11:2083–2087
95. Liu X, Lu CH, Willner I (2014) Switchable reconfiguration of nucleic acid nanostructures by stimuli-responsive DNA machines. Acc Chem Res 47(6):1673–1680
96. Modi S et al (2009) A DNA nanomachine that maps spatial and temporal pH changes inside living cells. Nat Nanotechnol 4:325–330
97. Mao C et al (1999) A nanomechanical device based on the B-Z transition of DNA. Nature 397(6715):144–146
98. Zhou M et al (2010) A light-driven DNA nanomachine for the efficient photoswitching of RNA digestion. Angew Chem Int Ed Engl 49(12):2167–2170
99. Yurke B et al (2000) A DNA-fuelled molecular machine made of DNA. Nature 406(6796):605–608
100. Bath J, Green SJ, Turberfield AJ (2005) A free-running DNA motor powered by a nicking enzyme. Angew Chem Int Ed Engl 44(28):4358–4361
101. Aldaye FA, Palmer AL, Sleiman HF (2008) Assembling materials with DNA as the guide. Science 321(5897):1795–1799
102. Aldaye FA, Sleiman HF (2007) Dynamic DNA templates for discrete gold nanoparticle assemblies: control of geometry, modularity, write/erase and structural switching. J Am Chem Soc 129(14):4130–4131
103. Park SY et al (2008) DNA-programmable nanoparticle crystallization. Nature 451(7178):553–556
104. Tyagi S, Kramer FR (1996) Molecular beacons: probes that fluoresce upon hybridization. Nat Biotechnol 14(3):303–308
105. Sajja S et al (2018) Dynamic behavior of RNA nanoparticles analyzed by AFM on mica/air interface. Langmuir. https://doi.org/10.1021/acs.langmuir.8b00105
106. Zakrevsky P et al (2017) Preparation of a conditional RNA switch. Methods Mol Biol 1632:303–324
107. Xie Z et al (2010) Logic integration of mRNA signals by an RNAi-based molecular computer. Nucleic Acids Res 38(8):2692–2701
108. Xie Z et al (2011) Multi-input RNAi-based logic circuit for identification of specific cancer cells. Science 333(6047):1307–1311
109. Soukup GA, Breaker RR (1999) Nucleic acid molecular switches. Trends Biotechnol 17(12):469–476
110. Rinaudo K et al (2007) A universal RNAi-based logic evaluator that operates in mammalian cells. Nat Biotechnol 25(7):795–801
111. Penchovsky R, Breaker RR (2005) Computational design and experimental validation of oligonucleotide-sensing allosteric ribozymes. Nat Biotechnol 23(11):1424–1433
112. Gupta K et al (2015) Bolaamphiphiles as carriers for siRNA delivery: from chemical syntheses to practical applications. J Control Release 213:142–151
113. Gupta K et al (2015) Oxime ether lipids containing hydroxylated head groups are more superior siRNA delivery agents than their nonhydroxylated counterparts. Nanomedicine 10:2805–2818
114. Kim T et al (2013) In silico, in vitro, and in vivo studies indicate the potential use of bolaamphiphiles for therapeutic siRNAs delivery. Mol Ther Nucleic Acids 2:e80
115. Byun H-M et al (2004) Erythrocyte ghost-mediated gene delivery for prolonged and blood-targeted expression. Gene Ther 11:492–496
116. Hung ME, Leonard JN (2016) A platform for actively loading cargo RNA to elucidate limiting steps in EV-mediated delivery. J Extracell Vesicles 5:31027

117. Smalheiser NR (2007) Exosomal transfer of proteins and RNAs at synapses in the nervous system. Biology Direct 2:35
118. Ramachandran S, Palanisamy V (2012) Horizontal transfer of RNAs: exosomes as mediators of intercellular communication. Wiley Interdiscip Rev RNA 3:286–293
119. Valadi H et al (2007) Exosome-mediated transfer of mRNAs and microRNAs is a novel mechanism of genetic exchange between cells. Nat Cell Biol 9:654–659
120. Ha D, Yang N, Nadithe V (2016) Exosomes as therapeutic drug carriers and delivery vehicles across biological membranes: current perspectives and future challenges. Acta Pharmaceutica Sinica. B 6:287–296
121. Pardridge WM (2012) Drug transport across the blood-brain barrier. J Cereb Blood Flow Metab: Official Journal of the International Society of Cerebral Blood Flow and Metabolism 32:1959–1972
122. Petrocca F, Lieberman J (2011) Promise and challenge of RNA interference-based therapy for cancer. J Clin Oncol: Official Journal of the American Society of Clinical Oncolology 29:747–754
123. Shen H, Sun T, Ferrari M (2012) Nanovector delivery of siRNA for cancer therapy. Cancer Gene Ther 19:367–373
124. Alvarez-Erviti L et al (2011) Delivery of siRNA to the mouse brain by systemic injection of targeted exosomes. Nat Biotechnol 29:341–345
125. El-Andaloussi S et al (2012) Exosome-mediated delivery of siRNA in vitro and in vivo. Nat Protoc 7:2112–2126
126. Wahlgren J et al (2012) Plasma exosomes can deliver exogenous short interfering RNA to monocytes and lymphocytes. Nucleic Acids Res 40:e130
127. Shtam TA et al (2013) Exosomes are natural carriers of exogenous siRNA to human cells in vitro. Cell communication and signaling: CCS 11:88
128. Banizs AB et al (2014) In vitro evaluation of endothelial exosomes as carriers for small interfering ribonucleic acid delivery. Int J Nanomed 9:4223–4230
129. Pitt JM et al (2014) Dendritic cell-derived exosomes as immunotherapies in the fight against cancer. J Immunol (Baltimore, Md.: 1950) 193:1006–1011
130. Schmidt O, Teis D (2012) The ESCRT machinery. Curr Biol 22:R116–R120
131. Wollert T, Hurley JH (2010) Molecular mechanism of multivesicular body biogenesis by ESCRT complexes. Nature 464:864–869
132. Trajkovic K et al (2008) Ceramide triggers budding of exosome vesicles into multivesicular endosomes. Science (New York, N.Y.) 319:1244–1247
133. Villarroya-Beltri C et al (2014) Sorting it out: regulation of exosome loading. Semin Cancer Biol 28:3–13
134. Fader CM et al (2008) Induction of autophagy promotes fusion of multivesicular bodies with autophagic vacuoles in k562 cells. Traffic (Copenhagen, Denmark) 9:230–250
135. Villarroya-Beltri C et al (2016) ISGylation controls exosome secretion by promoting lysosomal degradation of MVB proteins. Nat Commun 7:13588
136. De Jong OG et al (2014) Extracellular vesicles: potential roles in regenerative medicine. Front Immunol 5:608
137. Colombo M, Raposo G, Théry C (2014) Biogenesis, secretion, and intercellular interactions of exosomes and other extracellular vesicles. Annu Rev Cell Dev Biol 30:255–289
138. Théry C et al (2001) Proteomic analysis of dendritic cell-derived exosomes: a secreted subcellular compartment distinct from apoptotic vesicles. J Immunol (Baltimore, Md.: 1950) 166:7309–7318
139. Ogawa Y et al (2008) Exosome-like vesicles with dipeptidyl peptidase IV in human saliva. Biol Pharm Bull 31:1059–1062
140. Sokolova V et al (2011) Characterisation of exosomes derived from human cells by nanoparticle tracking analysis and scanning electron microscopy. Colloids Surf B, Biointerfaces 87:146–150
141. Böing AN (2014) Single-step isolation of extracellular vesicles by size-exclusion chromatography. J Extracell Vesicles 3

142. Hong CS et al (2014) Isolation and characterization of CD34+ blast-derived exosomes in acute myeloid leukemia. PLoS ONE 9:e103310
143. Lee C et al (2012) Exosomes mediate the cytoprotective action of mesenchymal stromal cells on hypoxia-induced pulmonary hypertension. Circulation 126:2601–2611
144. Musante L et al (2012) Biochemical and physical characterisation of urinary nanovesicles following CHAPS treatment. PLoS ONE 7:e37279
145. Wubbolts R et al (2003) Proteomic and biochemical analyses of human B cell-derived exosomes. Potential implications for their function and multivesicular body formation. J Biol Chem 278:10963–10972
146. Caby M-P et al (2005) Exosomal-like vesicles are present in human blood plasma. Int Immunol 17:879–887
147. Kim G et al (2012) Noble polymeric surface conjugated with zwitterionic moieties and antibodies for the isolation of exosomes from human serum. Bioconjug Chem 23:2114–2120
148. Balaj L et al (2015) Heparin affinity purification of extracellular vesicles. Scientific Reports 5:10266
149. Chen C et al (2010) Microfluidic isolation and transcriptome analysis of serum microvesicles. Lab Chip 10:505–511
150. Davies RT et al (2012) Microfluidic filtration system to isolate extracellular vesicles from blood. Lab Chip 12:5202–5210
151. Wang Z et al (2013) Ciliated micropillars for the microfluidic-based isolation of nanoscale lipid vesicles. Lab Chip 13:2879–2882
152. He M et al (2014) Integrated immunoisolation and protein analysis of circulating exosomes using microfluidic technology. Lab Chip 14:3773–3780
153. Kanwar SS et al (2014) Microfluidic device (ExoChip) for on-chip isolation, quantification and characterization of circulating exosomes. Lab Chip 14:1891–1900
154. Vaidyanathan R et al (2014) Detecting exosomes specifically: a multiplexed device based on alternating current electrohydrodynamic induced nanoshearing. Anal Chem 86:11125–11132
155. Nakai W et al (2016) A novel affinity-based method for the isolation of highly purified extracellular vesicles. Sci Rep 6:33935
156. Wan Y et al (2017) Rapid magnetic isolation of extracellular vesicles via lipid-based nanoprobes. Nat Biomed Eng 1
157. Lu X, Kang Y (2007) Organotropism of breast cancer metastasis. J Mammary Gland Biol Neoplasia 12:153–162
158. Costa-Silva B et al (2015) Pancreatic cancer exosomes initiate pre-metastatic niche formation in the liver. Nat Cell Biol 17:816–826
159. Hoshino A et al (2015) Tumour exosome integrins determine organotropic metastasis. Nature 527:329–335
160. Jung Y et al (2008) Hematopoietic stem cells regulate mesenchymal stromal cell induction into osteoblasts thereby participating in the formation of the stem cell niche. Stem cells (Dayton, Ohio) 26:2042–2051
161. Melo SA et al (2015) Glypican-1 identifies cancer exosomes and detects early pancreatic cancer. Nature 523:177–182

Fullerenes for Cancer Therapy and Bioimaging

Xuejiao J. Gao, Xiaomei Shen, Gengmei Xing and Xingfa Gao

Abstract Chemical derivatives of fullerene and endohedral metallofullerenes (EMFs) not only keep some of the physicochemical properties of their parent molecules, but can also have water solubility and biocompatibility that are not owned by their parent molecules. These brilliant properties provide these materials potentials in biomedical applications. Here, we review the atomistic-level structural models for fullerene derivatives, and their applications as magnetic resonance imaging (MRI) contrast and cancer therapy agents. Because atomistic level structures are the base for their physicochemical properties, we first review the structural models of fullerene derivatives, which are applicable to fullerene derivatives synthesized in different conditions. Second, we review the widely explored applications of the Gd-containing EMF derivatives as the safe and efficient MRI contrast agents, thanks to their special geometries and electronic structures. Reportedly, fullerene-based materials, like $Gd@C_{82}(OH)_{22}$ nanoparticles, possess intrinsic antitumor activities. The recent advance in using fullerene derivatives for cancer therapy through different pathways, including photodynamic therapy (PDT), reactive oxygen species (ROS) scavenging, drug delivery, adjuvant activity for cancer vaccines, and reduction of the nutrient supply to tumor cells, are further summarized. Finally, we briefly discuss the challenges and perspectives for the future biomedical applications of fullerene derivatives.

Keywords Cancer therapy · Electronic structures · Fullerenols · First-principles calculations · Structural models

---

X. J. Gao · X. Shen · X. Gao (✉)
College of Chemistry and Chemical Engineering, Jiangxi Normal University, Nanchang 330022, China
e-mail: gaox@ihep.ac.cn

G. Xing
CAS Key Laboratory for Biomedical Effects of Nanomaterials and Nanosafety, Institute of High Energy Physics, Chinese Academy of Sciences, Beijing 100049, China

G. Gonçalves and G. Tobias (eds.), *Nanooncology*, Nanomedicine and Nanotoxicology, https://doi.org/10.1007/978-3-319-89878-0_12

1 Introduction

Fullerene, which is constructed entirely of sp^2–C, is the third allotrope of carbon atoms in addition to graphite and diamond. In 1985, H. W. Kroto, R. E. Smalley, and R. F. Curl first discovered fullerene C_{60} [1], which led to their Nobel Prize in chemistry. Fullerenes are cage-like hollow carbon clusters and have cavities approximately 1 nm in diameter that can cage atoms, ions, or molecules to form endohedral fullerenes. The encapsulation of metal species inside the hollow fullerene cages produces novel structures termed as endohedral metallofullerenes (EMFs). La@C_{60} was the first experimentally synthesized EMF that was detected by Heath et al. shortly after the discovery of C_{60} [2].

The biomedical application is a developing field that provides great opportunities for fullerenes and EMFs, since relatively small dosages are required for high efficiency. In addition, the nontoxicity of the carbon cage makes fullerenes and EMFs potentially feasible for medical applications [3, 4]. In EMFs, the internal metals are effectively isolated from the biological environment, which maintains the high stability and low toxicity of EMFs and gives them a prominent advantage over the metal chelate complexes commercially used in radio medicine and diagnostic radiology.

Fullerenes and EMFs have absolutely hydrophobic surfaces and thus have poor water solubility. Chemical modifications can introduce hydrophilic groups, such as amino acids [5], carboxylic acids [6], and polyhydroxyl groups [7], onto the surfaces of fullerenes and EMFs, producing derivatives with improved water solubility and biocompatibility. Fullerenes, EMFs, and their derivatives make up the fullerene-based materials. In the past three decades, various fullerene-based nanomaterials with different functional groups in different sizes have been synthesized and widely explored for biomedical applications [5–47].

Bioimaging and cancer therapy are two main aspects of the biomedical applications of fullerene-based nanomaterials. Compared with pristine fullerene and EMFs, their derivatives have improved water solubility and biocompatibility and thus are more widely explored. Hydroxylated and malonic acid-substituted fullerenes and EMFs (fullerenols and carboxyfullerenes) are two major groups of water-soluble fullerene derivatives, which have been found to have biomedical significance [5–20]. These two groups of fullerene derivatives have been demonstrated to be efficient in bioimaging [48, 49] and cancer therapy. Many mechanisms for their antitumor activities have been proposed, including photodynamic therapy (PDT) [24–29], reactive oxygen species (ROS) scavenging [5–20], drug delivery [30–40], adjuvant activity for cancer vaccines [41–44], reduction of the nutrient supply to tumor cells [45–47], and others.

In this chapter, in three sections, we will focus primarily on the geometric structures of fullerene derivatives, their biomedical applications as bioimaging agents and their antitumor mechanisms. Finally, we will discuss challenges and perspectives involved in their further exploration.

2 The Geometrical Structures of Fullerene and Metallofullerene Derivatives

Fullerenols and metallofullerenols are the main derivatives of fullerene and EMFs. The hydroxylation reactions of fullerenes in different chemical conditions, e.g., in acidic [50–52] and in basic [53–55] solution, have been reported. Fullerenols and metallofullerenols contain many OH groups. For instance, $C_{60}(OH)_x$ ($x = 12$, 22–26, 36–40, and 44) [56–58] and $C_m(OH)_n$ ($m = 60$ and 82, $n = 16$, 22 and 40) [59–62] have been experimentally synthesized. Gd@C_{2v}-$C_{82}(OH)_{22}$ nanoparticles have also been synthesized and widely explored for their antitumor effects and bioimaging applications. However, the amorphous nature of the experimentally synthesized fullerenols makes it difficult to purify them and characterize their precise structures. Computational methods can be applied as a useful tool to predict the geometric structures of fullerenols and metallofullerenols. A number of fullerenol models with different numbers of OH groups have already been predicted. In this section, we will summarize models of fullerenols and metallofullerenols that are predicted based on different rules.

2.1 Fullerenol Models

The purification and characterization of fullerenols are difficult. Although a number of fullerenols $C_{60}(OH)_x$ (x = 12, 22–26, 36–40, and 44) [56–58] and $C_m(OH)_n$ (m = 60 and 82, n = 16, 22 and 40) [59–62] have been experimentally synthesized, only one X-ray crystal structure of $C_{60}(OH)_8$ was obtained by Gan et al. in 2010 [63]. Theoretical method is a useful tool to predicate the precise structures of fullerenols. Based on computational calculations, some fullerenol models have been proposed. Rodríguez-Zavala and Guirado-López reported that at low addition coverages (up to 7 OH groups), OH groups preferred to gather on one pole of the C_{60} cage, forming an island, and that at higher addition coverages (8 to 14 OH groups), a second OH island forms on the opposite pole of C_{60}. [64, 65] Guirado-López and Rincón also reported that OH groups of C_{60} fullerenols preferred to distribute aggregatively, forming OH islands of different sizes to obtain low-energy configurations [66]. For fullerenols with more than 20 OH groups, e.g., $C_{60}(OH)_{24}$, Yang et al. constructed ten isomers with different OH addition patterns and calculated their relative energies. They found that the isomer with the 24 OH groups located on the equator of the cage has the lowest energy [67]. A common feature of these low-energy structures is the aggregation of OH groups on fullerene cages.

The above models are predicted according to the low-energy criterion. The low-energy criterion can be stated as follows: the most likely species to exist stably are those with lower energies. The low-energy criterion is widely applied to evaluate the stabilities of pristine fullerenes and EMFs in forms such as the isolated pentagon rule (IPR) [68], the maximum pentagon separation rule [69–71], and the maximum

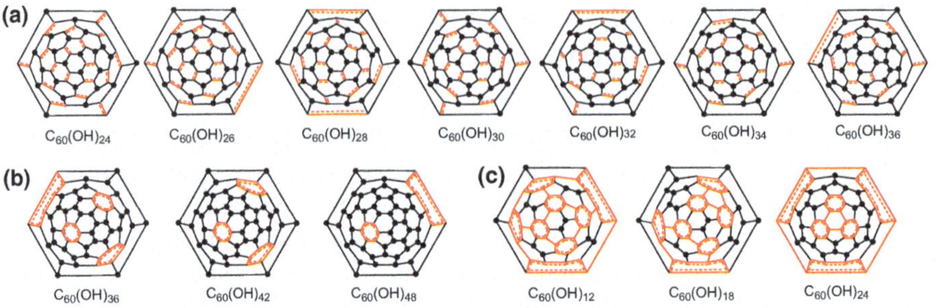

Fig. 1 Schlegel diagrams for C_{60} fullerenols. **a** Fullerenols with 24–36 hydroxyl groups whose sp^2 substructures are isolated double bonds. **b** Fullerenols with 36–48 hydroxyl groups whose sp^2 substructures are isolated benzene rings. **c** Fullerenols with 12–24 hydroxyl groups, where sp^2 substructures are full Clar's structures. (Reproduced with permission from [75]. ©2016 American Chemical Society.)

aromaticity criterion [72]. However, unlike the fullerenes and EMFs, which are synthesized at high temperatures (higher than 1200 °C) and thus have enough energy to undergo structural relaxation and thereby reach the global minimum on the potential energy surface (PES), the fullerenols are commonly synthesized based on prepared fullerene and EMFs under mild conditions [73]. Thus, the wide-gap criterion should also be considered when predicting the structures of fullerenols. The wide-gap criterion can be stated as follows: molecules with larger highest occupied molecular orbital—lowest unoccupied molecular orbital (H–L) energy gaps are more inert to chemical reactions and thus more likely to be separated experimentally.

Based on the wide-gap criterion, Gao et al. have proposed an isolated aromatic patch (IAP) rule to predict structures with high chemical stabilities for fullerenols and metallofullerenols. The IAP rule can be stated as follows: only fullerenols or metallofullerenols whose sp^2—C atoms form aromatic patches isolated by sp^3—C atoms have large H–L energy gaps and thus are relatively experimentally separable [74, 75]. Based on the IAP rule, Gao et al. contrasted a number of models of C_{60} and C_{2v}-C_{82} fullerenols $C_{60}(OH)_n$ ($n = 12$–48) and $C_{2v}(9)$-$C_{82}(OH)_x$ ($x = 14$–58). Figures 1 and 2 show the Schlegel diagrams for the structures of C_{60} and C_{2v}-C_{82} fullerenols, respectively.

The above fullerenol models contain only hydroxyl groups. In fact, the hydroxylation reactions are conducted in acidic or basic conditions, and the species of functional groups on the fullerenes cage are related to the chemical reactions. Based on the mechanisms of their synthesis reactions under different chemical conditions, Gao et al. identified the species and distribution of the functional groups on fullerenol surfaces and proposed a precise fullerenol model [76, 77]. They reported that epoxy, carbonyl, carboxyl, and hemiacetal structures also exist on the fullerenol cage in addition to hydroxyl groups (Fig. 3).

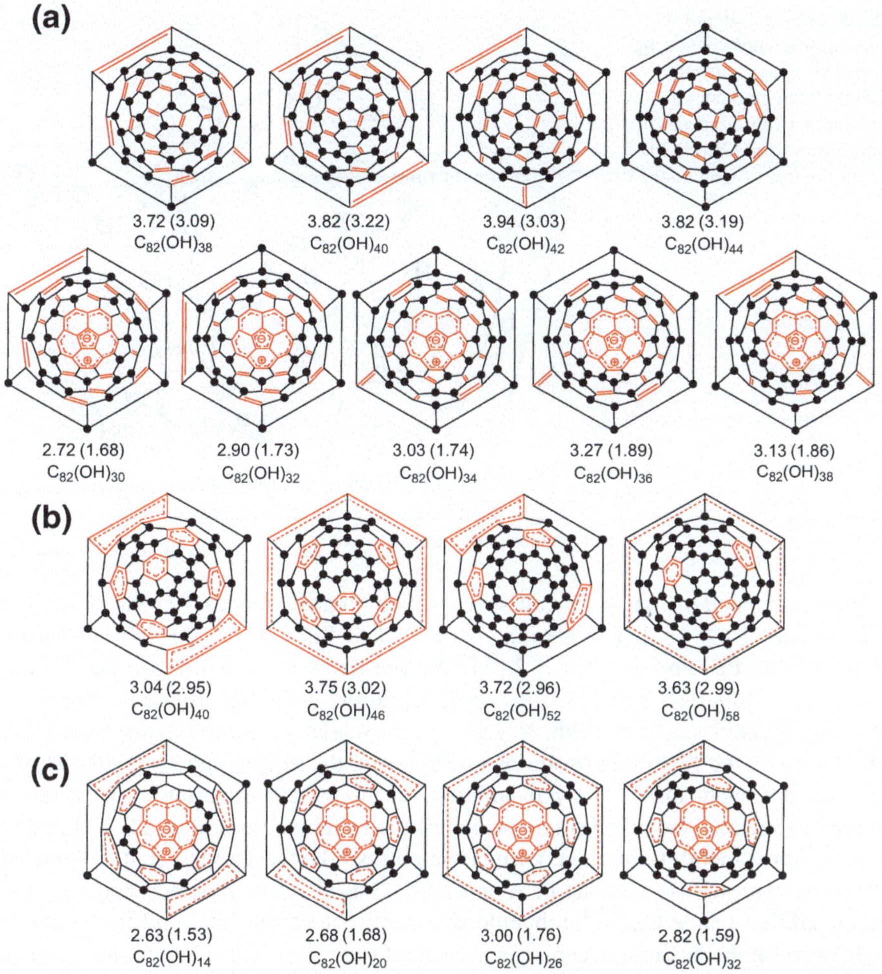

Fig. 2 Schlegel diagrams for C_{82} fullerenols. The H–L gaps and the first excitation energies (FEEs), in parentheses, for each structure are under its diagram. The FEE and H–L gap of $C_{2v}(9)$-C_{82} are 0.75 and 0.12 eV, respectively. (Reproduced with permission from [75]. ©2016 American Chemical Society.)

2.2 *Metallofullerenol Models*

Unlike fullerene, metallofullerene contains a metal atom, and thus the intramolecular interaction between the inner metal and the outer carbon cage should be considered. The electronic structures of EMF can be described using the ionic model $M^{x+}@C_{2n}{}^{x-}$ according to previous studies that have consistently demonstrated that in EMFs, the metals donate electrons to the carbon cages. The number of transferred electrons

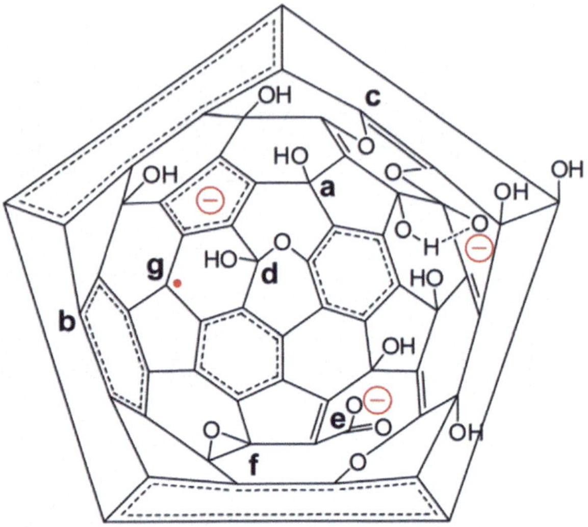

Fig. 3 Schlegel diagrams for C_{60} fullerenol containing hydroxyl, epoxy, carbonyl, carboxyl and hemiacetal structures. (Reproduced with permission from [76], ©2014 Royal Society of Chemistry.)

is mainly dependent on the metal species. For alkaline metals Li, Na, K, or Cs, x is usually 1 [78–82]. For alkaline-earth metals Be, Ca, Sr, Ba and lanthanide elements Sm, Eu, Tm, or Yb, x is 2 [83–88]. For Sc, Y, La, Gd, La–Nd, Gd–Er, Lu, or U, x is 3 [89–92]. The positions of metal atoms inside the fullerene cages are also investigated. Early in 1998, Nagase and co-workers reported that for the EMF $M^{x+}@C_{2n}{}^{x-}$, the favorable positions of M^{x+} were the electrostatic potential (ESP) minima sites inside the C_{2n}^{x-} [93]. In addition, the inner metal is commonly considered to prefer to reside above a hexagonal ring of the fullerene instead of a C–C bond to maximize the metal–fullerene coordination number. However, Gao et al. conducted DFT calculations for a series of mono-EMFs based on lanthanides and C_{60} and predicted that inside C_{60}, a lanthanide will either sit on the 6/6 bond in a divalent state or sit on the hexagonal center in a trivalent state [94]. This phenomenon can be understood based on the IAP rule, which states that the metal atom donates electrons to the adjacent sp^2–C patch following the $4n + 2$ rule to maintain aromaticity [74].

Metallofullerenols are metallofullerene multiple adducts (MMAs). Similar to empty fullerenols, metallofullerenols are difficult to purify and separate. Although $Gd@C_{82}(OH)_x$ ($x \approx 22$) is well known for its intrinsic antitumor activity with low toxicity in vivo and in vitro and for its superior bioimaging function as an MRI contrast agent, little is known about its molecular structures. It is easy to understand that molecules with large first excitation energies (FEEs) are chemically stable and most likely to survive during experimental synthesis. On this basis and on the basis of the quantum confinement effect, Gao et al. used the IAP rule to quickly predict whether MMAs have large FEEs: only MMAs whose sp^2–C atoms form aromatic patches isolated by sp^3–C atoms have large FEEs. Figure 4 illustrates the IAP rule used to construct $C_{2n}{}^{x-}$ hosting cages for MMAs [74].

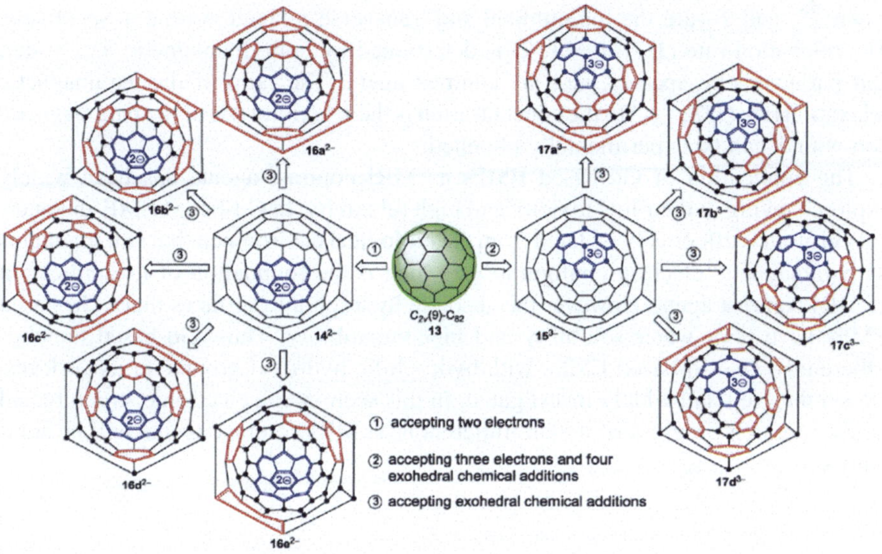

Fig. 4 Constructing IAP-obeying hosting cages C_{2n}^{x-} for MMAs. Anionic and neutral patches are shown in blue and red colors, respectively. The black dots represent sp^3–C. (Reproduced with permission from [74], ©2015 Elsevier Ltd.)

3 Biomedical Applications of EMFs for Bioimaging

Magnetic resonance imaging (MRI) is a mature technique that is commonly used in clinical practice to obtain pictures of organs and structures inside the body using a magnetic field and pulses of radio wave energy. MRI is usually used to find problems such as tumors, bleeding, injury, blood vessel diseases, or infection. Contrast agents are often used to drastically improve the sharpness of MRI through changing the proton relaxation times in tissues and body cavities. Gd-based chelate complexes, such as Gd-DTPA (DTPA: diethylenetriamino-pentaacetic acid), are the most commonly used MRI contrast agents. The paramagnetism of Gd^{3+} is the foundation of its high efficiency as an MRI contrast agent. The electronic configuration of Gd^{3+} is [Xe]$4f7$. The seven unpaired $4f$ electrons with the same spin states are dipoles and thus have magnetic moments. The magnetic susceptibility of an electron is approximately 660 times that of a proton. When Gd-DTPA is applied during MRI, the dipole–dipole interaction between proton and electron can dramatically enhance the proton relaxation effect and thus shorten the proton relaxation times T_1 and T_2, which can improve the relaxivity r_i, which is defined by the following general equation:

$$\frac{1}{T_i,\,\mathrm{obs}} = \frac{1}{T_i,\,\mathrm{H_2O}} + \frac{1}{T_i,\,\mathrm{para}} = \frac{1}{T_i,\,\mathrm{H_2O}} + r_i[\mathrm{M}]$$

where T_1 and T_2 are the longitudinal and transverse relaxation time, respectively. The relaxation rate ($1/T_i$, i = 1, 2) is determined by both diamagnetic, i.e., water, and paramagnetic species, i.e., *the* contrast agent. The ratio of the paramagnetic relaxation rate ($1/T_{i,\,para}$) to the concentration is the relaxivity (r_i) of the paramagnetic compound, can be experimentally obtained.

The application of Gd-based EMFs as MRI contrast agents has been widely explored owing to their low toxicity and high relaxivity. In Gd-based EMF, the internal Gd atom is effectively isolated from the biological environment but still maintains its paramagnetic electron configuration, which is the foundation of its application as MRI contrast agent. However the absolute hydrophobic surfaces make Gd-based EMFs have poor water solubility and biocompatibility. Thus, Gd-based metallofullerenols, i.e. Gd-based EMFs with hydrophilic hydroxyl groups on its surfaces, are synthesized and widely investigated. In this section, the electronic structure and recent research progress of metallofullerenols as MRI contrast agents will be summarized.

3.1 The Electronic Structures of Metallofullerenes

The foundation of the function of Gd-based metallofullerenols as MRI contrast agents is their paramagnetic electron configurations. Previous experimental studies have established that the metal atoms inside the EMF cage donate electrons to the outer cages, leading to the formation of the specific ionic structures $M^{x+}@C_{2n}{}^{x-}$. The numbers of electrons transferred from the metal to the cage depend on the metal species. For Gd-based EMFs or metallofullerenols, the Gd atom usually donates three valence electrons to the carbon cage, leading to its [Xe]4*f*7 electronic configuration. Using density functional theory (DFT), Gao et al. investigated the geometric and electronic structures for $Gd@C_{60}$ [94]. Their results demonstrated that the Gd atom prefers to reside on a hexagon ring. According to orbital analysis, all the 5*d* and 6*s* orbitals of Gd are empty, which indicates that the three valence electrons are transferred to the carbon cage and that the $Gd@C_{60}$ has an electronic structure of $Gd^{3+}@C_{60}{}^{3-}$. Figure 5 shows the selected molecular orbitals of $Gd@C_{60}$. The seven 4*f* orbitals of Gd, numbered (1–7), are half occupied by spin-up electrons (Fig. 5a). Among the three transferred electrons, two of them make a pair, and the extra spin-up electron singly occupies a molecular orbital, giving rise to a net spin of 7/2 and −1/2 on the Gd and C_{60} cage, respectively. The intramolecular interaction between the inner Gd atom and the outer cage makes the whole Gd EMF paramagnetic and thus can be used as MRI contrast agent. The chemical decoration of EMFs, including hydroxylation, will not interrupt the electron donation from metals to fullerene cages and can notably improve their water solubility and biocompatibility. Thus, Gd-based metallofullerenols, such as $Gd@C_{82}(OH)_x$ ($x = 22$ and 40), and other water-soluble derivatives, such as $Gd@C_{82}O_6(OH)_{16}(NHC_2\text{–}H_4CO_2H)_8$, are widely studied as MRI contrast agents.

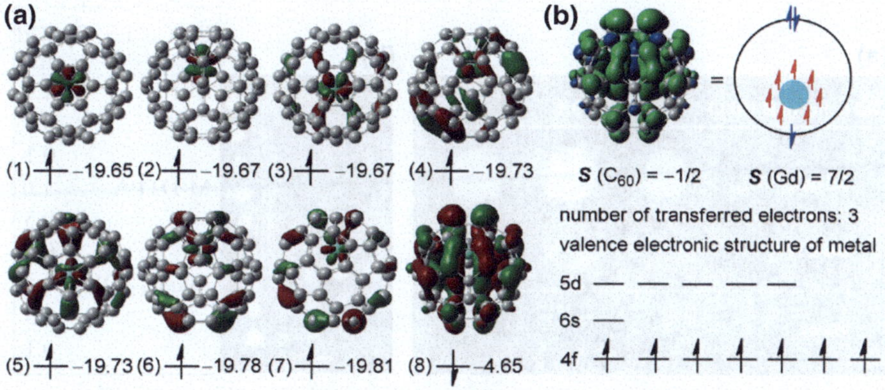

Fig. 5 **a** Selected molecular orbitals of Gd@C_{60}. The seven 4*f* orbitals of Gd, numbered (1–7), are half occupied by spin-up electrons. **b** The electronic configuration of Gd@C_{60}. The net spin on Gd and C_{60} are 7/2 and −1/2, respectively. (Reproduced with permission from [94], ©2014 Royal Society of Chemistry.)

3.2 *Applications of Metallofullerenes as MRI Contrast Agents*

Early in 1996, the application of EMFs as MRI contrast agents was proposed, and such applications have been extensively studied during the past two decades [59, 62, 95–107]. Thus far, metallofullerenols [59–61, 108, 109], lutetium-based trimetallic nitride-templated (TNT) EMFs [99, 104, 110–115], carboxylated EMF derivatives [96, 98], amino acid derivatives of Gd-EMFs [105, 116], organophosphonate-functionalized EMs [117], and other fancy EMF structures [118, 119] have been synthesized and investigated as MRI contrast agents. Mechanisms for their extremely high relaxivity r_l are also proposed [120].

In 2001, Shinohara et al. synthesized Gd-metallofullerenols Gd@$C_{82}(OH)_{40}$ and evaluated their paramagnetic properties for MRI in vivo and in vitro. The in vitro water proton relaxivity r_1 of Gd@$C_{82}(OH)_{40}$ (81 mM^{-1} s^{-1}) is almost 20-fold higher than that of the commercial MRI contrast agent Gd-DTPA (3.9 m M^{-1} s^{-1}) at 1.0 T (Fig. 6a). This unusually high r_1 of Gd@$C_{82}(OH)_{40}$ makes it possible to achieve the highest signal enhancement at much lower concentration [59]. Later, in 2003, metallofullerenols M@$C_{82}(OH)_n$ (M = La, Ce, Gd, Dy, and Er) were systematically synthesized and characterized as MRI contrast agents by the same group of Shinohara to get more information for the proton relaxation mechanism [108]. According to their results, the r_1 and r_2 values of these metallofullerenols are in the range of 0.8–73 and 1.2–80 m M^{-1} s^{-1}, respectively, which are much higher than those of the corresponding free ions and their DTPA chelates (Fig. 6b). The phantom NMR images of Gd@$C_{82}(OH)_{40}$, M@$C_{82}(OH)_n$ (M = La, Ce, Gd, Dy, and Er) as well as those lanthanoid-free ions and DTPA chelates are illustrated in Fig. 6a, b. Compared

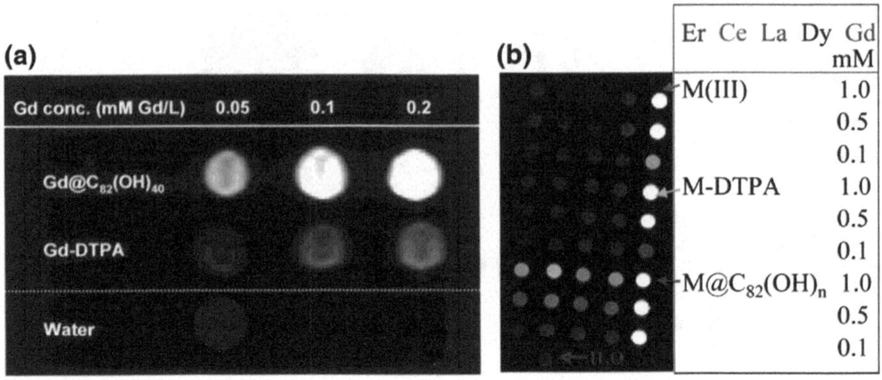

Fig. 6 **a** T_1-weighted MRI of Gd@$C_{82}(OH)_{40}$, Gd-DTPA and pure water phantom at the concentration of 0.05, 0.1, and 0.2 mmol Gd/L, respectively, by 4.7 T Unity INOVA. (Reproduced with permission from [59], ©2001 American Chemical Society.) **b** Phantom NMR images of M@$C_{82}(OH)n$ (M = La, Ce, Gd, Dy, and Er) as well as their corresponding ions and DTPA complexes solutions at 1.0, 0.5, and 0.1 mmol metal/L. (Reproduced with permission from [108], ©2003 American Chemical Society.)

with the Gd-free ions and its DTPA chelates, the Gd-based fullerene materials present distinctly strong contrast enhancement at the same concentration.

Hydroxyl-intense metallofullerenols, like Gd@$C_{82}(OH)_n$ with $n > 36$, usually have open-caged structures, which leads to the release of highly toxic Gd ions. Thus, to maintain the integrity and stability of metallofullerenols, Gd@$C_{82}(OH)_x$ materials with lower OH numbers, such as Gd@$C_{82}(OH)_{16}$ [60] and Gd@$C_{82}(OH)_{22\pm2}$ [61, 109], are also synthesized and characterized as MRI contrast agents. Gd@$C_{82}(OH)_{16}$ was reported to have high proton relaxivity, even higher than that of Gd-DTPA [60]. The bioimaging efficiency of Gd@$C_{82}(OH)_{22\pm2}$ were evaluated in mice by Zhao et al. and their results demonstrated that Gd@$C_{82}(OH)_{22\pm2}$ exhibited 12-fold enhancement of r_1 than that of Gd-DTPA in vitro and in vivo [61, 109]. The MRI efficiencies of other water-soluble Gd@C_{82} derivatives were also evaluated. Wang et al. synthesized Gd@$C_{82}O_6(OH)_{16}(NHC_2\text{–}H_4CO_2H)_8$, Gd@$C_{82}O_6(OH)_{16}(NHCH_2CH_2C$ OantiGFP$)_5$, and Gd@$C_{82}O_2(OH)_{16}(C(PO_3Et_2)_2)_{10}$, which all possess greatly enhanced relaxivity r_1 compared with that of Gd-DTPA [116–118].

Carboxylated EMF derivatives are also used as MRI contrast agents. Carboxylated EMFs have advantages over hydroxylated ones. Metallofullerenols can lead to the aggregation of erythrocytes in blood while the carboxylated EMFs have favorable biodistribution in vivo and have comparable MRI efficiencies at the same time. In 2003, Bolskar et al. synthesized the first carboxylated EMF Gd@C_{60} [C–$(COOH)_2]_{10}$ with a higher relaxivity r_1 (4.6 mM^{-1} s^{-1} at 20 MHz and 40 °C) than Gd-DTPA [96]. In 2005, Merbach et al. also reported similar relaxivities r_1 of Gd@C_{60} [C–$(COOH)_2]_{10}$ (6.8–24.0 m M^{-1} s^{-1}). In addition, they reported that the addition of salts, such as phosphate and sodium halides, could lead to better dispersion of both Gd@C_{60} $[C(COOH)_2]_{10}$ and Gd@$C_{60}(OH)_{27}$ in aqueous [62]. Wilson et al. also proved that the aggregation of $Gd@C_{60}$ $[C(COOH)_2]_{10}$ and $Gd@C_{60}(OH)_{27}$ could be disrupted by salt addition and their high relaxivities root in the rapid exchange of water molecules with the bulk [120].

The applications of TNT EMFs $M_3N@C_{2n}$ [99, 104, 110–115] as MRI contrast agents are also widely studied. Compared with mono-EMFs, TNT EMFs $M_3N@C_{2n}$ have higher stability and the encapsulation of three Gd^{3+} ions per molecule can lead to higher imaging efficiency. In 2014, Zhang et al. reported the preparation and characterization of $Gd_3N@C_s$-$C_{84}O_6(OH)_{28}$ and $Gd_3N@I_h$-$C_{80}O_{11}(OH)_{21}$ [104]. $Gd_3N@C_s$-$C_{84}O_6(OH)_{28}$ has a higher ^1H relaxivity than that of $Gd_3N@I_h$-$C_{80}O_{11}(OH)_{21}$ in low (0.47 T), medium (1.4 T), and high (9.4 T) magnetic fields. The higher relaxivity of $Gd_3N@C_s$-$C_{84}O_6(OH)_{28}$ was ascribed to its higher hydroxyl content and aggregate size. Experimentally, the yield of $Gd_3N@C_{80}$ was relatively low thus Gd-containing mixed-metal TNT EMFs were intended to be synthesized. In 2007, two water-soluble hydroxylated derivatives of Gd, Sc-mixed TNT EMFs, i.e., $Sc_2GdN@C_{80}O_{12}(OH)_{26}$ and $ScGd_2N@C_{80}O_{12}(OH)_{26}$ were synthesized and their relaxivity r_1 were evaluated to be 20.7 and 17.6 mM^{-1} s^{-1} [112], which are lower than that of pure Gd-based TNT EMFs but still higher than that of Gd-DTPA.

Some other types of $Gd_3N@C_{80}$-based water-soluble MRI contrast agents were also synthesized by different groups. In 2006, Fatouros et al. prepared the new functionalized TNT EMF species $Gd_3N@C_{80}$ [DiPEG5000$(OH)_x$] where DiPEG5000 represents poly(ethylene glycol) malonate. They demonstrated that the $Gd_3N@C_{80}$ [DiPEG5000$(OH)_x$] nanoparticle can dramatically enhance relaxivity both in vitro and in vivo. The measured relaxivity r_1 are 102, 143, and 32 mM^{-1} s^{-1} at 0.35, 2.4, and 9.4 T, respectively, which are markedly higher than that of gadodiamide. The T1-weighted MRI images of aqueous solutions of $Gd_3N@C_{80}$ [DiPEG5000$(OH)_x$] and gadodiamide at different concentrations are compared. The corresponding concentrations for the $Gd_3N@C_{80}$ [DiPEG5000$(OH)_x$] agent were at least 30 times smaller [111]. In 2008, MacFarland et al. synthesized hydrochalarones which are a series of novel derivatives of $Gd_3N@C_{80}$. The addition of glycol methyl ethers, ranging from monoethylene glycol to hexaethylene glycol, made hydrochalarones water soluble. The general structure of the hydrochalarones is $Gd_3N@C_{80}$-R_x, where $R = [N(OH)(CH_2CH_2O)_nCH_3]_x$ where $n = 1, 3, 6$ and $x = 10$–22. Hydrochalarone-6 had the highest r_1 of 205 mM^{-1} s^{-1}, which is markedly higher than that of Gd-DTPA (3.8 mM^{-1} s^{-1}) [121]. In 2009, Dorn et al. synthesized a new TNT EMF, $Gd_3N@C_{80}(OH)_{\sim 26}(CH_2CH_2COOM)_{\sim 16}$ (M = Na or H), in high yield by a facile method and determined the relaxivity r_1 to be 207 mM^{-1} s^{-1} at 2.4 T, which is 50 times larger than that of Magnevist [113]. In 2010, the same group reported the synthesis of water-soluble PEG-functionalized and hydroxylated $Gd_3N@C_{80}$, i.e. $Gd_3N@C_{80}$ [DiPEG$(OH)_x$], whose molecular weight depends on the PEG content (350–5000 Da). The measured relaxivities r_1 for the 350/750 Da PEG derivatives are 237/232 mM^{-1} s^{-1} (2.4 T), which are approximately 60 times higher than that of Gd-DTPA [114].s

The phospholipid bilayers of cellular surfaces are negatively charged. Thus, the fullerene nanoparticles with positive charges will bind more efficiently with the cellular surface than those with negatively charged carboxyl or hydroxyl groups.

Thus, amino acid functionalization is considered as another way to make EMFs or TNT EMFs water soluble and biocompatible because the cage surface bearing positively charged amino groups $-NH^{3+}$. In 2006, the amino acid derivatives of Gd-based metallofullerenes $Gd@C_{82}O_m(OH)_n(NHCH_2CH_2COOH)_l$ ($m \approx 6$, $n \approx 16$ and $l \approx 8$) (AAD-EMFs) were synthesized by Wang et al. The measured r_1 of AAD-EMFs is 9.1 mM^{-1} s^{-1} (1.5 T) which is higher than that of Gd-DTPA (5.6 mM^{-1} s^{-1}) but lower than that of Gd-based metallofullerenols (23.1 mM^{-1} s^{-1}) under the same experimental conditions [116]. In 2015, Dorn et al. reported the preparation of an amino acid-functionalized TNT EMF characterized as $Gd_3N@C_{80}O_{12}(OH)_{10}(NH2)_7(NO_2)_2$ [105]. Their results demonstrated that these positively charged nanoparticles exhibited enhanced charge attraction for GBM cellular endocytosis compared with carboxyl-surface functionalized nanoparticles. The r_l values of $Gd_3N@C_{80}O_{12}(OH)_{10}(NH2)_7(NO_2)_2$ are 88, 101, and 30.5 mM^{-1} s^{-1} at 0.47, 1.4, and 9.4 T, respectively.

Along with the rapid developments in the application of EMFs as MRI contrast agents, multifunctional EMFs nanomaterials with fancy structures have been proposed to realize the integration of diagnosis and treatment. Very recently, Chen et al. developed a radionuclide-^{64}Cu-labeled doxorubicin-loaded polydopamine (PDA)–Gd EMF core–satellite nanotheranostic agent (termed CDPGM) for MR/photoacoustic (PA)/positron emission tomography (PET) multimodal imaging-guided combination cancer therapy. In this agent, the near-infrared (NIR)-absorbing PDA acts as a platform for the assembly of different moieties; $Gd_3N@C_{80}$ acts as a satellite anchoring it on the surface of PDA. The CDPGM nanoparticles show good biocompatibility, strong NIR absorption, high relaxivity ($r_1 = 14.06$ mM^{-1} s^{-1}), low risk of release of Gd ions, and NIR-triggered drug release. CDPGM can efficiently accumulate in the tumor and the tumor is completely eliminated by NIR laser irradiation [122].

In this section, we have reviewed the electronic structures of Gd-based EMFs, which are the foundation for their applications as MRI contrast agents. Additionally, research progress of mono-EMFs or TNT EMFs acting as MRI contrast agents has been summarized. Table 1 lists the relaxivities r_1 of all reported Gd-based EMFs in comparison with those of the Gd-DTPA.

3.3 Applications of Metallofullerenes as X-ray Contrast Agents

The Hounsfield Unit (HU) scale or computed tomography (CT) numbers is a quantitative scale for describing radiodensity and commonly applied to X-ray CT data. HU is defined by the general equation

$$HU = 1000 \times (\mu - \mu_{water})/\mu_{water}$$

Table 1 Relaxivities r_1 of endohedral fullerenes in comparison with that of the commercial MRI contrast agent Gd-DTPA

Fullerene materials	r_I (m M^{-1} s^{-1})	*T* Magnetic field	Year	References
$Gd@C_{82}(OH)_{40}$	67	0.47	2001	[59]
	81	1.0		
	31	4.7		
$Gd@C_{60}(OH)_{27}$	14.1–83.2	1.4	2005	[62]
$Gd@C_{82}(OH)_{16}$	19.3	4.7	2007	[60]
$Gd@C_{60}(OH)_{22\pm2}$	37.7 (pH 2)	4.7	2008	[61]
	61 (pH 7)			
$Gd@C_{82}O_{14}(OH)_{14}(NH_2)_6$	47.0	0.5	2013	[123]
	41.0	1.5		
	39.5	3		
	27.4	7.1		
$ScGd_2N@C_{80}O_{12}(OH)_{26}$	20.7	14.1	2007	[112]
$Gd_3N@C_{80}$ [DiPEG5000(OH)x]	102	0.35	2006	[111]
	143	2.4		
	32	9.4		
$Sc_2GdN@C_{80}O_{12}(OH)_{26}$	17.6	14.1	2007	[112]
$Gd_3N@C_{80}$-hydrochalarone	205	0.47	2008	[121]
$Gd_3N@C_{80}(OH)_{26}(CH_2CH_2COOM)_{16}$ (M = Na or H)	154	0.35	2009	[113]
	207	2.4		
	76	9.4		
$Gd_3N@C_{80}$ [DiPEG(OH)$_x$]	107–227	0.35	2010	[114]
	139–237	2.4		
	41.9–68.2	9.4		
$Gd_3N@C_{80}O_{11}(OH)_{21}$	137	0.47	2014	[104]
	140	1.4		
	58	9.4		
$Gd_3N@C_{84}O_6(OH)_{28}$	170	0.47	2014	[104]
	173	1.4		
	63	9.4		
$Gd@C_{60}$ [C(COOH)$_2$]$_{10}$	6.8–24.0	1.4	2003	[96]
$Gd@C_{82}O_6(OH)_{16}(NHC_2H_4CO_2H)_8$	8.1	0.35	2006	[116]
	9.1	1.5		
$Gd@C_{82}O_2(OH)_{16}(C(PO_3Et_2)_2)_{10}$	37	0.35	2008	[117]
	38.9	2.4		
	19.9	9.4		
$Gd@C_{82}O_6(OH)_{16}(NHCH_2CH_2CO$antiGFP$)_5$	12	0.35	2008	[116, 118]
$Gd_3N@C_{80}O_{12}(OH)_{10}(NH_2)_7(NO_2)_2$	88 ± 1	0.47	2015	[105]
	101 ± 1	1.4		
	30.5	9.4		
CDPGM	14.06	7	2017	[122]
Gd-DTPA	3.8	0.47	1999	[95]
	3.9	1.0		
	3.8	4.7		

where μ is the attenuation coefficient, and μ_{water} is the attenuation of water. Different materials have different HUs. For instance, the HU of water and fat are 0 and –100 respectively. The HU of bones is commonly larger than 500. The soft tissues have HU ranging from 0 to 100. However, the HU differences between soft tissues are small, especially, the HU differences between normal and pathologic tissues are small, which greatly challenges the contrast resolution of CT scanners. Thus, exogenously delivered contrast media, i.e., radiocontrast agents, are applied to enhance the distinguishability of X-ray CT, radiography, and fluoroscopy. Typically, the commonly used radiocontrast agents in the clinic are iodine or barium compounds.

The application of Gd chelates as a radiocontrast agent for X-ray imaging has been investigated. However, the attenuation of Gd-DTPA is less than that of iodinated contrast agent under the same single energy and concentrations. In addition, the maximum safe administration dose for Gd-DTPA (0.3 mmol/kg) is significantly lower than that of iodinated contrast agents (3–4 mmol/kg). Thus, the application of Gd-DTPA for X-ray imaging is very limited. In contrast to Gd chelates, EMFs can separate the inner metal atom from the environment, leading to its low toxicity, and thus can be considered as safe contrast agents for X-ray imaging. In 2002, $Lu_3N@C_{80}$ was reported by Dorn et al. to possess application potential as a good X-ray contrast [110]. Additionally, they proposed that the mixed-metal TNT EMFs, $Lu_{3-x}M_xN@C_{80}$, where M is Gd or Ho and x =0–2, can serve as multifunctional contrast agents for both X-ray and MR imaging, which can be simultaneously applied for diagnosis.

4 Biomedical Applications of Fullerenes for Cancer Therapy

In 2005, Zhao et al. first reported the intrinsic high antitumor activities of $Gd@C_{82}(OH)_{22}$ nanoparticles against H22 hepatoma in mice with low toxicity in vivo and in vitro [124]. Since then, the antitumor activities of different fullerene materials targeting different tumor cells have been extensively studied. The absolutely hydrophobic surfaces of fullerenes and EMFs hinder their direct applications in the biomedical field, and thus the fullerene materials discussed in this section are commonly their water-soluble derivatives.

The biological environments of organisms are complicated, and the biomedical effects of fullerenes are diverse. On the one hand, fullerenes can be used for PDT because they can be photoexcited from the ground state to the excited states, which will subsequently transfer energy to peripheral oxygen forming singlet oxygen [24–29], a vital example of ROS. On the other hand, it is reported that fullerene-based materials are also effective antioxidants with a superior capacity to scavenge ROS and thus can be used to treat ROS-related diseases [5–20], including cancer. In addition to the intrinsic antitumor activities of fullerene derivatives, they can also serve as drug delivery agents [30–40] and adjuvants to cancer vaccines [41–44] for cancer

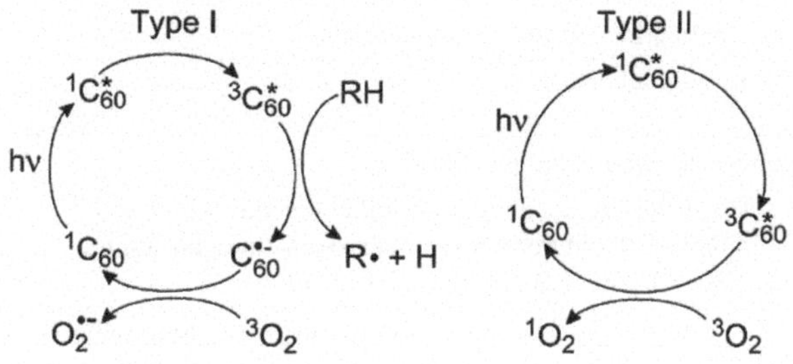

Fig. 7 Schematic representation of Type I (charge transfer) and Type II (energy transfer) photochemical mechanisms (*denotes excited singlet and triplet states of C_{60}). (Reproduced with permission from [25], © 2008 Elsevier Ltd.)

therapy. Fullerenes can also treat cancer via reducing the nutrient supply to tumor cells [45–47].

Table 2 summarizes the biomedical applications of fullerene derivatives relating to cancer therapy. In addition, based on extensive studies, different mechanisms have been proposed for the antitumor activities of fullerenes [125]. In this section, we will focus on the antitumor effects of fullerenes and summarize the proposed mechanisms: (i) fullerenes treat cancer via photodynamic therapy; (ii) fullerenes treat cancer as antioxidants and neuroprotective agents; (iii) fullerenes treat cancer as drug delivery agents; (iv) fullerenes treat cancer as adjuvants for cancer vaccines; and (v) fullerenes treat cancer via reducing the nutrient supply to tumor cells.

4.1 Fullerenes Treat Cancer via Photodynamic Therapy

Fullerenes can be photoexcited from the ground state to excited states, which will subsequently transfer energy to peripheral oxygen to form singlet oxygen ($^1O_2$) or superoxide anion ($O_2^{\cdot -}$)(Fig. 7). These ROS are highly reactive thus can oxidize biological targets like protein, lipid, DNA, etc., leading to cell damage [24–29]. Therefore, when fullerene is accumulated in the tumor tissue, its photodynamic effect that generates singlet oxygen under the irradiation with light should suppress the increase of tumors.

In 1997, Ikada et al. modified C_{60} with polyethylene glycol (PEG) to obtain the C_{60}-PEG conjugate, which can preferentially accumulate in a subcutaneous tumor and suppress the increase in tumor mass with exposure to visible light [26]. In 2006, Yamakoshi et al. synthesized a water-soluble C_{60}-N-vinylpyrrolidone (NVP) copolymer and indicated that C_{60}-NVP can generate ($O_2^{\cdot -}$) under visible light to break the specific DNA and achieve photodynamic therapy [24]. In 2008, Trajkovic et al. investigated the mechanism of the ROS generation/quenching mediated by C_{60}-based derivatives. They found that changing the number, species, and addition pattern of the addends on the fullerene cages will lead to a change in their ROS produc-

Table 2 Biomedical applications of fullerene materials

Fullerene materials	Biological effects	Year	References
Fullerenes treat cancer via photodynamic therapy			
C_{60}-PEG	C_{60}-PEG can accumulate in subcutaneous tumor and suppress the increase in tumor mass with exposure to visible light. (424 μg kg^{-1}, 107 J cm^{-2})	1997	[26]
C_{60}-NVP	C_{60}-NVP can generate $O_2^{\cdot-}$ under visible light to break the specific DNA and achieve photodynamic therapy	2006	[24]
C_{60}-Glc	C_{60}-Glc could induce phototoxicity against various cancer cells but not against normal fibroblast cells.	2010	[27]
FC4S	FC4S can induce cytotoxicity with light irradiation mediated by singlet oxygen	2016	[29]
Fullerenes treat cancer as antioxidants and neuroprotective agents			
$C_{60}(C(COOH)_2)_2$	These three derivatives could reduce intracellular ROS production. The relative activities order is $Gd@C_{82}(OH)_{22}$ > $C_{60}(OH)_{22}$ > $C_{60}(C(COOH)_2)_2$	2009	[6, 7]
$C_{60}(OH)_{22}$			[7]
$Gd@C_{82}(OH)_{22}$			[7]
Fullerenes treat cancer as immune system regulators			
Polyhydroxy-C_{60}, *N*-ethyl-polyamino-C_{60}	Inhibit the IgE-dependent mediator release and the IgE-induced elevation in cytoplasmic ROS levels. (2.5 μg kg^{-1} in mice)	2007	[11]
$C_{60}(OH)_{20\pm2}$	Fullerenol could attenuate neutrophilic lung inflammation induced by quartz via blocking ROS generation. (0.01, 0.1, 1 mg kg^{-1})	2008	[126]
$C_{60}(OH)_{20}$	Induce TNF-α-mediated cellular immunity to inhibit the tumor growth. (0.76, 0.53 mg kg^{-1})	2009	[22]
$Gd@C_{82}(OH)_{22}$	$Gd@C_{82}(OH)_{22}$ nanoparticles are a strong immunomodulator of the activation of T cells and macrophages. (0.76, 0.53 mg kg^{-1})	2009	[21]
$Gd@C_{82}(OH)_{22}$	It could induce DC maturation and activate T1-biased immune response. (0.5 mg kg^{-1})	2010	[127]
$C_{60}(OH)_{24}$	Pretreatment could protect against ionizing-radiation-induced mortality possibly via enhancing immune function, decreasing oxidative damage, and improving mitochondrial function. (40 mg kg^{-1})	2010	[128]
Fullerene mixture (C_{60}: 60%, C_{70}: 25%, C_{75}: 15%)	It could selectively induce CD11b+cells (e.g., macrophages)-mediated splenic inflammation at day 6 post administration. (0.2, 2 mg kg^{-1})	2011	[129]

(continued)

Table 2 (continued)

Fullerene materials	Biological effects	Year	References
Fullerenes treat cancer as drug delivery agents			
C_{60}-PTX	C_{60}-PTX conjugate has a significant anticancer activity in vitro	2005	[31]
C_{60}-DOX	C_{60}-DOX conjugate is distributed mostly in the cytoplasm	2010	[30, 38]
FA-CD-C_{60}-CBP	FA-CD-C_{60} significantly increased the intracellular uptake and release of CBP, thereby providing higher cytotoxicity against the HeLa cells with high expression of folate receptor	2016	[39]
C_{60}-DOX-NGR	C_{60}-DOX-NGR has excellent stability in physiological solutions and active tumor-targeting capacity	2016	[40]
C_{60}+CIS	C_{60}+CIS complex has higher toxicity toward tumor cell lines in vitro when compared to CIS alone	2017	[33, 34]
C_{60}-DTX	The C_{60}-DTX conjugate was able to deliver the drug at the desired pH to the interiors of the cancer cells with enhanced efficacy	2017	[35]
DOXO-C_{82}-cRGD	The DOXO-C_{82}-cRGD nanoparticles showed great preferential cellular uptake in non-small lung cancer cells as compared with DOXO-C_{82} and particularly, DOXO	2017	[37]
Fullerenes treat cancer as adjuvants for cancer vaccines			
$C_{60}(OH)_{20}$	It could serve as vaccine adjuvant mainly via triggering multiple TLR signaling pathways to enhance the immunogenicity of HIV-1 DNA antigen. (0.04, 0.2, 1 mg kg^{-1})	2013	[44]
Fullerenes treat cancer via reducing the nutrient supply to tumor cells			
$C_{60}(OH)_{20}$	$C_{60}(OH)_{20}$ can markedly reduce the expression of platelet endothelial cell adhesion molecule and the vessel density in tumors	2010	[45]

ing/quenching ability and the type of ROS produced/quenched [25]. In 2010, Morita et al. examined the effect of PDT with glycoconjugated fullerene (C_{60}-Glc) [27]. They found that PDT with C_{60}-Glc could induce phototoxicity against various cancer cells but not against normal fibroblast cells. In addition, C_{60}-Glc-induced apoptosis was blocked by NaN_3, which indicates that the cytotoxicity of C_{60}-Glc resulted from the generation of $^1O_2$. In 2016, Yu et al. reported the PDT with hexa(sulfo-*n*-butyl) [60] fullerene (FC4S) against sarcoma. FC4S can induce cytotoxicity to both human fibrosarcoma cells and murine sarcoma 180 cells with light irradiation because of the production of singlet oxygen [29].

4.2 Fullerenes Treat Cancer as Antioxidants and Neuroprotective Agents

Oxidative stress, i.e., the generation of ROS, has been proven to be involved in the progression for many diseases such as arthritis, cancer, amyotrophic lateral sclerosis, cardiovascular disease, and a number of neurodegenerative disorders [130, 131]. ROS include superoxide radical anion ($O_2^{\cdot -}$), hydroxyl radical (HO·), singlet oxygen ($^1O_2$), and hydrogen peroxide (H_2O_2). ROS can oxidize biological targets like protein, lipid, and DNA, which may lead to cancer, atherosclerosis, ischemia, inflammation, and liver injury [132–134]. The microenvironments of cancer or inflammatory cells are under increased oxidative stress than those of the normal cells. The high level of ROS can promote cellular proliferation, mutation, and genetic instability [135–137]. Thus, antioxidative enzymes, like superoxide dismutase (SOD), or chemical species that can scavenge ROS provides opportunities for cancer therapy.

It is well documented that fullerene-based materials possess unique capacities to scavenge ROS, thereby acting as antioxidants [138–141]. Hydroxylated and carboxylated fullerenes are two major water-soluble fullerene derivatives that possess free-radical scavenging activities [5–20]. A series of research works by Zhao and colleagues and Chen et al. have well established the ability of fullerene materials to modulate oxidative stress [6, 7, 45, 142–150]. In 2009, using cerebral microvessel endothelial cells (CMECs) as an in vitro model of the blood–brain barrier (BBB), Lao et al. found that $C_{60}(C(COOH)_2)_2$ nanoparticles could selectively enter oxidized CMECs rather than normal cells and greatly inhibit the apoptosis of CMECs induced by H_2O_2, which is related to their modulation of the JNK pathway [6]. The cytoprotective functions of these fullerene derivatives $C_{60}(C(COOH)_2)_2$, $C_{60}(OH)_{22}$, and $Gd@C_{82}(OH)_{22}$ were also studied by the same research group. They found that all these three derivatives could protect cells against H_2O_2-induced oxidative damage, stabilize the mitochondrial membrane potential, and reduce intracellular ROS production. The relative activities order is $Gd@C_{82}(OH)_{22} > C_{60}(OH)_{22} > C_{60}(C(COOH)_2)_2$ [7].

Inflammation is the body's response to tissue insult. Based on the above reports that fullerene derivatives possess unique capacities to scavenge ROS, it is proposed that fullerenes may inhibit inflammation. In 2007, Ryan et al. investigated the interactions between fullerene derivatives and inflammatory-related cells, human mast cells (MCs) and peripheral blood basophils (PBBs). They showed that polyhydroxy-C_{60} and *N*-ethyl-polyamino-C_{60} could significantly inhibit the IgE-dependent mediator release and lower the cytoplasmic ROS levels. Also the allergic responses, including the histamine release and body temperature are negatively regulated [11]. Subsequently in 2009, Dellinger et al. confirmed the inhibitory effect of fullerene derivatives on inflammation induced by phorbol 12-myristate 13-acetate (PMA) [151]. Another study reported that C_{60} fullerenols could significantly attenuate neutrophilic lung inflammation via blocking ROS-induced inflammation. [126]

Another important system related to ROS is the immune system. The immune system is a defensive barrier to the protection of the organism against external stimuli and invasion. Undoubtedly, fullerene materials will encounter the immune system after absorption. It is important to understand the interaction between fullerenes and the immune system for a comprehensive and accurate understanding of the cancer therapy applications of fullerenes. Fullerene-based materials are reported to have efficient ROS-scavenging capabilities, and ROS are involved in the cell signaling pathways that are related to the immune system. For instance, ROS can induce macrophage activation, leading to the release of cytokines and other inflammatory responses in adipose tissues [152]. Therefore, it is reasonable to explore the interaction between fullerene materials and the immune system regarding this aspect.

In 2005, Chen et al. first reported the antitumor activity of the star material $Gd@C_{82}(OH)_{22}$. Intrinsically, the $Gd@C_{82}(OH)_{22}$ nanoparticles were nontoxic to tumor cells and had a low accumulation in the tumor tissue. In order to understand the antitumor activity of $Gd@C_{82}(OH)_{22}$, they conducted a histopathological evaluation and found that there was lymphocyte infiltrated in the tumor tissues after the treatment of $Gd@C_{82}(OH)_{22}$. In other words, instead of directly damaging tumor cells, $Gd@C_{82}(OH)_{22}$ may manifest antitumor effects via recruitment of immune cells and regulate immune responses. To uncover the underlying mechanisms for the effects of fullerene materials on the immune system, the same group systematically investigated the interaction process between macrophages and fullerene materials including the uptake and subcellular distribution of metallofullerenols in macrophages. Taking advantages of synchrotron-based scanning transmission X-ray microscopy (STXM) with high spatial resolution of 30 nm, they found that a large amount of $Gd@C_{82}(OH)_{22}$ was taken up by peritoneal macrophages after peritoneal injection into mice for nearly 1 week and that the derivatives were internalized in a time-dependent manner. Meanwhile, they found that both $C_{60}(OH)_{22}$ and $Gd@C_{82}(OH)_{22}$, especially $Gd@C_{82}(OH)_{22}$, can significantly active macrophages to produce pro-inflammatory cytokines like IL-1β. They proposed two signal pathways for the secretion of IL-1β after fullerenol exposure: on one hand, fullerenols stimulate macrophages to express pro-IL-1β via TLRs/MyD88 pathway and subsequently activate NF-κB; on the other hand, fullerenols activate NLRP3 inflammasome under the assistance of relative factors including K^+ efflux and $P2X_7$ receptor, which is required for processing pro-IL-1β into mature IL-1β. This work can help get an insight into the underlying mechanisms of fullerene derivatives as good immunoregulatory agents with low toxicity [145].

4.3 Fullerenes Treat Cancer as Drug Delivery Agents

Paclitaxel (PTX, Taxol), carboplatin (CBP), cisplatin (CIS), and DOX have revolutionized cancer treatment in the past decade and are recognized as great advances in oncological medicine. The standard delivery modalities of intravenous infusion result in multiple side effects, and targeting of the drug to specific areas within the body can result in better efficacy and lower toxicity. Fullerene and its derivatives are star materials for cancer therapy not only owing to their intrinsic antitumor activities but also to their applications as drug delivery agents. The employment of fullerenes for drug delivery is still in an early stage of development. The design and synthesis of multifunctionalized fullerene systems that can cross cell membranes and efficiently deliver active molecules is an attractive challenge that involves multidisciplinary strategies. Promising results have emerged in recent years, bringing fullerenes again to the forefront of interest. Herein, the state of the art of this emerging field is presented and illustrated with some of the most representative examples.

In 2005, Wilson et al. designed and synthesized the first C_{60}-PTX conjugate slow-release system for liposome aerosol delivery to the lung [31]. The PTX was covalently attached to the C_{60} via an ester bond, permitting the paclitaxel to retain its pharmacological activity. Hydrolysis of the ester bond released the drug. Moreover, it was demonstrated that the C_{60}-PTX conjugate had significant anticancer activity in vitro when administered with a liposome aerosol formulation even though the IC_{50} was 1.6 times higher than that of the analogous formulation containing the drug alone [31]. In 2010, taking advantage of the optical properties of DOX, Wang et al. examined the uptake and distribution of the fullerene-DOX conjugate in cells [30]. They found that compared with free DOX, which is predominantly distributed in the cell nucleus, the C_{60}-DOX conjugate is distributed mostly in the cytoplasm. In 2015, Panchuk et al. evaluated the effect of C_{60}-DOX on the growth and metastasis of Lewis lung carcinoma in mice and performed primary screening of the potential mechanisms of action of the C_{60}-Dox complex. They reported that the volume of tumors treated with the C_{60}-Dox complex was 1.4 times less than that in control untreated animals. The potential mechanisms of the antitumor effect of C_{60}-Dox complex include both its direct action on tumor cells by inducing cell death and increased stress sensitivity and an immunomodulating effect [38]. In 2016, Zhang et al. synthesized the folic acid (FA)-cyclodextrin (CD)-C_{60} conjugate as a carrier for tumor-targeted drug delivery to enhance the anticancer effect of carboplatin (CBP). FA-CD-C_{60} significantly increased the intracellular uptake and release of CBP, thereby providing higher cytotoxicity against HeLa cells with high expression of folate receptor [39]. Wang et al. synthesized C_{60}-DOX-NGR nanoparticles. In this complex, DOX was covalently conjugated to C_{60} nanoaggregates via a ROS-sensitive thioketal linker (C_{60}-DOX), and then a hydrophilic shell (NGR) was attached to the outer surface of C_{60}-DOX, giving C_{60}-DOX-NGR excellent stability in physiological solutions in addition to active tumor-targeting capacity [40]. In 2017, the self-organization of C_{60} and CIS in aqueous solution was investigated by Prylutska et al. Their results revealed the complexation between the two compounds. The genotoxicity of C_{60} fullerene, CIS, and their complex was evaluated in vitro. The results showed that C_{60} did not induce DNA strand breaks and did not influence the genotoxic activity of CIS in vitro. The C_{60} in the C_{60}+CIS complex did affect the cell death mode in treated resting lymphocytes from healthy persons and reduced the fraction of necrotic cells [33]. In a subsequent study, they showed that the C_{60} + CIS complex had higher toxicity toward tumor cell lines in vitro than does CIS alone [34]. Misra et al. functionalized C_{60} with glycine to make it water soluble and conjugated the C_{60}-Gly with the anticancer drug docetaxel (DTX), forming C_{60}-DTX. The C_{60}-DTX conjugate was able to deliver the drug at the desired pH to the interiors of the cancer cells with enhanced efficacy and substantial erythrocyte compatibility [35]. Zhao et al. developed an attractive method to prepare a C_{82}-based drug delivery system modified with doxorubicin (DOXO) and a cyclic RGD motif (DOXO-C_{82}-cRGD) for potential clinical use in treating solid tumors in lung cancer. The in vitro results demonstrated that DOXO-C_{82}-cRGD nanoparticles showed great preferential cellular uptake into non-small lung cancer cells compared with that of DOXO-C_{82} and particularly compared with that of DOXO. Moreover, the cell viability results suggested that DOXO-C_{82}-cRGD had significant cytotoxicity to cells at a low dose [37].

Despite the above promising achievements in research on fullerenes as drug delivery agents, further research is needed, especially to better understand the possible harmful effects and the fate of fullerenes after delivering their cargos. It is clear that fullerenes can offer new opportunities in the upcoming generations of drug delivery systems.

4.4 Fullerenes Treat Cancer as Adjuvants for Cancer Vaccines

Preventive or therapeutic cancer vaccines have vast significance for cancer treatment. Owing to their positive effects on the uptake of antigen and/or modulating immune cells, especially for antigen-presenting cells (APCs), vaccine adjuvants are essential to further the immunogenicity of tumor-associated/specific antigens [153]. As fullerene derivatives could modulate immune cells, especially for macrophages and APCs, they are considered as a potential candidate of vaccines for crucial infectious diseases and cancers.

It is reported that fullerene derivatives $C_{60}(OH)_{20}$ have dual functions: on one hand, they have distinct self-assembly property [154] to encapsulate the antigen during self-assembly; on the other hand, they can improve the immunogenicity of the antigen via modulating the immune cells [44]. Meanwhile, fullerenol could significantly facilitate the phagocytosis capability of peritoneal macrophages of immunized mice and trigger the generation of effector memory $CD8^+$ T cells (TEM), which are vital for protection during the early stage of infection. Through high-throughput screening of toll-like receptor (TLR) signaling pathways using gene knockout mice, a possible mechanism of fullerenol as a potent adjuvant has been proposed. When encountering with fullerenols, which serve as vaccine adjuvants, diverse TLR signaling pathways (e.g., TLR2, TLR4, TLR5, TLR8, TLR9) had been triggered to activate DC maturation and finally enhance the immunogenicity of the vaccine antigen. Therefore, fullerenol may act as a potential adjuvant candidate for vaccines in clinical practice.

4.5 Fullerenes Treat Cancer via Reducing the Nutrient Supply to Tumor Cells

The tumor metastasis, instead of the tumor itself, is a dominated factor of cancer mortalities. Chemical species that are highly effective in inhibiting tumor metastasis are promising chemotherapeutics for cancer therapy. Tumor cells need nutrients and energies to grow and immigrate. Cut off the nutrient supply to tumor cells could efficiently inhibit its growth and metastasis. In 2010, Chen et al. reported that in a mouse breast cancer model, $C_{60}(OH)_{20}$ could negatively modulate the oxidative stress and markedly reduce the expression of platelet endothelial cell adhesion molecule and the vessel density in tumors. Subsequently, the nutrient supply to tumor cells is inhibited [45].

5 Conclusion and Perspectives

In this chapter, we have summarized the latest advances in the use of fullerenes for bioimaging and cancer therapy. First, we focused on the geometric structures of fullerene derivatives. The structural problem is one of the most important issues in obtaining a comprehensive understanding of the biological effects of fullerene materials because the structure determines the physicochemical properties, which are vital to its applications. The experimental purification and identification of the precise structures of fullerene-based nanomaterials are difficult because of the gigantic number of its isomers. Theoretical methods are an important tool to predict the structures of fullerene materials. Based on the synthesis conditions and the fact that molecules with large HOMO–LUMO gaps are most likely to survive, some models of fullerene derivatives have been proposed. Second, we have summarized the applications of fullerene materials as bioimaging agents. Metallofullerenols are widely studied as MRI contrast agents owing to their low toxicity and high relaxivity. The paramagnetic electron configurations of Gd-based metallofullerene are the foundation for their high relaxivity. In addition, the geometric structures, in which the internal Gd atom is effectively isolated from the biological environment, provide EMFs with high stability and low toxicity. Third, we summarize the application of fullerene materials as antitumor agents for cancer therapy. Five possible mechanisms have been summarized based on previous studies: (i) fullerenes treat cancer via photodynamic therapy; (ii) fullerenes treat cancer as antioxidants and neuroprotective agents; (iii) fullerenes treat cancer as drug delivery agents; (iv) fullerenes treat cancer as adjuvants for cancer vaccines; and (v) fullerenes treat cancer via reducing the nutrient supply to tumor cells. Based on the above achievements, fullerene materials are promising nanomedicine candidates with feasible uses in the treatment of diseases.

Based on the abovementioned results, several key issues urgently need to be addressed. (i) The development of new methods to synthesize, purify, and characterize fullerene derivatives is an important issue. (ii) The structure–activity relationship needs to be extensively investigated. For instance, the effects of the physicochemical properties on the interactions of fullerene nanomaterials with the immune system need to be extensively investigated. (iii) The toxicology profiles as well as the biodegradation behaviors of fullerene nanomaterials need to be extensively studied, although some important pieces of information have been obtained. It is necessary to adopt new techniques or methods to evaluate the interactions between fullerene materials and the tumor cells and uncover the underlying mechanisms. For instance, molecular dynamics simulation may provide some information about how these materials interact with biomacromolecules. Intense cooperation among research groups worldwide working in different disciplines will also be very helpful in addressing this issue within a short time. In summary, with our rapidly growing understanding of the geometric and electronic structures as well as the underlying mechanisms of the antitumor activities of fullerene nanomaterials, the rational design of fullerene nanomaterials with high efficiency and low toxicity for applications of interest can be realized in the future.

Acknowledgements This work was financially supported by the National Natural Science Foundation of China (21371118, 41430644).

References

1. Kroto HW et al (1985) C60: Buckminsterfullerene. Nature 318:162–163
2. Heath JR et al (1985) Lanthanum complexes of spheroidal carbon shells. J Am Chem Soc 107:7779–7780
3. Wilson JL et al (1999) Metallofullerene drug design. Coord Chem Rev 190:199–207
4. Wilson JL (1999) Medical applications of fullerenes and metallofullerenes, vol 8. Electrochemical Society, Pennington, NJ, ETATS-UNIS
5. Yang J et al (2007) The use of fullerene substituted phenylalanine amino acid as a passport for peptides through cell membranes. Org Biomol Chem 5:260–266
6. Lao F et al (2009) Fullerene nanoparticles selectively enter oxidation-damaged cerebral microvessel endothelial cells and inhibit JNK-related apoptosis. ACS Nano 3:3358–3368
7. Yin J-J et al (2009) The scavenging of reactive oxygen species and the potential for cell protection by functionalized fullerene materials. Biomaterials 30:611–621
8. Yin J-J et al (2008) Inhibition of tumor growth by Endohedral Metallofullerenol nanoparticles optimized as reactive oxygen species scavenger. Mol Pharmacol 74:1132–1140
9. Fang L et al (2009) Fullerene derivatives protect endothelial cells against NO-induced damage. Nanotechnology 20:225103
10. Norton SK et al (2012) Epoxyeicosatrienoic acids are involved in the C70 fullerene derivative–induced control of allergic asthma. J Allerg Clin Immunol 130:761–769
11. Ryan JJ et al (2007) Fullerene nanomaterials inhibit the allergic response. J Immunol 179:665–672
12. Application of fullerenes in nanomedicine: an update. Nanomedicine 2013, *8*, 1191–1208
13. Beuerle F et al (2007) Cytoprotective activities of water-soluble fullerenes in zebrafish models. J Exp Nanosci 2:147–170
14. Witte P et al (2007) Water solubility, antioxidant activity and cytochrome C binding of four families of exohedral adducts of C60 and C70. Org Biomol Chem 5:3599–3613

15. Lotharius J et al (1999) Distinct mechanisms underlie neurotoxin-mediated cell death in cultured dopaminergic neurons. J Neurosci 19:1284–1293
16. Dugan LL et al (1996) Buckminsterfullerenol free radical scavengers reduce excitotoxic and apoptotic death of cultured cortical neurons. Neurobiol Dis 3:129–135
17. Gharbi N et al. (2005) [60]Fullerene is a powerful antioxidant in vivo with no acute or subacute toxicity. Nano Lett 5:2578–2585
18. Dugan LL et al (1997) Carboxyfullerenes as neuroprotective agents. Proc Natl Acad Sci 94:9434–9439
19. Dugan LL et al (2001) Fullerene-based antioxidants and neurodegenerative disorders. Parkinsonism Relat Disord 7:243–246
20. Lin AMY et al (1999) Carboxyfullerene prevents iron-induced oxidative stress in rat brain. J Neurochem 72:1634–1640
21. Liu Y et al (2009) The effect of Gd@C82(OH)22 nanoparticles on the release of Th1/Th2 cytokines and induction of TNF-α mediated cellular immunity. Biomaterials 30:3934–3945
22. Liu Y et al (2009) Immunostimulatory properties and enhanced TNF-α mediated cellular immunity for tumor therapy by C60 (OH) 20 nanoparticles. Nanotechnology 20:415102
23. Wu G et al (2016) Fullerenes and their derivatives as inhibitors of tumor necrosis factor-α with highly promoted affinities. J Mol Model 22:161
24. Iwamoto Y, Yamakoshi Y (2006) A highly water-soluble C60-NVP copolymer: a potential material for photodynamic therapy. Chem Commun. 4805–4807
25. Markovic Z, Trajkovic V (2008) Biomedical potential of the reactive oxygen species generation and quenching by fullerenes (C60). Biomaterials 29:3561–3573
26. Tabata Y et al (1997) Photodynamic effect of polyethylene glycol–modified fullerene on tumor. Jpn J Cancer Res 88:1108–1116
27. Otake E et al (2010) Effect and mechanism of a new photodynamic therapy with glycoconjugated fullerene. Photochem Photobiol 86:1356–1363
28. Abrahamse H, Hamblin MR (2016) New photosensitizers for photodynamic therapy. Biochem J 473:347–364
29. Yu C et al (2016) Photodynamic therapy with hexa(sulfo-n-butyl) 60 fullerene against sarcoma in vitro and in vivo. J Nanosci Nanotechnol 16:171–181
30. Liu J-H et al (2010) Fullerene-conjugated doxorubicin in cells. ACS Appl Mater Interfaces 2:1384–1389
31. Zakharian TY et al (2005) A fullerene–paclitaxel chemotherapeutic: synthesis, characterization, and study of biological activity in tissue culture. J Am Chem Soc 127:12508–12509
32. Montellano A et al (2011) Fullerene C60 as a multifunctional system for drug and gene delivery. Nanoscale 3:4035–4041
33. Prylutska S et al (2017) A nanocomplex of C-60 fullerene with cisplatin: design, characterization and toxicity. Beilstein J Nanotechnol 8:1494–1501
34. Prylutska S et al (2017) C-60 fullerene enhances cisplatin anticancer activity and overcomes tumor cell drug resistance. Nano Res 10:652–671
35. Misra C et al (2017) Improved cellular uptake, enhanced efficacy and promising pharmacokinetic profile of docetaxel employing glycine-tethered C-60-fullerenes. Mater Sci Eng C-Mater Bio Appl 76:501–508
36. Guan M et al (2016) Fullerene/photosensitizer nanovesicles as highly efficient and clearable phototheranostics with enhanced tumor accumulation for cancer therapy. Biomaterials 103:75–85
37. Zhao L et al (2017) A novel fullerene-based drug delivery system delivering doxorubicin for potential lung cancer therapy. J Nanosci Nanotechnol 17:5147–5154
38. Prylutska SV et al. (2015) Complex of C-60 fullerene with doxorubicin as a promising agent in antitumor Therapy. Nanoscale Res Lett. 10
39. Liu Y et al (2016) Folic acid functionalized gamma-cyclodextrin C-60, a novel vehicle for tumor-targeted drug delivery. J Biomed Nanotechnol 12:1393–1403
40. Shi J et al (2016) Fullerene (C-60)-based tumor-targeting nanoparticles with "off-on" state for enhanced treatment of cancer. J Control Release 235:245–258

41. Turabekova M et al (2014) Immunotoxicity of nanoparticles: a computational study suggests that CNTs and C60 fullerenes might be recognized as pathogens by Toll-like receptors. Nanoscale 6:3488–3495
42. Kawai T, Akira S (2010) The role of pattern-recognition receptors in innate immunity: update on Toll-like receptors. Nat Immunol 11:373–384
43. Pulendran B, Ahmed R (2006) Translating innate immunity into immunological memory: implications for vaccine development. Cell 124:849–863
44. Xu L et al (2013) Morphologically virus-like fullerenol nanoparticles act as the dual-functional nanoadjuvant for HIV-1 vaccine. Adv Mat 25:5928–5936
45. Jiao F et al (2010) Studies on anti-tumor and antimetastatic activities of fullerenol in a mouse breast cancer model. Carbon 48:2231–2243
46. Meng H et al (2010) Potent angiogenesis inhibition by the particulate form of fullerene derivatives. ACS Nano 4:2773–2783
47. Sun C et al (2016) C-60(OH)(22): a potential histone deacetylase inhibitor with anti-angiogenic activity. Nanoscale 8:16332–16339
48. Lyu Y et al (2016) Intraparticle molecular orbital engineering of semiconducting polymer nanoparticles as amplified theranostics for in vivo photoacoustic imaging and photothermal therapy. ACS Nano 10:4472–4481
49. Shi J et al (2016) A multi-functional tumor theranostic nanoplatform for MRI guided photothermal-chemotherapy. Pharm Res 33:1472–1485
50. Chiang LY, et al. (1992) Multihydroxy Additions Onto C-60 fullerene molecules. J Chem Soc-Chem Commun. 1791–1793
51. Chiang LY et al (1992) Versatile nitronium chemistry for C-60 fullerene functionalization. J Am Chem Soc 114:10154–10157
52. Chiang LY et al (1993) Evidence of hemiketals incorporated in the structure of fullerols derived from aqueous acid chemistry. J Am Chem Soc 115:5453–5457
53. Li J et al. (1993) C-60 Fullerol formation catalyzed by quaternary ammonium hydroxides. J Chem Soc-Chem Commun. 1784–1785
54. Gonzalez KA et al (2002) Synthesis and in vitro characterization of a tissue-selective fullerene: vectoring C-60(OH)(16) AMBP to mineralized bone. Biorg Med Chem 10:1991–1997
55. Wang S et al (2005) Novel and efficient synthesis of water-soluble 60 fullerenol by solvent-free reaction. Synth Commun 35:1803–1808
56. Kokubo K et al (2008) Facile synthesis of highly water-soluble fullerenes more than half-covered by hydroxyl groups. ACS Nano 2:327–333
57. Kokubo K et al (2011) Facile and scalable synthesis of a highly hydroxylated water-soluble fullerenol as a single nanoparticle. Nano Res 4:204–215
58. Saitoh Y et al (2011) Super-highly hydroxylated fullerene derivative protects human keratinocytes from UV-induced cell injuries together with the decreases in intracellular ROS generation and DNA damages. J Photochem Photobiol B 102:69–76
59. Mikawa M et al (2001) Paramagnetic water-soluble metallofullerenes having the highest relaxivity for MRI contrast agents. Bioconjug Chem 12:510–514
60. Zhang J et al (2007) Synthesis and in vivo study of metallofullerene based MRI contrast agent. J Radioanal Nucl Chem 272:605–609
61. Xing G et al (2008) The strong MRI relaxivity of paramagnetic nanoparticles. J Phys Chem B 112:6288–6291
62. Laus S et al (2005)Destroying gadofullerene aggregates by salt addition in aqueous solution of Gd@$C_{60}(OH)_x$ and Gd@$C_{60}[C(COOH_2)]10$. J Am Chem Soc 127:9368–9369
63. Zhang G et al (2010) Facile synthesis of isomerically pure fullerenols and formation of spherical aggregates from $C_{60}(OH)_8$. Angew Chem Int Ed 49:5293–5295
64. Rodríguez-Zavala J, Guirado-López R (2004) Structure and energetics of polyhydroxylated carbon fullerenes. Phys Rev B 69:075411
65. Rodríguez-Zavala JG, Guirado-López RA (2006) Stability of highly OH-covered C60 fullerenes: role of coadsorbed O impurities and of the charge state of the cage in the formation of carbon-opened structures. J Phys Chem A 110:9459–9468

66. Guirado-López RA, Rincón ME (2006) Structural and optical properties of highly hydroxylated fullerenes: stability of molecular domains on the C60 surface. J Chem Phys 125:154312
67. He H et al (2011) The structural stability of polyhydroxylated C60(OH)24: Density functional theory characterizations. Comput Theor Chem 974:16–20
68. Kroto HW (1987) The stability of the fullerenes Cn, with n = 24, 28, 32, 36, 50, 60 and 70. Nature 329:529–531
69. Campanera JM et al (2005) General rule for the stabilization of fullerene cages encapsulating trimetallic nitride templates. Angew Chem Int Ed 44:7230–7233
70. Valencia R et al. (2007) Large fullerenes stabilized by encapsulation of metallic clusters. Chem Commun. 4161–4163
71. Valencia R et al (2008) Understanding the stabilization of metal carbide endohedral fullerenes M2C2@C82 and related systems. J Phy Chem A 112:4550–4555
72. Garcia-Borràs M et al (2013) Maximum aromaticity as a guiding principle for the most suitable hosting cages in endohedral metallofullerenes. Angew Chem Int Ed 52:9275–9278
73. Wang ZZ et al (2015) Syntheses, structures and antioxidant activities of fullerenols: knowledge learned at the atomistic level. J Cluster Sci 26:375–388
74. Gao X et al (2016) Isolated aromatic patches as a rule to select metallofullerene multiple adducts with high chemical stabilities. Carbon 96:980–986
75. Gao XJ et al (2016) Improved description for the structures of fullerenols C60(OH)n (n = 12–48) and C2v(9)–C82(OH)x (x = 14–58). J Phys Chem C 120:11709–11715
76. Wang Z et al (2014) A precision structural model for fullerenols. Chem Science 5:2940–2948
77. Wang Z et al (2015) Oxidation-induced water-solubilization and chemical functionalization of fullerenes C60, Gd@C60 and Gd@C82: atomistic insights into the formation mechanisms and structures of fullerenols synthesized by different methods. Nanoscale 7:2914–2925
78. Yaghobi M et al (2009) Optical and structural properties of the endohedral complexes M@C60 (M=Cs, Li, and Na). J Mol Struct (Thoechem) 905:48–50
79. Decleva P et al (1999) Theoretical study of resonances in the metal core photoionization of M@C 60 (M=Li, Na, K). J Phys B: At Mol Opt Phys 32:4523
80. Brocl/awik E, Eilmes A (1998) Density functional study of endohedral complexes M@C60 (M=Li, Na, K, Be, Mg, Ca, La, B, Al): Electronic properties, ionization potentials, and electron affinities. J Chem Phys 108:3498–3503
81. Jingnan L et al (1994) Structural properties of the endohedral complex Na+@C 60. J Phys: Condens Matter 6:L253
82. Stepniak F et al (1993) Electrical transport in Na, K, Rb, and Cs fullerides: phase formation, microstructure, and metallicity. Phys Rev B 48:1899–1906
83. Inoue T et al (1999) XAFS study on Eu@C60. J Synchrotron Radiat 6:779–780
84. Suzuki M et al (2012) Single-crystal X-ray diffraction study of three Yb@C82 isomers cocrystallized with NiII(octaethylporphyrin). J Am Chem Soc 134:18772–18778
85. Okazaki T et al (2000) Isolation and spectroscopic characterization of Sm-containing metallofullerenes. Chem Phys Lett 320:435–440
86. Lu J et al (1999) Electronic structures of endohedral Sr@C60, Ba@C60, Fe@C60 and Mn@C60. Mod Phys Lett B 13:97–101
87. Pichler T et al (1998) The metallofullerene Tm@C82: isomer-selective electronic structure. Appl Phys A 66:281–285
88. Wang LS et al (1993) The electronic structure of Ca@C60. Chem Phys Lett 207:354–359
89. Kubozono Y et al (2003) Crystal structure and electronic transport of $\mathrm{Dy}@{\mathrm{C}}_{82}$. Physical Review B 67:115410
90. Ding J, Yang S (1996) Isolation and Characterization of Pr@C82 and Pr2@C80. J Am Chem Soc 118:11254–11257
91. Ding J et al (1996) Isolation and characterization of a new metallofullerene Nd@C82. Chem Phys Lett 261:92–97
92. Suzuki T et al (1993) Electrochemical properties of La@C82. J Am Chem Soc 115:11006–11007

93. Kobayashi K, Nagase S (1998) Structures and electronic states of M@C82 (M=Sc, Y, La and lanthanides). Chem Phys Lett 282:325–329
94. Gao X et al (2015) Divalent metals can reside on bonds in fullerenes. Dalton Trans 44:9561–9568
95. Caravan P et al (1999) Gadolinium(III) chelates as MRI contrast agents: structure, dynamics, and applications. Chem Rev 99:2293–2352
96. Bolskar RD et al (2003) First soluble M@C-60 derivatives provide enhanced access to metallofullerenes and permit in vivo evaluation of Gd@C–60 C(COOH)(2) (10) as a MRI contrast agent. J Am Chem Soc 125:5471–5478
97. Lu X et al (2004) Studies on the relaxivities of novel MRI contrast agents two water-soluble derivatives of Gd@C-82. Chem J Chin Univ-Chin 25:697–700
98. Sitharaman B et al (2004) Gd@C-60 C(COOH)(2) (10) and Gd@C-60(OH)(x): Nanoscale aggregation studies of two metallofullerene MRI contrast agents in aqueous solution. Nano Lett 4:2373–2378
99. Chaur MN et al (2007) Gd3 N@C-2n (n = 40, 42, and 44): remarkably low HOMO-LUMO gap and unusual electrochemical reversibility of Gd3 N@C-88. J Am Chem Soc 129:14826–14829
100. Dunsch L, Yang S (2007) Metal nitride cluster fullerenes: their current state and future prospects. Small 3:1298–1320
101. Takano Y et al (2009) Anisotropic magnetic behavior of anionic Ce@C-82 carbene adducts. J Am Chem Soc 131:9340–9346
102. Grushko YS et al (2010) MRI-Contrasting system based on water-soluble fullerene/Gd-metallofullerene mixture. Fuller Nanotub Carb Nanostruct 18:417–421
103. Chen N et al (2012) Sc2S@C-s(10528)-C-72: a dimetallic sulfide endohedral fullerene with a non isolated pentagon rule cage. J Am Chem Soc 134:7851–7860
104. Zhang J et al (2014) Gd3 N@C-84(OH)(x): A new egg-shaped metallofullerene magnetic resonance imaging contrast agent. J Am Chem Soc 136:2630–2636
105. Li T et al (2015) A new interleukin-13 amino-coated gadolinium metallofullerene nanoparticle for targeted MRI detection of glioblastoma tumor cells. J Am Chem Soc 137:7881–7888
106. Li T et al (2016) A new interleukin-13 amino-coated gadolinium metallofullerene nanoparticle for targeted MRI detection of glioblastoma tumor cells (vol 137, pg 7881, 2015). J Am Chem Soc 138:1723
107. Li T, Dorn HC (2017) Biomedical applications of metal-encapsulated fullerene nanoparticles. Small. 13
108. Kato H et al (2003) Lanthanoid endohedral metallofullerenols for MRI contrast agents. J Am Chem Soc 125:4391–4397
109. Qu L et al (2006) Study of rare earth encapsulated carbon nanomolecules for biomedical uses. J Alloy Compd 408:400–404
110. Iezzi EB et al (2002) Lutetium-based trimetallic nitride endohedral metallofullerenes: new contrast agents. Nano Lett 2:1187–1190
111. Fatouros PP et al (2006) In vitro and in vivo imaging studies of a new endohedral metallofullerene nanoparticle. Radiology 240:756–764
112. Zhang E-Y et al (2007) Preparation and characterization of two new water-soluble endohedral metallofullerenes as magnetic resonance imaging contrast agents. J Phys Chem B 111:14223–14226
113. Shu C et al (2009) Facile preparation of a new gadofullerene-based magnetic resonance imaging contrast agent with high H-1 relaxivity. Bioconjug Chem 20:1186–1193
114. Zhang J et al (2010) High relaxivity trimetallic nitride (Gd3 N) metallofullerene MRI contrast agents with optimized functionality. Bioconjug Chem 21:610–615
115. Adiseshaiah P et al (2013) A novel gadolinium-based trimetasphere metallofullerene for application as a magnetic resonance imaging contrast agent. Invest Radiol 48:745–754
116. Shu CY et al (2006) Synthesis and characterization of a new water-soluble endohedral metallofullerene for MRI contrast agents. Carbon 44:496–500
117. Shu C-Y et al (2008) Organophosphonate functionalized Gd@C-82 as a magnetic resonance imaging contrast agent. Chem Mater 20:2106–2109

118. Shu C-Y et al (2008) Conjugation of a water-soluble gadolinium endohedral fulleride with an antibody as a magnetic resonance imaging contrast agent. Bioconjug Chem 19:651–655
119. Cui R et al (2015) Novel carbon nanohybrids as highly efficient magnetic resonance imaging contrast agents. Nano Research 8:1259–1268
120. Laus S et al (2007) Understanding paramagnetic relaxation phenomena for water-soluble gadofullerenes. J Phys Chem C 111:5633–5639
121. MacFarland DK et al (2008) Hydrochalarones: a novel endohedral metallofullerene platform for enhancing magnetic resonance imaging contrast. J Med Chem 51:3681–3683
122. Wang S et al. (2017) core-satellite polydopamine-gadolinium-metallofullerene nanotheranostics for multimodal imaging guided combination cancer therapy. Adv Mater (Deerfield Beach, Fla.)
123. Zheng J-P et al (2013) Multifunctional gadofulleride nanoprobe for magnetic resonance imaging/fluorescent dual modality molecular imaging and free radical scavenging. Carbon 65:175–180
124. Chen C et al (2005) Multihydroxylated[Gd@C82(OH)22]n Nanoparticles: antineoplastic activity of high efficiency and low toxicity. Nano Lett 5:2050–2057
125. Wang J et al (2014) Therapeutic applications of low-toxicity spherical nanocarbon materials. NPG Asia Mater 6:e84
126. Roursgaard M et al (2008) Polyhydroxylated C60 fullerene (fullerenol) attenuates neutrophilic lung inflammation in mice. Basic Clin Pharmacol Toxicol 103:386–388
127. Yang D et al (2010) Nanoparticles, [Gd@C(82)(OH)(22)](n), induces dendritic cell maturation and activates Th1 immune responses. ACS Nano 4:1178–1186
128. Cai X et al (2010) The polyhydroxylated fullerene derivative C60(OH)24 protects mice from ionizing-radiation-induced immune and mitochondrial dysfunction. Toxicol Appl Pharmacol 243:27–34
129. Ding N et al (2011) Intratracheal administration of fullerene nanoparticles activates splenic CD11b + cells. J Hazard Mater 194:324–330
130. Valko M et al (2007) Free radicals and antioxidants in normal physiological functions and human disease. Int J Biochem Cell Bio 39:44–84
131. Halliwell B, Gutteridge JM (2015) Free radicals in biology and medicine. Oxford University Press, USA
132. Stadtman ER, Berlett BS (1997) Reactive oxygen-mediated protein oxidation in aging and disease. Chem Res Toxicol 10:485–494
133. Loft S, Poulsen HE (1996) Cancer risk and oxidative DNA damage in man. J Mol Med 74:297–312
134. Aust SD et al (1993) Free radicals in toxicology. Toxicol Appl Pharmacol 120:168–178
135. Hileman EA et al (2001) Superoxide dismutase: an emerging target for cancer therapeutics. Expert Opin Ther Targets 5:697–710
136. Pelicano H et al (2004) ROS stress in cancer cells and therapeutic implications. Drug Resist Updates 7:97–110
137. Toyokuni S et al (1995) Persistent oxidative stress in cancer. FEBS Lett 358:1–3
138. Lai H-S et al (2000) Free radical scavenging activity of fullerenol on the ischemia-reperfusion intestine in dogs. World J Surg 24:450–454
139. Bisaglia M et al (2000) C3-fullero-tris-methanodicarboxylic acid protects cerebellar granule cells from apoptosis. J Neurochem 74:1197–1204
140. Tsai MC et al (1997) Polyhydroxylated C60, fullerenol, a novel free-radical trapper, prevented hydrogen peroxide- and cumene hydroperoxide-elicited changes in rat hippocampus in-vitro. J Pharm Pharmacol 49:438–445
141. Krusic PJ et al (1991) Radical reactions of C$_{60}$. Science 254:1183–1185
142. Cong W et al (2015) Evaluation of the influence of fullerenol on aging and stress resistance using Caenorhabditis elegans. Biomaterials 42:78–86
143. Tang J et al (2016) Polyhydroxylated fullerenols regulate macrophage for cancer adoptive immunotherapy and greatly inhibit the tumor metastasis. Nanomed Nanotechnol Biol Med 12:945–954

144. Pan Y et al (2015) Gd–Metallofullerenol nanomaterial suppresses pancreatic cancer metastasis by inhibiting the interaction of histone deacetylase 1 and metastasis-associated protein 1. ACS Nano 9:6826–6836
145. Chen Z et al (2014) Polyhydroxylated metallofullerenols stimulate IL-1β secretion of macrophage through TLRs/MyD88/NF-κB pathway and NLRP3 inflammasome activation. Small 10:2362–2372
146. Chen Z et al (2012) Applications of functionalized fullerenes in tumor theranostics. Theranostics 2:238–250
147. Li W et al (2011) The inhibition of death receptor mediated apoptosis through lysosome stabilization following internalization of carboxyfullerene nanoparticles. Biomaterials 32:4030–4041
148. Zhou G et al (2010) Subcellular distribution of polyhydroxylated metallofullerene Gd@C82(OH)22 in different tissues of tumor-bearing mice. J Nanosci Nanotechnol 10:8597–8602
149. Jiao F et al (2010) Modulation of oxidative stress by functionalized fullerene materials in the lung tissues of female C57/BL mice with a metastatic lewis lung carcinoma. J Nanosci Nanotechnol 10:8632–8637
150. Chen C (2016) Biomedical applications and toxicology of carbon nanomaterials, Wiley
151. Dellinger A et al (2009) Fullerene nanomaterials inhibit phorbol myristate acetate-induced inflammation. Exp Dermatol 18:1079–1081
152. Forman HJ, Torres M (2002) Reactive oxygen species and cell signaling. Am J Respir Crit Care Med 166:S4–S8
153. Li H et al (2011) Alpha-alumina nanoparticles induce efficient autophagy-dependent cross-presentation and potent antitumour response. Nat Nano 6:645–650
154. Brant JA et al (2007) Fullerol cluster formation in aqueous solutions: Implications for environmental release. J Colloid Interface Sci 314:281–288

Carbon Nano-onions for Bioimaging and Cancer Therapy Applications

Adalberto Camisasca and Silvia Giordani

Abstract Carbon nano-onions (CNOs) are a versatile class of carbon nanomaterials that recently aroused great interest in the scientific community. Several production methods have been published, where thermal annealing of detonation nanodiamonds (DNDs) is the elected method that ensures high yields along with a low cost. Thanks to their peculiar properties and to the possibilities for functionalizing their surface, CNOs are employed in different fields. In particular, fluorescently labeled CNOs are promising candidates in biomedical applications, such as scaffolds for cellular targeting, imaging, and drug delivery due to their good biocompatibility both in vitro and in vivo. In this chapter, we will present the unique properties of CNOs and the different methods used for their synthesis. Additionally, we will review the potential biological applications of CNOs, and their in vitro and in vivo biosafety assessments.

Keywords Carbon nano-onion · Imaging · Nanomedicine · Fluorescence

1 Introduction

The discovery of a new carbon allotropic form, C_{60} fullerene, in 1985 by H. Kroto, R. Curl, and R. Smalley [1], has constituted a turning point in the classical view of carbon-based structures, composed of diamond and graphite. From that moment, carbon nanomaterials (CNMs) started arousing great interest in the scientific community due to their intriguing properties suitable for a large number of applications.

A. Camisasca · S. Giordani
Nano Carbon Materials, Istituto Italiano di Tecnologia (IIT), via Livorno 60, 10144 Turin, Italy

A. Camisasca
Department of Chemistry and Industrial Chemistry, University of Genoa, via Dodecaneso 31, Genoa 16145, Italy

S. Giordani (✉)
Department of Chemistry, University of Turin, via Giuria 7, 10125 Turin, Italy
e-mail: silvia.giordani@iit.it; s.giordani@unito.it

G. Gonçalves and G. Tobias (eds.), *Nanooncology*, Nanomedicine and Nanotoxicology, https://doi.org/10.1007/978-3-319-89878-0_13

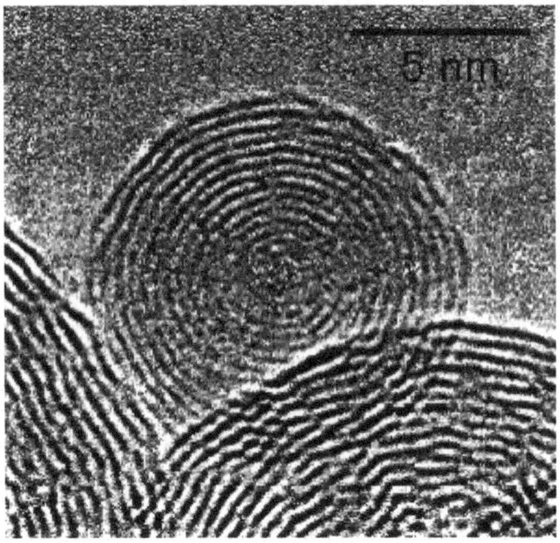

Fig. 1 HREM image of CNOs generated by electron irradiation. Reprinted with permission from [9]. Copyright 1995 Elsevier Ltd.

This unexpected excitement led, in the last 30 years, to the discovery of a variety of different CNMs, including carbon nanotubes (CNTs) [2] and graphene [3, 4], enriching the carbon family [5].

Among the different carbon forms, multi-shell fullerenes also known as carbon nano-onions (CNOs), or onion-like structures (OLCs), have recently received growing attention due to their peculiar physicochemical characteristics, such as low density, large surface area to volume ratio and a graphitic multilayer morphology [6].

CNOs are members of the fullerenes family and exhibit a spherical structure composed of multiple sp^2-hybridized carbon shells forming the particular cage-in-cage structure (Fig. 1). The spacing between two adjacent shells is approximately equal to 3.4 Å, slightly different to the distance between two [220] graphitic planes (3.35 Å) [7]. The innermost shell usually consists of a C_{60} fullerene, but hollow and metallic cores have been also reported [8].

Ideally, the carbon atoms which compose the graphitic outer layers increase following the formula $60n^2$, where n is the number of layers [10]. However, the real shape is typically far from the theoretical one. The presence of amorphous or sp^3-hybridized carbon atoms induces defectiveness in the structure and thus CNOs usually exhibit a quasi-spherical or polyhedral shape. Furthermore, the size, which typically ranges from few to tens of nanometers, and the number of concentric graphitic layers depend on the synthetic procedure used for their production [8].

CNOs were first reported by S. Iijima in 1980; he noted with a high-resolution transmission electron microscope (HRTEM) the presence of onion-like structured carbon in the amorphous carbon films, prepared by vacuum evaporation method at 10^{-6} Torr and 4000 K [11]. However, only a decade later, CNOs started obtaining attention thanks to D. Ugarte.

In 1992, by irradiating amorphous carbon with an intense electron beam, he observed the in situ formation of concentric spherical graphitic particles with well-distinct nucleation centers. The spherical particles formed through an irradiation-stimulated graphitization mechanism, which induced the graphitization and the following curling of the amorphous carbon, exhibited a diameter of about 45 nm and about 70 graphitic shells with an interlayer distance close to that of bulk graphite [9, 12].

In the perspective to obtain a large-scale fabrication method, a large number of synthetic strategies for the production of CNOs were developed in the last years, including arc discharge [13–15], thermal annealing of detonation nanodiamonds (DNDs) in vacuum [16, 17] and inert atmosphere (i.e., argon [18], nitrogen [19], hydrogen [20] helium [21, 22], chemical vapor deposition [23, 24], ball-milling [25, 26], ion implantation [27, 28], plasma treatment [29], and laser ablation [30]. The as-mentioned producing methods are discussed in detail in the next paragraph.

2 Synthesis Methods of Carbon Nano-Onions

Since Ugarte reported a reproducible way to obtain CNOs [12], an intense scientific research was performed by many groups with the intent to develop a scalable method for the production of CNOs with controllable size and shape, with high-yield and competitive production cost.

After Ugarte's experiment [12], the production of CNOs by electron beam irradiation was investigated by many research groups. Since 1995, Xu et al. conducted a series of experiments involving electron beam irradiation of amorphous carbon, demonstrating the formation of giant CNOs from the conversion of amorphous carbon film under Al nanoparticles (Fig. 2) [31].

The formation process is divided into three stages: in the first, the nucleus and the first flake of fullerene are formed under the Al nanoparticle. In the second step, the fullerene flakes grow into irregular ellipsoidal shells with a large hollow interior; finally, in the final stage, the formation of quasi-spherical shells is observed by eliminating the inner hollow. The Al nanoparticles act as nucleation sites promoting the formation of CNOs with 10–54 graphitic shells and diameter ranging from 4 to about 38 nm [31].

Thermal annealing of detonation nanodiamonds has emerged as the common method for the synthesis of small carbon nano-onions, due to the large number of advantages, such as high yield, high purity, narrow size distribution, and low cost; so far, this is the only technique that has the potential for industrial application due to the yield close to 100%. DNDs with average diameter of 5 nm are typically used as precursor, producing homogeneous, and small carbon nano-onions (typically below 10 nm) approximately consisting of 5–8 carbon shells (Fig. 3).

The high temperatures reached during the annealing process help overcome the kinetic barrier associated with graphitization. Thermodynamically, the most stable form of carbon under normal conditions is the graphite and for this reason, the formation of planar graphite should be promoted with respect to a multi-shell structure, as in the case of carbon nano-onions. This can be explained by the increase of the

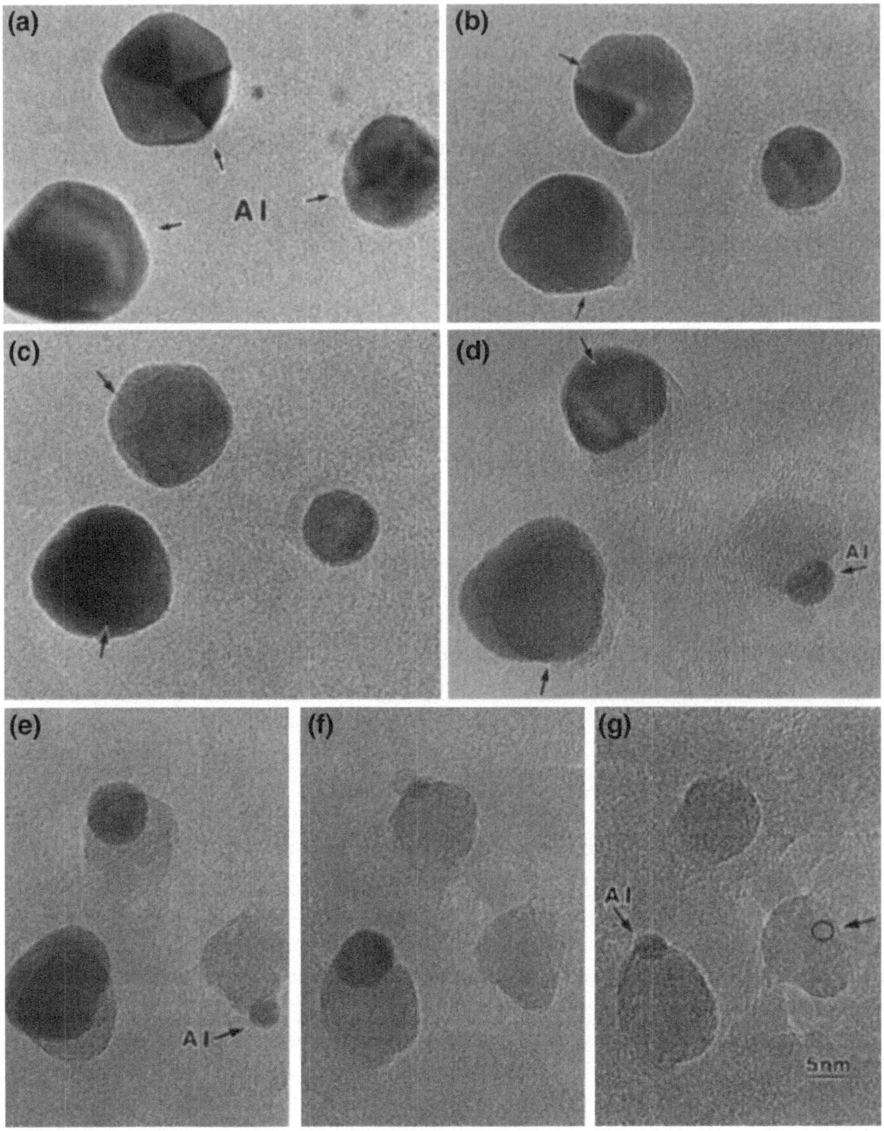

Fig. 2 HRTEM images of the formation of CNOs under Al nanoparticles induced by electron irradiation at different time: **a** 0 s, **b** 540 s, **c** 900 s, **d** 1260 s, **e** 1800, **f** 2100 s and **g** 2460 s. Reprinted with permission from [31]. Copyright 1998 Elsevier Ltd.

contribution of the surface energy of the particle to the total energy in the nanoscale. The curling of graphitic sheets to form curved structure provides to reduce significantly the surface energy of the nanoparticles, removing the dangling bonds present at the edges of the carbon atoms in the graphitic sheets [32].

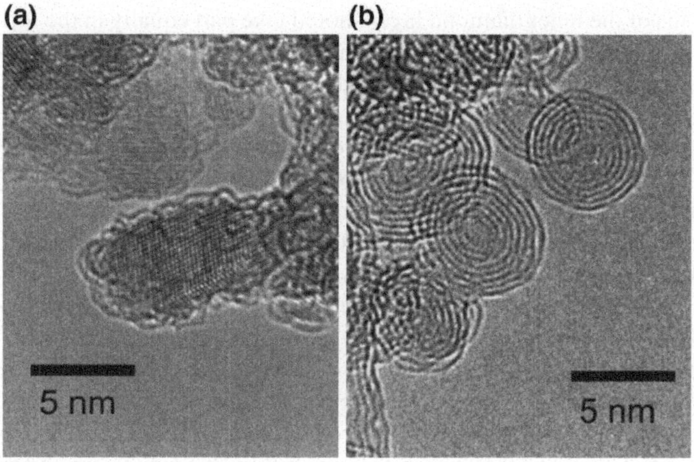

Fig. 3 HRTEM images of **a** diamond nanoparticles and **b** spherical carbon onions. Adapted with permission from [17]. Copyright 2001 American Institute of Physics

Fig. 4 Zipper-like scheme proposed for the formation of CNOs from nanodiamonds. Reprinted with permission from [33]. Copyright 1999 American Institute of Physics

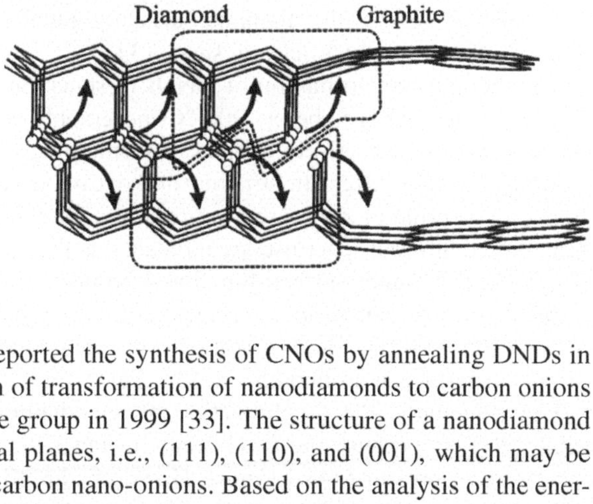

In 1994, Kuznetsov first reported the synthesis of CNOs by annealing DNDs in vacuum [16]. The mechanism of transformation of nanodiamonds to carbon onions has been reported by the same group in 1999 [33]. The structure of a nanodiamond contains three different crystal planes, i.e., (111), (110), and (001), which may be involved in the formation of carbon nano-onions. Based on the analysis of the energetics of the exfoliation of graphite sheets, the transformation of the (001) diamond planes to form the (002) graphitic planes is more unlikely than that of (111) and (110) planes due to the more atomic rearrangement required to form graphitic planes and the much higher cleavage energy.

Based on standard semiempirical method (MNDO) calculations of two-layer cluster models and physical arguments, the (111) diamond planes transform to the (002) graphitic planes with higher rate than that of diamond (110) planes; therefore, the preferential exfoliation of the (111) diamond plane is observed.

Based on these results, they proposed a zipper-like scheme (Fig. 4) in which three (111) diamond planes transform into two (002) graphitic sheets as the graphite–diamond interface migrates into the bulk of the diamond crystal.

The atoms of the inner diamond layer should take part equally in the formation of both graphite sheets, ensuring the same length for the two growing graphitic sheets. Each of the marked groups of atoms provides for the formation of an additional row of graphite sixfold rings. The process probably proceeds via successive insertion of two carbon atoms per two sixfold ring of a growing sheet with the intermediate formation and subsequent reconstruction of eightfold rings [33].

The formation process of carbon nano-onions includes the formation of graphite fragments, the connection and the curvature of graphite sheets between diamond (111) planes and closure of graphite layers.

Graphitization preferentially begins at diamond (111) planes, which are composed of zigzag hexagonal rings that can be easily rearranged into graphite sheet; therefore, graphite fragments with different numbers of carbon atoms exfoliate from the external surface of any diamond (111) plane and surround diamond particles.

These fragments are rearranged by introducing pentagonal and other polygonal rings to form a closed shell. The graphite fragments link and tangle around the surface of the diamond particle to eliminate dangling bonds, and then generate closed graphite shells to diminish surface energy. The inner diamond maintains the original shape and dwindles little by little in the course of transformation. Consequently, graphitic layers enclose around the diamond surface gradually and carbon nano-onions are similar to the original particles in shape [34].

The thermal transformation of DNDs to spherical CNOs is a multistep process (Fig. 5). At temperature below 900 °C, no graphitization occurs and carbon nano-onions cannot be effectively formed. Annealing performed below this temperature does not noticeably affect the structure and size of the nanodiamonds and results in the thermo-desorption of non-carbon impurities. Thermal desorption data and Fourier transformed infrared spectroscopy indicate that the desorption of water, detachment of oxygen-containing surface functional groups, and the decomposition of acidic groups (carboxyl, anhydride, and lactone groups) yielding CO_2 and CO gases are completed at 900 °C. The last non-carbon contaminant (hydrogen) leaves the nanodiamond surface at around 850 °C [35, 36].

Experimentally, the graphitization of nanodiamonds should occur at least at 1100–1200 °C because the C–C bonds in inner diamond require enough kinetic energy to be broken (348 kJ/mol). However, due to the presence of defects and higher kinetic energy of the surface with respect to the inner part, graphitic fragments are already formed at 900 °C and increase with increasing annealing temperature, gradually connecting up into curved graphitic sheets.

At temperatures between 900 and 1100 °C, carbon nano-onions are formed but still a small amount of nanodiamonds can be detected within the core of the particles, suggesting that graphitization of nanodiamonds begins from the particle surface toward the center. Only above 1400 °C, nanodiamonds are completely transformed into carbon nano-onions. The main product obtained by this synthetic method is a perfectly spherical multilayered structure with graphitic shells that ranges from several to 12 [34].

Parameters such as annealing atmosphere, temperature and holding time are of critical importance when thermal annealing is selected as synthesis method, because

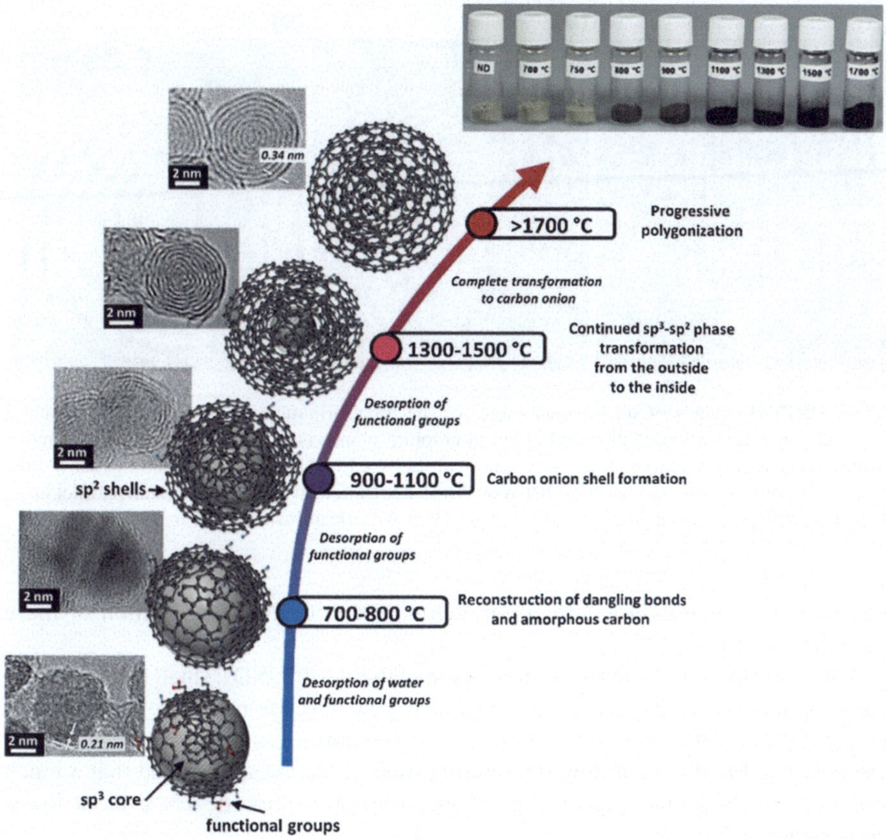

Fig. 5 Schematic representation of the transformation from nanodiamonds to carbon onions by annealing. Reprinted with permission from [8]. Copyright 2016 The Royal Society of Chemistry

from these factors depend shape and number of graphitic layers, and thus the physicochemical properties [37]. The synthesis atmosphere is a very important factor when producing CNOs. In vacuum, thermal annealing of DNDs produces mainly spherical CNOs at temperature higher than 1600 °C, while at temperature higher than 1900 °C, the main product is composed of polyhedral CNOs [38].

In an inert atmosphere such as argon, spherical CNOs are produced at 1500 °C and increasing the annealing time leads to a growing polygonization of the particles and the development of ordered graphene-like sheet structures [39]. Furthermore, the atmosphere in which NDs are annealed is important also in terms of purity of the product. It has been experimentally demonstrated that annealing in vacuum produces mainly CNOs, while, in argon, few-layer graphene nanoribbons are formed as by-products. This is due to the gaseous carbon oxides formed during the decomposition of the surface functional groups during the annealing; these gases create a partial

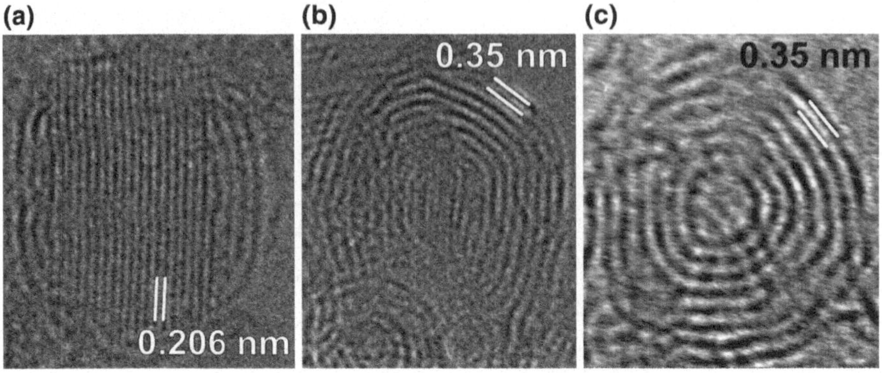

Fig. 6 HRTEM images of the various stages of the transformation of NDs to CNOs. **a** Initial stage: The transformation of diamond [111] to graphite planes is evident. **b** Intermediate stage: a diamond core is seen within nested nearly spherical shells. **c** Final stage: carbon onion with a hollow interior. The numbers are the spacings between the planes indicated by the highlighted parallel lines. Reprinted with permission from [33]. Copyright 1999 American Institute of Physics

pressure which causes a local carbon etching resulting in the formation of these graphitic structures [40].

Another important factor is the dimension of the starting NDs: small nanodiamond particles undergo to the graphitization process at lower temperature compared to nanoparticles with bigger size due to thermodynamic reason (i.e., higher surface free energy). Finally, regarding the holding time, it has been observed that a much prolonged heating time leads to an enhanced graphitization degree even at lower temperature [37].

In Fig. 6, the main stages of the transformation from nanodiamond to carbon nano-onions are shown. Figure 6a shows the presence of graphitic sheets on three sides of the nanodiamond, confirming that the exfoliation occurs preferentially from the diamond (111) surfaces. The exfoliated planes tend to merge with the nanodiamond surface and sometimes with each other to avoid the formation of dangling bonds. Figure 6b, c shows structures at more advanced stages of graphitization. In Fig. 6b, a residual nanodiamond core is still observed in the concentric spherical structure. Figure 6c illustrates a later stage of the transformation in which a carbon nano-onion is formed, enclosing a hollow fullerene-like sphere [33].

Arc discharge is a traditional method for the synthesis of fullerenes [41] and thus suitable for CNOs. The main drawback of this method is the inevitable production of carbonaceous impurities, such as CNTs and amorphous carbon, together with the desired product, requiring thus purification processes.

In 2001, Sano et al. reported the large-scale synthesis of high-purity spherical carbon nano-onions by an arc discharge in deionized water [13], providing high yield without the necessity of a vacuum equipment; the apparatus used for the synthetic procedure is reported in Fig. 7a. Under a constant and relatively low discharge current of 30 A, a discharge voltage of 16–17 V was applied between two graphite electrodes

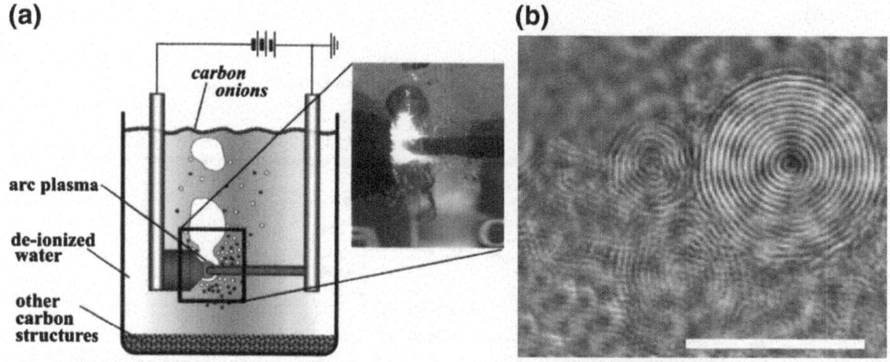

Fig. 7 **a** Schematic of the apparatus used for arc discharge in water with a digital image of the discharge. Reprinted with permission from [14]. Copyright 2002 American Institute of Physics; **b** HRTEM image of carbon nano-onions floating on the water surface after their production. Scale bars, 10 nm. Reprinted with permission from [13]. Copyright 2001 Macmillan Magazines Ltd.

submerged in deionized water. The as-synthetized carbon nano-onions (Fig. 7b), with average size of 25–30 nm and composed of 15–30 shells, were recovered as floating powder on the water surface, separated from the other products through natural segregation, thus ensuring high-purity material [13].

The same group proposed a model to explain the mechanism of formation of carbon nano-onions in a water arc [14], schematically shown in Fig. 8.

Between the two electrodes, a hot plasma zone is established; this produces the vaporization of the surrounding liquid, forming a gas bubble, which surrounds the plasma zone. At the gas–liquid interface, the reaction of C atomic vapor and H_2O produces CO and H_2 as given below

$$C + H_2O \rightarrow CO + H_2$$

The extremely sharp temperature gradient in the gas bubble from the hot plasma region ($T \approx 4000$ K) to the gas–water interface ($T = 373$ K) causes the rapid solidification of the vaporized carbon; this condenses when in contact with the surrounding cold aqueous environment, producing carbon nano-onions.

The type of product formed depends on which part of the cold zone occurs the quenching of the carbon vapor. In the zone (I), where the quenching occurs within the ion current adjacent to the hot plasma zone, elongated structures such as CNTs are preferentially produced due to the epitaxial growth which occurs in this zone. In the zone (II), instead, a three-dimensional (3D) isotropic growth occurs because of the absence of an axis of symmetry, thus producing CNOs [14].

Recently, Borgohain et al. reported the synthesis of carbon nano-onions prepared via optimized underwater arc discharge [15]. Under new experimental conditions, they produced high-quality carbon nano-onions (20–25 nm in size) with minimal impurities and narrow polydispersity (Fig. 9a). Furthermore, they developed an

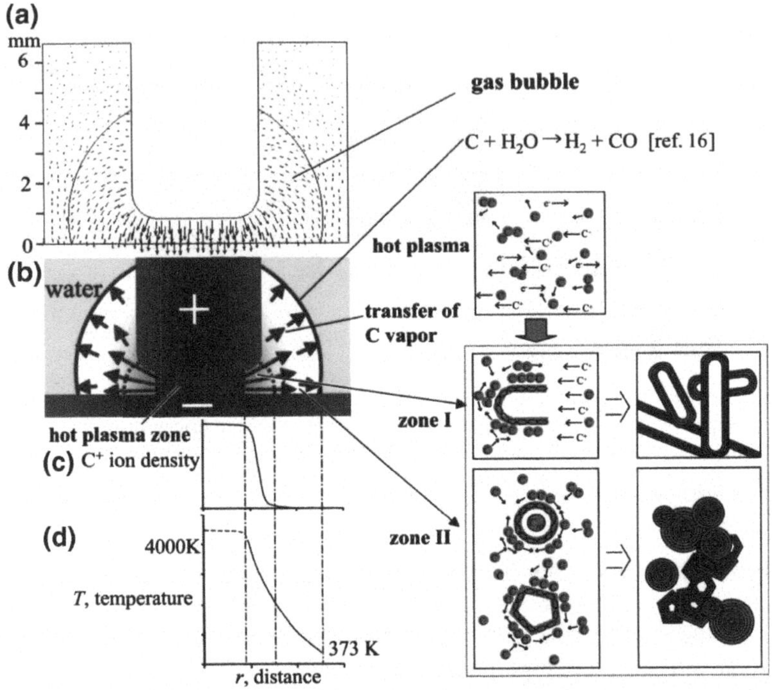

Fig. 8 Proposed formation mechanism of CNOs in a water arc. **a** Relative electric field strength between a rod anode (17 V) and a flat cathode (ground) in a gas bubble surrounded by water. **b** Direction of thermal expansion from plasma to the water interface. **c** Qualitative ion density distribution. **d** Temperature gradient. The formation of elongated nanoparticles in zone (I) and onions in zone (II) is also shown schematically. Reprinted with permission from [14]. Copyright 2002 American Institute of Physics

efficient purification method to screen out carbonaceous and metal impurities from carbon nano-onions (Fig. 9b) [15].

Chemical vapor deposition (CVD) is a convenient method for producing CNOs due to the low cost, high yield, and direct control on the synthesis. In a typical CVD process, hydrocarbons such as methane are decomposed with the aid of a catalyst to form both hollow and metal-encapsulated CNOs [42, 43].

In 2006, He et al. reported the production of carbon nano-onions via chemical vapor deposition (CVD) at a low temperature (600 °C) by the catalytic decomposition of CH_4 on a Ni/Al catalyst [24]. The carbon onions consist of several concentric carbon layers surrounding either a hollow core or a nickel particle with sizes ranging from 5 to 50 nm (Fig. 10). The shape of the carbon onions depends on the shape of the nickel particle: when the nickel particle has a quasi-spherical structure, the graphite layers roll around the spherical particle and form a similarly spherical carbon onion; otherwise, when the nickel particle exhibits a polyhedral shape, this results in a polyhedral carbon onion. They suggest that the formation of carbon nano-onions

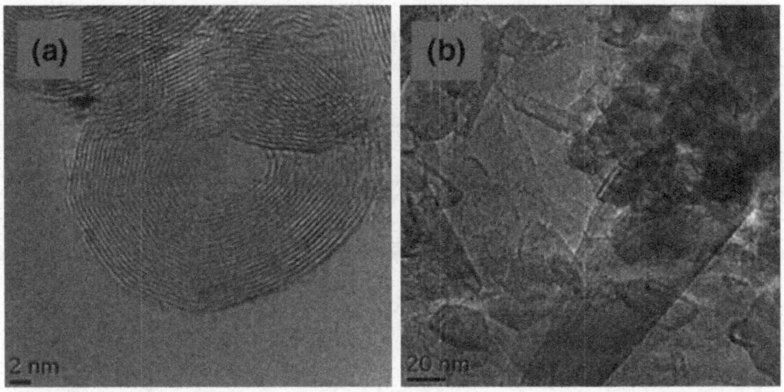

Fig. 9 HRTEM images of **a** purified CNOs and **b** carbonaceous impurities. Adapted from [15]. Copyright 2013 Elsevier Ltd.

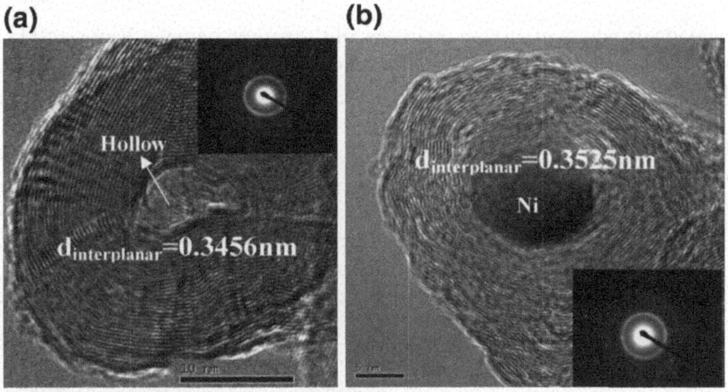

Fig. 10 High-resolution TEM images of **a** hollow and **b** Ni-encapsulated carbon onions. Adapted from [24]. Copyright 2005 Elsevier Ltd.

occurred directly from the gas phase and the Ni particles acted as catalyst, inducing curvature in graphite layers, remaining trapped in the carbon shells [24].

He et al. proposed a mechanism for the formation of metal-filled and hollow carbon onions (Fig. 11) [44]. In the first case (Fig. 11a), the growth processes involve the adsorption of carbon vapor on the nickel nanoparticle, followed by the formation of graphitic layers around its surface (VS mechanism). In the case of CNOs with an empty core (Fig. 11b), the process is the same but because of the nanometer-size of the particles, the evaporation of small nickel nanoparticles occurs, forming CNOs with a hollow core (VL mechanism) [44].

Chen et al. reported the high-yield synthesis of high-purity carbon nano-onions via radio frequency plasma-enhanced chemical vapor deposition (PE-CVD) by using a mixture of CH_4 and H_2 and a cobalt-based catalyst [23]. The product contains only spherical nanoparticles with size ranging from a few to several tens of nanometers (Fig. 12).

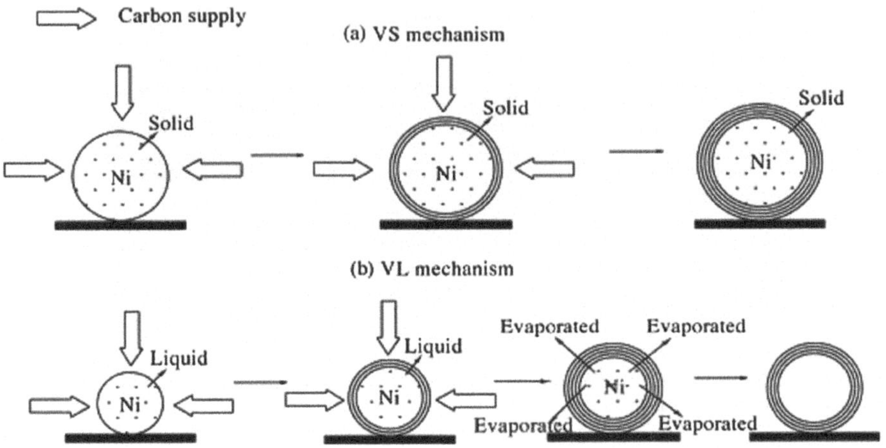

Fig. 11 Schematic depiction of growth processes of the carbon onion. **a** Vapor–solid (VS) mechanism and **b** vapor–liquid (VL) mechanism. Reprinted with permission from [44]. Copyright 2005 Elsevier Ltd.

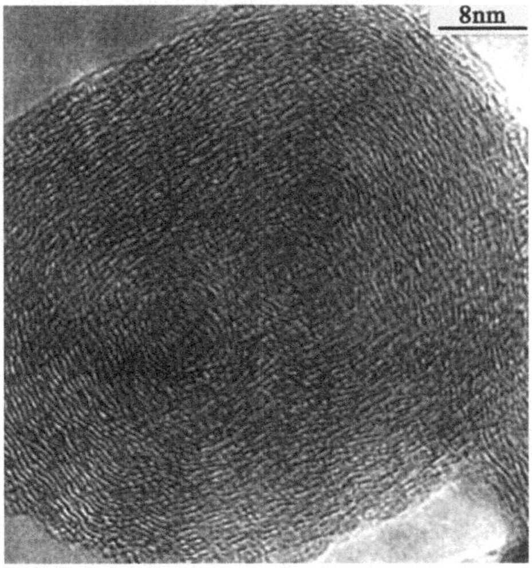

Fig. 12 HRTEM image of a typical onion formed on catalyst, containing several cores on its center. Reprinted with permission from [23]. Copyright 2001 Elsevier Ltd.

The low temperatures required in the radio frequency plasma-enhanced CVD method proved that CNOs could be produced without the help of high-energy levels.

In the process, the formation of many cages in successive stages from the core to the surface is essential. The nucleus is a dome formed by one part of a C_{60} molecule and its growth by addition of hexagons and pentagons leads to the formation of a first cage. This first cage acts as a substrate on which further layers may nucleate and grow into a second, third, etc., spherical cage.

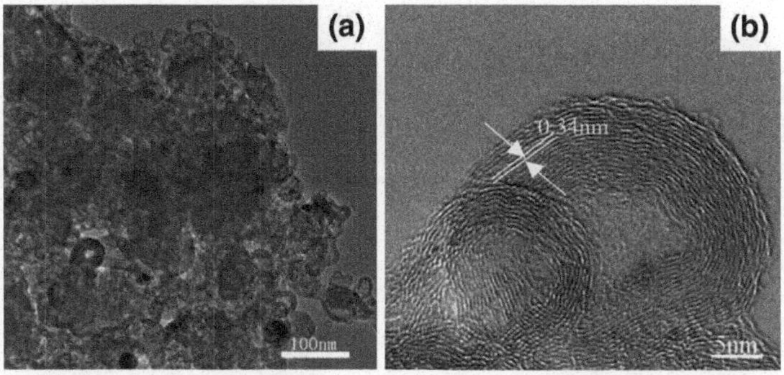

Fig. 13 TEM images of the products from coal: **a** overall views and **b** the partially magnified HRTEM images of (**a**). Reprinted with permission from [29]. Copyright 2006 Elsevier Ltd.

With the increase of cage surface area, the new secondary surface may involve pentagons, hexagons, and heptagons, which deposit with various degrees of epitaxial coherence, resulting in the formation of wavy carbon sheets.

In the classical pyrolytic CVD process, the hydrocarbon is adsorbed and decomposed on the front of the metal particles and then the carbon layers form on the back of the metal particles, resulting encapsulated by carbon layers.

In the case of radio frequency plasma-enhanced CVD, the hydrocarbon is adsorbed at the surfaces of the catalytic particles and, in the meantime, the process of glow discharge leads to CH_4 decomposition, resulting in the formation of carbon rings, which tend to nucleate on the surfaces of the catalyst. Furthermore, in the presence of atomic hydrogen in the plasma, free carbon valences on the ends of carbon layers can be stabilized by the formation of C–H bonds.

The product obtained by this method is pure and can be easily separated from the catalytic particles, which are not encapsulated in the nanoparticles [23].

Radio frequency plasma is another method for the production of CNOs due to the low cost of the precursors (typically coal and carbon black).

Du et al. reported the synthesis of carbon nano-onions in high yields from coal and a mixture of coal and graphite by radio frequency plasma [29]. In both cases, the particles display a clear polyhedral or quasi-spherical morphology with a hollow center, having an average diameter ranging from 10 to 35 nm and high purity (Fig. 13).

The particles synthesized from coal and graphite display higher degree of graphitization, more regular morphology, smoother and cleaner surfaces, whereas coal-derived particles show a relatively low degree of graphitization, and the graphite sheets are wavy. They proposed that graphite sheets act as the building blocks during the growth: over ten thousand carbon atoms were needed only in the innermost shell and enough regular aromatic fragments cannot be provided only by the decomposition of coal within the extremely short time, but the graphite added to the coal sample can fulfill the work and help to form regular cage structure.

Furthermore, the growth of the carbon nano-onions is supposed to begin at the innermost shell and progress toward the surface [29].

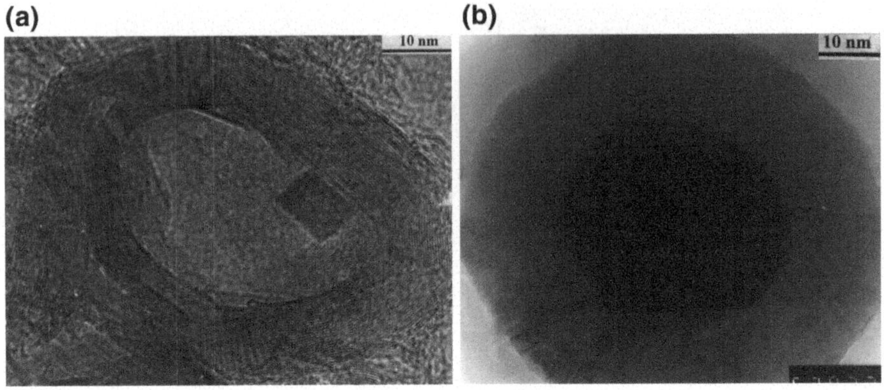

Fig. 14 HRTEM images of the closed-shells nanostructures with **a** a hollow core, **b** an inner metal particle. Reprinted with permission from [25]. Copyright 2000 Elsevier Ltd.

In 1999, Huang et al. reported the synthesis of highly curled carbon nanostructures during high-energy ball-milling of graphite. The direct bending of the flat sp^2 sheets under the heavy mechanical deformation was due to the introduction of strain or lattice defects in the structure, with bending that fall in the range of a few degrees to less than 180° [26].

The following year, Chen et al. proposed a two-step model for the as-mentioned production method [25]. In the first step, the flat graphite sheets are bent or curved by ball-milling and many defects and dangling bonds are present in the structure.

At a later time, when the ends of the two curved graphite sheets are very close to each other, the high-energy ball-milling provides enough energy to graphitize for imperfect graphitic structure and to allow closure of the network.

Anyway, different carbon nanostructures are produced via ball-milling, including nanoparticles which consist of an inner metal core and outer graphitic shells, making this method useless to produce high-purity carbon nano-onions (Fig. 14) [25].

In 1995, Cabioc'h et al. reported, for the first time, the synthesis of carbon nano-onions via carbon ion implantation onto copper substrates [45] and in the following years, the same groups published different work on this topic.

In 2000, they reported the carbon ion implantation performed onto copper and silver substrates and demonstrated the potential of the technique in adjusting the size and controlling the microstructure of the carbon onions.

By varying synthesis parameters such as temperature and ion flux, they obtained particles with size that ranges from 3 up to 30 nm (Fig. 15). Furthermore, they discussed the nucleation and growth mechanisms of the carbon onions during the ion implantation process and concluded that the formation of carbon nano-onions occurred inside the volume of the metallic substrate silver or copper due to the immiscibility of carbon in the metal. However, this method showed the disadvantage of the presence of heteroatoms in the structure [27].

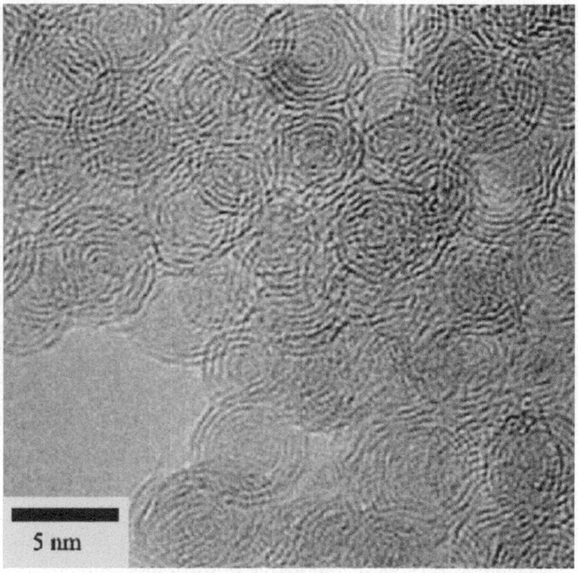

Fig. 15 HRTEM micrograph of carbon onions synthesized by ion implantation in a silver thin film deposited on a silica substrate. Reprinted with permission from [27]. Copyright 2000 Elsevier Ltd.

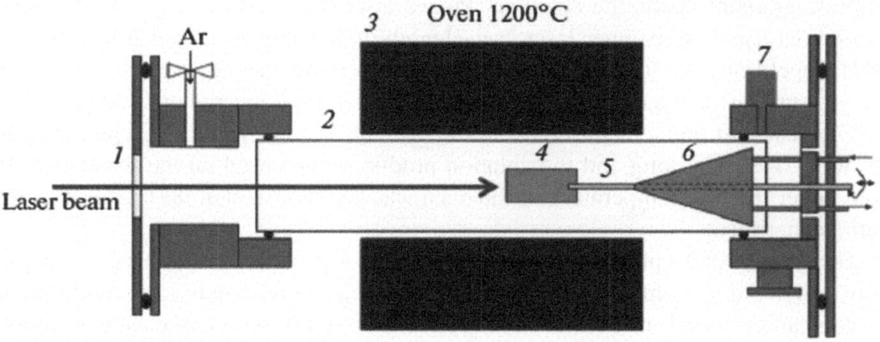

Fig. 16 Schematics of the laser ablation chamber design for KrF excimer laser. Chamber schematics: 1—quartz laser window; 2—quartz tube; 3—electrical oven; 4—target; 5—graphite transfer rod; 6—cold finger; 7—vacuum gauge. Reprinted with permission from [30]. Copyright 2014 Allerton Press, Inc.

In 2014, Dorobantu et al. reported the production of carbon nano-onions by laser ablation of commercial pure graphite targets [30]; the schematization of the chamber used is shown in Fig. 16.

The laser ablation chamber consists of a quartz tube (2), 60 mm in diameter, mounted inside a hinged tube furnace (3). The temperature of the furnace can be varied from 30 °C up to 1200 °C and the quartz tube is O-ring sealed to ensure pressure control from 10^{-3} Torr up to atmospheric pressure. The ablation gas is entering the chamber just after the quartz window (1), controlled by a flowmeter (7).

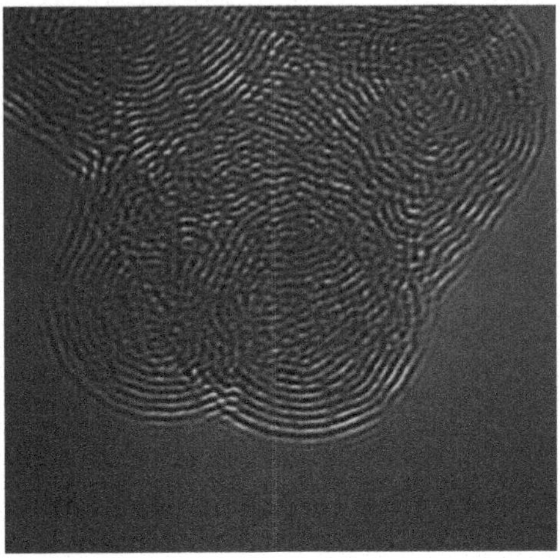

Fig. 17 HRTEM image of clustered CNOs. Reprinted with permission from [30]. Copyright 2014 Allerton Press, Inc.

The flow can be varied from 0 to 300 L/h and the length of the oven has been increased to 675 mm, ensuring a quite constant temperature zone or a smaller temperature gradient inside the chamber. Pulsed laser vaporization experiments were carried out using an excimer laser (wavelength of 248 nm, 25 ns pulse length, and 10 Hz repetition rate) focused on a pure graphite commercial target surface, keeping the pressure inside the reactor at 7 Torr, while the argon flow at about 300 L/h.

The observed plume resulting from the laser interaction with the graphite target (4) was 20–30 mm long and the ablation products condensed on the water-cooled cold finger (6). The temperature of the oven was kept constant at 900 °C during the entire experiment.

The ablation of the pure graphite target produces well-defined carbon nano-onions, both individual and clustered through an amorphous carbon matrix; no amorphous carbon can be found inside the shell, resulting in highly pure carbon nano-onions (Fig. 17). The as-synthetized carbon nano-onions are spherically shaped and have a concentric shell structure formed by 10–20 shells with a regular and very small (less than 5 nm) hollow core.

The size typically ranges from 10 to 25 nm and the graphitic interlayer distance was found to be 0.35 nm, which is close to the ideal graphitic interlayer spacing [30].

Several other ways to synthetize carbon nano-onions were published so far. Bystzejewski et al. reported the catalyst-free synthesis of carbon nano-onions by thermolysis of a sodium azide (NaN_3) and hexachlorobenzene (C_6Cl_6) mixture. The production was performed in a modified stainless steel calorimetric bomb under argon or air atmosphere where the reagents, after mixing in a mechanical shaker, were loaded in a quartz crucible inside the chamber. The by-products were completely removed using a simple purification process. This method allowed the production of carbon nano-onions with an average size between 30 and 100 nm along with impurities such as amorphous carbon nanoparticles and NaCl, removed after a purification step. The formation of carbon nano-onions was likely caused by a shock wave, a rapid increase of pressure, during thermolysis, which induced the coalescence of phenyl radicals [46].

In 2012, Choucair et al. reported the gram-scale synthesis of pure carbon nano-onions by flash pyrolysis of a polyaromatic hydrocarbon (naphthalene) vapor onto a glass or ceramic substrate. The product was then purified by heating at mild temperatures (ca. 200 °C) under vacuum, resulting in large quantities of carbon nano-onions with an average size of about 50 nm and composed of 50–54 shells, free of both impurities and other carbon allotropes [47].

In another work, Du et al. reported the production of carbon nano-onions with a size ranging from 60 to 90 nm via thermal reduction of a mixture of glycerin and magnesium at 650 °C in a stainless steel autoclave [48].

In 2003, Hou et al. reported the formation of carbon nano-onions under shock condition by using SiC as a source of carbon yielding nanoparticles with 10–30 nm in average size and several to 20 graphitic layers [49].

In 2007, Yan et al. reported the production at low temperature (below 200 °C) of graphitic carbon nanostructures by solution-phase synthesis methods (solvothermal synthesis and hot injection) by using commercial mesophase pitches (MPs) as a carbon precursor. This process allowed the production of carbon nano-onions with diameters of 30–80 nm along with small carbon nanoparticles (3–6 nm in diameter) and carbon nanopores (with a length of hundreds of nm and a width of 3–20 nm) [50].

Ghosh et al. reported the production of CNOs by pyrolysis of wood wool at 600 °C in a muffle furnace under a flow of nitrogen/oxygen (95:5) gas mixture for 2 h [51–53].

Dubey et al. reported the gram-scale synthesis of CNOs by burning common carbonaceous sources such as camphor and polystyrene foam in the presence of insufficient amounts of oxygen in the atmosphere [54].

Another way to produce carbon nano-onions is by counterflow diffusion flame synthesis onto a catalytic Ni substrate, published by Hou and coworkers in 2008. In details, they focused on the influence of $C_2H_4/CH_4/N_2$ mixed fuel in terms of ethylene and methane concentrations and oxygen concentrations on the synthesis of carbon nano-onions. The results showed that only carbon nano-onions (and no carbon nanotubes) were synthesized with an oxygen concentration of 50%, ethylene concentration of 5% and methane concentration between 15 and 45%. Additionally, they showed that the methane concentration strongly influenced the diameters of carbon nano-onions: the average diameters were increased ranging from 5 to 60 nm when methane concentration was increased from 15 to 45%, opening the possibility to control the yield and size by varying the methane concentration [55].

As clearly highlighted from the above discussion, depending on the production methods, CNOs differ on their morphological characteristics and this lead to nanoparticles that own unique physicochemical properties.

In 2007, Echegoyen et al. reported a comparison between carbon nano-onions synthetized in two different ways in terms of physicochemical properties and reactivity. Small CNOs (average size of 5 nm and composed of 6–8 shells) were produced by thermal annealing of DNDs, while the arching of graphite underwater led to the formation of larger carbon nano-onions (20–30 shells) with size ranging between 15 and 25 nm. Nanodiamond-derived CNOs exhibit smaller radii and consequently higher curvature; this results in a greater reactivity than the larger carbon nano-onions because it is known that fullerene reactivity is partially dictated by the degree of surface strain: the higher the curvature, the greater the reactivity.

Raman spectroscopy suggested a more defective outer structure for the small CNOs, while thermogravimetric analysis (TGA) higher thermal stability. According to this, differences in terms of reactivity and subsequent functionalization are expected. The authors proved that the smaller carbon nano-onions were successfully functionalized by [2 + 1] cycloaddition, free-radical addition, and oxidative reactions, whereas the bigger carbon nano-onions did not react under the same experimental conditions. This demonstrated the higher reactivity of the nanodiamond-derived CNOs due to the larger curvature as well as the higher surface-to-volume ratio [21].

3 Bioimaging Applications of Carbon Nano-onions

All carbon nanomaterials typically display poor solubility in both aqueous and organic solvents, due to their highly hydrophobic nature. The high tendency to form aggregates, promoted by strong intermolecular interactions, such as van der Waals forces, can seriously affect their applicability in different fields. Thus, the introduction of surface functional groups on the surface of CNMs is very important to make them suitable for applications; in particular, the improvement of the solubility and the dispersion properties are important factors for biomedical applications

In the last years, several chemical functionalizations of CNOs introducing different functional groups were reported, including both covalent and non-covalent pathways, schematized in Fig. 18 [56].

The first covalent functionalization of CNOs was published in 2003 by Prato et al. [57]. The authors reported an azomethine ylide addition reaction on CNOs, derived from the raw soot produced by arc discharge, performed in a mixed solution of amino acid and paraformaldehyde refluxing in toluene. ^{1}H NMR spectroscopy, MALDI mass spectroscopy, and UV-vis-NIR spectroscopy confirmed the successful functionalization of CNOs and HRTEM studies revealed an average size between 60 and 300 nm. The modified CNOs displayed enhanced solubility in many organic solvents, such as chloroform and dichloromethane [57]. Since this work, different functionalization pathways were published, including 1,3-dipolar cycloaddition [58], oxidation [21, 58–60], cyclopropanation [21], radical addition [21], fluorination [61], radical addition of diazonium compounds [62], and many others.

The development of several functionalization pathways to incorporate different functional groups and molecules, such as fluorophores onto the surface, led to the possibility to use these CNMs in a wide range of applications including electronics [63], sensing [64], and biology [65–67].

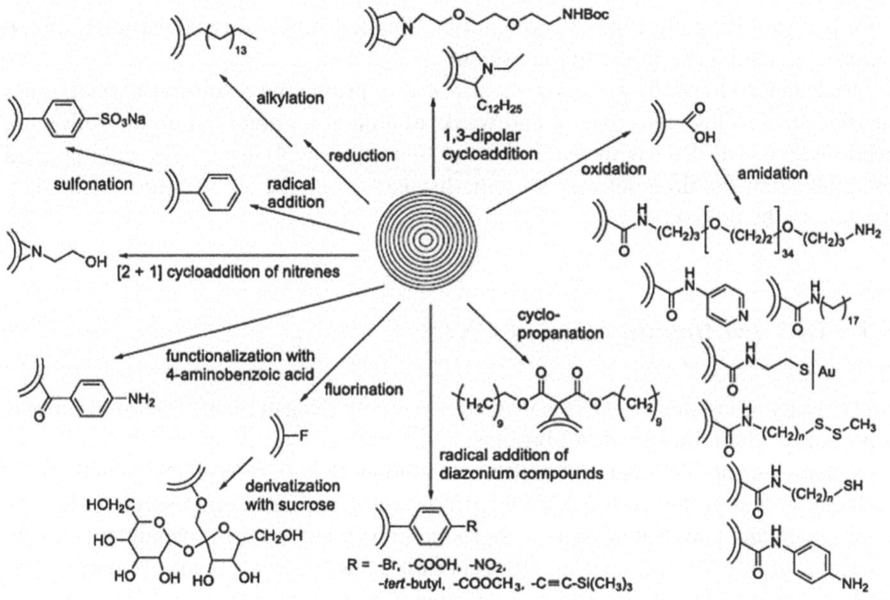

Fig. 18 Covalent functionalization pathways for CNOs. Reprinted with permission from [56]. Copyright 2014 Bartelmess and Giordani; licensee Beilstein-Institut

Among the different CNMs, fullerenes [68, 69], carbon nanotubes (CNTs) [70, 71], carbon dots [72, 73], and graphene [74] have been the subject of the most intense research in the biological field as imaging agents and drug delivery vectors.

Besides their unique physical and chemical properties, CNMs present the advantage of a small size, which ranges from a few nanometers up to a few hundred nanometers. This size is comparable to that of common biological macromolecules, such as proteins, enzymes, and DNA, making CNMs very suitable for biological applications. Furthermore, the possibility to use these nanomaterials as vehicle for specific drugs may pave the way for the development of innovative approaches for drug delivery [66].

CNOs have been successfully applied in several fields including electronics [8, 37, 75, 76], catalysis [77, 78], and tribology [79, 80]. In the last years, many efforts have been made to explore CNOs for biological and biomedical applications [81].

The development of nanosystems capable to carry therapeutic agents for specific targeting holds great potential in nanomedicine, in particular in the treatment of cancer.

In this framework, CNOs present unique physicochemical properties, making them suitable as nanoplatforms for intracellular transport. Their small size allows them to be easily carried in the circulatory system; the high surface area along with the several decoration methods available nowadays allow CNOs to be easily modified on their surface and be functionalized with fluorophores, or multiple ligands, giving the possibility for the targeted transport of therapeutic agents [65].

When exploring this kind of applications, detailed studies on toxicity are required before a potential use in the living system.

To date, CNO-based systems have shown very promising results both in vitro and in vivo, such as low cytotoxicity and weak inflammatory potential; in the following sections, we will discuss in particular the literature involving CNOs as biological imaging agent and drug delivery system, discussing first the in vitro studies and then moving to the in vivo.

3.1 In Vitro Investigation of CNOs

The toxicity assessment of CNOs was carried out by Ding et al. in 2005 in a comparative study where human skin fibroblast cell populations were exposed to different concentrations of CVD-derived multiwall carbon nanotubes (MWCNTs) and multiwall carbon nano-onions (MWCNOs), produced by arc discharge method [82]. In a dose-dependent way, both CNMs were found to be toxic in a dose-dependent way, in the exposed skin fibroblasts (Fig. 19), generating mRNA level changes and thus seriously affecting the cellular functions in maintenance, growth, and differentiation. However, a material-dependent cytotoxicity of MWCNOs and no innate immune responses on the cells were observed [82].

The same MWCNOs, with an average size of around 30 nm, were tested by Xu et al. on human umbilical vein endothelial cells (HUVECs) at different concentrations [83]. At low concentration, MWCNOs showed good cytocompatibility, while, at higher concentrations, significant toxic effect was observed (Fig. 20) suggesting a dose-dependent effect of MWCNOs on cell survival, which leads to the inhibition of the cell growth. Furthermore, MWCNOs showed genotoxic effects and the DNA damage on HUVECs [83].

In 2010, Luszczyn et al. reported the covalent functionalization of small CNOs with biomolecules availing of the avidin-biotin interactions [59]. The cytotoxicity of pristine and functionalized CNOs (oxidized and PEGylated derivatives) was inves-

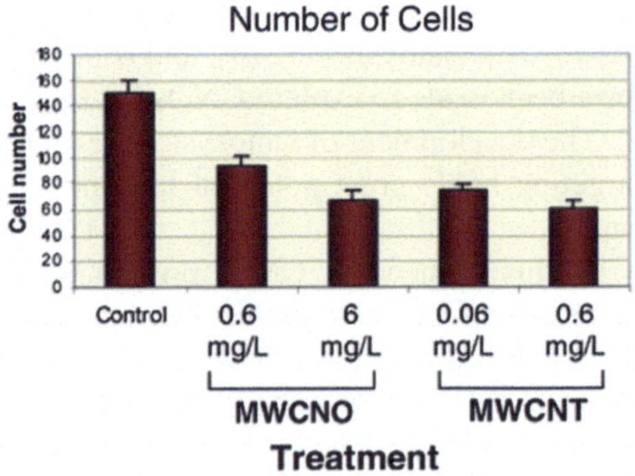

Fig. 19 Cell viability measurements after treatment of skin fibroblast cells with carbon nanomaterials at cytotoxic doses. Adapted with permission from [82]. Copyright 2005 American Chemical Society

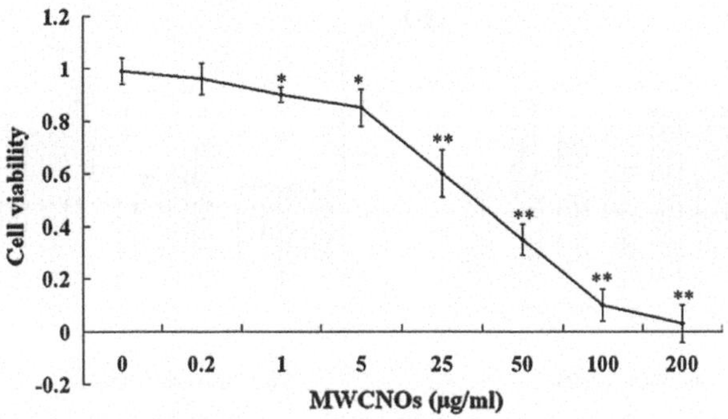

Fig. 20 The cell viabilities of MWCNOs on HUVECs at 24 h treated with various concentrations of MWCNOs. Reprinted with permission from [83]. Copyright 2011 Wiley Periodicals, Inc.

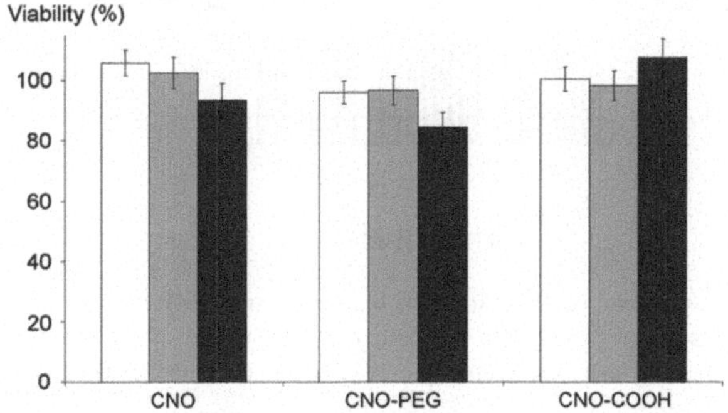

Fig. 21 Viability of fibroblasts after being exposed to three concentrations (white bar: 30 mg mL^{-1}, gray bar: 300 mg mL^{-1}, black bar: 3000 mg mL^{-1}) of pristine CNOs and modified CNOs (CNO-PEG and ox-CNO). Reprinted with permission from [59]. Copyright 2010 Wiley-VCH Verlag 4870 GmbH&Co. KGaA, Weinheim

tigated after exposure of the different CNMs, at three different concentrations, to rat dermal fibroblasts for 4 h. All the tested CNOs did not induce any toxic effects at the lower concentration (almost 100% viability), while just a slight reduction (about 15%) was observed at the highest concentration (Fig. 21). This was the first evidence of the non-cytotoxicity of CNOs synthetized by thermal annealing [59].

Another comparison study of the cellular toxicity between MWCNTs and CNOs was published by Kang et al. in 2015 [84]; in this work, the CNMs were produced by CVD and laser irradiated flame aerosol method, respectively. Both nanomaterials were tested on a human bronchial epithelial cell line, assessing the effects of different parameters, such as morphology, metal catalysts, hydrophilicity, and ROS, on cytotoxicity. CNOs showed higher cytocompatibility (Fig. 22a) compared to MWCNTs. These results were ascribed to the different morphologies of the carbon nanoparticles, suggesting that this parameter plays an important role in these processes. Furthermore, even in the presence of an antioxidant, CNOs displayed a lower generation

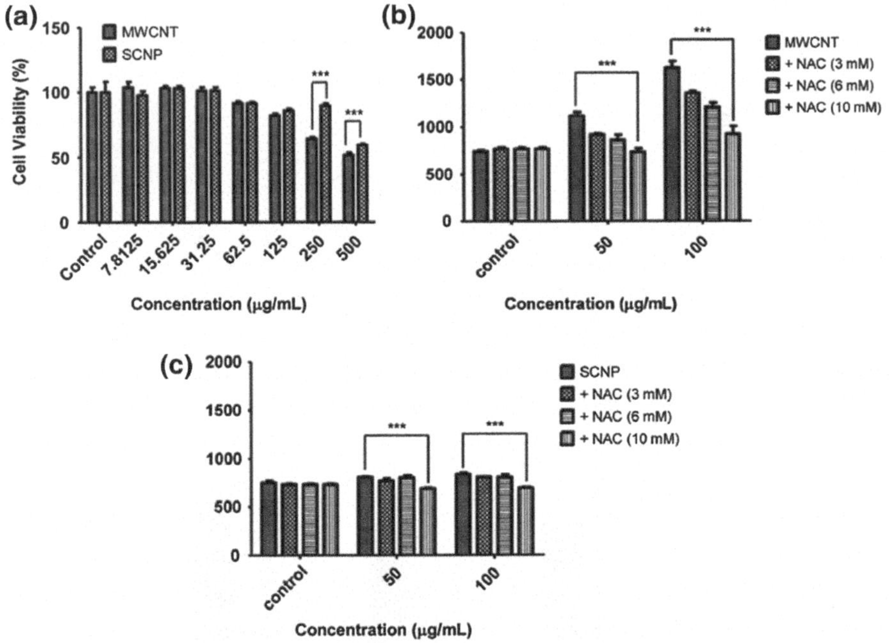

Fig. 22 **a** WST-1 assay on 16HBE14o-cells incubated with MWCNTs and SCNPs for 24 h. ROS formation measurement in 16HBE14o-cells using DCFH-DA for **b** MWCNTs and **c** SCNPs. Adapted with permission from [84]. Copyright 2015 Springer Science+Business Media Dordrecht

of ROS (Fig. 22b) and a higher cell viability than MWCNTs (Fig. 22c), suggesting that the cytotoxicity of MWCNTs was mainly arising from ROS generated by cell–particle interaction [84].

Recently, Tripathi et al. reported the use of CNOs as imaging agent in *Escherichia coli* and *Pseudomonas putida* cells [85]. CNOs were produced via pyrolysis of vegetable ghee and, after an oxidation process, were found to exhibit an intrinsic and tunable photoluminescence, allowing for their use as imaging fluorescent probes (Fig. 23). Furthermore, the same nanoparticles were used for the selective sensing of glucose based on the fluorescence "turn-off/turn-on" mechanism of CNOs, showing promising results in terms of response time along with a good detection limit [85].

In the last 5 years, our groups reported several examples of CNO-based fluorescent nanoprobes for biological imaging and drug delivery; CNOs tested in our works were synthetized by thermal annealing of DNDs. Different surface modification pathways were employed, in order to enhance the solubility and to attach specific functional molecules such as fluorophores, allowing for the development of system capable for both imaging and targeting purposes [81].

In 2013, we investigated the effects of raw, purified and the functionalized single-walled carbon nanotubes (SWCNTs) and CNOs on the inflammatory and NLRP3 inflammasome activation properties [86]. Purified single-walled carbon nanotubes

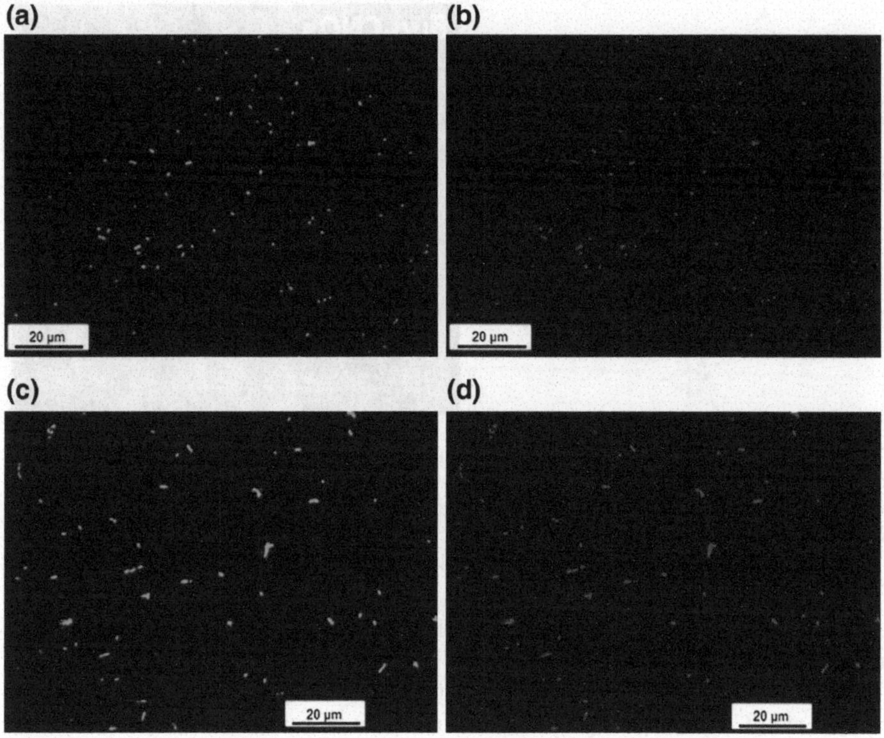

Fig. 23 Fluorescence microscopic images of *E. coli* (**a**) and (**b**) and *P. putida* (**c**) and (**d**) under 488 nm (green) and 562 nm (red) band pass filters, respectively. Reprinted with permission from [85]. Copyright 2016 The Royal Society of Chemistry

(p-SWCNTs) [87] and carbon nano-onions (p-CNOs) [58] were oxidized following the one-pot Tour reaction [62, 88, 89] and then coupled with a fluoresceinamine by amide reaction to yield the fluorescently labeled derivatives (fluo-CNTs/-CNOs). The fluoresceinamine-functionalized derivatives were efficiently internalized by immortalized bone marrow derived mouse macrophages (iBMM) and mouse bone marrow derived dendritic cells (BMDCs), as suggested by the green fluorescence in Fig. 24; furthermore, p-SWCNTs,and p-CNOs did not induce any dendritic cell maturation and any secretion of the inflammatory cytokines.

SWCNTs and CNOs exhibited the ability, in the presence of TLR ligands, to promote the release of the inflammatory cytokines IL-1 β by BMDCs and to induce the activation of the NLRP3 inflammasome; CNOs exhibited an in vitro and in vivo low inflammatory potential. Moreover, compared to SWCNTs, CNOs showed a higher uptake by antigen-presenting cells and were more efficiently targeted to draining lymph nodes. We also demonstrated that the chemical functionalization had an impact on the inflammatory properties, showing that the functionalization signifi-

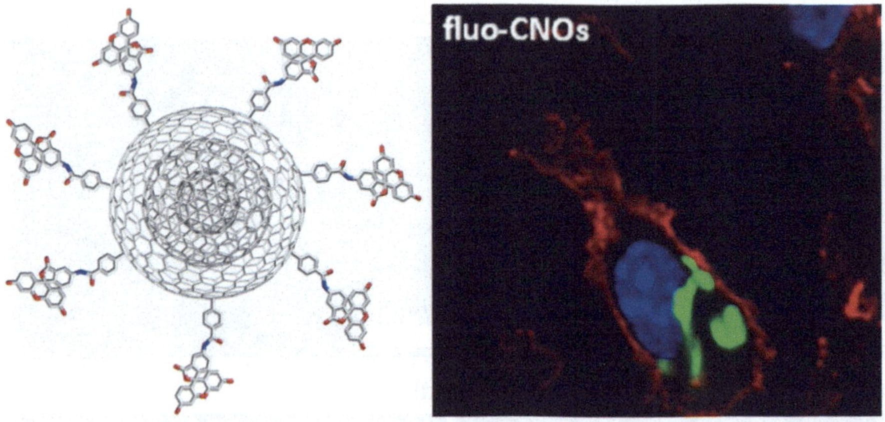

Fig. 24 Fluoresceinamine-functionalized CNOs (fluo-CNOs): structure (left) and confocal image (right) of C57BL/6 BMDCs incubated with fluo-CNOs and stained with wheat germ agglutinin-Alexa Fluor594 (red) and nuclei stain Hoechst (blue). Reprinted with permission from [81]. Copyright 2017 Elsevier Ltd.

cantly reduced the in vitro release of IL-1 β and the in vivo recruitment of innate inflammatory cells [86].

In 2014, we reported the development of a high-resolution imaging platform, composed of CNOs and a highly fluorescent boron dipyrromethene dyes (BODIPY-CNOs), [90]. The boron dipyrromethene (BODIPY) derivative was covalently attached by ester reaction to the surface of benzoic acid CNOs (benz-CNOs), obtained by using the modified *Tour* reaction previously described [62, 88, 89]. The in vitro toxicity of our nanoprobe was tested *on* MCF-7 human breast cancer cells, exposed to different concentrations of CNO derivatives. Cellular viability tests (Fig. 25a, b) did not show any toxic effects up to 72 h of exposure. Confocal live imaging showed an excellent internalization of BODIPY-CNOs inside the treated cells and, furthermore, co-localization experiments, by using a lysotracker probe, demonstrated that BODIPY-CNOs localized in lysosomal vesicles, as suggested by the yellow co-localization signals in Fig. 25c [90].

Another CNO-based fluorescent probe was reported in 2015 by our group, with enhanced water solubility [60]. The chemical oxidation, carried out in a 3M solution of nitric acid, led to the introduction of a large number of carboxylic acid functionalities onto the surface of p-CNOs. The oxidized CNOs (oxi-CNOs) were then coupled with fluoresceinamine dyes to produce fluoresceinamine functionalized CNOs (fluo-CNOs) with high water solubility and good emission properties in cellular media. The cellular viability of the CNOs derivatives was tested on HeLa cells and no significant cytotoxicity was observed for all the tested concentrations (Fig. 26a, b) after 72 h of incubation. Confocal microscopy showed an efficient uptake of fluo-CNOs by HeLa cells (Fig. 26c–e), with an intense green emission coming from the fluorescent CNO derivatives internalized by endocytosis [60].

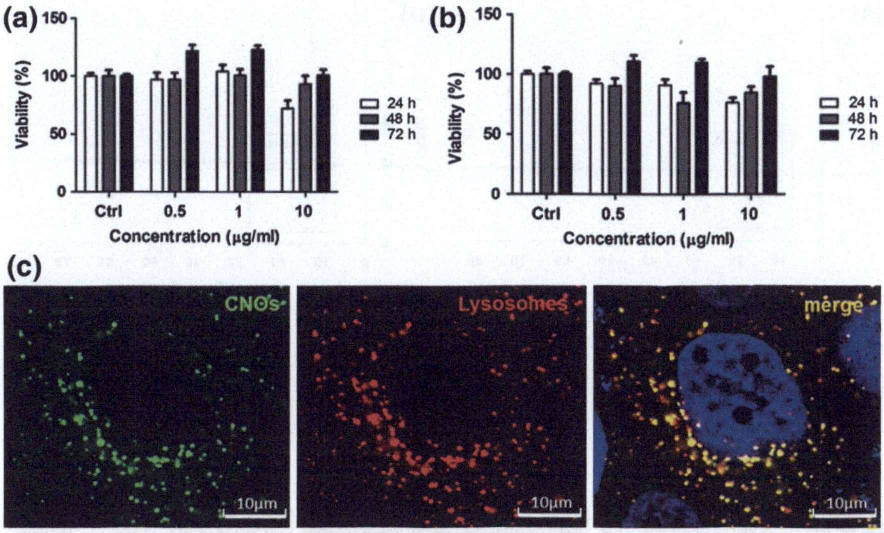

Fig. 25 Cellular viability of MCF-7 cells treated for 24, 48, and 72 h after exposure to different CNOs. **a** with 0.5, 1, and 10 μg mL^{-1} benz-CNOs and **b** with 0.5, 1, and 10 μg mL^{-1} BODIPY-CNOs. **c** Cellular uptake of BODIPY-CNO nanoparticles visualized by confocal microscopy imaging in MCF-7 cells incubated for 48 h with 10 μg mL^{-1} BODIPY-CNOs. Green fluorescent BODIPY-CNOs (left), lysosomes stained with Lysotracker Red probe (middle) and merged images (right). Adapted with permission from [90]. Copyright 2014 The Royal Society of Chemistry

In 2015, we reported the dual functionalization of CNOs with imaging and targeting ligands [91]. Starting from benzoic acid CNO derivatives (CNO-1), a multiple covalent functionalization strategy was employed to attach fluorescein (FITC) and folic acid (FA) units on the surface of CNOs, as imaging and targeting units, respectively. The successful realization of the CNOs conjugates was proved by several characterization techniques such as XPS, ToF-SIMS, TGA, FT-IR, and fluorescence spectroscopies. The cytocompatibility of the different CNOs derivatives, including CNO-1, CNOs with FITC and FA (CNO-4) and CNOs with only FITC (CNO-5), was tested in HeLa and KB cells and the results showed good viability for both cell lines (Fig. 27).

Confocal live cell imaging showed an efficient cellular uptake of the fluorescent CNO derivatives by HeLa cells and their localization in lysosomes. This was furthermore confirmed by correlative imaging analysis (CLEM), which showed the presence of large aggregates of CNOs inside late-endosome/lysosome vesicles (Fig. 28). Finally, in order to establish the effect of the conjugated folic acid on cell targeting, KB cells, which overexpress folate receptors, were incubated with both CNOs conjugates. While confocal images of CNO-5 showed low fluorescence signal, CNO-4 bearing FA was more efficiently internalized by the cells, accumulating in the vesicles and displaying a widely spread emission; this suggested a folate-mediated endocytosis pathway for the uptake of CNOs by cancer cells. This evidence was also confirmed

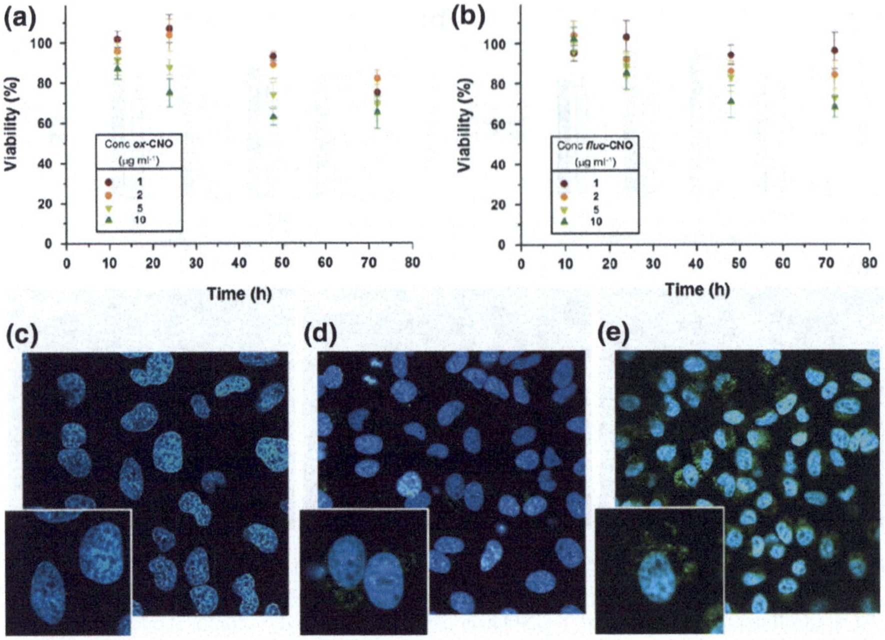

Fig. 26 Cellular viability of HeLa cells after exposure to 1, 2, 5 and 10 μg ml^{-1} ox-CNO (**a**) and fluo-CNO (**b**). Confocal image of HeLa cells after incubation for 24 h with 1 (**c**), 5 (**d**) and, 10 (**e**) μg ml^{-1} of fluo-CNO. Adapted with permission from [60]. Copyright 2015 IOP Publishing Ltd.

by minimal internalization of CNO-4 in the presence of free folic acid in solution, which is the consequence of the competition for the folate receptors on the cell membranes [91].

In 2015, we reported the first fluorescent CNO nanoprobe based on non-covalent interactions between CNOs and a synthesized pyrene–BODIPY molecule (3) [92]. This functionalization approach exploited the π–π stacking interactions between the pyrene unit of 3 and the extended aromatic system on the surface of p-CNOs. The cellular viability on HeLa cells showed low cytotoxicity for the CNOs hybrids at all the concentration tested up to 72 h of incubation (Fig. 29a). p-CNO/3 were efficiently uptaken by cells, as suggested by the strong green fluorescence detected in the perinuclear region (Fig. 29c); furthermore, a low CNOs hybrids internalization was observed when incubating cells at 4 °C (inset of Fig. 29c) [92].

In the visible region, tissue and water induce phenomena of absorption and scattering, leading to the attenuation of the signal. For this reason, in the last years, many researchers have been involved in the development of imaging agents emitting in the near infrared region (NIR, 650–1450 nm). In this "imaging windows", tissue exhibits minimal absorption, resulting in deeper penetration in biological tissues than visible light [66].

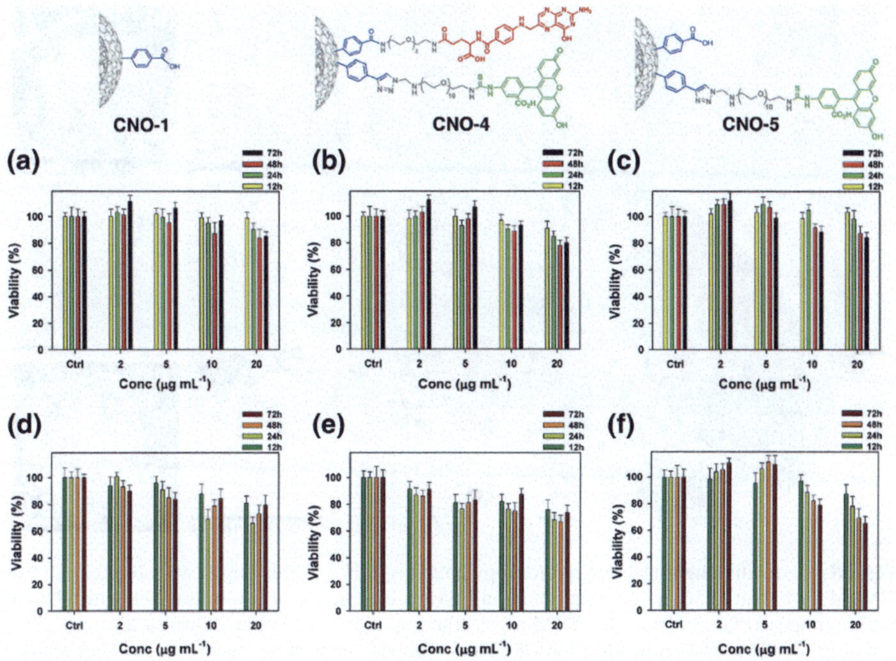

Fig. 27 **a–c** Cellular viability of HeLa cells, and **d–f** KB cells after exposure to 2, 5, 10 and 20 mg mL^{-1} of CNO-1 (**a**, **d**), CNO-4 (**b**, **e**) and CNO-5 (**c**, **f**). Adapted with permission from [91]. Copyright 2015 Wiley-VCH Verlag GmbH & Co. KGaA, Weinheim

With the aim to increase the imaging properties of CNO-based fluorescent probes, we reported, in 2014, the in vitro investigation on HeLa cells of an imaging system based on NIR active boron difluoride azadipyrromethene methoxy-substituted-functionalized CNOs (CNO-1a) [93]. Cytotoxicity studies showed no significant to effects up to concentrations of 100 µg/mL for 72 h of incubation (Fig. 30a). The NIR system showed an excellent internalization of CNOs and a very strong NIR emission, in particular in the perinuclear region (Fig. 30b). Furthermore, the hydroxyl-substituted NIR-BODIPY functionalized carbon nano-onions exhibited an emission on/off reversibility modulated in response to pH; the on/off switching ability was retained in vitro, where, only at acidic pH, a strong NIR emission was observed [93].

An interesting bioimaging system with emission in the far-red/NIR region was reported in 2015 [94]; the bright nanoprobe based on CNOs was obtained through the surface functionalization of CNOs with a highly fluorescent bromo-terminated BODIPY alkyl ether dyes (Fig. 31, left side). Different characterization techniques, including thermogravimetric analysis, Raman, absorption and fluorescence spectroscopies, confirmed the successful development of a promising biological imaging agent. In particular, the excellent performance in terms of fluorescence was confirmed by means of confocal microscopy. After the dispersion of the samples in a

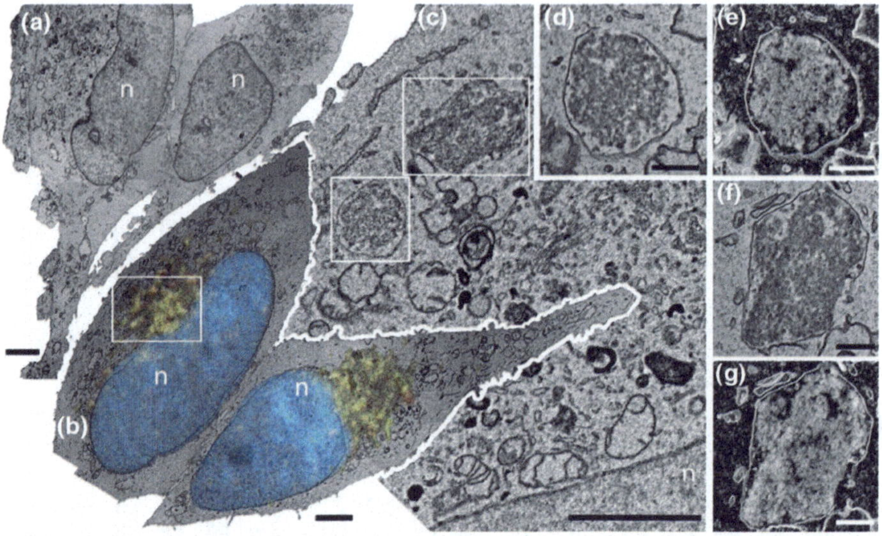

Fig. 28 Correlative light electron microscopy analysis of HeLa cells doped with 10 μg mL^{-1} of CNO-4 for 12 h at 37 °C. **a** Low magnification TEM reconstruction of two cells doped with CNO-4. **b** Superimposed fluorescence and TEM reconstruction of the same cells shown in **a** (imaged in a different section). The co-localization of CNO-4 (green fluorescence) within lysosomes stained with Lysotracker (red) results in yellow fluorescence. **c** Higher magnification TEM image of the boxed area in **b**. Note, the two endosomes containing CNO-4. **d–g** Higher magnification images of the boxed areas from C imaged, respectively, in defocused TEM (**d**, **f**) and in high angular annular dark field (HAADF) scanning TEM (**e**, **g**). The nuclei are labeled with "n". Scale bars: **a**, **b** 4 mm, **c** 2.5 mm, **d–g** 0.5 mm. Adapted with permission from [91]. Copyright 2015 Wiley-VCH Verlag GmbH & Co. KGaA, Weinheim

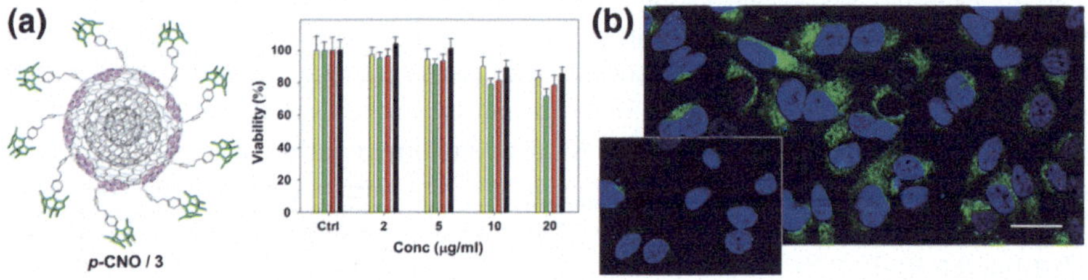

Fig. 29 **a** Cellular viability of HeLa wt cells after exposure to different concentrations (2, 5, 10, 20 μg mL^{-1}) of p-CNO/3. **b** Confocal images of fixed HeLa cells incubated for 30 min at 37 °C with 10 mg mL^{-1} of p-CNO/3. The insets show the corresponding uptake of p-CNO/3 by HeLa cells incubated at 4 °C. Adapted with permission from [92]. Copyright 2015 The Royal Society of Chemistry

drop of water on a polystyrene slide, the 3D z-stacking images (Fig. 31, right side) showed a bright red fluorescence throughout the whole sample area [94].

Recently, we reported a fluorescent on/off modulated nanoprobes based on the covalent functionalization of CNOs with a π-extended distyryl-substituted boron

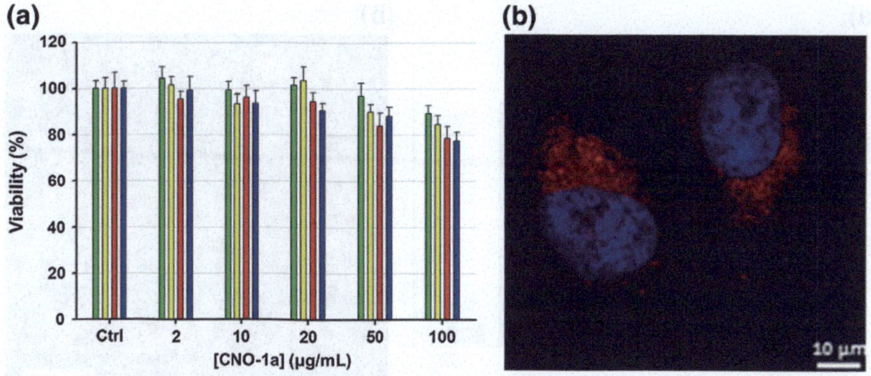

Fig. 30 **a** Viability of HeLa cells at 12 h (green bar), 24 h (yellow bar), 48 h (red bar) and 72 h (blue bar) after exposure to increasing doses (2; 10; 20; 50; 100 $\mu g\ mL^{-1}$) of CNO-1a. **b** confocal microscopy images of CNO-1a (red color), internalized in HeLa Kyoto cells in the case of the confocal microscopy image. Adapted with permission from [93]. Copyright 2014 The Royal Society of Chemistry

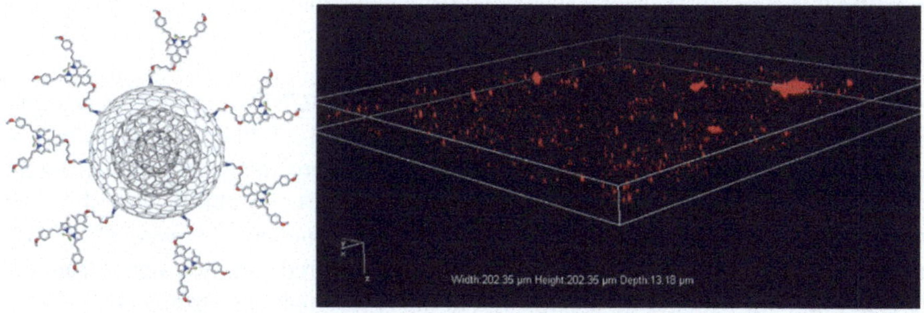

Fig. 31 Far red-NIR-BODIPY functionalized CNOs: structure (left) and 3D Laser confocal microscopy z-stacking image (right) of BODIPY-CNOs aggregates deposited on polystyrene, which illustrates the intense red fluorescence of the CNOs conjugates. Excitation at 647 nm. Reprinted with permission from [81]. Copyright 2017 Elsevier Ltd.

dipyrromethene (BODIPY) dye [95]. The emission switching properties, due to the protonated/non-protonated form of the dimethylamino group attached to the BODIPY core, are activated by an acidic environment, allowing for the development of pH-dependent probes for diagnostic applications. HeLa cells were incubated with different concentrations of the fluorescent nanoprobes, showing minimal systemic cytotoxicity (Fig. 32a). The activation of the emission properties in acidic environment was confirmed by confocal microscopy. At physiological pH (7.4), no fluorescence was detected. On the other hand, a widely spread red emission was observed in acidic pH (4.5) (Fig. 32b), confirming the pH-switchable properties of these nanoprobes [95].

Another very interesting work was recently published in 2017 by our groups [96]. In this paper, we synthetized a far-red emitting BODIPY dye with high quantum

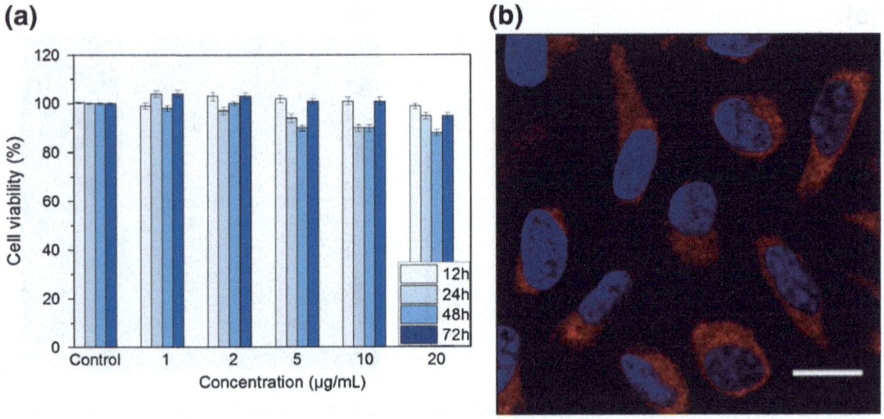

Fig. 32 **a** Cellular viability of HeLa cells treated with different concentrations (1, 2, 5, 10 and 20 μg mL^{-1}) of fluo-CNOs for 12, 24, 28 and 72 h in DMEM (pH 7.4). **b** Cellular uptake and localization of fluo-CNOs in HeLa cells in acidic conditions (PBS, pH 4.5) observed by confocal fluorescence microscopy after incubation for 48 h. Scale bars = 20 μm. Adapted with permission from [95]. Copyright 2017 Lettieri et al.; licensee Beilstein-Institut

yield and we covalently attached it to the surface of CNOs to create new high-resolution bioimaging agents, namely fluo-CNOs, which exhibited high fluorescence and dispersibility in aqueous media. MCF-7 cells were incubated with different concentration of fluo-CNOs and no adverse effects on the cell viability were observed up to 72 h of incubation (Fig. 33a). These fluorescent nanoprobes were efficiently uptaken by the cells, as confirmed by the widely spread red emission signal from the cells in Fig. 33b, with partial localization in the lysosomal compartments.

3.2 *In Vivo Investigation of CNOs*

In vitro studies demonstrated that CNOs are biocompatible, showing no toxic effects on tested cells viability and a good cellular uptake. With the aim to use these carbon nanomaterials in biomedical applications, careful in vivo studies are required, testing toxicity and biodistribution in living organisms. The group of Sarkar reported different in vivo assessment on CNOs prepared by pyrolysis. CNOs were synthetized from wood wool followed by an oxidation process with concentrated nitric acid, resulting in water-soluble and fluorescent CNOs with emission across the visible/NIR region. In 2011, Sakaer et al. performed in vivo imagings of the entire life cycle of *Drosophila melanogaster* upon oral ingestion of the fluorescent nanoparticles mixed with the dietary food [51]. The toxicity study showed no adverse effects induced by CNOs from pupal to adult stages (Fig. 34a, b), confirming that CNOs did not affect the activity of *Drosophila melanogaster*. The oral ingestion of CNOs allowed

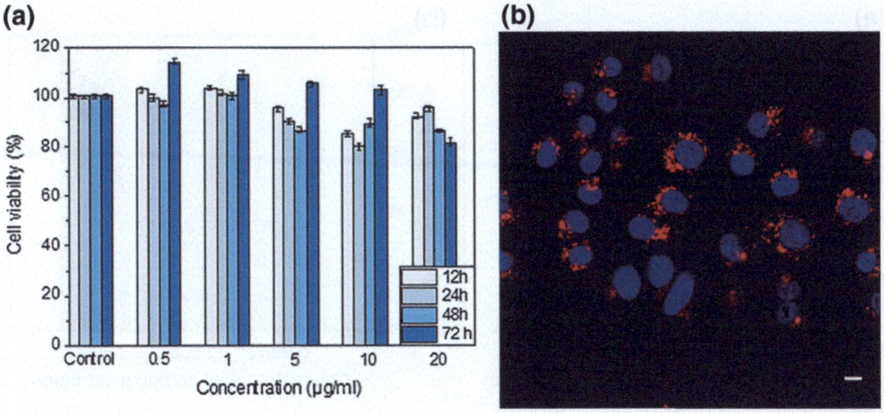

Fig. 33 **a** Cellular viability of MCF7 cells treated with different concentrations (0.5, 1, 5, 10 and 20 μg mL^{-1}) of fluo-CNOs for 12, 24, 28 and 72 h. **b** Confocal image of living MCF7 cells after incubation for 24 h with 20 mg mL^{-1} of fluo-CNOs. Scale bars = 10 mm. Reprinted with permission from [96]. Copyright 2017 The Royal Society of Chemistry

mapping of the growth from egg to adulthood by means of a confocal microscopy (Fig. 34c). Furthermore, CNOs were readily removed from the body, as confirmed by the fact that the excretion gradually becomes nonfluorescent [51].

Subsequently, Sarkar and coworkers used the same nanoparticles to demonstrate the ability of CNOs as in vivo imaging in *Escherichia coli* (*E. coli*) and in *Caenorhabditis elegans* (*C. elegans*), hermaphrodite worm [53]. After exposure to *E. coli* cells, no effects were found in terms of bacterial growth and numbers of colonies formed; moreover, CNOs enabled the imaging of *E. coli* cells, thanks to their broad emission properties. No toxic effects were observed on the cell viability and egg laying capacity of treated *C. elegans*; furthermore, despite the autofluorescence of *C. elegans*, the presence of CNOs allowed the imaging of the gut, which does not normally show any fluorescence [53].

Recently, the same group reported the possible use of CNOs as drug delivery system in the brain [97]. The reduced CNOs size, achieved after fragmentation, enabled the crossing of these fluorescent nanoparticles through the blood–brain barrier (BBB) into the brain of CADASIL mice in vivo as well as through the glioblastoma multiform (GBM) induced mice. After CNOs injection, in vivo imaging of the brain at different time point was performed (Fig. 35i). After 3 days, CNOs were excreted, as suggested by the disappearance of the red fluorescence (Fig. 35ii) [97].

In 2015, our group reported the first evaluation of the possible eco-toxicological effects induced by CNOs on the development of the freshwater polyp *Hydra vulgaris* [98]. The animals were exposed to increasing concentrations of pristine CNOs (p-CNOs), benzoic acid CNOs (benz-CNOs), pyridine CNOs (py-CNOs) and methylpyridinium iodide CNOs (py+-CNOs). No morphological and behavioral abnormalities were observed, event at the highest dose tested. All CNOs conjugates were efficiently internalized by macropinocytosis, as suggested by the dark spots

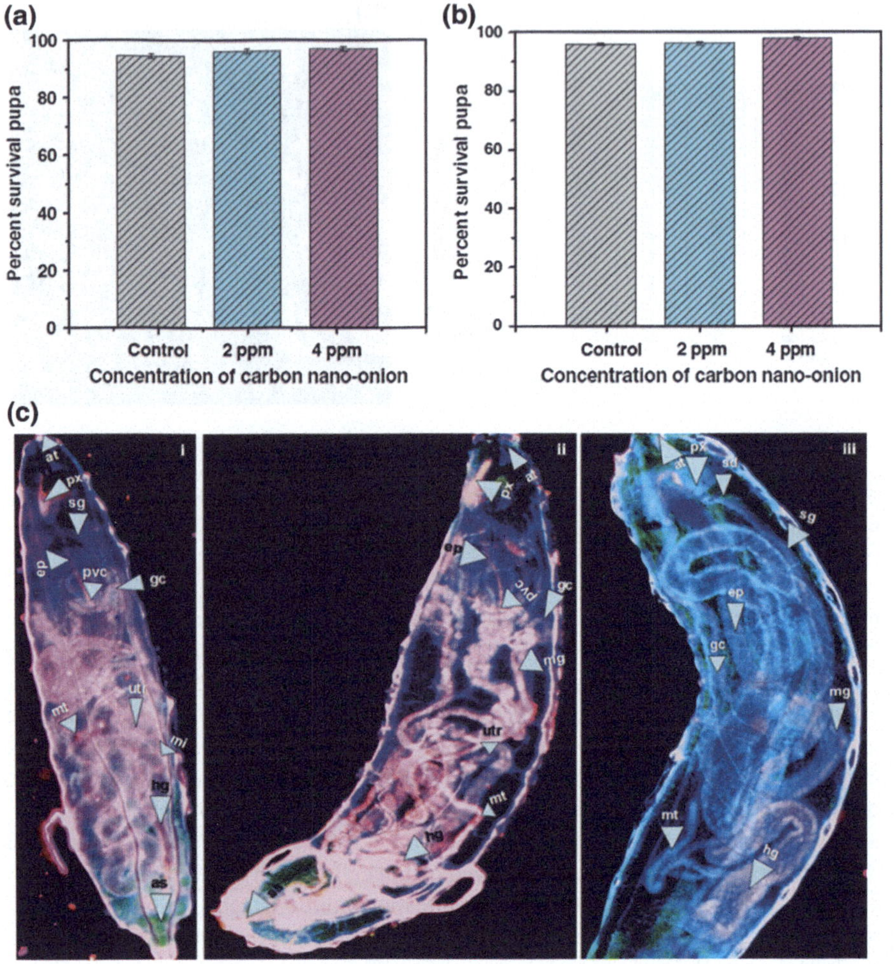

Fig. 34 The effect of water-soluble carbon nano-onions on the viability and growth of nano-onion-fed Drosophila in comparison with controls. **a** Comparison of the survival rates of the pupal stage fed with 2 and 4 ppm of water-soluble carbon nano-onions and on control food. **b** Comparison of the survival rates (30 pairs) of the adult stage fed with and without water-soluble nano-onions. **c** Images (merges of 385, 488, and 561 nm filters) of *Drosophila melanogaster* at the larval stages showing organ details: (i) first instar larva, (ii) second instar larva, and (iii) third instar larva. Organ labels: at = atrium; px = harynx; sd = salivary duct; sg = alivary gland; ep = esophagus; pvc = proventiculus; gc = gastric ceaca; mg = midgut; mi = mid intestine; utr = ureter; mt = malpighian tubule; hg = hind gut; as = anus. Adapted with permission from [51]. Copyright 2011 Wiley-VCH Verlag GmbH & Co. KGaA, Weinheim

visible in the body, head, and tentacles of animals (Fig. 36). Different effects on *Hydra* were examined, including cell apoptosis in vitro and regeneration and

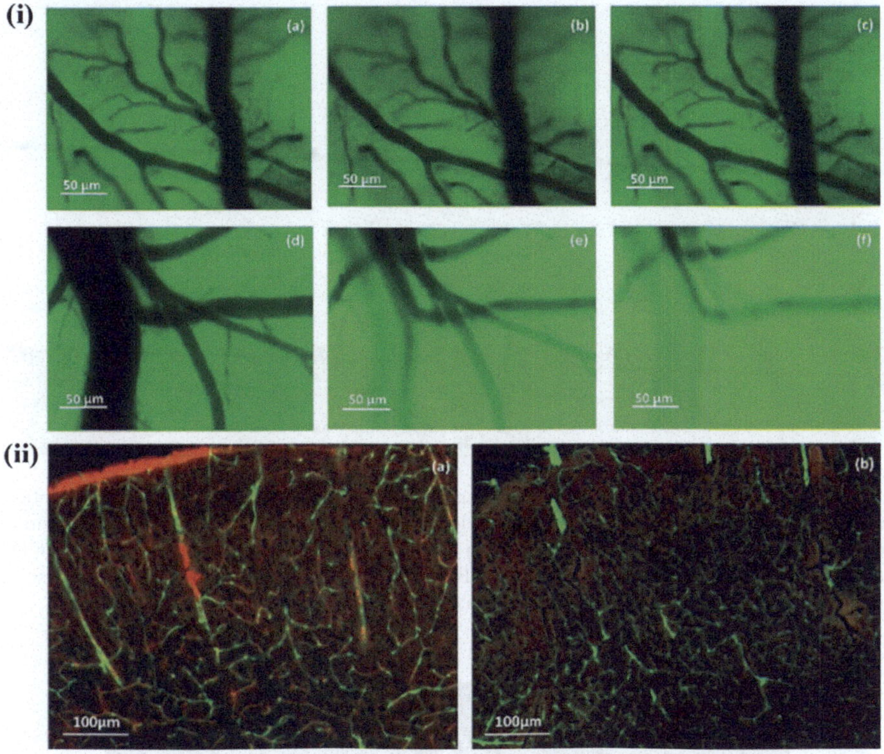

Fig. 35 (i) In vivo image of brain of control mouse after **a** first few second, **b** 14th second and **c** 35th second and in vivo image of brain of mouse injected with CNOs after **d** first few second, **e** 14th second and **f** 35th second; (ii) Labeling of floating section of brain cortex (100 mm) under fluorescent microscope with two color channels (red CNO and green, autofluorescence lectin stained vessels): **a** mouse sacrificed after an hour after CNOs treatment, **b** mouse sacrificed after 3 days to show the clearance of CNOs. Adapted from [97]. Copyright 2016 The Royal Society of Chemistry

reproductive capabilities and the results clearly showed that CNOs did not cause toxic effects over a long time scale [98].

Recently, we reported a careful assessment of the biological impact of functionalized CNOs in a vertebrate model [99]. The toxic effects induced by benz-CNOs and BODIPY-CNOs [90] were investigated on zebrafish (*Danio Rerio*) during the development, after exposure of embryos at different concentrations of the CNOs. No adverse effects on the survival and hatching rates were observed, as well as on the heartbeat rate and frequency of movements; furthermore, no delay on the zebrafish development was reported. Finally, the in vivo biodistribution of BODIPY-CNOs in zebrafish larvae was studied by means of inverted selective plane illumination microscopy (iSPIM); BODIPY-CNOs presented a ubiquitous distribution in the whole zebrafish [99] (Fig. 37).

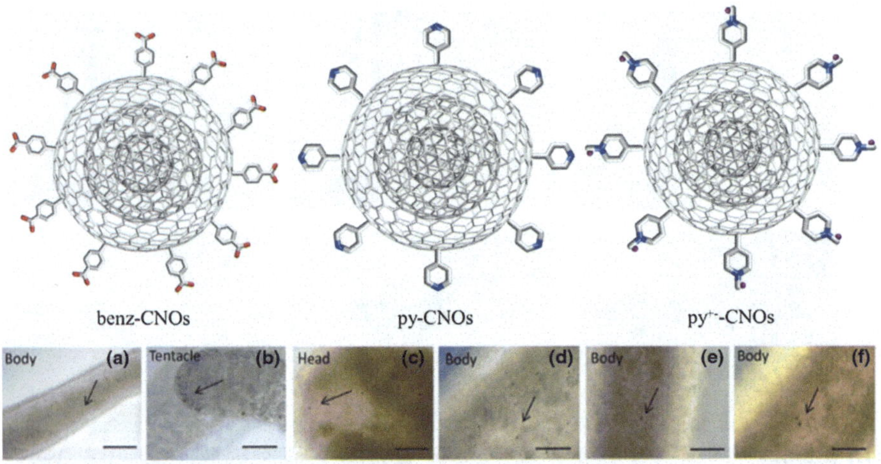

Fig. 36 Structure of benz-CNOs, py-CNOs and py$^{+\cdot}$-CNOs (top) and respective in vivo uptake and biodistribution in Hydra. (bottom) **a–b** Bright field images of benz-CNOs in the body and in tentacle; scale bars: 500 μm in (**a**) and 50 μm in (**b**); **c–d** Bright field images of py-CNOs in the head and in the body; scale bars: 200 μm; **e–f** Bright field images of py$^{+\cdot}$-CNOs in the body; scale bars: 200 μm. Reprinted with permission from [81]. Copyright 2017 Elsevier Ltd.

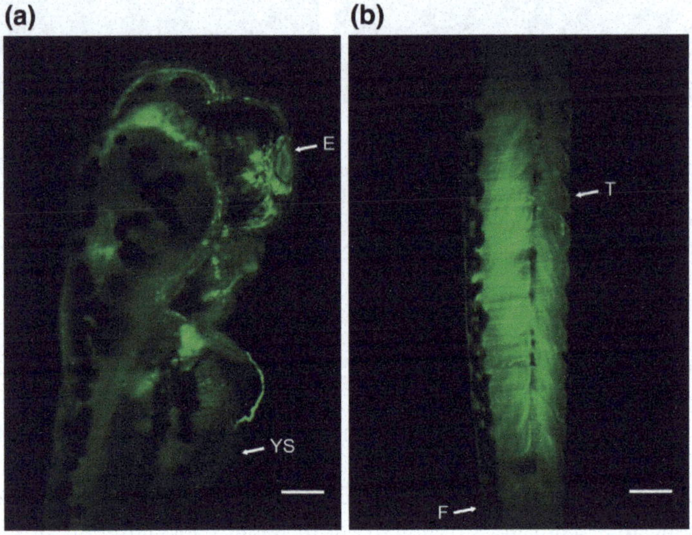

Fig. 37 Maximum intensity projections of the superior part (**a**) and tail (**b**) of treated larvae (100 μg/mL of BODIPY-CNOs). Exposure time: 200 ms, step size: 0.7 μm. Scale bars 100 μm. E, eye; YS, yolk sac; T, tail; F, finfold. Reprinted with permission from [99]

4 Conclusions

In this book chapter, we discussed in detail the current literature involving the use of carbon nano-onions in biological applications. Different synthetic pathways have been developed for its synthesis and, nowadays, thermal annealing is the method of choice because it ensures the production of CNOs in bulk quantities along with a low cost, thus allowing for a detailed understanding of its properties. Furthermore, the increasing development of surface functionalization methods enabled the possibility to explore CNOs in a large variety of applications. Among the different applications reported so far, one of the most promising is in the field of biology; the small size along with a high surface area permits the attachment of different functional groups, thus allowing for the conjugation of both diagnostic and therapeutic agents.

Several reports have demonstrated the suitability of functionalized CNOs as imaging and targeting agents, thanks to their in vitro and in vivo low toxicity, weak inflammatory potential, and high biocompatibility.

In particular, the high biocompatibility of functionalized CNOs in zebrafish can be used to predict their behavior on other vertebrates, due to the high similarity between humans and zebrafish. However, additional studies are still required before the effective use of CNOs as platform in nanomedicine, although these CNMs have already shown different promising properties for this application.

Acknowledgements The authors wish to thank Dr. S. Lettieri and Dr. M. d'Amora for proofreading.

References

1. Kroto HW, Heath JR, O'brien SC et al (1985) C60: buckminsterfullerene. Nature 318:162–163
2. Iijima S, Ichihashi T (1993) Single-shell carbon nanotubes of 1-nm diameter. Nature 363:603–605
3. Novoselov KS, Geim AK, Morozov SV et al (2004) Electric field effect in atomically thin carbon films. Science 306:666–669
4. Geim AK, Novoselov KS (2007) The rise of graphene. Nat Mater 6:183–191
5. Delgado JL, Herranz M, Martín N (2008) The nano-forms of carbon. J Mater Chem 18:1417
6. Molina-Ontoria A, Chaur MN, Plonska-Brzezinska ME et al (2013) Preparation and characterization of soluble carbon nano-onions by covalent functionalization, employing a Na–K alloy. Chem Commun 49:2406–2408
7. Tomita S, Burian A, Dore JC et al (2002) Diamond nanoparticles to carbon onions transformation: X-ray diffraction studies. Carbon 40:1469–1474
8. Zeiger M, Jäckel N, Mochalin VN et al (2016) Review: carbon onions for electrochemical energy storage. J Mater Chem A 4:3172–3196
9. Ugarte D (1995) Onion-Like graphitic particles. Carbon 33
10. Xu BS (2008) Prospects and research progress in nano onion-like fullerenes. New Carbon Mater 23:289–301
11. Iijima S (1980) Direct observation of the tetrahedral bonding in graphitized carbon black by high resolution electron microscopy. J Cryst Growth 50:675–683
12. Ugarte D (1992) Curling and closure of graphitic networks under electron-beam irradiation. Nature 359:707–709

13. Sano N, Wang H, Chhowalla M et al (2001) Synthesis of carbon 'onions' in water. Nature 414:506–507
14. Sano N, Wang H, Alexandrou I et al (2002) Properties of carbon onions produced by an arc discharge in water. J Appl Phys 92:2783–2788
15. Borgohain R, Yang J, Selegue JP et al (2014) Controlled synthesis, efficient purification, and electrochemical characterization of arc-discharge carbon nano-onions. Carbon 66:272–284
16. Kuznetsov VL, Chuvilin AL, Butenko YV et al (1994) Onion-like carbon from ultra-disperse diamond. Chem Phys Lett 222:343–348
17. Tomita S, Sakurai T, Ohta H et al (2001) Structure and electronic properties of carbon onions. J Chem Phys 114:7477–7482
18. Chen J, Deng SZ, Chen J et al (1999) Graphitization of nanodiamond powder annealed in argon ambient. Appl Phys Lett 74:3651–3653
19. Xie FY, Xie WG, Gong L et al (2010) Surface characterization on graphitization of nanodiamond powder annealed in nitrogen ambient. Surf Interface Anal 42:1514–1518
20. Aleksenski AE, Baïdakova MV, Vul' AY et al (2000) Effect of hydrogen on the structure of ultradisperse diamond. Phys Solid State 42:1575–1578
21. Palkar A, Melin F, Cardona CM et al (2007) Reactivity differences between carbon nano onions (CNOs) prepared by different methods. Chem Asian J 2:625–633
22. Mykhailiv O, Lapinski A, Molina-Ontoria A et al (2015) Influence of the synthetic conditions on the structural and electrochemical properties of carbon nano-onions. Chem Phys Chem 16:2182–2191
23. Chen XH, Deng FM, Wang JX et al (2001) New method of carbon onion growth by radio-frequency plasma-enhanced chemical vapor deposition. Chem Phys Lett 336:201–204
24. He CN, Zhao N, Du X et al (2006) Low-temperature synthesis of carbon onions by chemical vapor deposition using a nickel catalyst supported on aluminum. Scripta Mater 54:689–693
25. Chen XH, Yang HS, Wu GT et al (2000) Generation of curved or closed-shell carbon nanostructures by ball-milling of graphite. J Cryst Growth 218:57–61
26. Huang JY, Yasuda H, Mori H (1999) Highly curved carbon nanostructures produced by ball-milling. Chem Phys Lett 303:130–134
27. Cabioc'h T, Jaouen M, Thune E et al (2000) Carbon onions formation by high-dose carbon ion implantation into copper and silver. Surf Coat Technol 128:43–50
28. Thune E, Cabioc'h T, Gue´Rin P et al (2002) Nucleation and growth of carbon onions synthesized by ion-implantation: a transmission electron microscopy study. Mater Lett 54:222–228
29. Du AB, Liu XG, Fu DJ et al (2007) Onion-like fullerenes synthesis from coal. Fuel 86:294–298
30. Dorobantu D, Bota PM, Boerasu I et al (2014) Pulse laser ablation system for carbon nano-onions fabrication. Surf Eng Appl Electrochem 50:390–394
31. Xu BS, Tanaka S-I (1998) Formation of giant onion-like fullerenes under Al nanoparticles by electron irradiation. Acta Mater 46:5249–5257
32. Butenko YV, Šiller L, Hunt MRC (2013) Carbon onions. In: Carbon nanomaterials, 2nd edn. CRC Press
33. Kuznetsov VL, Zilberberg IL, Butenko YV et al (1999) Theoretical study of the formation of closed curved graphite-like structures during annealing of diamond surface. J Appl Phys 86:863–870
34. Zou Q, Wang M, Li Y et al (2009) Fabrication of onion-like carbon from nanodiamond by annealing. Sci China Ser E-Tech Sci 52:3683–3689
35. Butenko YV, Kuznetsov VL, Paukshtis EA et al (2006) The thermal stability of nanodiamond surface groups and onset of nanodiamond graphitization. Fullerenes Nanotubes Carbon Nanostruct 14:557–564
36. Obraztsova ED, Fujii M, Hayashi S et al (1998) Raman identification of onion-like carbon. Carbon 36:821–826
37. Zheng Y, Zhu P (2016) Carbon nano-onions: large-scale preparation, functionalization and their application as anode material for rechargeable lithium ion batteries. RSC Adv 6:92285–92298
38. Tomita S, Fujii M, Hayashi S et al (1999) Electron energy-loss spectroscopy of carbon onions. Chem Phys Lett 305:225–229

39. Bogdanov K, Fedorov A, Osipov V et al (2014) Annealing-induced structural changes of carbon onions: high-resolution transmission electron microscopy and Raman studies. Carbon 73:78–86
40. Zeiger M, Jäckel N, Weingarth D et al (2015) Vacuum or flowing argon: what is the best synthesis atmosphere for nanodiamond-derived carbon onions for supercapacitor electrodes? Carbon 94:507–517
41. Alekseyev NI, Dyuzhev GA (2003) Fullerene formation in an arc discharge. Carbon 41:1343–1348
42. Zhao N, Cui Q, He C et al (2007) Synthesis of carbon nanostructures with different morphologies by CVD of methane. Mater Sci Eng A 460–461:255–260
43. He CN, Tian F, Liu SJ et al (2008) Characterization and magnetic property of carbon coated metal nanoparticles and hollow carbon onions fabricated by CVD of methane. Mater Lett 62:3697–3699
44. He CN, Zhao N, Shi C et al (2006) Carbon nanotubes and onions from methane decomposition using Ni/Al catalysts. Mater Chem Phys 97:109–115
45. Cabioc'h T, Riviere JP, Delafond J (1995) A new technique for fullerene onion formation. J Mater Sci 30:4787–4792
46. Bystrzejewski M, Rummeli MH, Gemming T et al (2010) Catalyst-free synthesis of onion-like carbon nanoparticles. New Carbon Mater 25:1–8
47. Choucair M, Stride JA (2012) The gram-scale synthesis of carbon onions. Carbon 50:1109–1115
48. Du J, Liu Z, Li Z et al (2005) Carbon onions synthesized via thermal reduction of glycerin with magnesium. Mater Chem Phys 93:178–180
49. Kobayashi T, Sekine T, He H (2003) Formation of carbon onion from heavily shocked SiC. Chem Mater 15:2681–2683
50. Yan Y, Yang H, Zhang F et al (2007) Low-temperature solution synthesis of carbon nanoparticles, onions and nanoropes by the assembly of aromatic molecules. Carbon 45:2209–2216
51. Ghosh M, Sonkar SK, Saxena M et al (2011) Carbon nano-onions for imaging the life cycle of *Drosophila melanogaster*. Small 7:3170–3177
52. Sonkar SK, Roy M, Babar DG et al (2012) Water soluble carbon nano-onions from wood wool as growth promoters for gram plants. Nanoscale 4:7670–7675
53. Sonkar SK, Ghosh M, Roy M et al (2012) Carbon nano-onions as nontoxic and high-fluorescence bioimaging agent in food chain—an in vivo study from unicellular *E. coli* to multicellular *C. elegans*. Mater Express 2:105–114
54. Dubey P, Tripathi KM, Sonkar SK (2014) Gram scale synthesis of green fluorescent water-soluble onion-like carbon nanoparticles from camphor and polystyrene foam. RSC Adv 4:5838
55. Hou S-S, Chung D-H, Lin T-H (2009) High-yield synthesis of carbon nano-onions in counterflow diffusion flames. Carbon 47:938–947
56. Bartelmess J, Giordani S (2014) Carbon nano-onions (multi-layer fullerenes): chemistry and applications. Beilstein J Nanotechnol 5:1980–1998
57. Georgakilas V, Guldi DM, Signorini R et al (2003) Organic functionalization and optical properties of carbon onions. J Am Chem Soc 125:14268–14269
58. Rettenbacher AS, Elliott B, Hudson JS et al (2005) Preparation and functionalization of multilayer fullerenes (carbon nano-onions). Chem Eur J 12:376–387
59. Luszczyn J, Plonska-Brzezinska ME, Palkar A et al (2010) Small noncytotoxic carbon nano-onions: first covalent functionalization with biomolecules. Chem Eur J 16:4870–4880
60. Frasconi M, Maffeis V, Bartelmess J et al (2015) Highly surface functionalized carbon nano-onions for bright light bioimaging. Methods Appl Fluoresc 3:044005
61. Liu Y, Vander Wal RL, Khabashesku VN (2007) Functionalization of carbon nano-onions by direct fluorination. Chem Mater 19:778–786
62. Flavin K, Chaur MN, Echegoyen L et al (2009) Functionalization of multilayer fullerenes (carbon nano-onions) using diazonium compounds and "click" chemistry. Org Lett 12:840–843
63. Jariwala D, Sangwan VK, Lauhon LJ et al (2013) Carbon nanomaterials for electronics, optoelectronics, photovoltaics, and sensing. Chem Soc Rev 42:2824–2860

64. Baptista FR, Belhout SA, Giordani S et al (2015) Recent developments in carbon nanomaterial sensors. Chem Soc Rev 44:4433–4453
65. Hong G, Diao S, Antaris AL et al (2015) Carbon nanomaterials for biological imaging and nanomedicinal therapy. Chem Rev 115:10816–10906
66. Bartelmess J, Quinn SJ, Giordani S (2015) Carbon nanomaterials: multi-functional agents for biomedical fluorescence and Raman imaging. Chem Soc Rev 44:4672–4698
67. Baldrighi M, Trusel M, Tonini R et al (2016) Carbon nanomaterials interfacing with neurons: an in vivo perspective. Front Neurosci 10:250
68. Sayes CM, Fortner JD, Guo W et al (2004) The differential cytotoxicity of water-soluble fullerenes. Nano Lett 4:1881–1887
69. Bosi S, Da Ros T, Spalluto G et al (2003) Fullerene derivatives: an attractive tool for biological applications. Eur J Med Chem 38:913–923
70. Yang W, Thordarson P, Gooding JJ et al (2007) Carbon nanotubes for biological and biomedical applications. Nanotechnology 18:412001
71. Liu Z, Tabakman S, Welsher K et al (2009) Carbon nanotubes in biology and medicine: in vitro and in vivo detection, imaging drug delivery. Nano Res 2:85–120
72. Cao L, Wang X, Meziani MJ et al (2007) Carbon dots for multiphoton bioimaging. J Am Chem Soc 129:11318–11319
73. Zheng XT, Ananthanarayanan A, Luo KQ et al (2015) Glowing graphene quantum dots and carbon dots: properties, syntheses, and biological applications. Small 11:1620–1636
74. Chung C, Kim Y-K, Shin D et al (2013) Biomedical applications of graphene and graphene oxide. Acc Chem Res 46:2211–2224
75. Pech D, Brunet M, Durou H et al (2010) Ultrahigh-power micrometre-sized supercapacitors based on onion-like carbon. Nat Nanotechnol 5:651–654
76. Borgohain R, Li J, Selegue JP et al (2012) Electrochemical study of functionalized carbon nano-onions for high-performance supercapacitor electrodes. J Phys Chem C 116:15068–15075
77. Keller K, Maksimova NI, Roddatis VV et al (2002) The catalytic use of onion-like carbon materials for styrene synthesis by oxidative dehydrogenation of ethylbenzene. Angew Chem Int Ed 41:1885–1888
78. Su D, Maksimova NI, Mestl G et al (2007) Oxidative dehydrogenation of ethylbenzene to styrene over ultra-dispersed diamond and onion-like carbon. Carbon 45:2145–2151
79. Cabioc'h T, Thune E, Rivière JP et al (2002) Structure and properties of carbon onion layers deposited onto various substrates. J Appl Phys 91:1560–1567
80. Joly-Pottuz L, Vacher B, Ohmae N et al (2008) Anti-wear and friction reducing mechanisms of carbon nano-onions as lubricant additives. Tribol Lett 30:69–80
81. Camisasca A, Giordani S (2017) Carbon nano-onions in biomedical applications: promising theranostic agents. Inorg Chim Acta 468:67–76
82. Ding L, Stilwell J, Zhang T et al (2005) Molecular characterization of the cytotoxic mechanism of multiwall carbon nanotubes and nano-onions on human skin fibroblast. Nano Lett 5:2448–2464
83. Xu Y, Wang SY, Yang J et al (2013) Multiwall carbon nano-onions induce DNA damage and apoptosis in human umbilical vein endothelial cells. Environ Toxicol 28:442–450
84. Kang S, Kim J-E, Kim D et al (2015) Comparison of cellular toxicity between multi-walled carbon nanotubes and onion-like shell-shaped carbon nanoparticles. J Nanopart Res 17
85. Tripathi KM, Bhati A, Singh A et al (2016) From the traditional way of pyrolysis to tunable photoluminescent water soluble carbon nano-onions for cell imaging and selective sensing of glucose. RSC Adv 6:37319–37329
86. Yang M, Flavin K, Kopf I et al (2013) Functionalization of carbon nanoparticles modulates inflammatory cell recruitment and NLRP3 inflammasome activation. Small 9:4194–4206
87. Flavin K, Kopf I, Del Canto E et al (2011) Controlled carboxylic acid introduction: a route to highly purified oxidised single-walled carbon nanotubes. J Mater Chem 21:17881
88. Flavin K, Lawrence K, Bartelmess J et al (2011) Synthesis and characterization of boron azadipyrromethene single-wall carbon nanotube electron donor-acceptor conjugates. ACS Nano 5:1198–1206

89. Strano MS, Dyke CA, Usrey ML et al (2003) Electronic structure control of single-walled carbon nanotube functionalization. Science 301:1519–1522
90. Bartelmess J, De Luca E, Signorelli A et al (2014) Boron dipyrromethene (BODIPY) functionalized carbon nano-onions for high resolution cellular imaging. Nanoscale 6:13761–13769
91. Frasconi M, Marotta R, Markey L et al (2015) Multi-functionalized carbon nano-onions as imaging probes for cancer cells. Chem Eur J 21:19071–19080
92. Bartelmess J, Frasconi M, Balakrishnan PB et al (2015) Non-covalent functionalization of carbon nano-onions with pyrene–BODIPY dyads for biological imaging. RSC Adv 5:50253–50258
93. Giordani S, Bartelmess J, Frasconi M et al (2014) NIR fluorescence labelled carbon nano-onions: synthesis, analysis and cellular imaging. J Mater Chem B 2:7459–7463
94. Bartelmess J, Baldrighi M, Nardone V et al (2015) Synthesis and characterization of far-red/NIR-fluorescent BODIPY dyes, solid-state fluorescence, and application as fluorescent tags attached to carbon nano-onions. Chemistry 21:9727–9732
95. Lettieri S, D'amora M, Camisasca A et al (2017) Carbon nano-onions as fluorescent on/off modulated nanoprobes for diagnostics. Beilstein J Nanotechnol 8:1878–1888
96. Lettieri S, Camisasca A, D'amora M et al (2017) Far-red fluorescent carbon nano-onions as a biocompatible platform for cellular imaging. RSC Adv 7:45676–45681
97. Pakhira B, Ghosh M, Allam A et al (2016) Carbon nano onions cross the blood brain barrier. RSC Adv 6:29779–29782
98. Marchesano V, Ambrosone A, Bartelmess J et al (2015) Impact of carbon nano-onions on *Hydra vulgaris* as a model organism for nanoecotoxicology. Nanomaterials 5:1331–1350 (Basel)
99. D'Amora M, Rodio M, Bartelmess J et al (2016) Biocompatibility and biodistribution of functionalized carbon nano-onions (f-CNOs) in a vertebrate model. Sci Rep 6:33923

The manufacturer's authorised representative in the EU is Springer Nature Customer Service Centre GmbH, Europaplatz 3, 69115 Heidelberg, Germany. If you have any concerns regarding our products, please contact ProductSafety@springernature.com

Printed and bound by CPI Group (UK) Ltd, Croydon, CR0 4YY
05/10/2026
02228147-0004